Thermochemistry of Alloys
Recent Developments of Experimental Methods

NATO ASI Series

Advanced Science Institutes Series

A Series presenting the results of activities sponsored by the NATO Science Committee, which aims at the dissemination of advanced scientific and technological knowledge, with a view to strengthening links between scientific communities.

The Series is published by an international board of publishers in conjunction with the NATO Scientific Affairs Division

A Life Sciences	Plenum Publishing Corporation
B Physics	London and New York
C Mathematical	Kluwer Academic Publishers
and Physical Sciences	Dordrecht, Boston and London
D Behavioural and Social Sciences	
E Applied Sciences	
F Computer and Systems Sciences	Springer-Verlag
G Ecological Sciences	Berlin, Heidelberg, New York, London,
H Cell Biology	Paris and Tokyo

Series C: Mathematical and Physical Sciences - Vol. 286

Thermochemistry of Alloys
Recent Developments of Experimental Methods

edited by

H. Brodowsky

and

H.-J. Schaller
Institut für Physikalische Chemie,
Universität Kiel, Kiel, F.R.G.

Kluwer Academic Publishers

Dordrecht / Boston / London

Published in cooperation with NATO Scientific Affairs Division

Proceedings of the NATO Advanced Study Institute on
Thermochemistry of Alloys
Recent Developments of Experimental Methods
Kiel, F.R.G.
August 16–27, 1987

Library of Congress Cataloging in Publication Data

```
Thermochemistry of alloys : recent developments of experimental
  methods / edited by H. Brodowsky and H.-J. Schaller.
     p.   cm. -- (NATO ASI series. Series C, Mathematical and
  physical sciences ; vol. 286)
    Lectures and seminar papers given at a NATO advanced study
  institute held in Aug. 1987 in Kiel, Germany.
    "Published in cooperation with NATO Scientific Affairs Division."
    ISBN 0-7923-0434-9
    1. Alloys--Thermal properties--Congresses.  2. Thermochemistry-
  -Congresses.   I. Brodowsky, H. (Horst), 1933-   . II. Schaller, H.
  -J. (Hans-Jürgen), 1941-  III. Series: NATO ASI series.  Series C,
  Mathematical and physical sciences ; no. 286.
  TN690.T477   1989
  669'.9--dc20                                            89-36645
```

ISBN-13: 978-94-010-6953-3   e-ISBN-13: 978-94-009-1027-0
DOI: 10.1007/ 978-94-009-1027-0

Published by Kluwer Academic Publishers,
P.O. Box 17, 3300 AA Dordrecht, The Netherlands.

Kluwer Academic Publishers incorporates the publishing programmes of
D. Reidel, Martinus Nijhoff, Dr W. Junk and MTP Press.

Sold and distributed in the U.S.A. and Canada
by Kluwer Academic Publishers,
101 Philip Drive, Norwell, MA 02061, U.S.A.

In all other countries, sold and distributed
by Kluwer Academic Publishers Group,
P.O. Box 322, 3300 AH Dordrecht, The Netherlands.

Printed on acid free paper

CONTENTS

L:Lecture
S:Seminar

PREFACE

The thermochemistry of alloys has interested generations of scientists and the subject was treated in classical textbooks long ago, e.g. by Hume-Rothery, by Wagner, and by Kubaschewski and Alcock. Nevertheless, the appearance of new materials and the desire to improve traditional materials and metallurgical processes has kept up demand for more information on the thermodynamics of these systems.

The advent of computing power has created new opportunities to tie various aspects and properties together, such as phase diagrams and thermodynamic functions, that are in principle thermodynamically inter-related but were too cumbersome to work out before. The computer has also been a powerful tool in building and testing models that help to explain the underlying causes of non-ideal behavior. At the same time, these calculations have pinpointed areas, where additional and more accurate data are needed.

In the laboratory, new methods, improved materials, and sophistica-ted instrumentation have gradually changed the way in which experiments are done. Within the time span of perhaps thirty years, the development went from jotting down individual readings of data points to strip chart recording to automatic digital data acquisition.

Scholars and students active in the field of "Thermochemistry of Alloys" convened for a NATO Advanced Study Institute at Kiel in August 1987 to discuss these developments. This book collects most of the lectures and seminar papers given at the Institute.

Part of the Advanced Study Institute was devoted to the discussion of the current status of the electronic structure of alloys, of methods of estimating thermodynamic properties, and of practical ways to store empirical data for easy retrieval and application to specific problems.

The largest part was devoted to reviews and to the presentation of recent results of thermodynamic measurements on alloys by various methods, in particular by calorimetry, by free energy determinations with vapour pressure or e.m.f.measurements, and by chemical equilibration techniques.

A number of special lectures and seminars were concerned with liquid alloys, with metals in the glassy state, and with diffusion problems.

Finally, a number of applications were dealt with, in particular problems of corrosion of alloys in gaseous and in molten salt environ-ments, efficient formation of protective coatings by pack cementation, or problems arising at the electrical contacts between metals and semicon-ductors.

It is hoped that this collection of papers on the recent deve-lopments in alloy thermodynamics will be as welcome to the larger audience of students and practioners of metallurgy as it was to the participants of the ASI. On behalf of all the attendants, the directors wish to thank the NATO Scientific Affairs Division for making this meeting possible. Thanks are also due to the Government of the State of Schleswig-Holstein, to the University of Kiel, and to the Kennedy-Haus, Kiel, for substantial support.

Kiel, FR Germany
August 1988

Horst Brodowsky
Hans-Jürgen Schaller

MODELLING AND THEORIES OF ALLOY PHASE BEHAVIOR

R. E. Watson, J. W. Davenport, and M. Weinert
Department of Physics
Brookhaven National Laboratory
Upton, NY 11973 USA

L. H. Bennett
National Bureau of Standards
Gaithersburg, MD 20899 USA

ABSTRACT. It has been recognized since the early work of Hume-Rothery and others that many trends in alloy phase formation are readily understood in terms of physically plausible atomic parameters. For example, a substitutional alloy can only occur if there is not too great a difference in the sizes of the alloy constituents, so as not to require too great a cost in the elastic energy associated with deforming the lattice. This has led, in turn, to the introduction of so-called structural maps where two (or more) such atomic parameters are employed as the coordinates and well defined regions are observed to be associated with particular crystalline phases. These coordinates sometimes involve the difference in atomic parameters, such as the difference in the sizes of the constituent atoms, and sometimes involve an average, such as the average d-band occupancy of constituent transition element metals. An alternative approach to the emphasis on atomic parameters has been the consideration, as pioneered by Pearson, of how atoms are packed in some crystal structure and how this controls what the constituent atoms may be. Recently this has led to the utilization of Wigner-Seitz (sometimes called Voronoi or Dirichlet) constructs of the atomic cells in a crystal structure and, in turn, to the observation that sometimes two crystals which are nominally considered to have the same crystal structure according to normal crystallographic designation should, in fact, be considered to be different. The Wigner-Seitz cell constructs have also offered a framework for understanding trends in the magnetic and chemical properties of particular phases as well as making coordination between crystalline and glassy structures. Neither of the above approaches-- correlations with atomic parameters or with packing considerations-- provide numerical estimates of quantities of thermodynamic interest such as heats of formation. Such heats are being calculated with varying rigor and varying computational complexity ranging from model Hamiltonians employing atomic parameters to intricate electron band theory calculations. This chapter will attempt to provide the reader

1

H. Brodowsky and H.-J. Schaller (eds.), Thermochemistry of Alloys, 1–28.
© 1989 by Kluwer Academic Publishers.

with a sense of some of successes and some of the problems when
employing the above approaches to trace out trends in alloy phase
behavior. Because of space limitations, this review will be highly
selective.

I. ATOMIC PARAMETERS AND STRUCTURAL MAPS

For the most part, the elemental solids form in crystal structures which correlate with the number of their valence electrons. For
example, neglecting the magnetic solids, the sequence

$$hcp \rightarrow bcc \rightarrow hcp \rightarrow fcc \qquad (1)$$

occurs upon traversing the transition metal rows. With one modification:

$$hcp \rightarrow bcc \rightarrow tcp \rightarrow hcp \rightarrow fcc \qquad (2)$$

describes transition metal alloying, provided that strong compound
formation does not occur upon alloying, and the average d-band occupancy determines the crystal structure which occurs. Here tcp refers
to the topologically close-packed structures, such as the $A15(Cr_3Si)$,
σ and αMn phases, the first two being Frank-Kasper phases of which
more will be said in the next section. These tcp phases have atomic
sites in polyhedral environments which do not lend themselves to the
planar stacking of atoms and hence the systems are not ductile. This
is important in high strength steels and superalloys since these tcp
phases compete for the refractory transition metal alloy components,
such as V, Mo and W. The atomic sites have different numbers of
nearest neighbors and hence are of different size, i.e. differences

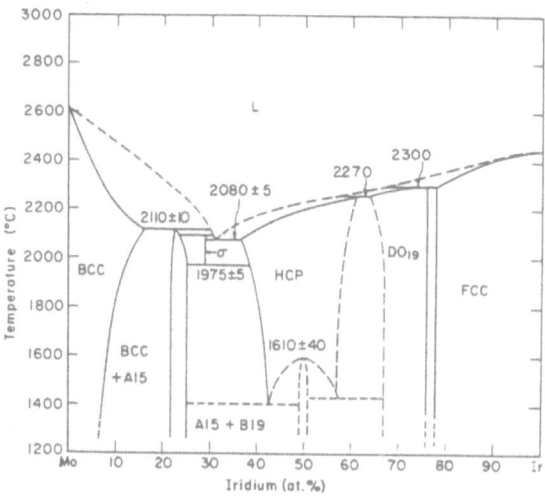

Fig. 1 The phase diagram of the Mo-Ir system.

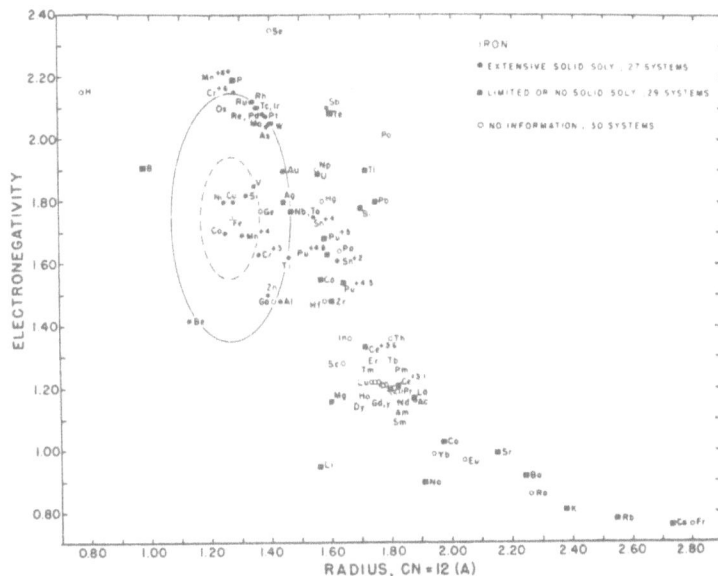

Fig. 2 A Darken-Gurry mapping of the elements with a solubility in either of the allotropic phases of iron greater or less than an arbitrarily chosen value of 5 atomic percent (ref. 3).

in atomic size of the constituent atoms are important to the formation of these phases. The αMn structure is closely related to, while not being[1] a Frank-Kasper phase. It too has sites of different size which accommodate Mn atoms of differing magnetic moment (Mn atoms with substantial magnetic moments have much larger volumes than do those with near zero moments). The Mn atoms in the α (and β) Mn structure(s) may truly be said to be forming alloys where it is alloy components of differing magnetic moment, rather than differing nuclear charge, which are involved . Sequence (2) is to be seen in the Mo-Ir phase diagram[2] illustrated in Fig. 1. Here one goes from bccMo → tcp → hcp → fccIr. Note that DO$_{19}$ structure is simply a (111) ordered atomic restacking of an fcc lattice where the Mo enters the Ir substitutionally in the terminal fcc phase.

While the average valence electron count may be important to what phase occurs, the ability to form a substitutional alloy requires that there neither be too great a difference in atomic size nor too great a strength of chemical bonding between the solute and the solvent, for otherwise ordered compounds are favored. This has led to the implementation of Darken-Gurry plots where one coordinate is the atomic radius and another the electronegativity--the idea of the latter being that a large electronegativity difference implies strong bonding. Such a plot, as obtained[3] by Waber et al. for Fe as the solvent, appears in Fig. 2. Here almost perfect separation is obtained between

4

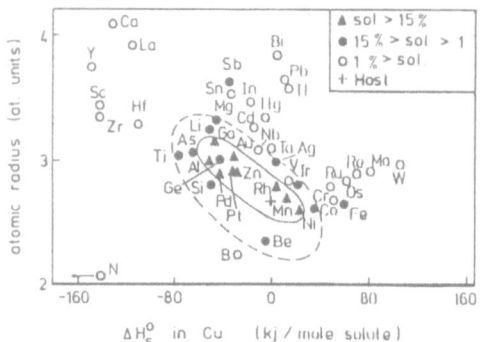

Fig. 3 A Darken-Gurry-type solubility map for impurities in Cu
employing elemental Wigner-Seitz radii and calculated heats of solution as the coordinates (ref. 4). Note that three ranges of solubility have been differentiated here as contrasted with the two of Fig. 2.

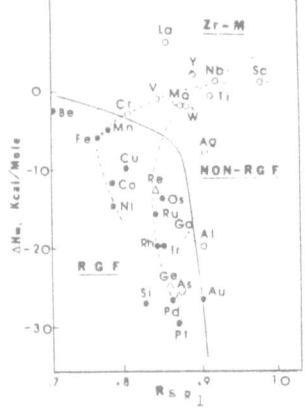

Fig. 4 A map differentiating ready glass formation in metallic alloys
where one coordinate is the heat of mixing and the other, the ratio of
small atom's to the larger's atomic radius (ref. 5). Glass formation
is generally favored providing $\Delta H_m < \sim -5$ Kcal/mole and the radius
ratio is less than ~ 0.86.

those alloys having more and those less than 5% solubility as defined
by an ellipse with its principle axes parallel to the map
coordinates. Lopez and Alonso have considered such solubility maps
where the electronegativity coordinate has been replaced by the heat
of solution and their result[4] for Cu alloys is shown in Fig. 3. Here
they consider three ranges of solubility, against the two in Fig. 2,
and obtain reasonable separation with roughly elliptical boundaries
whose axes, however, are no longer parallel to the map coordinates.

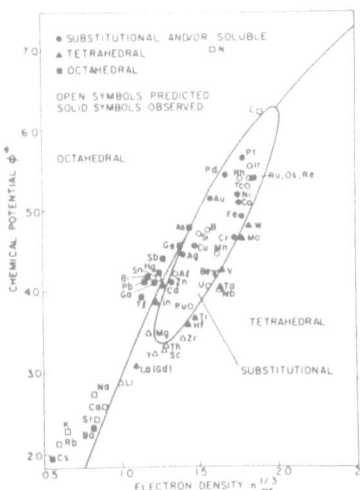

Fig. 5 A map, employing the Miedema parameters as coordinates for impurities entering Be metal substitutionally or at tetrahedral or octahedral intertices (ref. 6). Solid symbols are experimentally determined cases while the open symbols are predictions based on what region they fall within on the map. Small adjustments were made in Miedema's values of his parameters for Al, Ga, In, Ca and Cd so as to increase the success of the map.

Simple ideas of the relative size of the solutes and of the strength of their bonding with the host would seem to provide a reasonable basis for understanding the occurrence of substitutional alloys. Similar mappings, though ones without elliptical regions of "favor" have been applied[5] by Giessen and coworkers to the issue of what alloys readily form metallic glasses and their plot for Zr alloys appears in Fig. 4 where one coordinate is the heat of mixing and the other the ratio of the smaller to the larger atom's radius. The map is quite successful in defining two regions--one where metallic glasses readily form and one where they do not.

Kaufman and coworkers employed[6] rather different coordinates when considering Be alloys. Here impurities had been implanted into Be hosts and the issue was did they go in substitutionally or, instead, into tetrahedral or octahedral interstitial sites. The coordinates of their map, appearing in Fig. 5 are the chemical potential and the cube root of the interstitial charge density as employed[7] by Miedema in his model Hamiltonian for heats of formation. Separate well-defined regions are associated with the three differing types of impurity behavior.

St. John and Bloch adopted[8] a pseudopotential approach to struc tural mapping. They defined a quantum defect, $\hat{\ell}(\ell)-\ell$, in terms of the ionization energy of an atom,

$$E(n,\ell) = -Z^2/2[n + \hat{\ell}(\ell)-\ell]^2 \tag{3}$$

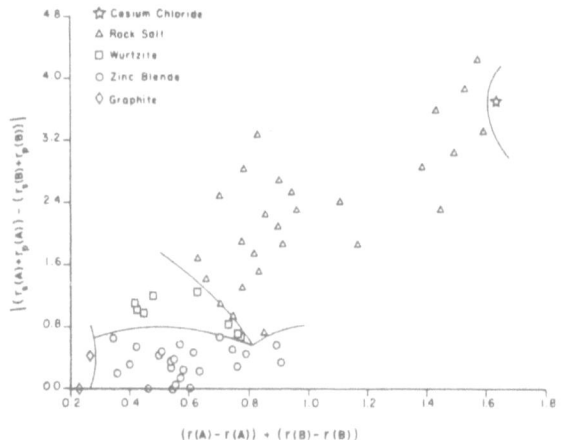

Fig. 6   A structural map for $A^N B^{8-N}$ octet compounds, having eight valence electrons per molecular unit, employing combinations of effective orbital radii, the abscissa effectively being a measure of relative s versus p bonding and the ordinate an electronegativity difference (ref. 8).

and in turn defined the maximum of an effective radial orbital

$$r_\ell \equiv \hat{\ell}(\hat{\ell} + 1)/Z \ . \tag{4}$$

Employing s and p r_ℓ they obtained the structural map shown in Fig. 6 for the so-called octet compounds formed of main group elements where the sum of the constituent's valence electrons equals eight. There is good separation between the various structures. Note that the abscissa involves the average of the elemental (r_p-r_s) while the ordinate involves the difference in (r_p+r_s).

One problem with the atomic parameters employed in such mappings is that they are not unrelated to one another. For example, in defining the chemical potential in terms of a single parameter for a metallic system that parameter would be the electron density in the interstitial region. The failure of the points in Fig. 5 to lie on a single smooth curve is a measure of the failure of such an oversimplified description of chemical potentials and the deviations from such a curve are presumably primarily due to crystal potential effects.

When dealing with maps for transition metal alloys it proves useful to employ one coordinate which is a measure of the d-band occupancy. Since the top of the d bands are more easily defined than the bottoms, which are heavily mixed with s-p character, we've found it

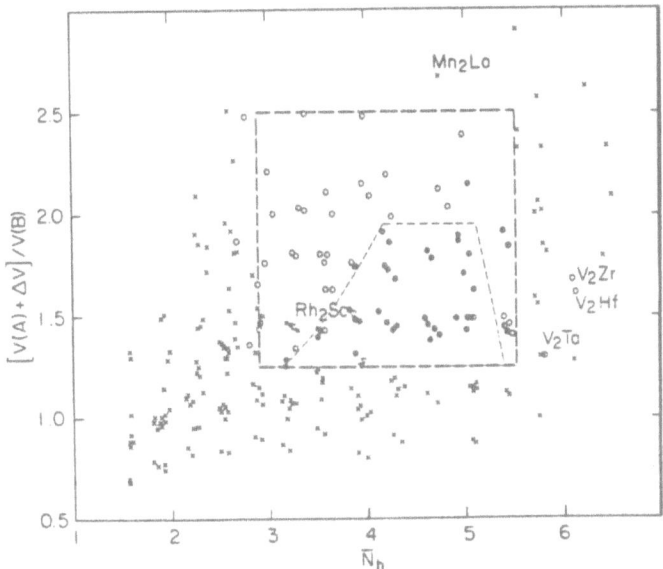

Fig. 7   The occurrence and nonoccurrence of transition metal AB_2 Laves phases as a function of the volume ratio $[V(A) + \Delta V]/V(B)$ and the average alloy d-band electron hole count (calculated at 1:2 composition).   Open circles denote $MgCu_2$ phases, solid circles the $MgZn_2$ and half-circles when both occur.   The x's indicate cases for which Laves phases have not been reported (ref. 9).

useful to define the occupancy in terms of the d hole count, $\bar{N}_h$, which is the number of unoccupied states per atom between the top of the d bands and Fermi level. Using the weighted average of the constituent atom's d hole counts, $\bar{N}_h$, as one coordinate, for the 1:2 AB_2 transition element-transition element alloys, a structural map[9], sampling whether these compounds do or do not form in Laves structures appears in Fig. 7.   The second coordinate is a measure of the volume difference between atoms A and B.   A volume difference is to be expected because the Laves phases are Frank-Kasper structures with markedly different site volumes at the two species of atomic sites. Now it so happens that these structures form with a molecular volume which is smaller than the sum of the elemental volumes, i.e. ΔV is negative where

$$\Delta V \equiv V(AB_2) - V(A) - 2V(B) \tag{5}$$

In the map this volume contraction has been attributed to the A atom which is larger, sloshier and more electropositive than the B.   The x's in the plot denote cases where either an AB_2 compound of different

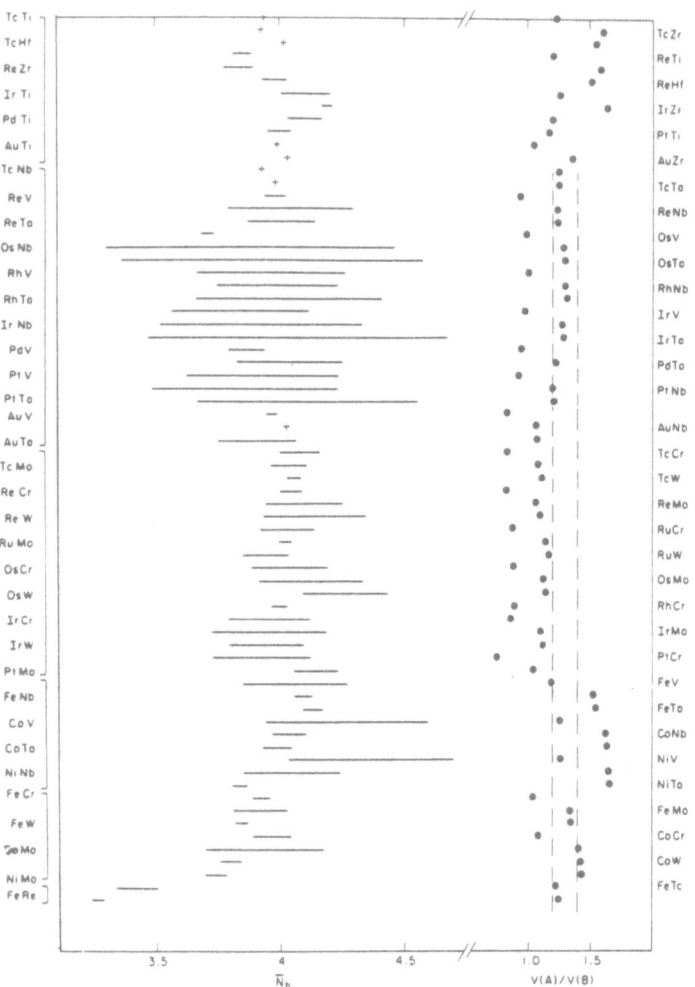

Fig. 8 Ranges of effective d-band hole count associated with binary non-Laves tcp phases of the transition metal alloys, except those of Mn (ref. 10). To the right are shown the ratios of the elemental metal volumes (the A atom is to the right and the B atom to the left in the alloy labels and in the periodic table). The dashed lines indicate a range 1.2 < V(A)/V(B) < 1.4 in which tcp phase formation appears to be especially favorable.

symmetry occurs or no compound of such composition has been reported. There is an extraordinary division between the Laves and non-Laves regions with rectangular bounds which contrast with the elliptical bounds seen in some of the earlier figures. The x's within the bounds are for systems where no compounds of AB_2 composition have been reported (and where Laves structures might well occur).

Equation 2 suggests that there may be a range of $\bar{N}_h$ in which the tcp phases, other than the Laves structures occur. This is the case[10] and a plot of the range of $\bar{N}_h$ for which this happens appears in Fig. 8. The elemental hole counts inferred from energy band calculations (and employed in Fig. 7) have been adjusted so as to provide a better defined range of $\bar{N}_h$. The main feature of this adjustment is to make the Ti, V and Cr columns of the transition elements more alike than they truly are (see Fig. 4 of Ref. 10 for a plot of the adjusted values versus their band theory counterparts). The tcp alloys occur with a varying range of N_h implying a varying range of stoichiometry since

$$\bar{N}_h(A_{1-x}B_x) \equiv (1-x)N_h(s) + xN_h(B) . \tag{6}$$

The ratios of the elemental volumes are plotted to the right. When a compound's ratio falls in the favored region, indicated by the dashed lines, it tends to have a wide range of N_h and also FeTc and FeRe with their apparently unfavorable N_h do have volume ratios in the "favorable" regime. A wide range in $\bar{N}_h$ arises in two ways: there may be more than one tcp phase involved but also there may be a range in stoichiometry in a given phase. A range in stoichiometry implies substitutional disorder, i.e. A atoms on B sites, or vice versa, or alternatively the introduction of vacancies. From Fig. 8 it would appear that such a range of stoichiometry is more likely when the relative sizes of the constituents is "right." Note that this favored region is not centered on a volume ratio of one.

There have been many mappings of compounds of the main group elements, of the transition elements and between main group and transition elements. With well chosen coordinates, maps with two coordinate provide reasonable separation between structures. Of recent interest is the work[11] of Pettifor where he has defined an _effective_ Mendeleev number for the elements where they are ordered according to chemical properties rather than nuclear charge. Using this quantity as both coordinates he has generated a two-dimensional map for the 1:1 compounds. The region involving the octet compounds of Fig. 6 shows similar good separation. One of the less successful regions includes the transition element-transition element systems and this is replotted in Fig. 9 using more recent crystallographic data[12] than was available to Pettifor (leading to changes in assignments of crystal structures for a few systems). Also, when more than one crystal structure was reported for some system, both are indicated in Fig. 9 without any attribution as to which is the most stable. Certain of the structures are associated with well defined regions of the map while others, such as FeB (the open diamonds) are scattered across it. There is one problem associated with structure attributions which deserves mention here. Consider the CsCl and CuAuI crystal structures which consist of alternate (100) layers of A and B atoms on the bcc

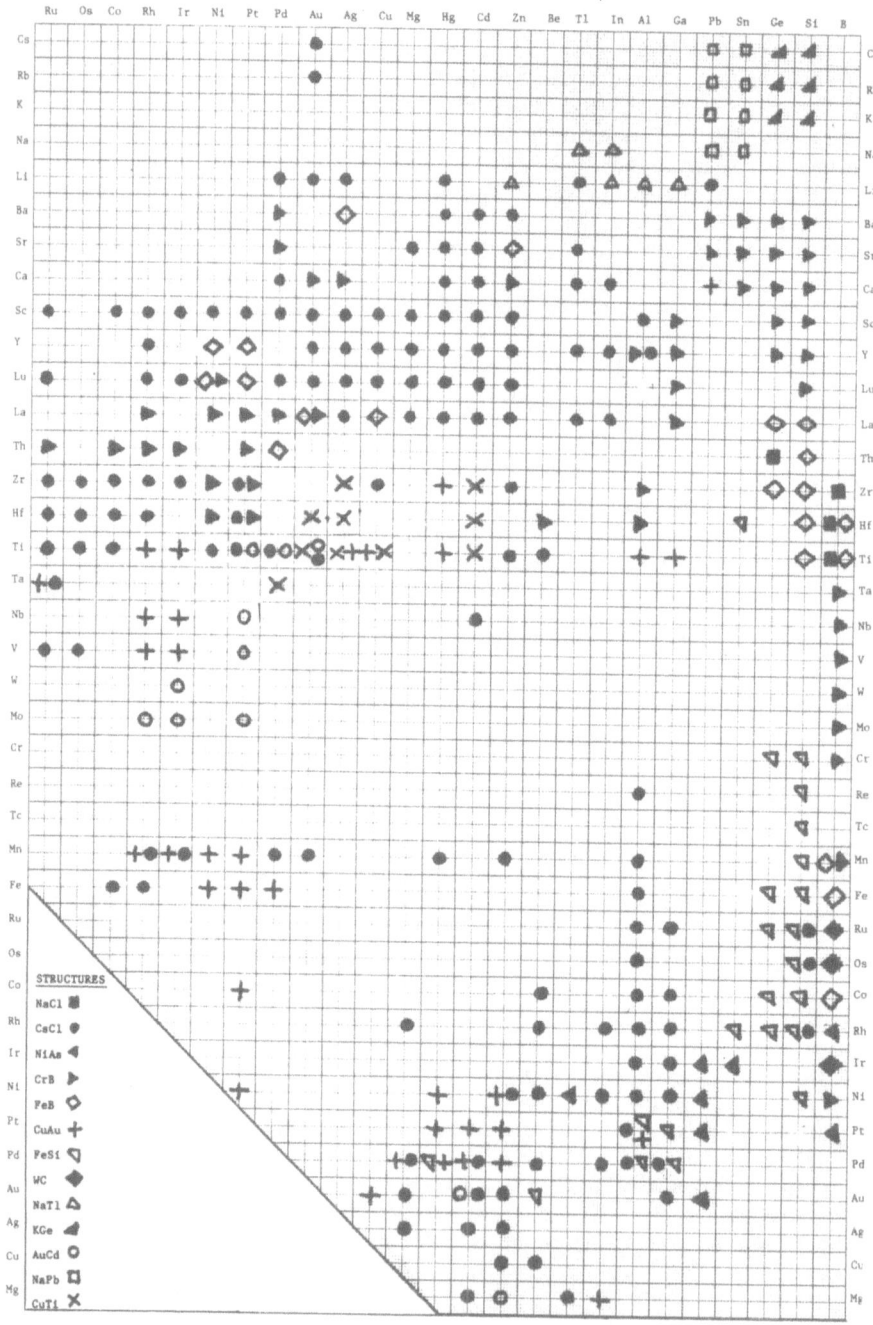

Fig. 9 A segment of the Pettifor map of 1:1 structures

and fcc lattices respectively. One structure can be taken into the other respectively by a tetragonal distortion: tetragonally distorting the CsCl structure until its c/a ratio equals √2 yields the cubic form of CuAuI. Now TiRh, TiIr and TaRu, which lie on the edge of a CsCl region in Fig. 9, have c/a, in the sense defined above, between 1.12 and 1.14, much closer to the c/a of 1.0 characteristic of CsCl than that of cubic CuAuI. Therefore, while they have the tetragonal symmetry characteristic of CuAuI (the 100 layering gives it tetragonal symmetry even if its three lattice vectors are equal), rather than the cubic of CsCl, the three compounds might be better considered to be bcc-like rather than fcc-like as implied by the CuAuI label. More will be said concerning this in the next section.

There has been considerable two-dimensional mapping of structures at 1:1 and other compositions which space does not allow to be reviewed here. These usually involve distinctly different parameters for the two coordinates and usually do better[13] than what is seen in Fig. 9. However, Pettifor's mapping with but a single intrinsic parameter, the effective Mendeleev number, is remarkable in how well it does.

One recent mapping effort deserves mention before leaving this section. There has been considerable interest in icosahedral phases which are neither crystalline nor glassy, having long range orientational but no long range periodic order. A number of these icosahedral phases involve ternary systems which also form in topologically close-packed crystal structures, such as $Al_6Mg_{11}Zn_{11}$. While there is some question as to whether the local atomic environments are the same in the crystalline and icosahedral phases, the occurrence of both led Villars and coworkers to construct[14] a structural map for these crystalline phases, the intent being to identify other systems which might order icosahedrally. Villars et al. found it necessary to employ a six-dimensional map in order to isolate the structure of interest and this, in our view, is unsatisfying. On the other hand, it would appear the mapping has provided a promising set of candidates to be scanned for icosahedral ordering.

II. CRYSTAL STRUCTURES AND WIGNER-SEITZ CELLS

Instead of considering atomic parameters and what crystal structures correlate with particular ranges of these parameters one can, instead, consider the crystal structures themselves. One can, for example, consider the packing of hard spheres in some structure, observe what spheres touch and how this touching changes as the lattice is distorted. Using such observations, over the years, Pearson has provided[15] considerable insight into the circumstances under which particular phases occur.

In a tour de force, Frank and Kasper observed[16] several systems where atoms were packed in particular twelve, fourteen, fifteen and sixteen-fold nearest neighbor environments and they then asked what crystal structures could be formed by the packing of such environments. This led to the definition of a number of hypothetical crystal

12

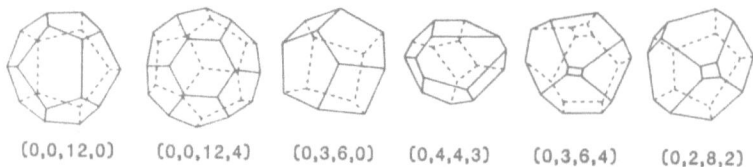

| (0,0,12,0) | (0,0,12,4) | (0,3,6,0) | (0,4,4,3) | (0,3,6,4) | (0,2,8,2) |

Fig. 10   Some Wigner-Seitz (i.e. Voronoi or Dirichlet) polyhedra appropriate to atomic sites in various crystal structures. The (0,0,12,0) and (0,0,12,4) are two of the four Frank-Kasper polyhedra while the (0,3,6,0) is one of the Bernal environments. See text for a definition of the labels.

structures, some of which have been subsequently seen and some of which may well exist though they have yet to be reported. These Frank-Kasper structures include the Laves, σ and Al5(Nb_3Sn) phases and constitute a major class of crystal structures which are important to transition metal alloy formation.

   Implicit in the above discussions is the question of what sort of local environment does a crystal structure provide a particular atomic site--how many nearest neighbors does the site have, what is the site symmetry and how many nearest neighbors are shared by a nearest neighbor pair? This question can be addressed by constructing Wigner-Seitz (or Voronoi or Dirichlet) atomic cells. In doing this, one wishes cell boundaries where the planes separating adjacent cells are positioned in proportion to the atomic radii of the atoms concerned. A clever, computationally efficient, algorithm has been devised[17],[18] for this which involves identifying the corners of the polyhedron. Some sample Wigner-Seitz polyhedra, encountered in crystal structures appear in Fig. 10. Here the notation (a,b,c,d...) indicates that there are a faces with 3 edges, b with 4, c with 5, d with 6, etc. where the number of edges, of course, indicates how many nearest neighboring sites are shared by the nearest neighbor pair responsible for the face. The Frank-Kasper phases involve two or more of the (0,0,12,0) and (0,0,12,4) which are shown and the (0,0,12,2) and (0,0,12,3) which are not. The (0,3,6,0) [along with the (0,2,8,0) and (0,4,4,0) which aren't shown] are environments originally defined by Bernal as the interstitial sites in glasses but which are common to small atom sites, such as boron, in many crystal structures.

   For some years there has been the question of whether αMn is a Frank-Kasper phase. The structure has 58 atoms in a unit cell in four different types of sites. Recently Wigner-Seitz cell constructs have shown[1] these to be (0,0,12,4), (0,0,12,4), (0,0,12,0), which are of the Frank-Kasper type and (0,1,10,3) which, with its four-fold facet, is not. Inspection of its Wigner-Seitz cells thus indicates that αMn is not a Frank-Kasper phase though it closely resembles them. Now Frank and Kasper originally designated their structure as "topologically close-packed." We think it appropriate to extend the tcp designation to structures, such as αMn, which share so many common features.

Pearson has employed[19] Wigner-Seitz cell constructs in order to resolve a more general crystallographic complication. There are many situations where systems, nominally having the same crystal structure, have such different c/a ratios or such different internal coordinates (which define exactly where atoms lie in the unit cell) that the local atomic environments are quite different, so that while the systems have common symmetry they have different bonding neighbors, hence different chemistry. Such a case arose concerning TiRh, TiIr and TaRu and Fig. 9. These systems are nominally in the CuAuI structure, which being a layered stacking of the fcc lattice would seem to imply 12 nearest atomic neighbors to each site. However, the c/a ratios of these compounds suggests that they are weakly distorted CsCl structures and that Wigner-Seitz cell constructs would indicate fourteen nearest neighbors which is characteristic of bcc lattices (such as CsCl's). Pearson has argued that in cases such as these, the symmetry designation incompletely defines the structure and that it is meaningful to make a more complete designation. We agree.

One traditional approach to the question of where the atoms are in a glassy solid has been the "quasicrystalline" model where the atoms in a glass are presumed to have the local environments encountered in the glass's crystalline counterpart. [The term "quasicrystal" has also been applied to the recently discovered icosahedral phases, but its usage here, for glassy environments, long predates this.] While it is unclear experimentally as to the extent that it's true, this view led to the inspection[20] of the crystalline transition metal rich-metalloid phases for those phases having "disorderly" structures. By disorderly we mean that the metal finds itself in a number of quite different local environments, implying a complicated radial distribution function (when averaged over the different sites). The inspection indicated that disorderly crystal structures tended to correlate with the ready occurrence of glass formation and the scan suggested some systems, which had hitherto not received much consideration, as ready glass formers. Further, it has been observed that the Ni rich-P glass system is polymorphic, that is, there are two distinct glasses at the same concentration, with distinct local structural configurations, whose occurrence depends upon how the glass is made. It has been also observed[21] that Ni rich-P systems form in crystalline phases where the P has distinctly different environments, one being of the Bernal type and others, which occur in $Ni_{12}P_5$, are not. This can be seen in Table 1. In addition, NMR and atomic density information make it possible to associate a particular crystalline environment with a particular glass. These observations suggested scanning other transition metal–metalloid systems for other cases where polymorphism might be favored. This has been done[21] for the phosphides, silicides and borides of the magnetic 3d elements as listed in Table 1. For the most part, the metalloid sites in the phosphides and borides take the form of the (0,4,4), (0,3,6) and (0,2,8) Bernal environments and of these only Cr-B looks like a contender for polymorphism. Quite different metalloid sites (implying quite different bonding?) occur for the silicides. Here all but Co-Si look like candidates for polymorphism with Ni-Si the most likely. It

Table 1

Voronoi polyhedra for the metalloid site in transition-metal rich phosphides, silicides and borides

Ni_3P	(0,3,6)	Ni_3Si	(0,12,0)	Ni_3B	(0,3,6)
$Ni_{12}P_5$	(0,3,8),(0,4,8)	δNi_2Si	(0,4,4,5)	Ni_2B	(0,2,8)
Ni_2P	(0,3,6)	θNi_2Si	(0,6,0,5)	mNi_4B_3	(0,3,6)
		Ni_3Si_2	(1,4,4,5,1) etc.	oNi_4B_3	(0,3,6)
Co_2P	(1,2,5,3)	Co_2Si	(0,5,2,8)	Co_3B	(0,3,6)
				Co_2B	(0,2,8)
Fe_3P	(0,3,6)	Fe_3Si	(0,6,0,8)	Fe_3B	(0,3,6)
ϵFe_3P	(0,3,6)	Fe_5Si_3	(0,2,8,1)	$Fe_3(B,P)$	(0,3,6)
Fe_2P	(0,3,6)			$[Fe_5B_2P]$	(0,3,6)]
				Fe_2B	(0,2,8)
Mn_3P	(0,3,6)	Mn_3Si	(0,6,0,8)	Mn_4B	(0,4,4)
Mn_2P	(0,3,6)	Mn_5Si_3	(0,2,8,1)	Mn_2B	(0,2,8)
		$Mn_{0.815}Si_{0.185}$	(0,0,12),(0,0,12,2)		
Cr_3P	(0,3,6)	Cr_3Si	(0,0,12,0)	Cr_2B	(0,2,8)
		Cr_5Si_3	(0,2,8),(0,2,8,2)	Cr_5B_3	(0,2,8), (0,8,0,2,2)

might be noted that the "different" sites tend to crop up when 5:3, 12:5 or 3:2 crystalline phases occur,, i.e. the very occurrence of such phases suggests the likelihood of polymorphism.

As already noted, an atomic environment as measured by the Wigner-Seitz construct not only counts the number of atoms in the nearest neighbor shell but also the number of nearest neighbors common to a nearest neighbor pair. Having 6 as compared with, say, 5 such neighbors will generally imply that the neighbors are spread out further radially from the bond line and thus, other things being equal, the two atoms on the bond line will lie closer together. Both the closer proximity and the sharing of more common neighbors should affect the chemical and magnetic coupling between sites. Frank and Kasper termed the 6-fold lines "major ligands" but granted the usage of the word ligand in chemistry they might be better termed "major lignates" or "major ligand lines" (and the 4-fold "minor" ligand lines). Nelson observed[22] that these may usefully be viewed as disclinations, that is as -72° disclinations introduced into icosahedra in a higher order space which is then projected down into three dimensions. Now the packing of equal sized atoms around a bond line is equivalent to packing regular tetrahedra with a common edge. It so happens that one can pack a noninteger number, five and a fraction, of such tetrahedra hence geometric considerations frustrate such packing. Nelson observed that the average number of common nearest neighbors in a Frank-Kasper structure comes very close to this noninteger number. Thus, the packing in this class of structures acts to relieve the frustration. It has also been shown[23] that the examination of such nets of disclinations offers insight into the supersymmetry associated with the three-dimensional space-group representations of centered crystals which have nonunique asymmetric units in their descriptions.

The new classes of 3d-rare earth hard magnets, with their "square" hysteresis loops of large area, tend to be tcp phases (though not strictly Frank-Kasper phases). There are important nets of major ligand lines in these systems which can be related to their magnetic

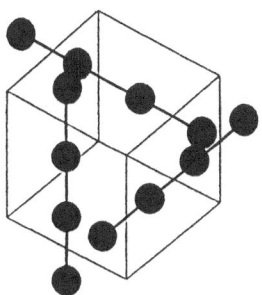

Fig. 11 The major ligand line (disclination) network in the Frank-Kasper A15 crystal (such as Nb_3Sn). Displayed is the cubic unit cell with only the majority, larger, atoms (2 per face) shown. The minority, smaller atoms are not shown and lie at the corners and the body center of the cube. The disclinations form separate chains parallel to the x, y and z axes. Here four atoms are shown for each of the chains on the near faces of the cube, the center two of each four being associated with the unit cell: the outside two lie in adjacent cells. The chains involve (0,0,12,2) Wigner-Seitz cells where the two disclinations connect a site to the neighbors on the chain. These cells are severely distorted, being compressed along the chain and quite large radially, normal to the chain. The small atom sites (not shown) have (0,0,12,0) cells and hence no disclinations.

magnetic properties[24-26]. In these magnets the 3d elements carry most of the magnetic energy while the rare earths are responsible for the anisotropy which is so necessary if these are to be useful magnets. It has been observed[24] that those 3d sites having the largest local magnetic moments lie on these major ligand lines and this has led to a search[25] for other structures, having transition metal atoms on such lines, which might be suitable hosts for hard magnetism. The disclination lines also serve[26] to define the crystal field axes at the rare earth sites and hence the rare earth anisotropies. One suspects that superconducting properties can sometimes be usefully related to disclination nets as well. Consider the A15 structure to which Nb_3Sn belongs. It is a Frank-Kasper structure with the disclination lines sketched in Fig. 11. These lines consist of separate chains of closely spaced atoms parallel to the x,y and z axes separately and, for years, these chains have been considered an essential feature of the superconductivity of these materials. However, second only to the A15's among traditional superconducting compounds are systems such as $NbC_xB_{(1-x)}$ which form in the NaCl structure which has no such ligand lines. The situation for the new high T_c superconductors is unclear.

The existence of an intimate relationship between magnetism, superconductivity and chemical activity was emphasized by B. T. Matthias over the years, i.e. one compound might be active in one

way and it or a near neighboring compound active in another. There
are even some correlations[27] between catalytic activity and bulk
magnetism. What has been suggested here is that in systems having
Frank-Kasper or tcp structures, the major ligand lines seem to have a
relevance to magnetic, superconducting and chemical properties; 5-fold
bond lines predominate and there are relatively sparse, chemically
important major ligand lines. This view breaks down when going to
lower coordinated systems where 4-fold minor ligand lines tend to
predominate as in covalently bonded compounds. Therefore, while
Wigner-Seitz constructs may be a useful tool in our overall understanding of crystal structures and how they form, tracing the major
ligand lines may be of much less universality in its usefulness. This
issue remains to be charted.

III. PREDICTIONS OF ALLOY HEATS OF FORMATION

Structural maps employing atomic parameters and observations
concerning the atomic packing associated with some crystal structure
may provide insights into the occurrence of particular phases in
particular alloy systems but they do not provide numerical estimates
of the thermodynamic quantities underlying the relative stabilities of
the competing phases. It so happens that the difference in energy
between such phases may be as small as 10^{-8} the energy of an
individual phase. Individual energies are not calculated to such an
accuracy but it is often possible to calculate separate total energies
to a sufficient common accuracy that the resulting calculated energy
difference between phases is of the requisite accuracy. There is a
wide range in the intrinsic accuracy of the calculated heats of
formation published in the literature and an even wider range, of the
order of 10^7, in the scale of the computational effort--ranging from
hand calculator efforts to hours on a supercomputer.

Obviously one employs the easiest means when it works but often
at the cost of also employing such means to cases where they are
unreliable or even yield wrong results. One of the "best buys"
because of its computational simplicity is the model Hamiltonian of
Miedema's which employs[7] as parameters the variables in Fig. 5.
Slightly more complicated is the d-band bonding model of Pettifor[28]
and Varma[29] for transition metal alloys where both the elemental
metals and compounds are given rectangular electron densities of
state. Here the essential parameters are the elemental band centers
of gravity, the band widths and the extent of band occupancy. This
scheme has been used[30] as an interpolation scheme for the 1:2 alloys
with the band parameters adjusted to fit experimental thermodynamic
data (just as Miedema's parameters were adjusted to yield agreement
with alloying trends). The results are superior to Miedema's as they
should be since they were developed and applied to a much smaller set
of alloys. Neither approach deals properly with the structure of the
phases--the numerical successes, such as they are, of the methods
relying on smooth trends in structural behavior across sequences of
systems. There is a whole array of more rigorous approaches to alloy

heats, one being the approach of Colinet and coworkers discussed elsewhere in these proceedings. In the remainder of this section we will
consider several of the assumptions and approximations common to some
of the more rigorous quantum mechanical estimates of the total energies in solids followed by three sample sets of results, of the
authors, to provide some sense to the current level of success.

Electron band theory employs a one-electron crystal potential and
solves for one-electron eigenvalues and eigenfunctions in this potential. These states are then occupied in increasing energy until the
proper number of states, per atom, is accommodated in the crystal.
This defines the Fermi energy, ϵ_F. From these occupied eigenfunctions a charge density and, in turn, a potential may be generated and
the process iterated. In doing this, a common potential is used for
occupied and unoccupied levels below and above ϵ_F. This works for
electron states which are sufficiently delocalized, having only a
small fraction of their probability at an individual atomic site.
Such a band theory is not appropriate when there are partially occupied sets of localized levels such as the open 4f shells of the rare-
earths. This does not appear to be an issue for the metals of concern
here.

Of the approximations normally employed in band theory
calculations, there is one intimately coupled to the above
assumption. It is the utilization of local density potentials. In
such a potential, interelectron terms other than the direct Coulomb
interaction are taken to depend on the density of all electrons at
that point in space where the electron sampling the potential is
located. Now exchange and correlation effects are inherently nonlocal
in nature but Hohenberg, Kohn and Sham have shown[31] that this local
density approximation is the correct leading term in an effective
potential in which one-electron equations are to be solved when
describing the ground state properties (which include the total
energy) of a system. There is yet to be a consensus as to the correct
leading correction term, to be used in computations to this local
density potential.

Now solids were considered shortly after the advent of quantum
mechanics and it was recognized that there are two regions in a solid,
one in the vicinity of an ion core where the potential is essentially
spherical (and which is most important to the total energy) and the
other in the interstitial region between ion cores. It was only
natural to mentally draw spheres around the cores. For ordered
crystalline solids, of concern here, these spheres are ordered in
periodic arrays, the electronic wavefunction is periodic and in
reciprocal space the pseudomomentum, k, associated with individual
one-electron states is a good quantum number. It is then natural to
divide the solution of the problem into two parts, one involving the
spheres and the other the region between. There are three obvious
choices for sphere size:

- One, shrink the spheres down to the sizes characteristic of the
  atomic cores and orthogonalize the interstitial wavefunctions,
  such as plane waves, to the atomic cores. The orthogonaliza-
  tion introduces terms into an effective one-electron potential

which will, in general, cause it to depend on the angular
momentum, ℓ, of the wavefunction even if the bare potential is
universal. The orthogonalized plane wave, OPW, and pseudopotential methods fall under this choice and the coordinates of
Fig. 6 are closely related to such methods.

- Two, expand the spheres so that, on average, they fill space.
  Thus some space will be twice occupied and an equal volume will
  be unoccupied. Then solve the quantum mechanics equations
  carefully inside the spheres. The interstitial region has been
  put out of existence. The augmented spherical wave, ASW,
  method is an example of such a scheme.
- Three, expand the spheres so that they touch, or almost touch.
  Solve the equations carefully inside the spheres and match
  these solutions to wavefunctions constructed from combinations
  of plane waves, or gaussians, or exponential functions etc. in
  the interstitial region. The APW, KKR, LAPW, LASTO, etc.
  methods do this.

The above methods involve solving one-electron equations which
have been derived variationally for the many-electron system and the
starting point is usually a common one-electron type of potential
which will be described below though it may look different if turned
into a pseudopotential.

Traditionally only the spherical term in the potential is
included when integrating for one-electron wave functions in the atomic spheres and only a constant term--a "muffin-tin" potential is
employed in the interstitial region. "Full potential" calculations
are increasingly being done where aspherical terms are included within
the atomic spheres and the interstitial potential, with its "ruffles",
is treated realistically. This is particularly important when the
atomic sites have low symmetry, such as at a surface or interface, or
when a structure is ill-packed with an unusually large fraction of the
crystal volume in the interstitial region. The cost in computing time
may be as much as an order of magnitude on going from muffin-tin to
full crystal potentials. Of course, the above mentioned ASW method
avoids the computational effort associated with the interstitial
region by having none. As will be seen shortly, the ASW method
sometimes yields heats of formation which are in accord with
calculations dealing explicitly with the interstitial region but this
is not always the case.

Other physical approximations are made in the course of band
theory calculations: the core electrons are or are not included in
the self-consistent process and relativistic one-electron equations
are or are not solved. Numerical approximations are also made since
there are expansions in reciprocal space and in spherical harmonics
which must be truncated somewhere: e.g., while there may not be
g-like ($\ell=4$) bands in a solid, there are g-like components in the
charge density at a site which may be viewed as being associated with
the tails of orbitals centered on adjacent sites. These high ℓ charge
components can be significant[32].

One motive for calculating heats of formation is to augment our
thermodynamic data base, particularly for metastable phases which are

inaccessible experimentally. There are, for example, phases which appear in ternary systems but not in the binaries forming their bounds yet it would be desirable when constructing the ternaries to have thermodynamic data for the binaries. This, of course, raises the question of whether the thermodynamic parameters entering a particular phase diagram construct are, in fact, strict thermodynamic quantities or, instead, effective parameters whose values depend, in part, on the model employed in the construct. Consider Fig. 1 with its terminal solution phases. In most constructs, the energy for the fcc terminal phase is taken to be

$$E(Mo_x Ir_{(1-x)}) = (1-x)E(fccIr) + xE(fccMo) \tag{7}$$

$$= (1-x)E(fccIr) + x\left[E(bccMo) + \Delta E(bcc \rightarrow fccMo)\right]$$

where ΔE is the structural energy difference involved in promoting bccMo into the fcc structure. Equation 7 neglects the loss or gain in energy associated with the fact that solute and solvent atoms are neighbors and bonded to one another. The intermediate hcp solution phase of Fig. 1 would be similarly treated,, i.e. at 1:1 we require

$$\frac{1}{2} \Delta E(bcc \rightarrow fccMo) > \frac{1}{2} \Delta E(bcc \rightarrow hcpMo) \tag{8}$$

$$+ \frac{1}{2} \Delta E(fcc \rightarrow hcpIr) < \frac{1}{2} \Delta E(fcc \rightarrow bccIr)$$

for the hcp phase to be stable relative to a bcc or fcc phase at this composition. The implication from the left hand inequality is

$$\left[\Delta E(bcc \rightarrow hcpMo) - \Delta E(bcc \rightarrow fccMo)\right] + \Delta E(fcc \rightarrow hcpIr) \tag{9}$$

$$= \Delta E(fcc \rightarrow hcpMo) + \Delta E(fcc \rightarrow hcpIr) < 0$$

and granted that the second term is positive, since fccIr is stable, this would imply that hcpMo is to be more stable than its fcc counterpart. For twenty years it has been recognized[33] that the structural ΔE found applicable to equations such as (7) in phase diagram constructs are much smaller than the ΔE resulting from band theory. This issue has become impossible to overlook with the recent generation of band calculations. The problem is illustrated in Fig. 12 for the hcp-fcc structural energy differences. The solid line depicts the values[34] of Kaufman and Bernstein standardly used in phase diagram constructs and the dashed line indicates ASW band theory results[35] of Skriver plus muffin-tin potential results of the present authors (here the two band theory treatments yield identical results). There is an order of magnitude discrepancy in the two sets of ΔE and a disagreement in sign for Ta and W implying no intermediate hcp phase, (which occurs for W-Ir as well as Mo-Ir, though not for Ta-Ir) within the

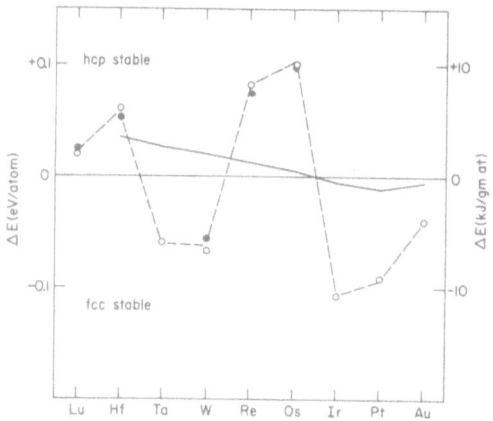

Fig. 12 fcc-hcp structural energy differences for the 5d transition
element metals Lu through Au. The open circles are Skriver's ASW
predictions (ref. 35) and the filled circles muffin-tin potential
results of the authors. The solid line is Kaufman and Bernstein's
estimate (ref. 24).

assumptions underlying Eq. 7. There have been several reconsidera-
tions[36],[37] of the experimental thermodynamic data with emphasis on
the bcc-fcc energy difference for W. These suggest that larger structural ΔE, than those traditionally used, are consistent with experiment though the "best" values are perhaps as much as a factor of two
smaller than the band theory estimates. This issue remains to be
resolved.

The heat of formation of a compound involves the difference
between the enthalpy of the compound and the weighted average of those
for the elemental solids. Now for the transition and noble metals,
the elemental solids are close-packed systems and one might expect
that an accurate heat of formation can be calculated if the compound
is similarly close-packed. That this is the case is suggested for the
Au-Hf systems in Fig. 13. The top panel shows the phase diagram while
the bottom displays heats of formation obtained with local density
muffin-tin potential calculations. Calculations were done for AuHf in
both the CsCl and CuTi structures[32]. While CsCl involves alternate
(100) layers of Cs and Cl atoms in the bcc lattice, CuTi involves a
double stacking, first two (100) layers of Cu and then two of Ti in
the same bcc lattice. The calculations correctly predict the CuTi
structure to be the more stable of the two phases. Consider Au_3Hf in
the Cu_3Ti structure. Its energy lies below the line drawn between
pure Au and Au_2Hf. Therefore, Au_3Hf is calculated to be stable relative to a two phase system involving these two. Similarly Au_2Hf's
heat lies below the line between Au_3Hf and AuHf, so it's stable and
the same situation holds across the sequence of alloys. The four compounds, taken in their correct structures, are calculated to be stable

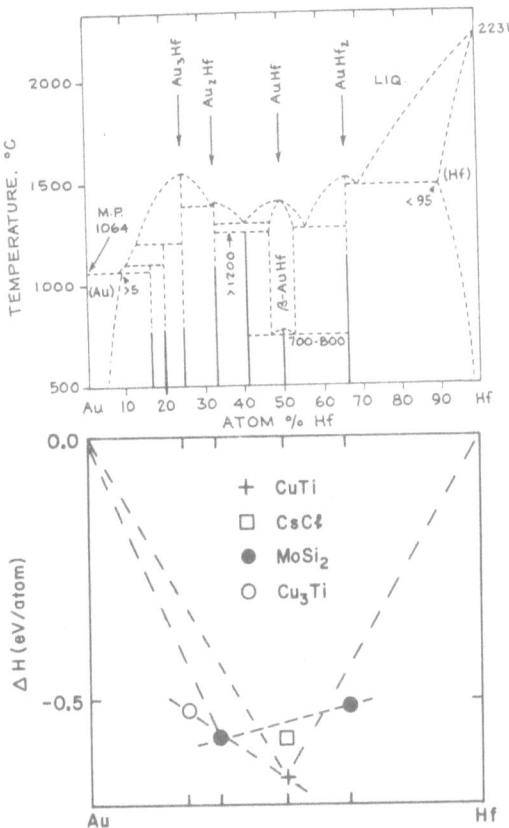

Fig. 13 The Au-Hf system. At the top is the phase diagram and at the bottom are calculated heats of formation for Au₃Hf, Au₂Hf, AuHf and AuHf₂ in the CuTi, CsCl, MoSi₂ and Cu₃Ti structures (with the exception of CsCl, these are the structures appropriate to the Au-Hf system).

with respect to one another and to the terminal phases. Au_2Hf is calculated to be the most weakly stable with respect to competition from its neighboring Au_3Hf and $AuHf$ phases and this is consistent with this compound displaying the weakest feature in the phase diagram. Thus, the essential features of the phase diagram are reproduced by the calculations. It should be emphasized that all the compounds considered here are close-packed. It might also be expected that total energy difference, between a compound and reference elemental solids or between compounds, would be most accurate when there are no severe differences in d and non-d charge counts at individual atomic sites (i.e., no significant charge flow). This condition is reasonably well met in Au-Hf.

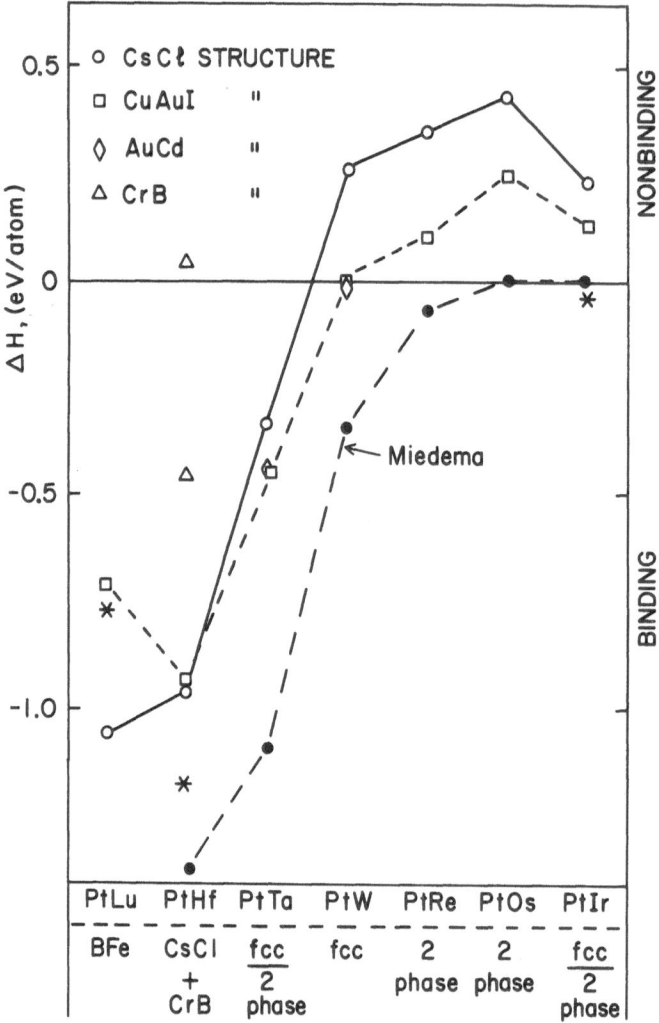

Fig. 14 Calculated heats of formation of ordered 1:1 Pt-5d element phases (ref. 38). The asterisks are experimentally obtained heats (enthalpies) and the panel at the bottom indicates the observed phase diagram behavior (the horizontal bars for PtTa and PtIr separate the differing high and low temperature phase behavior). Except for the Miedema values, the results were obtained with scalar relativistic, local density, muffin-tin potential total energy calculations.

Calculations[38] for Pt alloyed with its 5d neighbors, Lu through Ir, appear in Fig. 14. At the bottom is indicated the phase diagram behavior for the 1:1 alloys. The BFe structure involves a unit cell which was beyond the means of the computational machinery to deal with and has been neglected. The asterisks represent experimental heats data. Amongst the ordered phases, those for PtTa and PtW are calculated[38] to be suppressed by competition from ordered phases at other compositions, that is their calculated heats lie above the lines drawn between heats of adjacent phases in the sense of Fig. 12. No calculations were done for disordered phases which prevail here for PtW (PtTa and PtIr are two phase at low temperatures). The CsCl structure is correctly calculated to be stabler than CuAuI for PtHf but there are serious problems with PtHf in the CrB structure. This structure is ill-packed, comprised of close Cr-B pairs with large interstices along lines between pairs. The atomic sites are of low symmetry with nonspherical terms as low as $\ell=1$ in the site potential. The upper of the two plotted calculated heats was obtained with muffin-tin potentials with nonoverlapping atomic spheres. These spheres are smaller than usual due to the ill-packed nature of the structure. The more bound value for the CrB heat was obtained with a muffin-tin potential calculation where an empty atomic sphere was dropped into the interstice. This halves the error in this heat relative to those of the competing compounds. We expect that CrB's heat can be brought into line by employing full potential calculations. This issue will be investigated in the future. This set of alloys was shown because the calculation for PtHf in the CrB structure is, by far, the worst case we've encountered. Other than this, the heats represented here are consistent with phase diagram behavior and in semiquantitative agreement with the experimentally reported values (the disagreement for PtLu and PtHf is of the order of the experimental uncertainties).

Also shown in Fig. 14 are heats of formation calculated with Miedema's model Hamiltonian. They are consistently more bound than those from the band calculations and this is typically the case for transition metal alloys. The Miedema values are in as good agreement with experiment here although this is not the case for some transition metal alloy sequences and we would argue that the sign for PtRe and the magnitude for PtW is inconsistent with known phase diagram behavior.

Another sampling of heats of formation, calculated with muffin-tin potentials, is shown in Fig. 15. These are for ordered 1:1 compounds of Pd and Pt alloyed with the bottom ends of the 4d and 5d rows. Miedema values for the heats, which are not shown, lie below the band theory values much as in Fig. 14. Here, again, the band theory heats are in accord with the available experimental data, the asterisks. Note that the spread in the experimental data is of the order of the disagreements between theory and experiment. One also sees, for those cases where an ordered phase is reported to be the ground state, that the calculations have correctly predicted the stablest among the competing ordered phases—the one exception being PtHf in the ill-packed CrB structure. (As with Fig. 14, calculations were not done for the FeB structure.) There is an indication of a stronger tendency for compound formation for Pt alloyed with the 5d elements than for alloying between 5d and 4d and these, in turn, show stronger compound formation than Pd alloyed with the 4d. Two factors contribute to this. First, the d bandwidths of the 5d elemental

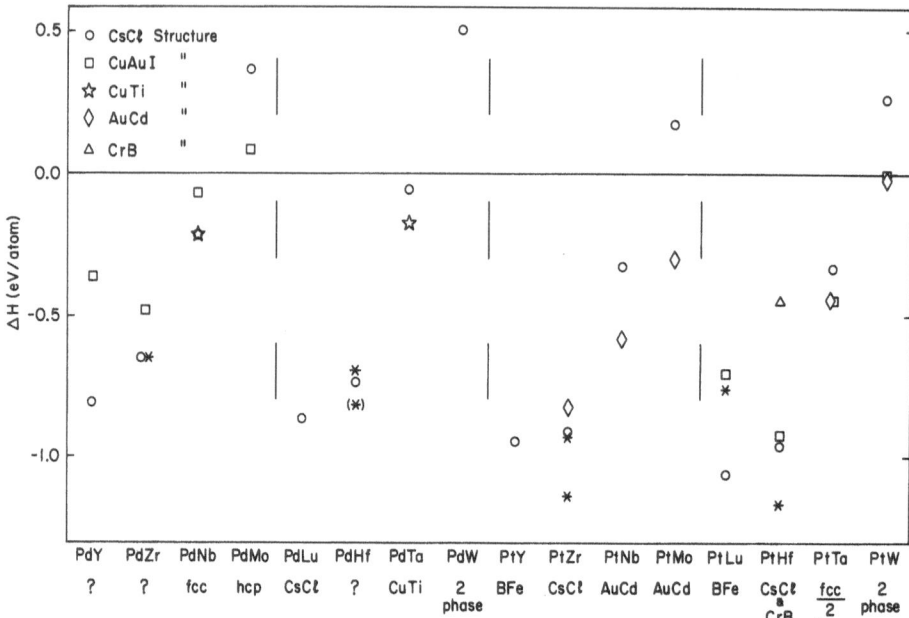

Fig. 15 Calculated heats (enthalpies) of formation of ordered 1:1 compounds of Pd and Pt alloyed with the bottom ends of the 4d and 5d metal rows (Y through Mo and Lu through W). The asterisks are experimentally obtained heats and the panel at the bottom indicates the observed phase diagram behavior. For the three systems with question marks (PdY, PdZr and PdHf) the most recent phase diagrams indicate the existence of 1:1 compounds but X-ray structure tables (ref. 12) do not list their structures. (From Fig. 9 one expects PdY most likely to be FeB or CsCl and PdZr and PdHf to be CrB, CsCl and/or CuTi.)

metals are greater than those in the 4d and all other things equal this could imply[27],[28],[29] greater bandwidths, hence stronger bonding, in the compounds as well. Second, the relative positions of the centers of gravity of the d bands of the separate alloy components affects[28],[29] the alloy's bandwidth, for the greater the separation, the greater that bandwidth and the resulting bonding energy. Pd (and Ag and La) is unusual in having d bands which lie unusually low with respect to the non-d bands. This affects both the d band occupancy and the above mentioned band separation. It is the interplay of these two factors which affects the bonding trends seen in Fig. 15.

In this section we have concentrated on trends in the structural heats of the elemental transition metals and heats of formation of ordered transition metal compounds which have been obtained by the present authors. There is, of course, other important work in the field such as that[39] of the Berkeley group on covalently bonded elemental solids and compounds. Considerable advances[40] have also been made for substitutionally disordered systems where both the enthalpy and the configurational entropy must be dealt with. The message to be taken away from this section is that total energy calculations can yield accurate heats of formation though these may sometimes require large computer investments. For example, PtHf in the CrB structure is not satisfactorily dealt with in the muffin-tin approximation and one must go to full potentials which are expected to remedy this case. The total energy calculations described here use local density theory based potentials which are held in bad repute by many chemists who deal with small molecular systems (but who offer no useful alternative for solid state calculations). Calculated cohesive energies, that is the difference in energy between free atoms and a solid, can be[41] as much as 1 eV per atom in error. This is unsatisfactory and arises from the very different nature of a solid with its constrained charge density and an atom with multiplet effects, not encountered in a metal, and a charge density tailing off into the vacuum. Calculated heats of adsorption of admolecules on solid surfaces are also unsatisfactory. The situation appears to be different for a heat of formation of a compound where the similarities between the compound and the reference solids leads to similar, cancelling errors in the total energies. Assuming this to be so, we are still not in a position where band theory can compete with experiment in the exploration of phase diagrams (though, of course, it is, when searching for low-lying metastable phases which may not be readily accesible to the experimenter). One problem is that the full exploration of competing phases requires consideration of structures with large unit cells and the scale of a calculation, employing current computational means, goes as the cube of the number of atoms in the unit cell. The σ and αMn structures which are important competing phases in transition metal alloys involve 30 and 58 atoms in the unit cell respectively, though the symmetry of the latter is body-centered implying that a calculation need deal with but 29 distinct atoms. Large unit cells of 50 to 100 atoms also offer a way to study the heats of systems which are structurally as well as substitutionally disordered. To compound the problem, many of these large unit cell systems are ill-packed with low site symmetries. For example, the Frank-Kasper σ and A15 (8 atoms/unit cell) phases involve the Frank-Kasper polyhedra which can be quite distorted. The major ligand lines of the A15 structure were illustrated in Fig. 11. These involve large atomic cells which are compressed severely along the ligand lines implying substantial $\ell=2$ components in the crystal potentials. Matheiss and Hamann have observed[42] that a proper accounting of the band structure of the A15's requires the use of full potentials and this will likely prove to be the case for the tcp phases in general. Efforts are under way to devise alternative band theory techniques where the scale of a calculation depends on a smaller power of the number of atoms. Even without this, the total energy calculations of today allow considerable exploration of the stabilities and metastabilities of alloy phases.

The work at Brookhaven National Laboratory supported by the Division of Materials Sciences, U.S. Department of Energy, under contract DE-AC02-76CH00016.

26

REFERENCES

1. R. E. Watson and L. H. Bennett, 'Alpha Manganese and the Frank-Kasper Phases', Scripta Metall. **19**, 535 (1985).
2. L. Brewer and R. H. Lamoreaux, 'Phase Diagrams', in Molybdenum: Physico-Chemical Properties of Its Compounds and Alloys, L. Brewer, Ed. (Int. Atomic Energy Agency, Vienna, 1980).
3. J. T. Waber, K. Gschneidner, Jr., A. C. Larson and M. Y. Price, 'Prediction of Solid Solubility in Metallic Alloys', Trans. Metall. Soc. AIME **227**, 717 (1963).
4. J. M. Lopez and J. A. Alonso 'A Comparison of Two Parameterizations of Solid Solubility in Alloys: Thermochemical Coordinates Versus Orbital Radii Coordinates', Physica **113B**, 103 (1982).
5. B. C. Giessen and S. H. Whang, 'Metallic Glass Formation Diagrams' in Alloy Phase Diagrams, L. H. Bennett, T. B. Massalski and B. C. Giessen, Eds., Materials Research Society Symposia Proceedings **19** (North Holland, NY, 1983).
6. E. N. Kaufman, R. Viarden, J. R. Chelikowsky and J. C. Phillips, 'Extension of Equilibrium Formation Criteria to Metastable Microalloys', Phys. Rev. Lett. **39**, 1671 (1977).
7. e.g. A. R. Miedema, 'On the Heat of Formation of Solid Alloys II', J. Less-Common Metall. **46**, 67 (1976).
8. J. St. John and A. N. Bloch, 'Quantum-Defect Electronegativity Scale for Nontransition Elements', Phys. Rev. Lett. **33**, 1095 (1974).
9. R. E. Watson and L. H. Bennett, 'Model Predictions of Volume Contractions in Transition-Metal Alloys and Implications for Laves Phase Formation', Acta. Metall. **32**, 491 (1984).
10. R. E. Watson and L. H. Bennett, 'Transition-Metal Alloy Formation: The Occurrence of Topologically Closed-Packed Phases', Acta. Metall. **32**, 477 (1984).
11. D. G. Pettifor, 'New Alloys from the Quantum Engineer', New Scientist, May 29, 1986, p. 48.
12. P. Villars and L. D. Calvert, Pearson's Handbook of Crystallographic Data for Intermetallic Phases (ASM, Metals Park, OH, 1985).
13. e.g. R. E. Watson and L. H. Bennett, 'Electron Factors in the Occurrence of Sigma and Structurally Related Transition Metal Alloy Phases', Scripta Metall. **12**, 1165 (1978) and R. E. Watson and L. H. Bennett, 'Transition Metals: d-Band Hybridization, Electronegativities and Structural Stability of Intermetallic Compounds', Phys. Rev. **B18**, 6439 (1978).
14. P. Villars, J. C. Phillips and H. S. Chen, 'Icosahedral Quasicrystals and Quantum Structure Diagrams', Phys. Rev. Lett. **57**, 3085 (1986).
15. W. P. Pearson, The Crystal Chemistry and Physics of Metals and Alloys, (Wiley, NY, 1972).
16. F. C. Frank and J. S. Kasper, 'Complex Alloy Structures Regarded as Sphere Packings', Acta Crystallogr. **11**, 184 (1958) and **12**, 483 (1959).

17. W. Fischer, E. Koch and E. Hellner, 'Zur Berechnung von Wirkungsbereichen in Strukturn Arorganischer Uerbindungen', N. Jb. Min. Mh. p. 227 (1971).

18. Also see, B. J. Gellatly and J. L. Finney, 'Characterization of Models of Multicomponent Amorphous Metals: The Radical Alternative to the Voronoi Polyhedron', J. Non-Crystall. Solids **50**, 313 (1982).

19. W. B. Pearson, 'Calculation in Establishing the Building Principles of the Crystal Structures of Intermetallic Phases' in Computer Modeling of Phase Diagrams, L. H. Bennett, Ed. (The Metallurgical Soc., Warrendale, PA, 1986).

20. R. E. Watson and L. H. Bennett, 'Disorderly Crystal Structures in Transition Metal Rich-Metalloid Alloys: Implications for Glass Formation', Scripta Metall. **17**, 827 (1983).

21. R. E. Watson and L. H. Bennett, 'The Quasicrystalline Structures of Transition Metal/Metalloid Glasses', J. Magn. Magnetic Mater. **54-57**, 295 (1986).

22. D. R. Nelson, 'Order, Frustration and Defects in Liquids and Glasses', Phys. Rev. **B28**, 5515 (1983).

23. L. H. Bennett and R. E. Watson, 'Symmetry and Supersymmetry in Crystals', Phys. Rev. **B35**, 845 (1987).

24. L. H. Bennett, R. E. Watson and W. B. Pearson, 'Topology of Local Atomic Environments: Implications for Magnetism and Superconductivity', J. Magn. Magnetic Mater. **54-57**, 1537 (1986).

25. R. E. Watson, M. Melamud and L. H. Bennett, 'Disclinations and Magnetism in Rare-Earth-Transition-Metal Hard Magnets', J. Appl. Phys. **61**, 3580 (1987).

26. M. Melamud, L. H. Bennett and R. E. Watson, 'Disclinations: Their Relation to the Anisotropies of Rare-Earth Hard Magnets', Scripta Metall. **21**, 573 (1987).

27. L. H. Bennett, A. J. McAlister and R. E. Watson, 'Interstitial Compounds', Physics Today **30**, 34 (1977).

28. D. G. Pettifor, 'Theory of the Heats of Formation of Transition-Metal Alloys', Phys. Rev. Lett. **42**, 846 (1979).

29. C. M. Varma, 'Quantum Theory of the Heats of Formation of Metallic Alloys', Solid State Commun. **31**, 295 (1979).

30. R. E. Watson and L. H. Bennett, 'Optimized Prediction for Heats of Formation of Transition-Metal Alloys II', CALPHAD 8, 307 (1984).

31. e.g. Theory of the Inhomogeneous Electron Gas, S. Lundqvist and N. H. March, Eds. (Plenum, NY, 1983).

32. R. E. Watson, J. W. Davenport and M. Weinert, 'Linear Augmented Slater-type Orbital Study of Au-5d Transition Metal Alloying', Phys. Rev. **B35**, 508 (1987).

33. Phase Stabililty in Metals and Alloys, P. S. Rudman, J. Stringer and R. I. Jaffee, Eds. (McGraw-Hill, NY (1967).

34. L. Kaufman and H. Bernstein, Computer Calculation of Phase Diagrams (Academic Press, NY, 1970).

35. H. L. Skriver, 'Crystal Structure from One-electron Theory', Phys. Rev. **B32**, 1909 (1985).

36. A. P. Miodownik, 'The Phase Stability of the Elements', in Computer Modeling of Phase Diagrams, L. H. Bennett, Ed. (The Metallurgical Soc., Warrendale, PA, 1986).

37. G. Grimvall, M. Thiessen and A. F. Guillermet, 'Thermodynamic Properties of Tungsten', Phys. Rev. (to appear).

38. R. E. Watson, J. W. Davenport and M. Weinert, 'A Linear Augmented Slater-type Orbital Study of Pt-5d Transition Metal Alloying', Phys. Rev. (to appear).

39. For a recent example see K. J. Chang and M. L. Cohen 'Ab initio Psuedopotential Study of Structural and High Pressure Properties of SiC', Phys. Rev. **B35,** 8196 (1987).

40. For a recent example see C. Sigli, M. Kosugi and M. M. Sanchez, 'Calculation of Thermodynamic Properties and Phase Diagrams of Binary Transition Metal Alloys', Phys. Rev. Lett. **57,** 253 (1986).

41. L. F. Mattheiss and D. R. Hamann, 'Linear Augmented-Plane-Wave Calculation of the Structural Properties of Bulk Cr, Mo and W', Phys. Rev. **33B,** 823 (1986).

42. L. F. Mattheiss and D. R. Hamann, 'Electronic Charge Density of V_3Si', Solid State Commun. **38,** 689 (1981).

THE "MACROSCOPIC ATOM" MODEL:
AN EASY TOOL TO PREDICT THERMODYNAMIC QUANTITIES

A.K Niessen and A.R. Miedema
Philips Research Laboratories
5600 JA Eindhoven
Netherlands

ABSTRACT   The model developed by Miedema and coworkers [1-4] to predict
enthalpy effects upon alloying uses the "macroscopic atom" as a basic
concept to describe the microscopic geometry of alloys. The assumption
that enthalpy effects are generated at the interface of dissimilar
neighbouring "macroscopic atoms" enables one to predict the enthalpy of
alloying. Not so well-known are the applications of "macroscopic atom"
model to predict other quantities, such as surface energies, the enthalpy
of formation of monovacancies and related topics. These quantities will
be reviewed in this paper, emphazing the ease to estimate their values
in a very simple way from a few experimental observations. This paper
is a condensed version of a manuscript for three chapters of a book to
be published by North-Holland Publishing Co. entitled "Cohesion in
Metals" [5].

1. THE "MACROSCOPIC ATOM" PICTURE

In problems of alloy stability we are dealing with formation energies,
that is, differences between the cohesive energies of alloys and their
constituents in the metallic state. Our basic assumption is that atoms
as they are imbedded in a pure metal can be chosen as a reference system.
Hence, many of the considerations that apply to the situation when two
macroscopic pieces of metal are brought into contact remain valid for
suitably defined "atoms in the metallic state". Energy considerations
are then made in terms of contact interactions that take place at the
interface between dissimilar atoms.
     A parameter of central importance in the description of interfacial
phenomena on an atomic scale is the electron density parameter, n_{ws}. This
quantity is defined as the electron density at the boundary of the
"macroscopic atomic" cell as derived for the pure elements in the metallic
state. An alloy or intermetallic compound is thought to be built up of
atomic cells, the electron density being kept unchanged as the cell is
removed from the metal. When dissimilar cells are brought into contact
in the alloy, there will be discontinuities in the electron density.
Elimination of such discontinuities requires energy, hence a positive

29

H. Brodowsky and H.-J. Schaller (eds.), Thermochemistry of Alloys, 29–54.

contribution to interfacial energies (heat of mixing) can be expected
determined by Δn_{ws}.

Since the effect will be a positive contribution to the energy of
alloying it must be an even (e.g. quadratic) function of the difference
in electron density of the two metals. The order of magnitude of the
proportionality constant can be estimated simply from the fact that in
the extreme case of a metal in contact with vacuum the mismatch in
electron density is to reproduce largely the pure metal surface energy.
In fig. 1 we show that there exists an approximate proportionality between
the surface energy of a solid metal at zero temperature, γ^0, and n_{ws}.

The values for n_{ws} can in principle be calculated easily for
non-transition metals; for these metals an acceptable assumption is that
the total electronic charge distribution in a metal crystal can be looked
upon as a superposition of the charge distributions of free atoms placed
at the lattice points. As it is essential to have n_{ws} values for transition
metals, too, these values are estimated in the following way. At first,
for non-transition metals there exists a rather accurate relation between
the so-calculated values for n_{ws} and the ratio of the experimental bulk
modulus K and the molar volume V of pure metals [6]

$$(n_{ws})^2 = \frac{K}{V}. \tag{1.1}$$

This relation may then be applied to find the values of n_{ws} for transition
metals. The empirical character of the procedure is underlined by leaving
the unit of n_{ws} as those of an empirical electron-density scale.

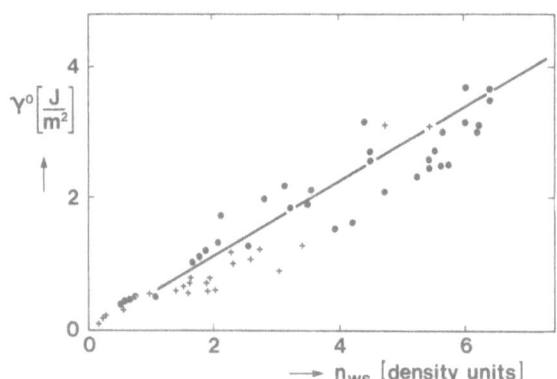

Fig. 1. The approximate proportionality between γ^0, the surface energy at zero temperature, and
the interatomic electron density n_{ws} for non-transition metals (+) and transition metals (•). The
values for γ^0 are taken from fig. 13 and those for n_{ws} from fig. 2.

For metallic alloys (metal-metal interfaces) there is another
contribution to the total formation energy. Whereas for van der Waals
substances this positive term is the only one that shows up in the heat
of mixing of liquids [2], for metallic alloys there must be an additional,
negative term in order to explain the existence of stable alloys and

intermetallic compounds. The physical background of such a negative term can easily be understood with metallic interfaces in mind. When two blocks of different metals are brought in contact, the charge redistribution will not be limited to the inside of each block, but there will be net charge transfer, governed by the difference in contact potential between the two metals. Charge will flow to places of lower potential energy, until the resulting dipole layer compensates the potential difference. Visualized on an atomic scale, this charge transfer corresponds to a negative, ionic contribution to the heat of formation. In order to describe ionicity we introduced the parameter ϕ^*. In the true spirit of the "macroscopic atom" picture, we should use the work function ϕ of the pure metallic elements, when discussing interfaces between dissimilar atoms. The asterisk in ϕ^* will remind us that the work function had to be readjusted, by amounts comparable to the experimental uncertainty of ϕ values, in order to arrive at a set of parameters relevant to alloying behaviour (the three parameters ϕ^*, $n_{WS}^{1/3}$ and $V^{2/3}$ are listed for all metals in fig. 2).

2. HEAT EFFECTS ON ALLOYING

From the definition of the adhesion energy

$$\Delta\gamma_{A\text{-}B}^{adhesion,\ T} = \gamma_A^T + \gamma_B^T - \gamma_{A\text{-}B}^{interface,\ T} \tag{2.1}$$

with (for the proportionality factor see refs. 7, 8)

$$\Delta\gamma_{A\text{-}B}^{adhesion,\ T} = -2\sqrt{\gamma_A^T\ \gamma_B^T} \tag{2.2}$$

it follows that the interfacial energy per unit area of contact equals

$$\gamma_{A\text{-}B}^{interface,\ T} = \left[\sqrt{\gamma_A^T} - \sqrt{\gamma_B^T}\right]^2. \tag{2.3}$$

From relation (2.3) it is clear that if we apply this result to non-ionic "interfaces" on an atomic scale, i.e. we consider a mole of atomic cells of species A fully surrounded by dissimilar cells of species B, we can obtain at zero temperature an expression for

$$\Delta\overline{H}_{A\ in\ B}^{interface\ (\gamma)} = c_0\ V_A^{2/3}\left[\sqrt{\gamma_A^0} - \sqrt{\gamma_B^0}\right]^2, \tag{2.4}$$

where the enthalpy effect of course scales with the molar surface area $V_A^{2/3}$ via the proportionality constant c_0. We prefer to replace γ_A^0 in relation (2.4) by $(n_{WS})_A$, making use of the approximate proportionality between these quantities. Hence,

$$\Delta\overline{H}_{A\ in\ B}^{interface\ (\gamma)} = Q''\ V_A^{2/3}\left[\sqrt{(n_{WS})_A} - \sqrt{(n_{WS})_B}\right]^2 \tag{2.5}$$

with Q'' as proportionality coefficient. Relation (2.5) can be rewritten as

$$\Delta\overline{H}_{A\ in\ B}^{interface\ (\gamma)} = Q\ \frac{V_A^{2/3}}{\left(n_{WS}^{-1/3}\right)_{av}}\left[\left(n_{WS}^{1/3}\right)_A - \left(n_{WS}^{1/3}\right)_B\right]^2, \tag{2.6}$$

where $(n_{ws}^{1/3})_{av}$ is the mean value of $(n_{ws}^{1/3})_A$ and $(n_{ws}^{1/3})_B$. For not too large differences in n_{ws} the two relations are equivalent [2] while $Q = 9/4\, Q''$.

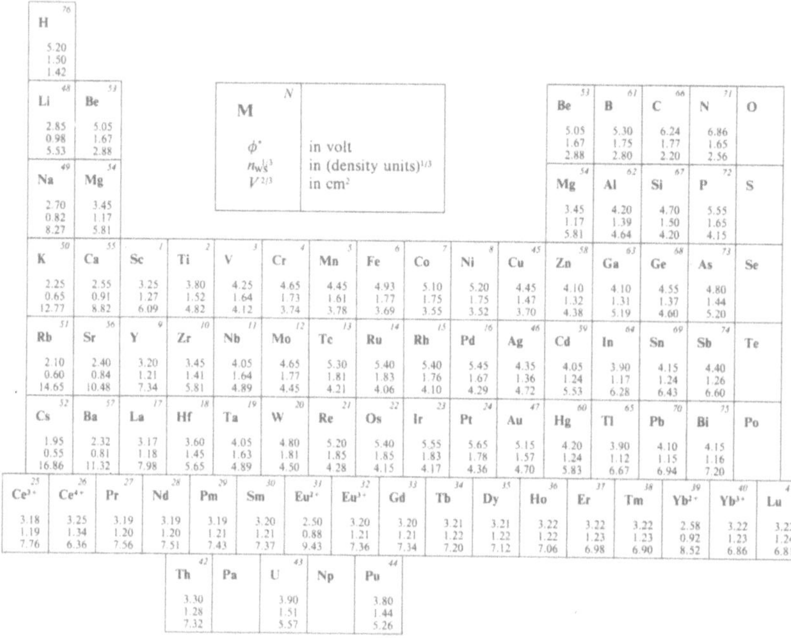

Fig. 2 — Legend: M ; ϕ^* in volt ; $n_{ws}^{1/3}$ in (density units)$^{1/3}$; $V^{2/3}$ in cm²

Each element block lists (top to bottom): ϕ^*, $n_{ws}^{1/3}$, $V^{2/3}$.

Element	ϕ^*	$n_{ws}^{1/3}$	$V^{2/3}$
H (76)	5.20	1.50	1.42
Li (68)	2.85	0.98	5.53
Be (53)	5.05	1.67	2.88
Na (49)	2.70	0.82	8.27
Mg (54)	3.45	1.17	5.81
K (50)	2.25	0.65	12.77
Ca (55)	2.55	0.91	8.82
Sc (1)	3.25	1.27	6.09
Ti (2)	3.80	1.52	4.82
V (3)	4.25	1.64	4.12
Cr (4)	4.65	1.73	3.74
Mn (5)	4.45	1.61	3.78
Fe (6)	4.93	1.77	3.69
Co (7)	5.10	1.75	3.55
Ni (8)	5.20	1.75	3.52
Cu (45)	4.45	1.47	3.70
Zn (58)	4.10	1.32	4.38
Ga (63)	4.10	1.31	5.19
Ge (68)	4.55	1.37	4.60
As (73)	4.80	1.44	5.20
Se			
Rb (51)	2.10	0.60	14.65
Sr (56)	2.40	0.84	10.48
Y (9)	3.20	1.21	7.34
Zr (10)	3.45	1.41	5.81
Nb (11)	4.05	1.64	4.89
Mo (12)	4.65	1.77	4.45
Tc (13)	5.30	1.81	4.21
Ru (14)	5.40	1.83	4.06
Rh (15)	5.40	1.76	4.10
Pd (16)	5.45	1.67	4.29
Ag (46)	4.35	1.36	4.72
Cd (59)	4.05	1.24	5.53
In (64)	3.90	1.17	6.28
Sn (69)	4.15	1.24	6.43
Sb (74)	4.40	1.26	6.60
Te			
Cs (52)	1.95	0.55	16.86
Ba (57)	2.32	0.81	11.32
La (17)	3.17	1.18	7.98
Hf (18)	3.60	1.45	5.65
Ta (19)	4.05	1.63	4.89
W (20)	4.80	1.81	4.50
Re (21)	5.20	1.85	4.28
Os (22)	5.40	1.85	4.15
Ir (23)	5.55	1.83	4.17
Pt (24)	5.65	1.78	4.36
Au (47)	5.15	1.57	4.20
Hg (60)	4.20	1.24	5.83
Tl (65)	3.90	1.12	6.67
Pb (70)	4.10	1.15	6.94
Bi (75)	4.15	1.16	7.20
Po			
Be (53)	5.05	1.67	2.88
B (61)	5.30	1.75	2.80
C (66)	6.24	1.77	2.20
N (71)	6.86	1.65	2.56
O			
Mg (54)	3.45	1.17	5.81
Al (62)	4.20	1.39	4.64
Si (67)	4.70	1.50	4.20
P (72)	5.55	1.65	4.15
S			
Ce³⁺ (25)	3.18	1.19	7.76
Ce⁴⁺ (26)	3.25	1.34	6.36
Pr (27)	3.19	1.20	7.56
Nd (28)	3.19	1.20	7.51
Pm (29)	3.20	1.21	7.43
Sm (30)	3.20	1.21	7.37
Eu²⁺ (31)	2.50	0.88	9.43
Eu³⁺ (32)	3.20	1.21	7.36
Gd (33)	3.20	1.21	7.34
Tb (34)	3.21	1.22	7.20
Dy (35)	3.21	1.22	7.12
Ho (36)	3.22	1.22	7.06
Er (37)	3.22	1.23	6.98
Tm (38)	3.22	1.23	6.90
Yb²⁺ (39)	2.58	0.92	8.52
Yb³⁺ (40)	3.22	1.23	6.86
Lu (41)	3.22	1.24	6.81
Th (42)	3.30	1.28	7.32
Pa			
U (43)	3.90	1.51	5.57
Np			
Pu (44)	3.80	1.44	5.26

Fig. 2

In order to include ionicity the form of the negative energy term involving ϕ^* must reflect, however, its origin as a dipole-layer energy:

$$\Delta\overline{H}_{A\ in\ B}^{\text{interface (ionic)}} = -P\,\frac{V_A^{2/3}}{\left(n_{ws}^{-1/3}\right)_{av}}\left[\phi_A^* - \phi_B^*\right]^2 \tag{2.7}$$

for a mole of A atoms, where $V_A^{2/3}$ represents the contact-surface area; P is a constant to be determined empirically, that contains the electronic unit of charge; $(n_{ws}^{1/3})_{av}$, enters the expression as a measure of electrostatic screening length, which determines the width of the dipole layer.

Now we can easily combine relations (2.6) and (2.7) and we obtain for $\Delta\overline{H}^\circ_{(A)\ in\ (B)}$, the heat of solution of liquid metal A in liquid metal B at infinite dilution

$$\Delta\overline{H}^\circ_{\{A\}\ in\ \{B\}} = \Delta\overline{H}_{A\ in\ B}^{\text{int}} \equiv \Delta\overline{H}_{A\ in\ B}^{\text{interface (ionic)}} + \Delta\overline{H}_{A\ in\ B}^{\text{interface }(\gamma)} \tag{2.8}$$

with

$$\Delta \overline{H}_{A \text{ in } B}^{\text{int}} = \frac{V_A^{2/3}}{\left(n_{WS}^{-1/3} \right)_{av}} \left[- P \left(\Delta \phi^* \right)^2 + Q \left(\Delta n_{WS}^{1/3} \right)^2 \right], \tag{2.9}$$

where $\Delta \overline{H}_{A \text{ in } B}^{\text{int}}$ contains our key expression for the sign of the heat of formation of binary alloys,

$$\Delta \overline{H}_{A \text{ in } B}^{\text{int}} \propto \left[- \left(\Delta \phi^* \right)^2 + \frac{Q}{P} \left(\Delta n_{WS}^{1/3} \right)^2 \right]. \tag{2.10}$$

The notation {A} and [A] will be used in this paper to indicate whether the liquid state or the solid state of A is meant. The concept that the energy effect upon alloying is generated solely at the contact surfaces between dissimilar atomic cells, i.e. the concept of interfacial energies, results consequently in the fact that $\Delta \overline{H}_{A \text{ in } B}^{\text{int}}$ does not vary with concentration as long as the dissolved atomic cells (A) remain fully surrounded by dissimilar cells (B).

For solid solutions the heat of solution can be different from that of liquid solutions, e.g. elastic energies originating from size mismatch need to be considered if the constituent metals differ in molar volume. For intermetallic compounds the elastic mismatch effect will be of no practical importance, since the compounds which exist in the equilibrium phase diagram will prefer those crystal structures that are favourable given the atomic sizes. Therefore, relation (2.10) directly applies to ordered intermetallic compounds AB_n if they are sufficiently rich in B such that A atoms are completely surrounded by B neighbours and $\Delta H_{AB_n}^{\text{for}}$, the enthalpy of formation of the ordered composition AB_n per mole A is equal to

$$\Delta H_{AB_n}^{\text{for}} = \Delta \overline{H}_{A \text{ in } B}^{\text{int}}. \tag{2.11}$$

A consequence of the simple structure of relations (2.8) to (2.11) is that, provided Q and P are truly constant for arbitrary choices of metals A and B, the sign of the heat of alloy formation is simply determined by the ratio (from eqn. 2.10)

$$W = \frac{\left| \Delta \phi^* \right|}{\left| \Delta n_{WS}^{1/3} \right|}. \tag{2.12}$$

For $W > Q/P$ the heat of alloying is negative, while in the opposite case the heat of alloy formation is positive.

The analysis of the sign of predicted and experimental heats of alloying is demonstrated in figs. 3 and 4 for binary systems involving two transition metals. Each symbol in the $|\Delta \phi^*|$ versus $|\Delta n_{WS}^{1/3}|$ diagram corresponds to a binary system. Data concerning $\Delta H_{\exp}^{\text{mix}}$ have been taken from Hultgren et al. [9] and phase-diagram information from Hansen et al. [10] and Moffat [11]. With the values of ϕ^* and $n_{WS}^{1/3}$ listed in fig. 2 a fine separation between + and • signs is obtained by drawing a straight line.

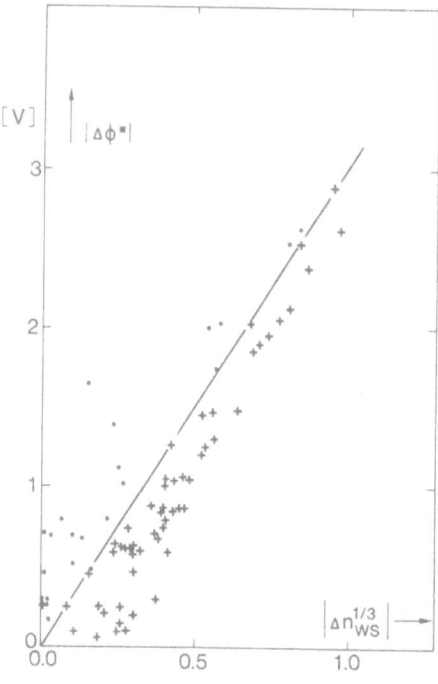

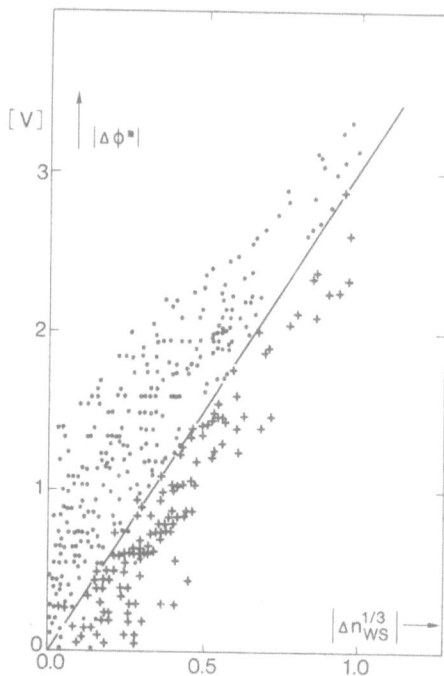

Fig. 3. Demonstration of the validity of relation (2.10) for liquid equiatomic alloys of transition metals, noble metals and the alkaline-earth metals: Ca, Sr, Ba. The actinide metals are represented by Th, U and Pu, while La, Nd, Gd, Eu and Yb represent the rare earth metals (all these metals have a clear d-character in their conduction-electron wave functions). The straight line separates the + and • signs. The deviating system with a + sign is the binary Fe - Mo. Each symbol corresponds to one binary system. The meaning of the signs is • : $\Delta H_{exp}^{mix} < 0$ for the equiatomic composition and + : $\Delta H_{exp}^{mix} > 0$ for the equiatomic composition, or the phase diagram shows in the liquid phase a central miscibility gap, while no compounds exist in the solid phase.

Fig. 4. The counterpart of fig. 3 for binary solid alloys of two transition metals. The various signs now have the following meaning: • in the binary system one or more compounds exist, which are stable at low temperatures, indicating that $\Delta H_{exp}^{for} < 0$ and a + sign if there are no compounds in the binary or if both terminal solubilities are smaller than 10 a/o, indicating that $\Delta H_{exp}^{for} > 0$. The slope of the line separating + and • signs is identical to the one in fig. 3. Hence, for both liquid alloys and solid compounds of two transition metals Q / P, i.e. the ratio of the proportionality constants in relation (2.10), equals $Q / P = 9.4 \ V^2 / (\text{density units})^{2/3}$. Notable discrepancies among binaries with a + sign instead of a • sign are Cr - Fe, Cr - U, Cu - Rh, Hf - U; binaries with a • sign instead of + sign are Au - Cr, Ca - Pr, Cr - Mn, Cu - Rh, Cr - Mo, Cu - Mn.

Since information on phase diagrams is easier to retrieve than numerical values of ΔH_{exp}^{for} the criteria for assigning + or • signs, as indicated in the caption of fig. 4, are fully based on phase-diagram information, and consequently they differ somewhat from the ones used in fig. 3. As we have already mentioned, in the solid phase not only compounds may exist but also solid solutions. The enthalpy effect for

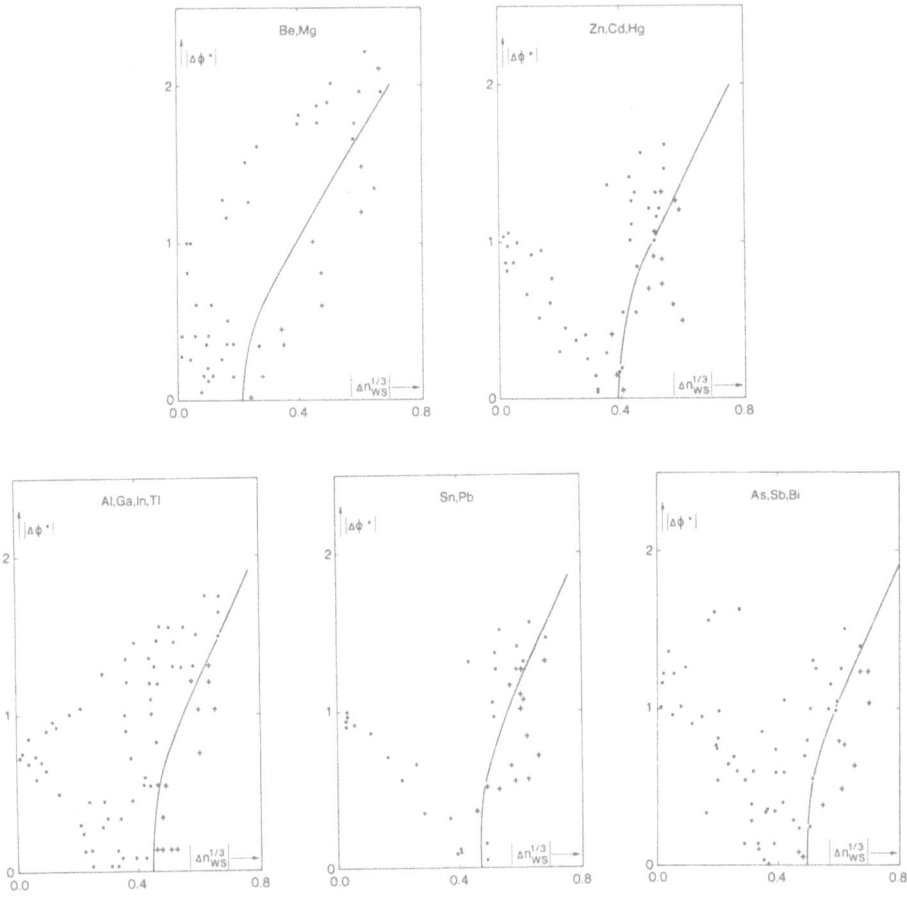

Fig. 5. The analysis as in fig. 4 for solid alloys of a transition metal and a non-transition metal. The intersection of the curve, separating + and · signs, with the horizontal axis increases clearly to higher $| \Delta n_{WS}^{1/3} |$ values with increasing number of valence electrons of the non-transition metal.

solid solutions differs from that for compounds because in the former case additional contributions are needed to predict the formation enthalpy. Consequently, unless compounds exist, the systems with substantial terminal solubilities at low temperatures are not taken into consideration for the determination of the sign.

The overall impression from figs. 3 and 4 is that the straight line with an identical slope separates the two signs very well indeed. In binary systems combining non-transition metals with transition metals there are complications. If the analysis is performed in the same way as in fig. 4 the demarcation line deviates from a straight line, and there appears to be a systematic deviation, which can be resolved most favourably in dependence on the number of conduction electrons of the non-transition metal; see fig. 5. It is still possible to separate regions of + and • signs by a demarcation line. However, the demarcation line now has a hyperbolic shape, very different from the straight line we had before. These demarcation lines are similar to the curves of equal energy derived from relation (2.9). Apparently an additional energy contribution is required. We have tried to analyse this additional enthalpy contribution along several lines. Neither from theoretical considerations, nor from purely empirical analyses, has a good understanding been achieved.

Since theoretical modelling of the third term does not exist, in contrast to the contributions owing to charge transfer and mismatch of electron density, we have decided to keep the model description as simple as possible. Henceforth, we shall use the formalism that is suggested as an outcome of the sign analyses and we modify relation (2.9) to read

$$\Delta \overline{H}_{A \text{ in } B}^{\text{int}} = \frac{V_A^{2/3}}{\left(n_{WS}^{-1/3} \right)_{\text{av.}}} \left[- P \left(\Delta \phi^* \right)^2 + Q \left(\Delta n_{WS}^{1/3} \right)^2 - R^* \right]. \tag{2.13}$$

It has been demonstrated [5] that heats of formation for alloys can be predicted, using the "macroscopic atom" model, for both liquid and solid phases. For alloys of two transition metals two counteracting terms have been found to be sufficient. Alloys of two non-transition metals in principle follow the same pattern, however an additional structure-dependent term (for semiconductors) cannot be ignored here. The heats of formation of alloys of a transition metal and a non-transition metal can be described only to a lesser degree of accuracy than for alloys of two transition metals by the introduction of an additional energy term. Although for a particular alloy the accuracy of the predicted enthalpy effect has its limitations, with respect to differences in predicted enthalpy effects for related alloys the model can be much more reliable. The model therefore deserves its strength, particularly, in relative considerations. In spite of all shortcomings, the chief merit of the model description remains, i.e. the simplicity with which it can produce values of formation enthalpies for arbitrary metal combinations.

3. SOLID SOLUTIONS

The prediction of enthalpy effects in the formation of solid solutions differs on various points from that for compounds. For 4d and 5d transition metals a detailed analysis has been given in refs. 12 and 13. The enthalpy effect for forming solid solutions of 3d metals [14,15] with other transition metals is more complicated due to the magnetic behaviour of some 3d metals and their alloys.

The first difference between compounds and solid solutions is that the degree of ordering of the atomic cells is very large in compounds but in a solid solution is assumed to approach randomness, in much the same way as in the liquid phase. Therefore the extent to which dissimilar atomic cells are each other's neighbours is smaller in solutions than in compounds. Consequently, the total chemical interaction is less in a solid solution, since the interaction only occurs at the interface of dissimilar cells.

Another difference is the existence in solid solutions of elastic energies due to size mismatch of the constituent atomic cells, whereas these energies are almost absent in compounds, since the compound chooses only those configurations in which the elastic energy is a minimum. The last difference to be mentioned is a structure dependence: a matrix with a bcc structure, for example, can be stabilized by one kind of atoms and destabilized by others.

On the basis of these ideas the enthalpy effect is predicted for solid solutions at large dilution for binary combinations of two 4d or 5d transition metals:

$$\Delta \bar{H}^\circ_{[S] \text{ in } [M]} = \Delta \bar{H}^{\text{int}}_{S \text{ in } M} + \Delta \bar{H}^{\circ\,\text{elastic}}_{[S] \text{ in } [M]} + \Delta \bar{H}^{\circ\,\text{structure}}_{[S] \text{ in } [M]}. \qquad (3.1)$$

where the chemical contribution has been taken equal to the predicted enthalpy of solution in the liquid phase. The elastic part can be estimated along the lines of classical elasticity from the distortions of a crystal lattice due to differences in size between the alloy components. The difference in size is not given by the difference in atomic size, as calculated from the molar volumes of the pure constituent metals, but by the difference in atomic size in the alloy, taking into account the change in volume [16] due to charge transfer upon alloying, see also §6.3. If we assume as a first approximation that the elastic constants do not change, both when they are used on an atomic scale or when the solute atoms are very dispersed, the elastic energy [13] per mole solute is given by

$$\Delta \bar{H}^{\circ\,\text{elastic}}_{[S] \text{ in } [M]} = E^{\text{elastic}} = \frac{2 K_S \, G_M \, (W_M - W_S)^2}{3 K_S \, W_M + 4 G_M \, W_S}. \qquad (3.2)$$

where W_M and W_S are the molar volumes of matrix and solute, respectively, corrected for charge transfer due to alloying, K_S is the bulk modulus of the solute and G_M the shear modulus of the matrix metal.

From the periodic system it is known that iso-electronic transition elements tend to have the same crystal structure in the solid phase. The structural contribution in eqn. 3.1 reflects this preference and depends on changes in energy relative to the number of valence electrons for each of the main crystallographic structures bcc, fcc or hcp. Theoretically the structure-dependent energy is expected to vary systematically with Z, the (average) number of valence electrons per atom, for solid solutions of two transition metals, too. The value of $\Delta \bar{H}^{\circ\,\text{structure}}_{[S] \text{ in } [M]}$ is predicted from eqn. 3.3, where $E_\sigma(Z)$ can be read from fig. 6:

$$\Delta \bar{H}^{\circ\,\text{structure}}_{[S] \text{ in } [M]} = (Z_S - Z_M) \frac{\partial}{\partial Z} E_{\sigma_M}(Z) + E_{\sigma_M}(Z_M) - E_{\sigma_S}(Z_S). \qquad (3.3)$$

The empirical curve [15] in fig. 6 (which is smoothed with respect to the one given in ref. 12) has been constructed from data of phase diagrams and thermodynamics supplemented by theoretical information.

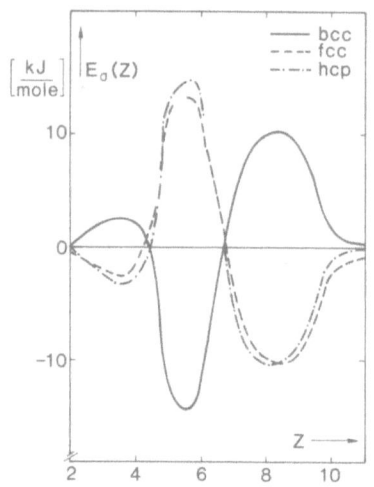

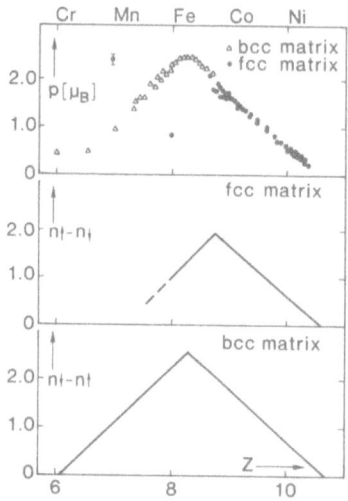

Fig. 6. The curve $E_\sigma = E_\sigma(Z)$ representing the relative lattice-stability energies.

Fig. 7. Above the magnetic moment P versus Z (Slater-Pauling curve [17]) and below the deduced $(n\uparrow - n\downarrow)$ curves for fcc and bcc crystal structures.

Predictions from eqs. 3.1 to 3.3 for combinations of non-magnetic transition metals correspond excellently with the terminal solubilities as observed in the respective phase diagrams.

Obviously the structural energy $E_\sigma(Z)$ as shown in fig. 6 cannot be applied readily to ferromagnetic materials: ferromagnetic Fe has a bcc structure and ferromagnetic Co has at low temperatures the hcp structure in contrast to the lowest values of $E_\sigma(Z)$ given for $Z = 8$ and $Z = 9$. Although we are unable to give the full energy balance between ferromagnetic and paramagnetic materials, the structural energy $E_\sigma(Z)$ can be predicted for ferromagnetic materials, too. In a magnetically ordered system the occupation of the states in the subbands with spin-up and spin-down is not the same and therefore the structural energy can be looked upon as the sum of two contributions, one representing the metal with spin-up and another that with spin-down. Of course, both contributions may be found from fig. 6. One of these matrices corresponds effectively to a matrix metal with Z_1 valence electrons and the other one to Z_2 electrons such that $(Z_1 + Z_2) = 2Z$, where Z denotes the total number of valence electrons, ($Z = 8$ for Fe and $Z = 9$ for Co). Assuming that the structural energy can also be predicted for non-integer values of Z similarly to what we assumed for paramagnetic metals, we can construct the structural energy for magnetic metals if either Z_1 or Z_2 is known.

As long as one of the matrices is effectively related to a metal with a completely filled d band (at $Z_1 = 10.5$), the difference in the average number of electrons per atom with spin-up and spin-down, $(n\uparrow - n\downarrow)$, will increase linearly with decreasing Z from zero at $Z = 10.5$. Not only $(n\uparrow - n\downarrow)$ will increase, but also the magnetic moment per atom.

With this in mind, a connection of $(n\uparrow - n\downarrow)$ with Z, which in fact represents the Slater-Pauling curve, becomes obvious. Therefore we take from now on this Slater-Pauling curve into account to determine $(Z_1 - Z_2)$, as will be explained in fig. 7. In the upper part of this figure the experimentally observed values of the atomic moment in ferromagnetic solid solutions with bcc and fcc structures are plotted versus Z. The value for fcc Fe is from ref. 18. If we deduce from this behaviour two separate curves for $(n\uparrow - n\downarrow) = (Z_1 - Z_2)$ versus Z for the two crystal structures the curves in the lower parts of fig. 7 are obtained.

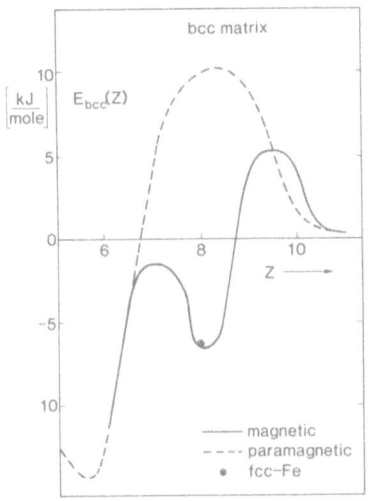 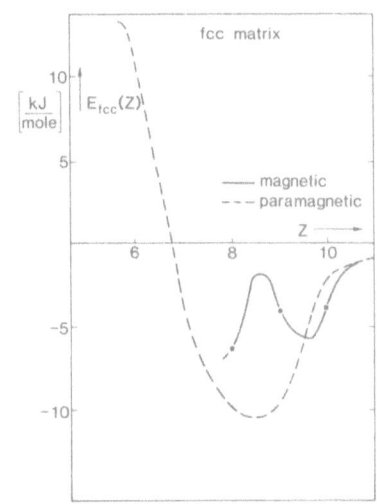

Fig. 8. The curve $E_\sigma = E_\sigma(Z)$ for a paramagnetic and a magnetic bcc matrix.

Fig. 9. The curve $E_\sigma = E_\sigma(Z)$ for a paramagnetic and a magnetic fcc matrix.

These curves show that $(n\uparrow - n\downarrow)$, and so $(Z_1 - Z_2)$, goes through an extremum with decreasing Z. With $(Z_1 - Z_2)$, evaluated according to these curves in fig. 7, we can construct $E_\sigma^{magn}(Z)$ for $\sigma = $ bcc and $\sigma = $ fcc, and even for $\sigma = $ hcp, if for this crystal structure $(n\uparrow - n\downarrow)$ behaves identically as for the bcc and fcc structures for $8.5 < Z < 10.5$. The constructed $E_\sigma^{magn}(Z)$ is shown in figs. 8-11; in figs. 8 and 9 $E_\sigma^{magn}(Z)$ with $\sigma = $ bcc or $\sigma = $ fcc in relation to the paramagnetic $E_\sigma(Z)$ values and in fig. 10 and fig. 11 the three $E_\sigma^{magn}(Z)$ curves, but now related to each other. The lowest values of $E_\sigma^{magn}(Z)$ for Fe $(Z = 8)$, for Co $(Z = 9)$ and for Ni $(Z = 10)$ are now in full agreement with the observed pure metal structures at low temperatures. Even the difference in $E_\sigma^{magn}(Z)$ between the magnetic phases for Co and Fe is small, in accordance, too, with the observed enthalpies of transition for the respective phases at high and low temperatures.

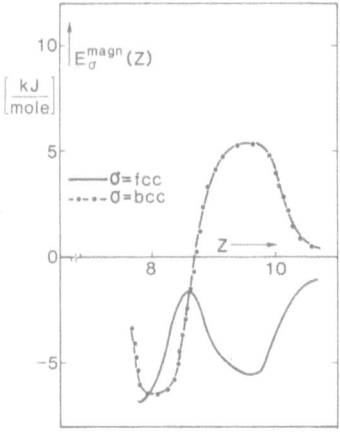 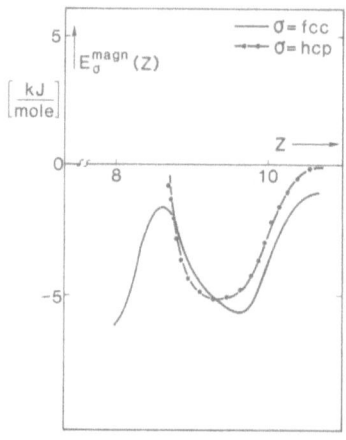

Fig. 10. The curves $E_\sigma = E_\sigma(Z)$ for a magnetic fcc matrix and a magnetic bcc matrix.

Fig. 11. The curves $E_\sigma = E_\sigma(Z)$ for a magnetic fcc matrix and a magnetic hcp matrix.

4. SOME APPLICATIONS OF $\Delta\overline{H}^{int}$

With a simple model based on the physical insight that energy effects upon alloying are generated at the common interface between dissimilar atomic cells, problems can easily be solved as will be shown in the following examples.

Recently there has been some interest in the preparation of metastable alloys of basically immiscible metals by means of laser or particle-beam heating, which involves rapid thermal quenching of high temperature liquid alloys. Here T^{mix}, the temperature above which two metals can be fully miscible, is of interest. This temperature can be estimated from the mean value of both heats of solution since this mean value may be used to approximate ΔG^{for}. From

$$\Delta G^{for} \simeq x\,(1\,-\,x)\left(\frac{\Delta\overline{H}^\circ_{\{A\}\,in\,\{B\}}\,+\,\Delta\overline{H}^\circ_{\{B\}\,in\,\{A\}}}{2}\right)\,-\,T\,\Delta S^{ideal}$$

one derives with

$$\left(\frac{\partial^2}{\partial x^2}\,\Delta G^{for}\right)_{T=T^{mix}} = 0 \tag{4.1}$$

$$T^{mix} \simeq \frac{\Delta\overline{H}^{int}_{A\,in\,B}\,+\,\Delta\overline{H}^{int}_{B\,in\,A}}{4\,R} \tag{4.2}$$

where T^{mix} is in 10^3 K and $\Delta\overline{H}^{int}_{A\,in\,B}$ and $\Delta\overline{H}^{int}_{B\,in\,A}$ in kJ per mole.

The adhesion of metals to sapphire (Al_2O_3) is of great technological interest. Experimentally [19] there is a strong adhesion of Ti to sapphire which points to chemical reactivity. At first sight this is difficult to explain because Al_2O_3 is more stable than TiO. However, as noted by Klomp [20] one should in fact consider the reaction

$$2\,Al_2O_3 \;+\; x\,Ti \;\rightarrow\; 6\,TiO \;+\; 4\,Al_{dissolved\ in\ Ti\cdot}$$

With the enthalpy of solution of Al in Ti of about $-$ 115 kJ per mole Al the reaction becomes practicable.

The stability of ternary alloys can be treated similarly as for binary alloys. Once the composition of the compound is known the degree to which a particular atomic cell is surrounded by each of the other two dissimilar atomic cells has to be estimated, given the crystal structure of the compound. If the environment factors f_B^A, f_C^A and f_C^B, representing the degree to which A atoms are surrounded in a ternary alloy by B and C neighbour atoms and the degree to which B atomic cells are in contact with C atomic cells, have been estimated, the enthalpy of formation for a ternary compound can be predicted,

$$\Delta H_{A_xB_yC_{1-x-y}}^{for} \;=\; x\,f_B^A\,\Delta\overline{H}_{A\ in\ B}^{int} \;+\; x\,f_C^A\,\Delta\overline{H}_{A\ in\ C}^{int} \;+\; y\,f_C^B\,\Delta\overline{H}_{B\ in\ C}^{int}. \qquad (4.3)$$

If transition metals form solid hydrides the entropy of the hydrogen gas is lost. This entropy change of about $-$ 130 J per degree K and per mole H_2 (NTP) counteracts the enthalpy of formation of a hydride ΔH_2 such that $\Delta G^{for} = 0$ at $T = 300$ K for $\Delta H^{for} = -$ 39 kJ per mole ΔH_2. The observed entropy change is indeed fairly constant for many different hydride systems. Hence, the main problem in predicting the stabilility of hydrides becomes the prediction of the enthalpy effect. The model used to predict the enthalpy of alloy formation cannot be applied readily to hydrogen in its standard gaseous state and an energy ΔH_H^{trans} is required to convert gaseous hydrogen into the metallic state. However, within the "macroscopic atom" model it is also possible to express the enthalpy of reaction of forming ternary hydrides by hydriding binary metal compounds in terms of ΔH^{for} of two hydrides and ΔH^{for} of the binary metal compound.

Rules for ternary solid solutions can be derived along similar lines to those we have given for binary solid solutions. We consider a solid solution of host metal A with two impurity metals, B and C, present in relatively small concentrations. In principle, the probability of finding a C atom as a nearest neighbour to a B atom differs from the "at random" statistics if the interaction energy for a B-C pair is non-zero which implies that $\Delta\overline{H}_{B\cdot C}^{pair}$ is non-zero, at zero temperature.

As pure metals the pure rare earth elements are trivalent except Eu and Yb which are divalent. Upon alloying the valence of Eu and Yb may vary from divalent to trivalent. This will be discussed here as an example of intra-atomic charge transfer. As pure metals Eu and Yb are divalent because of the extra stability given by a half-filled or a completely filled, but localized, 4f-shell, respectively. Breaking up this stable configuration and changing the 4f-electron configuration into the one commonly observed for rare earth metals requires at zero temperature an energy of conversion $\Delta H_{II\rightarrow III}^{trans}$. The enthalpy of alloying of Eu (or Yb) in the two valence modifications can easily be predicted to be

$$\Delta H^{\text{for}}(\text{ per mole Eu}^{\text{II}}) = f_B^{\text{Eu}^{\text{II}}}\, \Delta \overline{H}^{\text{int}}_{\text{Eu}^{\text{II}} \text{ in B}} \tag{4.4a}$$

for the divalent modification and

$$\Delta H^{\text{for}}(\text{ per mole Eu}^{\text{III}}) = f_B^{\text{Eu}^{\text{III}}}\, \Delta \overline{H}^{\text{int}}_{\text{Eu}^{\text{III}} \text{ in B}} + \Delta H^{\text{trans}}_{\text{Eu}^{\text{II}} \to \text{ Eu}^{\text{III}}} \tag{4.4b}$$

for the trivalent one from relation (4.4) upon inclusion of $\Delta H^{\text{trans}}_{\text{II} \to \text{III}}$.

The technique of X-ray-induced photoelectron spectroscopy allows accurate measurement of core-level binding energies in metals and in alloys. Due to electron redistribution effects these spectra show characteristics related to the cohesive energy. A scheme that can be used to calculate the difference in core-level binding energy in the metallic state with respect to the free atom, based on the approximation of the fully screened state, has been proposed by Mårtensson and Johansson [21]. The idea of a fully screened state is that upon removal of an electron from a core state there is a change in the outer electron distribution such that the outer electron distribution changes from the one characteristic for metal A to that characteristic for metal (A+1) which is the element with atomic number one higher than that of A. Experimental core-level binding energies therefore include

$$\Delta \overline{H}^{\text{int}}_{(A+1) \text{ in A}}$$

for pure metal A and

$$\Delta \overline{H}^{\text{int}}_{(A+1) \text{ in B}} - \Delta \overline{H}^{\text{int}}_{A \text{ in B}}$$

for a diluted solution of A in metal B. Since all other contributions to the core-level binding energies are identical for pure metal A and metal A diluted in metal B the core level shift $\Delta \overline{E}^{\text{core, A}}$ is

$$\Delta \overline{E}^{\text{core, A}}_{A \text{ in B}} = E^{\text{core, A}}(A \text{ diluted in B}) - E^{\text{core, A}}(A \text{ in pure A}) =$$

$$= + \Delta \overline{H}^{\text{int}}_{(A+1) \text{ in B}} - \Delta \overline{H}^{\text{int}}_{A \text{ in B}} - \Delta \overline{H}^{\text{int}}_{(A+1) \text{ in A}}. \tag{4.5}$$

5. INTERFACIAL ENERGIES

5.1. Introduction of interfacial energies

As we recall in this chapter, values of the surface energy of metals can be derived from the "macroscopic atom" model for those metals for which experimental information is lacking; also a complete set of enthalpies of formation of monovacancies in metals will be derived. Having at our disposal values of the heats of alloying, surface energies, enthalpies of formation of monovacancies, and in addition, having some insight into the limits of the applicability of the "macroscopic atom" model, a large variety of problems related to energy effects and charge redistribution effects in alloys have become accessible for simple reasoning.

The similarity between atomic cells and macroscopic metal blocks has been used in the "macroscopic atom" model to express $\Delta H^{\text{alloy}}_{A_x B_{1-x}}$, the heat of formation of an alloy with composition $A_x B_{1-x}$ per mole of A atoms, as the degree to which an atomic cell of metal A is surrounded by cells of metal B (f_B^A) times $\Delta \overline{H}_{A \text{ in } B}^{\text{int}}$

$$\Delta H_{A_x B_{1-x}}^{\text{for}} \text{ (per mole A)} = f_B^A \Delta \overline{H}_{A \text{ in } B}^{\text{int}},$$

where $\Delta \overline{H}_{A \text{ in } B}^{\text{int}}$ is a property defined for $(f_B^A = 1)$. This property is here, of course, equal to the heat of solution of A in B at large dilution. As far as possible, we here, too, prefer to express the relevant quantities as a quantity defined with $f_B^A = 1$ times f_B^A, the degree to which A atoms are surrounded by B atoms in the actual situation. Hence the Mössbauer isomer shift for atom A in a compound $A_x B_{1-x}$ will then be obtained from

$$\Delta E_{A \text{ in } A_x B_{1-x}}^{\text{isomer}} = f_B^A \Delta \overline{E}_{A \text{ in } B}^{\text{isomer}},$$

where $\Delta \overline{E}_{A \text{ in } B}^{\text{isomer}}$ denotes the isomer shift of A in AB_n with $n \gg 1$.

With regard to interfaces and interfacial energies an essential quantity is the surface area O_A of a mole of atomic cells of metal A. Since O_A will be proportional to the parameter $V_A^{2/3}$

$$O_A = c_0 V_A^{2/3}. \tag{5.1}$$

A representative value for the proportionality constant c_0 is

$$c_0 = 4.5 \times 10^8. \tag{5.2}$$

Since the value of c_0 will depend on the shape of the building block this value must be considered as an average. If the atomic cells are cubes then c_0 equals

$$c_0 = 6 N (1/N)^{2/3} = 6 N^{1/3} \simeq 5.1 \times 10^8, \tag{5.3a}$$

where N is Avogadro's number, and for spherically shaped cells

$$c_0 = 3 (4 \pi \frac{N}{3})^{1/3} \simeq 4.1 \times 10^8. \tag{5.3b}$$

5.2. Surface energy of pure metals

In the "macroscopic atom" model little difference is expected between $\gamma_{(A)}^0$, the surface energy of a solid metal at zero temperature, and $\gamma_{(A)}^0$, the (zero temperature) surface energy of this metal in the liquid phase, apart from the systematic difference induced by the average microscopic roughness of the metal-vacuum interface in both phases. The larger roughness apparent in the solid phase will increase the interfacial area between cell and vacuum, and with an identical interfacial energy per unit area, $\gamma_{(A)}^0$ will exceed $\gamma_{(A)}^0$. We shall demonstrate that the two surface energies $\gamma_{(A)}^0$ and $\gamma_{(A)}^0$ are indeed interrelated linearly.

At non-zero temperatures $\Delta G^{\text{surface}}$, the difference in Gibbs free energy between a mole of surface-layer atoms and a mole of bulk atoms is

$$\Delta G^{\text{surface}} = \Delta H^{\text{surface}} - T \Delta S^{\text{surface}}. \tag{5.4}$$

The entropy term originates from a lower Debye temperature Θ_s for atoms in the surface layer with respect to the Debye temperature Θ_b for bulk atoms and can be estimated at higher temperatures (near T^{fuse}) from the idealized (C_v versus T) curve for metals

$$\Delta S^{\text{surface}} = 3\,R\left[\ln\left(\frac{T}{\Theta_b}\right) - \ln\left(\frac{T}{\Theta_s}\right)\right] = 3\,R\,\ln\left(\frac{\Theta_b}{\Theta_s}\right), \tag{5.5}$$

where R denotes here the molar gas constant. A reasonable estimate for Θ_b/Θ_s may be 3/2 or 4/3. The value of $\Delta S^{\text{surface}}$ is then about R.

If microscopic and macroscopic surface energies are assumed to be proportional to each other $\Delta S^{\text{surface}}$ can also be derived from $\Delta G^{\text{surface}}$ by means of the "macroscopic atom" model and the experimentally observed surface energies. For liquid metal A $\Delta G_{\{A\}}^{\text{surface}}$ becomes proportional with γ^*, the atomic interfacial energy per unit area between metal and vacuum, at temperature T and with c_0 as defined in relation (5.1)

$$\Delta G_{\{A\}}^{\text{surface}} = f_{\text{vacuum}}^{\{A\}}\, c_0\left(\gamma^*\, V^{2/3}\right)^T \tag{5.6}$$

where V is the molar volume at temperature T and $f_{\text{vacuum}}^{\{A\}}$ represents the average fraction of the surface of the atomic cell in contact with vacuum. The entropy part $\Delta S_{\{A\}}^{\text{surface}}$ is calculated from relation (5.6) according to

$$\Delta S_{\{A\}}^{\text{surface}} = -\frac{\partial}{\partial T}\left(\Delta G_{\{A\}}^{\text{surface}}\right) = -f_{\text{vacuum}}^{\{A\}}\, c_0\,\frac{\partial}{\partial T}\left(\gamma^*\, V^{2/3}\right)^T =$$

$$= -f_{\text{vacuum}}^{\{A\}}\, c_0\left(\gamma^*\, V^{2/3}\right)^T\left[\frac{\partial}{\partial T}\ln(\gamma^*) + \frac{2}{3}\alpha\right]^T. \tag{5.7}$$

If γ^* is proportional to the experimental $\gamma_{\{A\}}^T$ relation (5.7) reflects the temperature dependence of $\gamma_{\{A\}}^T$ and V. Values of $\Delta S_{\{A\}}^{\text{surface}}$ can be estimated [22] reasonably well from experimental data for the expansion coefficent α and the temperature dependent $\gamma_{\{A\}}^T$ for a large number of liquid metals. It appears [22] that the values of $\Delta S_{\{A\}}^{\text{surface}}$ can be approximated by

$$\Delta S_{\{A\}}^{\text{surface}} = 5\times 10^{-8}\, f_{\text{vacuum}}^{\{A\}}\, c_0 \quad \text{(in J per K mole)} \tag{5.8}$$

for all metals within the experimental uncertainty. With $c_0 = 4.5\times 10^8$ and $f_{\text{vacuum}}^{\{A\}} = 1/3$, a reasonable value to indicate the part of the atomic cell surface in contact with vacuum, $\Delta S_{\{A\}}^{\text{surface}}$ is about 8 J per (mole K) and indeed nearly equal to the molar gas constant R. Clearly, the two approaches, relations (5.5) and (5.7), lead to nearly the same value of $\Delta S^{\text{surface}}$ and therefore relation (5.6) may be modified to read

$$\Delta G_{\{A\}}^{\text{surface}} = f_{\text{vacuum}}^{\{A\}}\, c_0\left(\gamma^*\, V^{2/3}\right)^0 - R\,T. \tag{5.9}$$

In the solid phase the approximation $\Delta S_{[A]}^{\text{surface}} \simeq R$ also appears to be correct within the range of widely scattering experimental data [23]. Taking for granted that

$$\Delta S_{[A]}^{\text{surface}} = \Delta S_{\{A\}}^{\text{surface}} = R \tag{5.10}$$

and assuming that the atomic interfacial energy is the same in both the solid and the liquid phase, then $\Delta G_{[A]}^{\text{surface}}$ can be expressed as

$$\Delta G_{[A]}^{\text{surface}} = f_{\text{vacuum}}^{[A]}\, c_0\left(\gamma^*\, V^{2/3}\right)^0 - R\,T =$$

$$= \beta\, f_{\text{vacuum}}^{\{A\}}\, c_0\left(\gamma^*\, V^{2/3}\right)^0 - R\,T \tag{5.11}$$

where $f_{\text{vacuum}}^{[A]}$ is increased by the factor β with respect to $f_{\text{vacuum}}^{(A)}$ due to the larger surface roughness of the interface (on a microscopic scale) between solid metal and vacuum.

For metal A the zero temperature value of $(\gamma^* \, V^{2/3})^0$ can be obtained by extrapolation of experimentally known values of either $(\gamma \, V^{2/3})_{[A]}^T$ or $(\gamma \, V^{2/3})_{[A]}^T$ to zero temperature, using the relations (5.9) and (5.11) if the appropriate values (e.g. for β) are taken. Experimental values of the surface energy for molten metals are known for nearly all metals, while those for the surface energy in the solid phase are available for about 20 metals, as reviewed in ref. 24.

Other values of $(\gamma^* \, V^{2/3})^0$ may be determined by considering different approaches, i.e. by starting from the enthalpy of vaporization or the dissociation energy of diatomic homonuclear particles (dimers). Since some of these quantities are more suited than others to be measured at a high temperature or under other severe environmental constraints a particular approach may be more appropriate for one metal than for the other metal. From these three values of $(\gamma^* \, V^{2/3})^0$ a recommended value is selected by taking into consideration the inherent difficulties of the three approximations.

In the "macroscopic atom" model surface energies of metals must be quantitatively related to the heat of vaporization of these metals since both quantities originate from the interface of metal atoms and vacuum. If the atomic interfacial energy at zero temperature is again equal to γ^* then ΔH_A^{vap}, the heat of vaporization of a mole of metal A at zero temperature, can be expressed to read

$$\Delta H_A^{\text{vap}} = c_0 \left(\gamma^* \, V^{2/3} \right)^0 \tag{5.12}$$

since the (evaporated) atomic cell is fully surrounded by vacuum ($f_{\text{vacuum}}^{[A]} = 1$) in this situation. From the experimental values one derives $c_0 = 5.1 \times 10^8$.

The dissociation energy D°_0 of a cluster is defined as the energy needed to dissociate a mole of clusters into free atoms. Within the "macroscopic atom" model it is straightforward to predict D°_0 for homonuclear diatomic molecules of metallic atoms since on dissociation of these dimers the surface area between the atomic cell of metal A and vacuum is increased by an amount proportional to $V_A^{2/3}$ times f_A^A, where f_A^A indicates the degree of contact of the two cells in the dimer. By analogy with relation (5.12) D°_0 is to read as

$$D^\circ_0 (\text{dimer}) = 2 \, c_0 \, f_A^A \left(\gamma^* \, V^{2/3} \right)^0. \tag{5.13}$$

For a large part of the elements the points are closely scattered around the line with a slope given by $c_0 \, f_A^A = 1.8 \times 10^8$.

From the recommended value of $(\gamma^* \, V^{2/3})^0$ a value for $\gamma_{[A]}^0$, the surface energy of a solid metal at zero temperature, can be obtained. The surface energy $\gamma_{[A]}^T$ of some metals has been determined [25] experimentally and for these metals from $(\gamma \, V^{2/3})_{[A]}^T$ a value of $(\gamma \, V^{2/3})_{[A]}^0$ can therefore be obtained from an extrapolation to zero temperature according to relation (5.9). If these values of $(\gamma \, V^{2/3})_{[A]}^0$ are compared to the recommended values of $(\gamma^* \, V^{2/3})^0$ then $(\gamma \, V^{2/3})_{[A]}^0$ is slightly larger than $(\gamma^* \, V^{2/3})^0$ where

$$\left(\gamma \, V^{2/3}\right)^0_{[A]} = \gamma^0_{[A]} \, V_A^{2/3} = \beta \left(\gamma^* \, V^{2/3}\right)^0 \qquad (5.14)$$

with $\beta = 1.1 \pm 0.1$.

The value of β being equal to 1.1 means an increase of about ten percent of $f^{[A]}_{vacuum}$ with respect to $f^{(A)}_{vacuum}$. An increase of this magnitude can be accepted as a reasonable value to indicate the microscopic roughness. With this empirical value of β values of $\gamma^0_{[A]}$ can be derived from the recommended value of $(\gamma^* \, V^{2/3})^0$ for nearly all pure metals. These values of $\gamma^0_{[A]}$ are given in fig. 13. For Sm, Tm and Pu we again give only upper limits since a change of the valence state which may indeed be expected for these metals will result in lower values of $\gamma^0_{[A]}$; see the discussion of this topic in § 6.1.

Moreover from the recommended values for $(\gamma^* \, V^{2/3})^0$ values for $\Delta H^{surface}_{[A]}$ may be derived from relation (5.11) upon assuming that the common situation in a surface layer is correctly described by $f^{[A]}_{vacuum} = \beta \, f^{(A)}_{vacuum} = 0.35$ and $c_0 = 4.5 \times 10^8$ (eqn. 5.2). The values of $\Delta H^{surface}_{[A]}$ derived in this way are given in fig. 13 together with the values of $\gamma^0_{[A]}$.

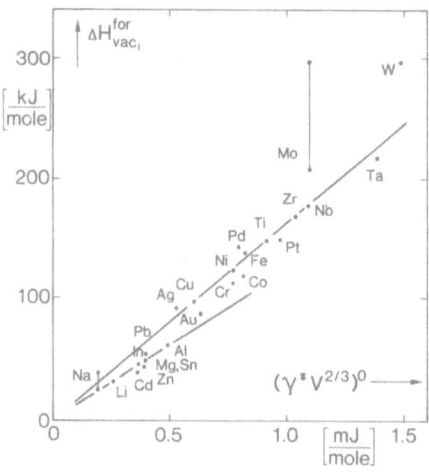

Fig. 12. The experimental values of the enthalpy of formation for a mole monovacancies $\Delta H^{for}_{vac_1}$ plotted versus the recommended values of $(\gamma^* \, V^{2/3})^0$ for a number of metals (collected in ref. 27). The straight line represents relation (5.17) with $C = 1.25 \times 10^8$ for non-transition metals and with $C = 1.64 \times 10^8$ for transition metals.

5.3. Formation enthalpy for monovacancies in pure metals

Solid state reactions are inconceivable without diffusion and for substitutional diffusion the presence of vacancies in the metal is essential. The concentration of monovacancies, c_{vac_1}, is related to $\Delta G^{for}_{vac_1}$

$$c_{vac_1} = \exp\left(\frac{-\Delta G^{for}_{vac_1}}{R\,T}\right) = \exp\left(\frac{\Delta S^{for}_{vac_1}}{R}\right) \exp\left(\frac{-\Delta H^{for}_{vac_1}}{R\,T}\right).$$

Similarly as for the creation of a surface-layer area, the entropy contribution $\Delta S_{vac_1}^{for}$ results from the change in lattice vibration frequencies when a vacancy is created. Therefore we expect that $\Delta S_{[A]}^{surface}$ and $\Delta S_{vac_1}^{for}$ are of the same order of magnitude, and from relation (5.10)

$$\Delta S_{vac_1}^{for} \simeq R \qquad (5.16a)$$

With respect to experimental information Seeger [26] quotes

$$\Delta S_{vac_1}^{for} = 0.6\,R \qquad (5.16b)$$

as the best value for both Al and Pb. The experimental data scatter strongly. Since $\Delta S_{vac_1}^{for}$ will vary only slightly for different metals the important material-dependent parameter will be $\Delta H_{vac_1}^{for}$ which, consequently is a central parameter in metallurgy.

Li	Be		M									Be	B	C	N	O
525	2700		$\gamma_{[M]}^0$	in mJ/m²								2700	3050			
40	109		$\Delta H_{[M]}^{surface}$	in kJ/mole								109	120			
40	90		$\Delta H_{vac_1}^{for}$	in kJ/mole								90	95			
Na	Mg											Mg	Al	Si	P	S
260	760											760	1160	1250	1100	
30	62											62	76	74	62	
32	50											50	63	60	50	
K	Ca	Sc	Ti	V	Cr	Mn	Fe	Co	Ni	Cu	Zn	Ga	Ge	As	Se
130	490	1275	2100	2550	2300	1600	2475	2550	2450	1825	990	1100	1000	850	
23	60	110	143	147	121	85	129	127	121	95	61	59	65	62	
24	65	115	150	155	120	90	140	135	130	100	50	48	55	50	
Rb	Sr	Y	Zr	Nb	Mo	Tc	Ru	Rh	Pd	Ag	Cd	In	Sn	Sb	Te
110	410	1125	2000	2700	3000	3150	3050	2700	2050	1250	740	675	675	535	
23	60	116	163	186	188	186	174	155	124	83	57	60	62	50	
24	63	125	170	185	200	195	185	165	135	90	41	46	50	40	
Cs	Ba	La	Hf	Ta	W	Re	Os	Ir	Pt	Au	Hg	Tl	Pb	Bi	Po
95	370	1020	2150	3150	3675	3600	3450	3000	2475	1500	575	575	600	490	
22	59	115	171	217	233	202	178	152	99	47	54	57	50	
23	62	120	180	225	250	230	210	190	160	100	40	44	48	40	

Ce	Pr	Nd	Pm	Sm	Eu	Gd	Tb	Dy	Ho	Er	Tm	Yb	Lu
1040	1080	1080	1100	<1100	450	1110	1130	1140	1150	1170	<1180	500	1225
114	115	115	115	<115	59	115	115	115	115	115	<115	60	118
120	120	120	120	<120	63	120	120	120	120	120	<120	65	125

Th	Pa	U	Np	Pu
1550		1900		<2000
153		150		<147
160		155		<155

Fig. 13. The values of γ^n, $\Delta H^{surface}$ and $\Delta H_{vac_1}^{for}$ of pure metals in relation of the position of these elements in the periodic table.

From the point of view of the "macroscopic atom" model a vacancy is a cavity in a metal of the size of a single atom. Hence, the enthalpy of formation needed for the formation of one mole of cavities is

$$\Delta H_{vac_1}^{for} = C \left(\gamma^* V^{2/3} \right)^0 \qquad (5.17)$$

where the proportionality constant C is expected to have about the same value as c_0 in eqn. 5.2 (with $f_A^{vac_1} = 1$ since a monovacancy is fully surrounded by atomic cells of metal A). Fig. 12 shows that there exists indeed a proportionality between $\Delta H_{\text{vac}_1}^{\text{for}}$ and $(\gamma^* \ V^{2/3})^0$ (recommended value) for all metals for which reasonably accurate experimental information on $\Delta H_{\text{vac}_1}^{\text{for}}$ is available [27]. However, the slope of the straight line is considerably smaller than expected and differs somewhat for transition metals ($C = 1.62 \times 10^8$) and for non-transition metals ($C = 1.25 \times 10^8$)

As demonstrated in fig. 12 the cavity model has led to a useful empirical relation for estimating vacancy formation enthalpies for all metals. An important feature is that this concept can also be used to estimate vacancy formation enthalpies for the different crystallographic sites in intermetallic compounds; see § 6.2.

6. APPLICATIONS OF KNOWN VALUES FOR γ^0 AND $\Delta H_{\text{vac}_1}^{\text{for}}$

The following examples of problems related to energy effects in metal physics and chemistry demonstrate the importance of having available (predicted) values for the surface energy at zero temperature for solid metals and the formation enthalpy of monovacancies.

6.1 Adsorption and segregation

From the "macroscopic atom" model the heat of adsorption of atoms of metal A on a surface of metal B is found easily. What really matters is the interfacial area between a mole of adsorbed A atoms and the surface formed by B atoms. Upon taking A atoms from the bulk and putting them on top of the substrate some B metal-vacuum interface has disappeared while an A metal-vacuum interface and an interface between the A and B metals have been created.

$$\Delta H_{\text{A on }[\text{B}]}^{\text{adsorption}} + \Delta H_{\text{A}}^{\text{vap}} = f_{\text{B}}^{\text{A}} \, \Delta \overline{H}_{\text{A in B}}^{\text{int}} +$$
$$+ \ c_0 \ V_{\text{A}}^{2/3} \ [\ - f_{\text{B}}^{\text{A}} \ \gamma_{\text{B}}^0 + (1 \ - \ f_{\text{B}}^{\text{A}}) \ \gamma_{\text{A}}^0 \ ]. \tag{6.1}$$

On the left-hand side $\Delta H_{\text{A}}^{\text{vap}}$ is present since the terms at the right-hand side of this relation refer to A atoms in the bulk as reference state while experimental adsorption enthalpies refer to free atoms (with $\Delta H^{\text{vap}} > 0$ and $\Delta H^{\text{adsorption}} < 0$).

The state of valence may vary for rare earth metals in alloy formation. An interesting question is whether the valence for bulk atoms and surface-layer or absorbed atoms on the other hand can differ. We shall first discuss a trivalent rare earth metal with a surface layer in which the rare earth metal atoms are in the divalent state. The change in energy per mole of rare earth metal atoms with respect to a surface formed by trivalent atoms can be estimated from the assumption that surface-layer atoms are surrounded for one third by bulk atoms, for one third by similar surface-layer atoms and for one third by vacuum.

$$\Delta H_{\text{III} \rightarrow \text{II}}^{\text{surface}} = \Delta H_{\text{II}}^{\text{surface}} - \Delta H_{\text{III}}^{\text{surface}} + \frac{1}{3} \, \Delta \overline{H}_{\text{II in III}}^{\text{int}}. \tag{6.2}$$

If the appropriate values [4] for Eu ($\Delta \overline{H}_{\text{II in III}}^{\text{int}}$ without inclusion of $\Delta H_{\text{Eu}}^{\text{trans}}$) are inserted in relation (6.2) one obtains

$$\Delta H_{Eu^{III} \to Eu^{II}}^{surface} \simeq -35 \text{ kJ per mole.} \tag{6.3}$$

Since bulk Eu is already stable in the divalent modification the result is uninteresting. However, the same value applies for example to Sm. Here the divalent state for surface-layer atoms on top of trivalent bulk Sm becomes more stable by 35 kJ per mole, which for a metal like Sm is decisive. A surface layer of divalent Sm has indeed been observed [28] by X-ray photoelectron spectroscopy.

In principle the description for the heat of adsorption of metals on metal substrates also applies to the enthalpy of adsorption of non-metallic species, e.g. Xe on metals. The difference between a metal adsorbate and Xe adsorbed on metal substrates is found in the order of magnitude of the various contributions in relation (6.1). The precision with which $\Delta \bar{H}_{A \text{ in } B}^{int}$, where A = Xe, can be predicted presents a difficulty. Hence, the problem here is treated differently. The enthalpy of adsorption can be given as (eqn. 2.1)

$$\Delta H_{A \text{ on } [B]}^{adsorption} = f_B^A c_0 V_A^{2/3} \Delta\gamma_{A-B}^{adhesion, 0} = -c'_A \sqrt{\gamma_{[B]}^0}. \tag{6.4}$$

Here c'_A is a quantity fully determined by properties of the noble gas.

Within the "macroscopic atom" model it is relatively easy to predict the atomic composition of the surface layer (c_A^l) once the bulk composition c_A is known. For dilute alloys

$$\frac{c_A^l}{c_A} = \exp\left(\frac{-\Delta H_{A \text{ in } B}^{surface \ segregation}}{R T}\right),$$

where the segregation enthalpy $\Delta H_{A \text{ in } B}^{surface \ segregation}$ is defined as the change in enthalpy upon transferring an atom of the minority constituent A from the bulk metal to the surface layer. Quantities relevant with respect to the prediction of the segregation enthalpy are the surface energies of solute metal and solvent metal (solutes with surface energies lower than those of the solvents will segregate) and the heat of solution (solutes with a positive enthalpy of solution tend to segregate, too). If f_B^A depicts the degree to which the atoms of the minority constituent A in the surface layer (fig. 3.21) are surrounded by dissimilar neighbours of the solid matrix metal B the enthalpy of segregation becomes

$$\Delta H_{[A] \text{ in } [B]}^{surface \ segregation} = \Delta H_{[A]}^{surface} - \Delta H_{[B]}^{surface} - (1 - f_B^A) \Delta \bar{H}_{A \text{ in } B}^{int}. \tag{6.5}$$

6.2 Vacancies, rapid diffusion and crystallization temperatures

From the model in which the energy of vacancy formation is related to the surface energy of a hole it is possible to derive the formation energy for vacancies in intermetallic compounds, too. Firstly we consider a compound AB_n with n so large that $f_B^A = 1$. If the atoms are of equal size, a vacancy at an A site is equivalent to a monovacancy in pure metal B. If the atoms differ in size a vacancy on an A site can still be looked upon as a cavity in B metal. By analogy to relation (5.17) one expects that the enthalpy of formation of such a vacancy will scale with $(V_A/V_B)^{2/3}$. In fact, for large A atoms this enthalpy will be larger: for a hole in metal B larger than the space occupied by a monovacancy in pure metal B the coefficent C will tend to increase, relation (5.17), towards the value of c_0 given in relation (5.2). This can be approximated by taking the scaling factor as $(V_A/V_B)^m$ with m > 2/3 . Since it is known that in a pure metal monovacancies generally attract each other, the exponent m must be smaller than unity. With m = 5/6 as a compromise between these two considerations we define

$$\Delta H^{for}_{vac_1 \text{ at A site in } AB_n} = \left(\frac{V_A}{V_B} \right)^{5/6} \Delta H^{for}_{vac_1 \text{ in } B} . \tag{6.6}$$

The consequences of this approach can be illustrated for compounds of La and Ni with n >> 1

$$\Delta H^{for}_{vac_1 \text{ at La site in } LaNi_n} = 320 \text{ kJ per mole vacancies}$$

and

$$\Delta H^{for}_{vac_1 \text{ at Ni site in } NiLa_n} = 40 \text{ kJ per mole vacancies.}$$

These values can be compared with the values of $\Delta H^{for}_{vac_1}$ in pure La and in pure Ni, which are both (accidentally) about 110 kJ per mole vacancies. A vacancy in Ni is relatively small in size while the surface energy for Ni is large; the opposite applies to La. In $LaNi_n$ a vacancy at a La site is a large cavity with the higher surface energy and correspondingly an extremely large enthalpy of vacancy formation. These values are simply first-order approximations since volume changes upon alloying are not included. The above values nevertheless explain why in compounds like $LaNi_5$ vacancies do not occur at La sites, while in compounds like $NiLa_3$ unusually large concentrations of vacancies at Ni sites will occur near the melting point. Within the fairly crude approximations made in § 4.1 and neglecting effects related to the heat of alloy formation the enthalpy of formation of a vacancy in a compound of arbitrary composition $A_x B_{1-x}$ at a B site, surrounded by A and B atoms for the respective fractions

$$f^{vac_1 \text{ at B site}}_A \quad \text{and} \quad f^{vac_1 \text{ at B site}}_B,$$

becomes with

$$f^{vac_1 \text{ at B site}}_A \simeq f^B_A \quad \text{and} \quad f^{vac_1 \text{ at B site}}_B \simeq f^B_B$$

$$\Delta H^{for}_{vac_1 \text{ at B site in } A_x B_{1-x}} = f^B_A \Delta H^{for}_{vac_1 \text{ at B site in } BA_n} + f^B_B \Delta H^{for}_{vac_1 \text{ in } B}. \tag{6.7}$$

In order to demonstrate relation (6.7) we consider ordered compounds in two binary systems. In fig. 14 we show the predicted $\Delta H^{for}_{vac_1}$ at the two different sites in ordered compounds in the binary system Y - Ni as well as in the binary Ti - Mn. The behaviour of $\Delta H^{for}_{vac_1}$ versus concentration differs for these two systems since

$$\gamma^0_{[Y]} < \gamma^0_{[Ni]}, \text{ while } \gamma^0_{[Ti]} > \gamma^0_{[Mn]}.$$

The present model is only a first approximation in the sense that effects of changes in volume upon alloying and of the heat of alloying have been neglected. However, since relation (6.7) points to a pronounced asymmetry, this first approximation is already quite relevant.

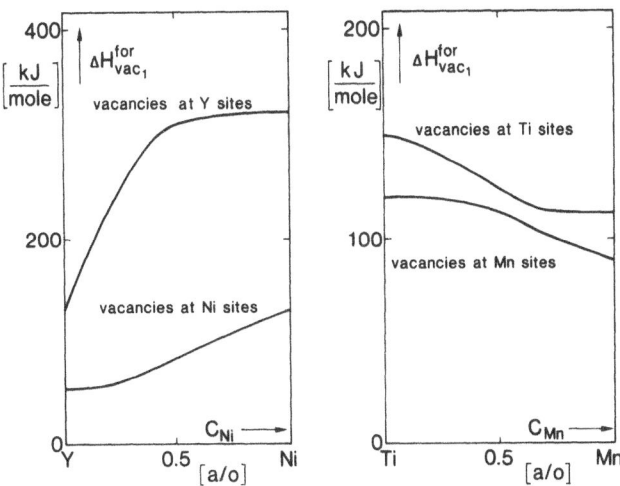

Fig. 14. The enthalpy of formation of monovacancies at the two different sites in a binary compound. The values of $\Delta H^{for}_{vac_1}$ vary strongly for these sites as demonstrated for ordered compositions in the Y - Ni system as well as in the Ti - Mn system.

A special case of solute-solute interaction in ternaries is the solute-vacancy interaction in a binary solid solution of B in A. If the two types of atomic cells are of the same size the interaction energy between solute and a vacancy is

$$\Delta \overline{H}^{pair}_{B\text{-}vac_1} = f^{vac_1}_B \left[\Delta H^{for}_{vac_1 \, in \, B} - \Delta \overline{H}^{int}_{B \, in \, A} - \Delta H^{for}_{vac_1 \, in \, A} \right]. \tag{6.8}$$

The requirement of equal volumes is a rigorous one. If B atoms are larger than A atoms, B atoms attract vacancies first of all because the elastic energy owing to the volume mismatch is relaxed when B gets a vacancy as neighbour, but more importantly, the excess volume of B atoms will very effectively reduce the volume of the cavity and the highly positive vacancy formation enthalpy is reduced accordingly. The strong attractive solute-vacancy interaction for metals like Sn in Al ($V_{Sn} = 16$ cm^3 and $V_{Al} = 10$ cm^3) has to be attributed to this effect.

In binary systems in which interstitial solubility tends to present a significant fraction of the solubility of solute B in matrix metal A one encounters the phenomenon of rapid diffusion. Generally solid-state diffusion requires vacancies and $\Delta H^{for}_{vac_1}$ constitutes a substantial fraction of the activation energy for diffusion. For interstitial solutes the activation energy can be much smaller, which, at low temperatures in particular, leads to unusually fast diffusion. Recently Bakker [29] has demonstrated that the data given in the literature on fast metal-impurity diffusion can be rationalized. For matrix metals Pr, U, In, Tl, Sn and Pb Bakker finds

$$\ln \left(\frac{D^A_{A \, in \, B}}{D^A_{self}} \right)_{T^{fuse}_B} = -12 \left(\frac{V^{eff}_A}{V_B} \right) + \text{const.}$$

where $D^A_{A \text{ in } B}$ and D^A_{self} are the impurity-diffusion and self-diffusion coefficients (at the melting point of B) and V^{eff}_A the effective volume of A in matrix metal B with

$$V^{\text{eff}}_A = V_A - \Delta \overline{V}^\circ_{A \text{ diluted in B}},$$

where $\Delta \overline{V}^\circ_{A \text{ diluted in B}}$ is given according to relation (6.12). The understanding of fast diffusion in metals is of main importance to a new method of preparing amorphous alloys. Schwarz and Johnson [30] have demonstrated that, starting for instance from layers of La and Au in contact, amorphous alloys are produced by diffusion of Au in La at relatively low temperatures. The basic requirements for the formation of amorphous alloys by interdiffusion are a large and a negative heat of mixing and a significant transport of one metal by diffusion at a temperature sufficiently low to prevent nucleation of crystalline phases. From the large number of systems for which interstitial solubility is feasible it is clear that the formation of amorphous alloys by interdiffusion can be a fairly general phenomenon.

Amorphous alloys have a disordered atomic arrangement comparable with that of liquid alloys. For diffusion-limited crystallization the central parameter will be an activation energy comparable with the enthalpy of formation of a cavity of the size of the mobile atom. This enthalpy is estimated similarly to that of a vacancy in compounds (eqn. 6.7). Hence,

$$\Delta H^{\text{for}}_{\text{cavity}} = \Delta H^{\text{for}} \text{ (hole of size of B in amorph } A_x B_{1-x} \text{)} =$$

$$= f^A_B \left(\frac{V_B}{V_A} \right)^{5/6} \Delta H^{\text{for}}_{\text{vac}_1 \text{ in A}} + (1 - f^A_B) \Delta H^{\text{for}}_{\text{vac}_1 \text{ in B}}. \tag{6.9}$$

Buschow [31] was able to demonstrate that the crystallization temperature for large families of amorphous alloys can be described by

$$T^{\text{xtal}} = 7.5 \, \Delta H^{\text{for}}_{\text{cavity}} \tag{6.10}$$

where T^{xtal} is in K if $\Delta H^{\text{for}}_{\text{cavity}}$ is in kJ per mole holes.

6.3 Charge-redistribution effects

In this section we shall indicate some consequences of the 'macroscopic atom" model, in particular, those related to charge transfer from an atomic cell to an adjacent dissimilar cell. With the "macroscopic atom" model the enthalpy effect upon alloying is related to differences in parameters defined for the pure constituent metals without specifying in detail in which way the discontinuities in electronegativity and the density of electrons are equalized during the alloying process. Here we make a connection between the amount of charge transferred and the magnitude of the ionic term in the enthalpy effect. Of the two terms in relation (2.9) the first term, $- P \, (\Delta \phi^*)^2$, is directly related to charge transfer; the $- P \, (\Delta \phi^*)^2$ term, also called the ionic term, can be looked upon as the energy gained if a charge of a magnitude proportional to $- \Delta \phi^*$ is transferred over a potential difference proportional to $\Delta \phi^*$. To this ionic term we shall relate the change in volume upon alloying of two transition metals and an ionic term in the Mössbauer isomer-shift. The discontinuity in n_{ws}, leading to the second term in relation (2.9), $+ Q (\Delta n_{ws}^{1/3})^2$, is formally assumed to be equalized by intra-atomic conversion of s-like and p or d-like electron states, in principle without a change in volume [16]. It will be shown that the equalization of n_{ws} does lead to a contribution of the Mössbauer isomer-shift.

In a binary system A - B. the change in volume per mole A on forming the compound A_xB_{1-x} can be expressed as

$$\Delta V_{A_xB_{1-x}} = f_B^A \, \Delta \bar{V}^\circ \text{ A diluted in B} \tag{6.11}$$

where $\Delta \bar{V}^\circ$ A diluted in B is the volume effect per mole A if A atoms are fully surrounded by dissimilar B atoms. We assume that the change in volume reflects the transfer of electrons from the outer regions of A cells with electron density $(n_{ws})_A$ to the outer regions of B cells, with $(n_{ws})_B$. The volume effect $\Delta \bar{V}^\circ$ A diluted in B will be proportional to differences in $(n_{ws})^{-1}$ as well as to the ratio of the contribution arising from $(\Delta \phi^*)^2$ in $\Delta \bar{H}_{A \text{ in } B}^{int}$ (.i.e. $\Delta \bar{H}_{A \text{ in } B}^{ionic}$), and $[(\phi^*)_A - (\phi^*)_B]$, leading to

$$\Delta \bar{V}^\circ \text{ A diluted in B} = P_V \, V_A^{2/3} \times$$

$$\times \left[(\phi^*)_A - (\phi^*)_B \right] \left[\frac{\left(\dfrac{1}{n_{ws}}\right)_A - \left(\dfrac{1}{n_{ws}}\right)_B}{\left(n_{ws}^{-1/3}\right)_{av}} \right], \tag{6.12}$$

where

$$P_V = 3 \text{ cm (density units)}^{2/3} \text{ per volt} \tag{6.13}$$

as the empirical constant [16], derived from experimentally observed deviations of Vegard's law.

An experimentally observable quantity which reflects charge-redistribution effects on an atomic scale is the isomer shift in Mössbauer transitions. The isomer shift is expected to be an indirect measure of changes in the density of s-like electrons at the nuclear site. An interpretation of the isomer shift for various types of alloys in terms of the atomic model was suggested by van der Woude and Miedema [32,33]. It is assumed that, similar to the heat of formation, the isomer shift contains two contributions. One term stems from inter-atomic charge transfer upon alloying, the same term which is responsible for the change in volume of intermetallic compounds; it is assumed that the s-like electron density at the nucleus varies linearly with the amount of electronic charge transferred to or from the corresponding atomic cell. The second contribution reflects the amount of intra-atomic redistribution of s, d and p electron states which takes place as a consequence of the mismatch of n_{ws} at the cell boundary of two dissimilar atoms. For instance, for a transition metal a change of d-type electron-states into s-type electron-states results in an increased density of electrons, both in the outside region of the atomic cell and at the nucleus. Again, for a series of ordered alloys in the binary system A - B, the variation of the isomer shift for an A atom reflects the degree to which the A cell is surrounded by dissimilar B cells,

$$\Delta E_{\text{A in A}_x\text{B}_{1-x}}^{\text{isomer}} = f_{\text{B}}^{\text{A}} \ \Delta \overline{E}_{\text{A diluted in B}}^{\text{isomer}} \tag{6.14}$$

where $\Delta \overline{E}_{\text{A diluted in B}}^{\text{isomer}}$ is the isomer shift in the compound AB_n with $n >> 1$ such that here $f_{\text{B}}^{\text{A}} = 1$ with

$$\Delta \overline{E}_{\text{A diluted in B}}^{\text{isomer}} = P'_{\text{M}} \frac{\Delta \overline{H}_{\text{A in B}}^{\text{ionic}}}{(\phi^*)_{\text{A}} - (\phi^*)_{\text{B}}} + Q_{\text{M}} \frac{(n_{\text{WS}})_{\text{B}} - (n_{\text{WS}})_{\text{A}}}{(n_{\text{WS}})_{\text{A}}}, \tag{6.15}$$

where the actual form of the second term in eqn. 6.15 is empirical and where P'_{M} and Q_{M} are constants.

REFERENCES

[1] A.R. Miedema and P.F. de Châtel in *Theory of Alloy Phase Formation*, Ed. L.H. Bennet (Met.Soc. AIME, Ohio, 1979).
[2] A.R. Miedema, P.F. de Châtel and F.R. de Boer, *Physica* **100B** (1980) 1.
[3] A.R. Miedema, F.R. de Boer and R. Boom in *Thermodynamics of Alloys* (North-Holland Publishing Co, Amsterdam, 1981)
[4] A.K. Niessen, F.R. de Boer, R. Boom, P.F. de Châtel, W.C.M. Mattens and A.R. Miedema, *CALPHAD* 7 (1983) 51.
[5] F.R. de Boer, R. Boom, W.C.M. Mattens, A.R. Miedema and A.K. Niessen, *Cohesion in Metals* (North-Holland Publishing Co, Amsterdam, to be published)
[6] A.R. Miedema, R. Boom and F.R. de Boer in *Crystal Structure and Chemical Bonding in Inorganic Chemistry*, Eds. C.J.M. Rooymans and A. Rabenau (North-Holland Publ.Co., Amsterdam, 1975).
[7] L.A. Girifalco and R.J. Good, *J. Phys. Chem.* 61 (1957) 904.
[8] F.M. Fowkes, *J. Phys. Chem.* 67 (1963) 2538.
[9] R. Hultgren, P.D. Desai, D.T. Hawkins, M. Gleiser and K.K. Kelley, *Selected Values of the Thermodynamic Properties of Binary Alloys* (Am. Soc. Metals, Ohio, 1973).
[10] M. Hansen and K. Anderko, *Constitution of Binary Alloys* (Mc Graw-Hill, New York, Toronto, London, 1958) R.P. Elliot, *ibid. First Supplement* (1965) F.A. Shunk, *ibid. Second Supplement* (1969).
[11] W.G. Moffat, *The Handbook of Binary Phase Diagrams* (General Electric Company, Schenectady, 1977 and updates).
[12] A.R. Miedema and A.K. Niessen, *CALPHAD* 7, (1983) 27.
[13] A.K. Niessen and A.R. Miedema, *Ber. Bunsenges. Phys. Chem.* **87** (1983) 717.
[14] A.K. Niessen and A.R. Miedema, *CALPHAD Conf. Liège* (1983).
[15] A.K. Niessen, F.R. de Boer, R. Boom and A.R. Miedema, *Physica B*, A review of Co-based alloys, to be published.
[16] A.R. Miedema and A.K. Niessen, *Physica* 114B (1982) 367.
[17] W. Kuemmerle and U. Gradmann, *Phys. Status Solidi (a)* **45** (1978) 171.
[18] U. Gradmann and H.O. Isbert, *J. Magn. Magn. Mater.* 15-18 (1980) 1109.
[19] S. Morozumi, M. Kikuchi and T. Nisheno, *J. Mat. Sci.* 16 (1981) 2137.
[20] J.T. Klomp, *Mat. Res. Soc. Symposium Proc.* **40** (1985) 381.
[21] N. Mårtensson and B. Johansson, *Solid State Comm.* 32 (1979) 791.
[22] A.R. Miedema and R. Boom, *Z. Metallkde* 69 (1978) 183.
[23] H. Jones and G.M. Leak, *Met. Sci. J.* 1 (1967) 211.
[24] A.R. Miedema, *Z. Metallkde* 69 (1978) 287.
[25] G. Lang, *Z. Metallkde* 67 (1976) 549; also in *Handbook of Chemistry and Physics* (CRS Press, Cleveland 1977).
[26] A. Seeger, *Crystal Lattice Defects* 4 (1973) 221.
[27] A.R. Miedema, *Z. Metallkde* 70 (1979) 345.
[28] G.K. Wertheim, *Mat. Sci. Eng.* 42 (1980) 85.
[29] H. Bakker, *J. Less-Common Met.* 105 (1985) 129.
[30] R.B. Schwarz and W.J. Johnson, *Phys. Rev. Lett.* 51 (1983) 415.
[31] K.H.J. Buschow, *Solid State Comm.* 43 (1982) 171.
[32] A.R. Miedema and F. van der Woude, *Physica* 100B (1980) 145.
[33] F. van der Woude and A.R. Miedema, *Solid State Comm.* 39 (1981) 1097.

AN ELECTRONIC THEORY OF BINARY TRANSITION METAL BCC PHASES

C. Colinet, A. Bessoud, A. Pasturel
Laboratoire de Thermodynamique
et Physico-Chimie Métallurgiques
ENSEEG, BP 75,
38402 Saint Martin d'Hères Cédex
France

ABSTRACT. To obtain phase diagrams, one needs certain physical parameters such as Gibbs energies of transformation of elements and Gibbs energies of mixing for both stable and completely metastable phases. In order to get insights into the physical origin of these parameters, it would be available to extract these parameters from solid-state physical calculations. For coherent phase diagrams of transition metal alloys, we show that energy estimations from band structure calculations and entropy estimations from Cluster Variation Method are sufficiently accurate to be considered as powerful input in phase diagram calculations.

1. INTRODUCTION

The study of phase diagrams, both experimental and theoretical, is one of the most important subjects of solid-state science. Although a great many binary phase diagrams have been determined experimentally, many systems are not yet well characterized. Therefore, some theoretical guidelines are necessary to understand and unify the experimental data allowing extrapolations from known phase diagrams to unknown ones. Nowadays, the theory is developed far enough to enable us to attempt a first principles calculation of the composition-temperature of phase diagram of a binary alloy : we start with a Hamiltonian describing the system of ions and electrons, without any other input than the elemental electronic data and we end at the phase diagram. Thus posed in all generality, the problem is difficult to be solved in practice. Indeed the precision in cohesive energies of pure constituents is just sufficient to select the simplest correct crystal structure, more complicated allotropic forms being not able to be treated. For compounds, there exists an infinity of possible crystal structures and it is not obvious to perform calculations for any given structure.

Yet, there is a class of phase diagram, the class of coherent phase diagram, which can be treated using theoretical approach. This class of phase diagram is characterized by equilibrium between phases which differ from one another solely by the distribution of elemental

55

H. Brodowsky and H.-J. Schaller (eds.), Thermochemistry of Alloys, 55–66.

atoms on the sites of a fixed lattice. The theoretical calculations are thus tractable because all the ordered ground states can be considered as perturbations of a common reference state, that of the disordered phase. Indeed, in this case, it suffices to minimize small free energy deviations rather that the full free energy. Method for calculating small energy and entropy deviations are now briefly described in the case of binary transition metal BCC phases.

2. ENERGY

As mentioned in the introduction, the free energy function of solid phases in binary alloys must include explicitely short- and long-range order. In general the free energy is given by :

$$F = \Sigma x_i F_i + \Delta Hf - T\Delta Sf \qquad (1)$$

where x_i and F_i are respectively the atomic composition and the free energy of pure element i, where ΔHf is the enthalpy of formation of alloy (including short range order SRO) and where ΔSf is the alloy entropy of formation (including also SRO).

The enthalpy of formation can be written as the sum of the random alloy enthalpy of formation (E_{rand}) plus the ordering energy (E_{ord}) which includes both short- and long- range order (1) :

$$\Delta Hf = E_{rand} + E_{ord} \qquad (2)$$

Friedel (2) has shown that the cohesive energy of pure transition metals can be understood in terms of a simple model for d-band bonding (tight-binding approximation). Detailed calculations (3) confirm the validity of this approach and this has motivated several model calculations (4,8) for random alloy enthalpy of formation, based on the changes in the d band with alloying. Our previous studies based on this tight-binding (T.B.) approximation (7,8) have shown that in order to achieve the accuracy in the random alloy enthalpy of formation of transition metal binary alloys, one must include off-diagonal disorder in the TB hamiltonian and carry out a self-consistent treatment of charge transfer.

Tight-binding approximation presents also the advantage to calculate simply the enthalpy of formation as a function of SRO and thus to get accurate value for the ordering energy, E_{ord}. The first approach proposed by Bieber et al (1) consist of calculating ΔHf for the alloy with SRO by extrapolation from the random system. This technique, known as the generalized perturbation method (GPM) is used to calculate localized interactions which contribute to ΔHf in addition to the non local coherent potential approximation energy of the random alloy. Moreover, for the case of the non-magnetic transition metals, the leading contributions to the ordering energy are given by pair interactions which extend to first nearest neighbors in the FCC lattice and to first and second nearest-neighbors in the BCC lattice (9). In this case, the ordering energy takes then the form :

3. ENTROPY

The crystal entropy can be separated into a vibrational and configurational part. In the following, we shall assume that the vibrational entropy change upon ordering is negligible compared to the change in configurational entropy. A relatively straight-forward and computationnaly efficient way of introducing SRO in the configurational entropy is by means of the cluster variation method (CVM) (12). In the CVM, the entropy is written as a function of the probabilities of arranging different atomic species on a set of lattice points included in one or several maximum clusters. The maximum cluster used in our study in the tetrahedron, containing both first and second neighbors in the BCC lattice. In the tetrahedron approximation, the entropy of a BCC disordered system is written for one mole of lattice sites :

$$S = - R \; (6 \sum_{ijkl} w_{ijhl} \ln w_{ijhP} - 12 \sum_{ijk} \ln t_{ijk}$$
$$+ 3 \sum_{ij} y_{ij}^{(2)} \ln y_{ij}^{(2)} + 4 \sum_{ik} y_{ik}^{(1)} \ln y_{ik}^{(1)} - \sum x_i \ln x_i) \quad (10)$$

where R is the perfect gas constant and w_{ijkl}, t_{ijk}, $y_{ij}^{(2)}$, $y_{ik}^{(1)}$ and x_i denote respectively the probability of finding tetrahedra, triangles, second-neighbor pairs, first-neighbor pairs and points in the configuration given by their subscripts (i equals A or B in a binary alloy).

It is now apparent that the configurational free energy F can be expressed as a functional in independent configurational variables (see equ.(6) and (10). As mentioned above, the equilibrium value of SRO parameter follows from the minimization at constant composition of the configurational free energy. In the case of the tetrahedron approximation, the minimization is conveniently carried out using the natural iteration method (NIM) developed by Kikuchi (13). The NIM equations used in the present work allow to obtain the cluster probabilities as well as the point and pair correlation functions at equilibrium (11). From these values, the thermodynamic properties of the alloy can be thus obtained.

To construct a phase diagram, it is more convenient to minimize the grand potential Ω given by :

$$\Omega = F - \mu \zeta_1 \qquad\qquad\qquad (11)$$

where μ is the effective chemical potential and ζ_1 the point correlation function.

In this work, this minimization is carried out with respect to the correlation functions ζ_n. To solve this minimization, we used the iterative method of Newton Raphson (NRM) (11). Compared to the NIM, the NRM is a quadratic iteration scheme and thus converges faster, if it does converge, than the NIM. However the initial values of the iteration calculations must be chosen carefully near the equilibrium values to obtain convergence when the NIM converges whatever the initial values may be. In our calculations the initial values have been determined by the free energy minimization.

$$E_{ord} = 1/2 \sum_k w_k V_k (\zeta_2^{(k)} - \zeta_1^2) \qquad (3)$$

where ζ_1 is the point correlation function, w_k, V_k and $\zeta_2^{(k)}$ are respectively the coordination number, the effective pair interaction and the pair correlation function for the k th-nearest neighbor defined by :

$$\zeta_2^{(k)} = y_{AA}^{(k)} + y_{BB}^{(k)} - 2 y_{AB}^{(k)} \qquad (4)$$

where $y_{ij}^{(k)}$ are the pair probabilities.

We can also introduce the Cowley SRO parameter which can be written as :

$$\sigma_k = (\zeta_2^{(k)} - \zeta_1^2) (1 - \zeta_1^2) \qquad (5)$$

and the expression of E_{ord} as a function of σ_k becomes :

$$E_{ord} = 1/2 (1 - \zeta_1^2) \sum_k w_k V_k \sigma_k \qquad (6)$$

In GPM, the corresponding effective pair interactions V_k are given approximately by :

$$V_k = - 1/2\pi . \text{Im} \int^{E_F} \Delta t^2 \sum_{\lambda,\mu} [G_k^{\lambda\mu} (z - \Sigma(z))]^2 \qquad (7)$$

In this relation, $\Delta t = t^A - t^B$ where t^i is the matrix associated to an atom i in the average medium (1). $G_k^{\lambda\mu} (z)$ is the interatomic Green function of the metal A taken between two kth neighbouring sites and $\Sigma(z)$ is the CPA self-energy.

An other approach to determine the effective pair interactions is the one proposed by Hawkins et al (10), based on the cluster Bethe lattice method (CBLM). Despite the fact that the CBLM relies on a topological approximation that replaces the real crystal lattice with a Cayley tree, this approach presents the great advantage to include SRO explicitly in the calculation of the electronic spectrum and internal energy. It is this approach which will be used in this paper (11). Within this scheme, the ordering energy is given by :

$$E_{ord} = \Delta E_{el}(x,\sigma) + \Delta E_{elect}(x,\sigma) \qquad (8)$$

where the first term ΔE_{el} is the variation of the one electron energy due to the modification of the electronic energy band upon ordering and the second term comes from the intraatomic electron-electron interactions and the interatomic electron-electron and ion-ion interactions and their variations with CSRO. The neighbor effective pair interactions are thus obtained by a parametrization of ΔHf as a function of SRO. The first and second nearest neighbor effective pair interactions are found to be fairly linear functions ~of concentration and they are approximated by the expression :

$$V_k (\zeta_1) = A_k + B_k \zeta_1 \qquad (9)$$

The parameters A_k and B_k being obtained by least squares fitting of the calculated enthalpy of formation.

4. RESULTS AND DISCUSSION

4.1. Heats of mixing of random alloys

In figure 1, we present calculated heats of mixing of the random alloys for Mo-Ta, Mo-Nb, Ta-W, Cr-Mo, Mo-W and Cr-W systems. The three first systems display negative energies while the three last one display positive values. As previously proposed by Colinet et al (7), this trend can be understood in terms of TB parameters, i.e., Δn_d the difference in the occupation of the elemental d band, ΔW_d the mismatch in d-bandwidths and ΔE_d the difference in the d band center. For the studied alloys, the sign of E_{rand} is the reflect of the competition between a positive contribution due to the d-bandwidth mismatch and a negative contribution due to the d-band mixing. However this hybridization contribution depends roughly on ΔN_d^2 (14). Consequently this term does not contribute to E_{rand} for the isoelectronic alloys, Cr-Mo, Mo-W and Cr-W and thus these three alloys segregate due to d bandwidth mismatch. The behaviour of Mo-Nb, Mo-Ta and Ta-W alloys is completely different due to the fact that the hybridization term becomes preponderant.

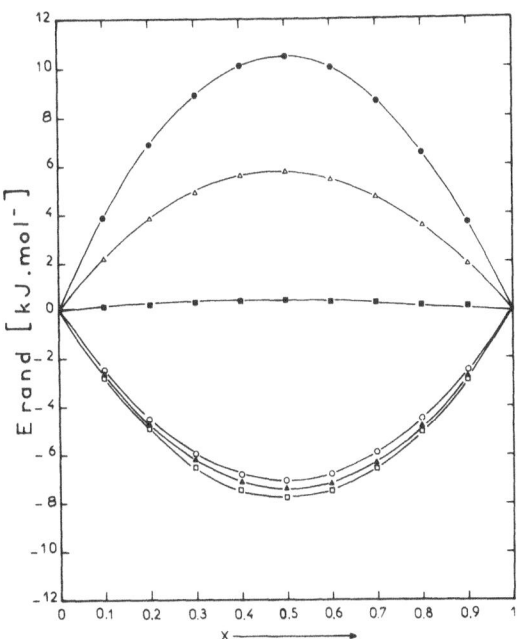

Figure 1 : Calculated energies of formation of the random $Cr_{1-x}Mo_x$ ($\longrightarrow\!\!\triangle\!\!\longrightarrow$), $Cr_{1-x}W_x$ ($\longrightarrow\!\!\bullet\!\!\longrightarrow$), $Mo_{1-x}Nb_x$ ($\longrightarrow\!\!\blacktriangle\!\!\longrightarrow$), $Mo_{1-x}Ta_x$ ($\longrightarrow\!\!\square\!\!\longrightarrow$), $Mo_{1-x}W_x$ ($\longrightarrow\!\!\blacksquare\!\!\longrightarrow$) and $Ta_{1-x}W_x$ ($\longrightarrow\!\!\circ\!\!\longrightarrow$) alloys as a function of composition x.

4.2. Calculations of the phase equilibrium

The first step to reach phase equilibrium calculations is the determination of the thermodynamic data at equilibrium SRO parameter values as indicated in the previous section. In order to show the validity of our calculated values, we present in figures 2 (a-b) and 3(a-b) a comparison of the calculated enthalpies, entropies and activities of both constituents with experimental measurements for Cr-Mo and Mo-Ta systems. A review of the phase equilibrium of Cr-Mo has recently been completed by Massalski et al (15), Jacob and Kumar (16) and Venkatraman and Neumann (17). Measurements of the chemical activity and of the enthalpy of formation have been reported by Laffitte and Kubaschewski (18), Kubaschewski and Chart (19) and recently by Jacob and Kumar (16). We can see that the agreement between calculated and experimental values is very satisfying, the positive value of ΔH and the positive deviation from ideality of the activity reflecting a positive excess free energy, indicative of a miscibility gap.

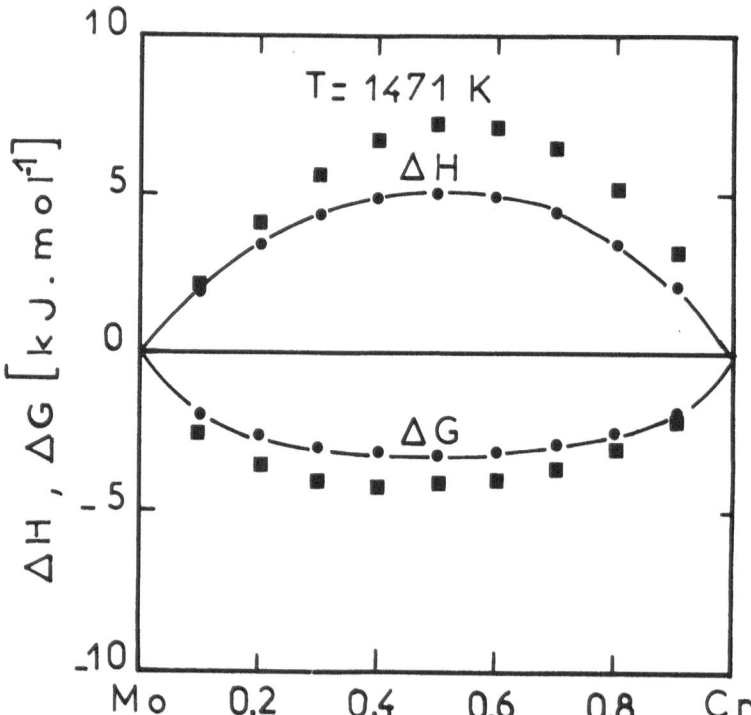

Figure 2(a) : Comparison of the calculated values,——●——, of the enthalpy (ΔH) and Gibbs free energy (ΔG) of formation with the experimental measurements,——■——, (18,19) for the Cr-Mo system.

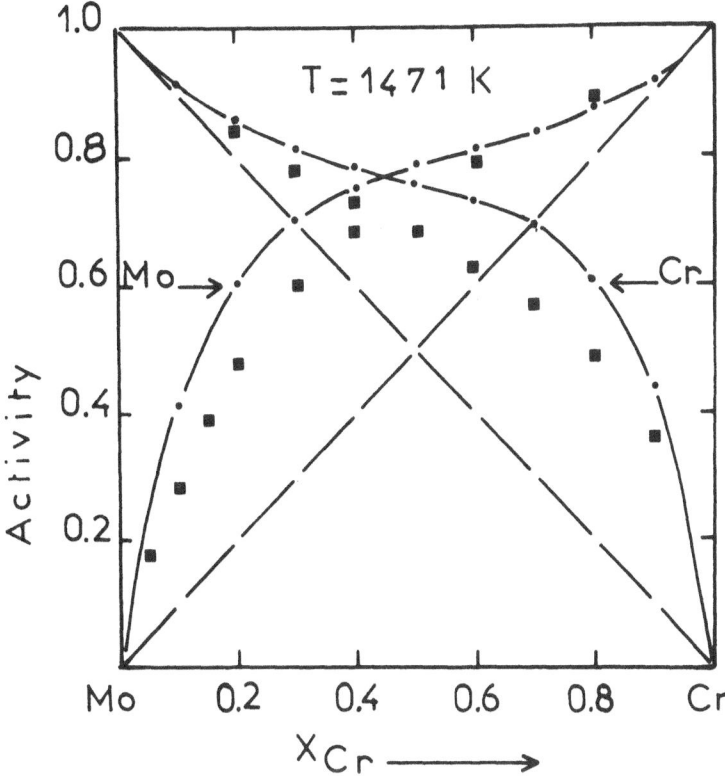

Figure 2(b) : Comparison of the calculated values, —•—, of activities with the experimental measurements, —■—, (18,19) for the Cr-Mo system.

For Mo-Ta system, the thermodynamic properties have been determined by Singhal and Worrell (20). In this system, the activities of niobium and molybdenum exhibit negative deviations from Raoult's law. As shown for the both presented systems and more generally for other systems (11), the close correspondence of experiment and theory indicates that the alloy free energy is predicted correctly in our approach.

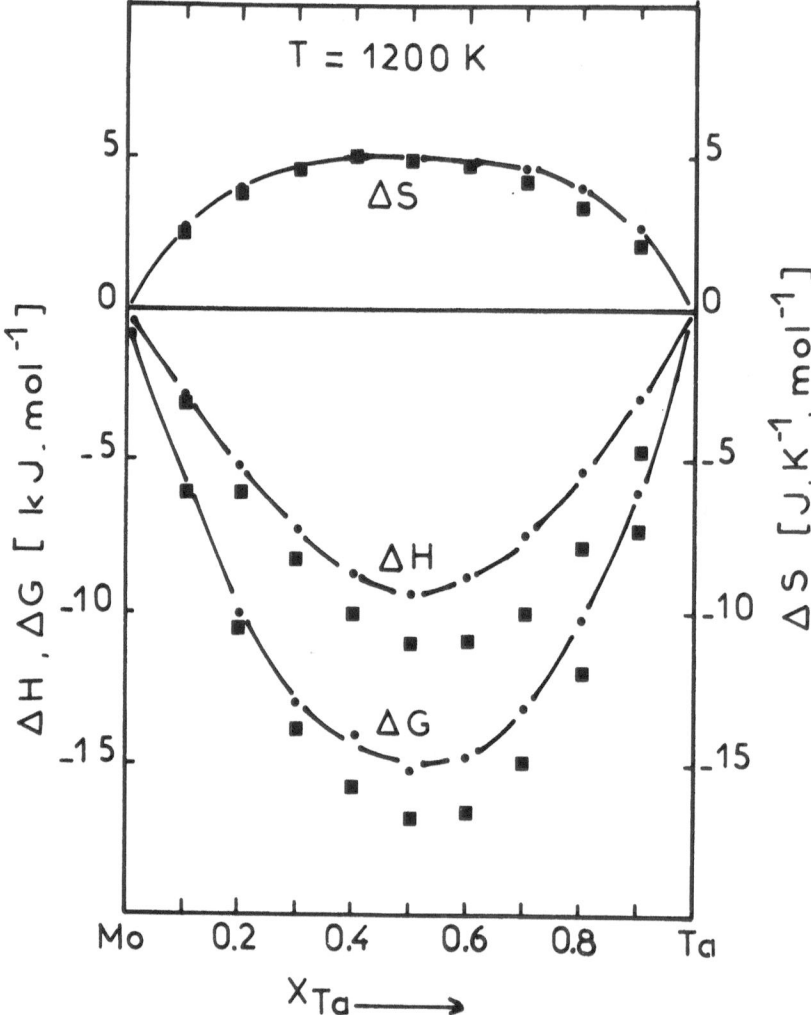

Figure 3(a) : Comparison of the calculated values, —•—, of the
enthalpy (ΔH), entropy and Gibbs free energy (ΔG) of formation with the
experimental measurements, —■— (20) for the Mo–Ta system.

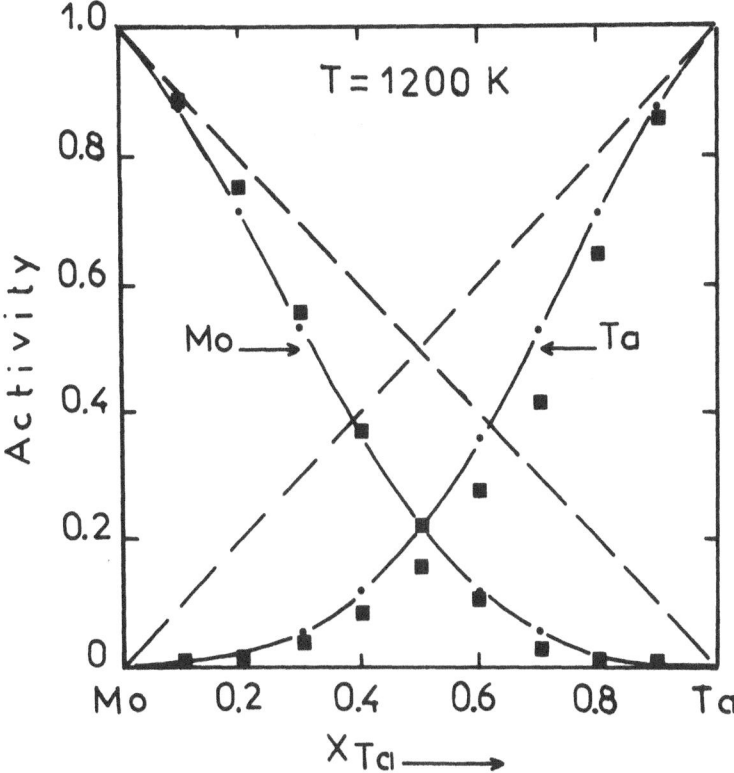

Figure 3(b) : Comparison of the calculated values, ——●——, of activities with the experimental measurements, ——■——, (20) for the Mo-Ta system.

The next step is the calculation of the solid-state equilibrium phase diagram. We focus our attention on Cr-W system for which experimental determinations exist. The phase equilibrium has been reviewed recently by Naidu et al (21). The experimental phase diagram reveals complete solid solution between 1950 K and the solidus : below 1950 K, a rather symmetric miscibility gap is found. In figure 4, we compare the calculated phase diagram with previous experimental results (Trzebiatowki et al. (22) ; Greenaway (23) ; Den Broeder (24) ;

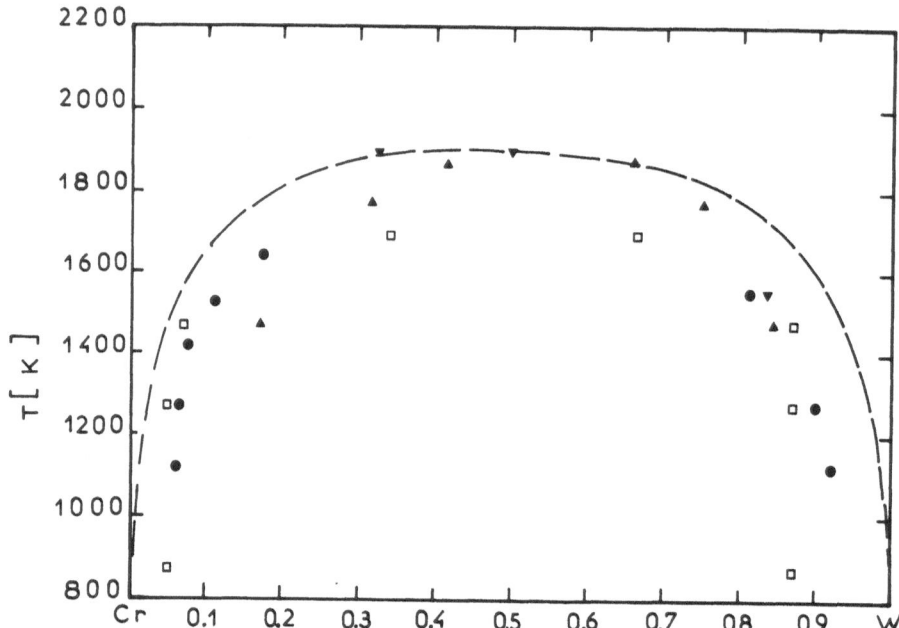

Figure 4 : Predicted miscibility gap for the Cr-W system. The experimental data are from Trzebiatowski et al (22), □ ; Grennaway (23), ● ; Den Broeder (24), ▽ and Margaria et al (25), △ .

Margaria et al (25)). As can be seen, good agreement with experiment is obtained and the experimentally determined critical temperature is quite well reproduced. Although the experimental phase boundary appears quite symmetric, there is a positive deviation on the Cr-rich side. In our treatment, this small asymmetry is due to the shape of the calculated ΔHf as a function of composition ; indeed it has been shown that in transition metal alloys (11), alloys with higher compositions of the metal with the larger bandwidth display less positive ΔHf values and thus are less strongly segregating.

5. CONCLUSION

We have shown that theoretical approaches are beginning to provide realistic calculations of phase diagrams. In this paper, we have presented results concerning coherent phase diagrams ; this class of phase diagram is characterized by equilibrium between phases that differ from one another solely by the distribution of atoms on the sites of a fixed lattice. In the particular case of transition metal alloys, we have used the CBLM-CVM method. Our results are in very good agreement with experiment ; segregation or not in these alloys, values of critical temperature, symmetry or asymmetry of the miscibility gap, ... etc, all these informations can be understood in terms of the electronic properties of the alloy constituent. The comprehension of all these phenomena can allow us to calculate unknown coherent phase diagram with a great confidence in our results.

6. REFERENCES

1　A. Bieber, F. Gautier, G. Treglia and F. Ducastelle, Solid State Commun., 39, 149 (1981) and references therein.

2　J. Friedel, in Phys of Metals : I. Electrons, ed. by J.M. Ziman (Cambridge University Press, Cambridge 1969), pp. 361-364.

3　A.R. Williams, C.D. Gelatt Jr., and J.F. Janak, in Theory of Alloy Phase Formation, ed. by L.H. Bennett (AIME, New York 1980) pp. 40-62.

4　D.G. Pettifor, Phys. Rev. Lett. 42, 846 (1979).

5　C.M. Varma, Solid State Commun. 31, 295 (1979).

6　R.E. Watson and L.H. Bennett, Phys. Rev. Lett. 43, 1130 (1979), Calphad 5, 25 (1981).

7　C. Colinet, A. Pasturel and P. Hicter, Calphad 9, 71 (1985).

8　A. Pasturel, C. Colinet and P. Hicter, Acta Metall. 32, 1061 (1984)

9　F. Gautier, in "High Temperature ALloys : Theory and Design" (ed. by J.O. Stiegler), Met. Soc. of AIME, pp. 163-181 (1984)

10　R.J. Hawkins, M.O. Robbins and J.M. Sanchez, Phys. Rev. B. 33, 4782 (1986).

11　C. Colinet, A. Bessoud and A. Pasturel, J. Phys. F., to be published.

12　R. Kikuchi, Phys. Rev. 81, 988 (1951).

13　R. Kikuchi and J. Murray, Calphad 9, 311 (1985).

14　D.G. Pettifor, in Physical Metallurgy, ed. by R.W. Cahn and P. Hassen (North Holland, Amsterdam 1983), Chap. 3.

15　T.B. Massalski, J.L. Murray and L.H. Bennett, Binary Alloy Phase diagrams, American Society for Metals, Metals Park, Ohio, 1986.

16　K.T. Jacob and B.V. Kumar, Z. Metallkde. 77, 207 (1986).

17　M. Venkatraman and J.P. Neumann, Bull. Alloy Phase Diag., 8, 216 (1987).

18　M. Laffitte and O. Kubaschewski, Trans. Faraday Soc. 57, 932 (1961).

19　O. Kubaschewski and T.G. Chart, J. Inst. Met. 93, 329 (1964-1965).

20 S.C. Singhal and W.L. Worrell, in Metallurgical Chemistry, Proceedings of the International Symposium held at Brunel University and National Physical Laboratory, 16–16, July 1971, edited by O. Kubaschewski (Her Majesty's Stationery Office, London 1972) p. 65.
21 S.V. Nagender Naidu, A.M. Sriramamurthy and P. Rama Rao, Bull. Alloy Phase Diag., 5, 289 (1984).
22 W. Trzebiatowski, H. Ploszek and J. Lobzowski, Anal. Chem., 19, 93 (1947).
23 H.T. Greenaway, J. Inst. Met. 80, 589 (1951–1952).
24 F.J.A. Den Broeder, Acta Metall. 20, 319 (1972).
25 T. Margaria, C. Allibert, I. Ansara and J. Driole, High Temp. High Pressures, 8, 451.

THERMODYNAMIC DATASETS FOR SYSTEMS WITH SOLUTION PHASES

Mats Hillert
Div. of Physical Metallurgy
Royal Institute of Technology
S-100 44 Stockholm,
Sweden

ABSTRACT. The concept of thermodynamic network, introduced in the discussion of evaluations of datasets for groups of pure substances, is applied to a discussion of similar problems concerning systems with several solution phases. The necessity of introducing mathematical models for solution phases is emphasized and an increase of the complexity of the network problem must be accepted. The simultaneous method of evaluation will then be limited to a network of fewer phases and the sequential method must be applied for larger systems. Examples are given of physical effects requiring special models.

INTRODUCTION

Extensive thermodynamic databases for pure substances have existed for a very long time. In contrast, datasets for systems with solution phases are still very small. The present report will concern alloy systems. In that field even a rather simple system usually contains several solution phases, a fact which creates great difficulties in the preparation of an extensive database. Such difficulties will now be discussed. As an introduction it may be instructive to examine earlier work on databases for substances.

THERMODYNAMIC DATA NETWORKS

Many of the problems one encounters when preparing an extensive database were reviewed by Garvin et al at the CODATA conference in Colorado 1976 (1). Already in the title they introduced an important concept "thermodynamic data network" and they used boron compounds for illustration, see Fig. 1. At the start of the evaluation there are a number of substances which are directly available to assessment through experimental information concerning themselves and other well known substances. In the diagram they are represented by a, b, c and d. However, there may also be a piece of information concerning two of them, in this example a and c or c and d. After completing the first assessment

H. Brodowsky and H.-J. Schaller (eds.), Thermochemistry of Alloys, 67–84.
© 1989 by Kluwer Academic Publishers.

68

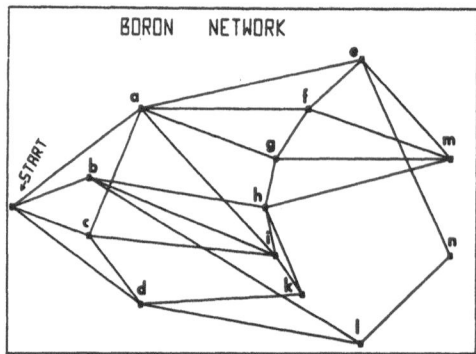

Figure 1. Thermodynamic network for enthalpy data on key compounds of boron. From ref. 1.

of these substances it is thus necessary to check that the results are consistent with this additional information. Normally, that will not be the case and consistency will not be achieved without some substantial iteration. When this stage has been completed successfully it is possible to consider another group of substances, e, f, g, h, i, j, k and l, the properties of which are known mainly through measurements concerning themselves and at least one from the first group. When more than one from the first group are concerned, for instance a, b and c in connection with i, it may be necessary to modify the descriptions already obtained for them and one must start a new iteration procedure. Again, the same happens when there is information concerning a combination of two or more substances in the second group.

Through all the information the boron substances are thus connected in a network which must be considered when their properties are assessed. In the old days, when the evaluations were made manually, the problem had to be solved as described here. This is called the sequential method and it usually involves a large amount of iteration. Using a computer we may today choose between the sequential method and a simultaneous treatment of all the pieces of information by an optimization procedure or a simultaneous solution of a large number of equations. This may be relatively easy if the equations can be linearized but is otherwise far from trivial.

THE CALPHAD METHOD

As another example of a network problem we may turn to the $CaO-Cr_2O_3$ system, Fig. 2. It contains a considerable number of well defined substances and for each of them there is direct information available. However, the phase diagram shows a number of three-phase equilibria. It is evident that one cannot be satisfied with the description obtained for each one of the substances before they are consistent with these equilibria. In fact, the metallurgist is usually even more concerned about the equilibria than the thermodynamic properties themselves and in most cases he is satisfied with having the phase diagram, as given in Fig. 2. That is why the method of evaluating thermodynamic properties by coupling information on phase diagrams and thermochemistry is called

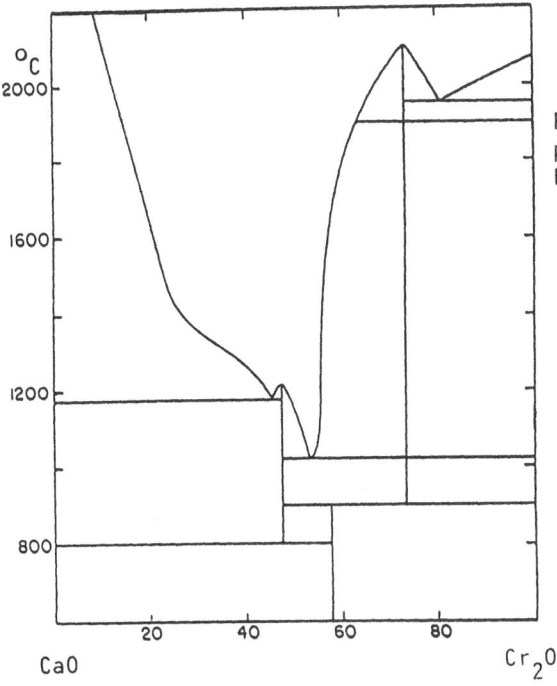

Figure 2. The CaO-Cr$_2$O$_3$ phase diagram according to Ford and White.

the CALPHAD (CALculation of PHAse Diagrams) method when applied to alloy systems. In principle it is not different from the evaluation work done by thermochemists without such a fancy name.

MATHEMATICAL MODELS

When the thermochemical evaluations were made manually, numerical methods could be applied and the results were obtained as tables. The results were primarily given as a table for the heat capacity, C_p, above 298 K and a set of values for the enthalpy of formation, $\Delta_f H$, the Gibbs energy of formation, $\Delta_f G$, and the entropy, S, at 298 K. Secondarily, the enthalpy, Gibbs energy and entropy have also been tabulated as functions of temperature although that information is available through the information given primarily. Analytical functions were also used, e.g. the following expression for the heat capacity

$$C_p = a + bT + cT^2 + dT^{-2}$$

It was thus possible to replace the tables by a set of parameters in the famous compilations by Kubaschewski and his collaborators (2). However, there might have been a feeling that such analytical descriptions are less accurate.

In order to take full advantage of computers it is now natural to favour the analytical approach and at the outset of an evaluation it is then necessary to choose a mathematical model for each phase. The loss

of accuracy due to the shortcomings, inherent in any model, may partly
be compensated by the increase in size of the network which a computer
can take into account. In any case, it will result in a better consistency within a given dataset and will allow a user to combine various
pieces of information with some confidence. The future dominance of
mathematical models in extensive datasets seems inevitable in spite of
the fact that tables are still being produced in large quantities.

THE PROCESS OF EVALUATION

The concept of network is also useful in a discussion of the process
of evaluating the properties of a single phase. It can be illustrated
by Fig. 3 which is an elaboration of a flow diagram presented by the
CODATA Task Group on Internationalization and Systematization of Thermodynamic Tables (ISTT) in a report from 1982 (3). From the flow diagram it is first evident that the computer cannot replace manual work
completely. For instance, the selection of relevant reports and the
extraction of information, relevant for the evaluation, must be made
manually by skillful assessors. Between levels 2 and 3 a computer can
be programmed to assist the assessor. There each type of data is evaluated. Between levels 3 and 4 one should introduce mathematical models
for each type of property by considering structural information, for
instance (from box 3c). Using a computer one can optimize their parameters to obtain the best fit to the information. It may turn out that
various pieces of information are conflicting and one should then go
back to previous levels. It may also turn out that a model is not
appropriate. Such decisions may be difficult to leave to the computer.
After sufficient iteration one may proceed and between levels 4 and 5
one should make a preliminary check of the consistency with respect to
thermodynamic relations between the various properties. Finally, between levels 5 and 6 one should make a complete assessment of all the
properties of the phase. It would probably be advantageous here to
apply a single mathematical model, representing all the properties of
the phase. That could be done by using an expression for the Gibbs
energy as function of T,P and composition or the Helmholtz energy as
function of T,V and composition. The diagram also gives the internal
energy as an alternative but it must be expressed as a function of S,V
and composition and S is not a very practical variable to work with.
    When evaluating the Gibbs or Helmholtz energy one may find it
necessary again to go back to the preceding levels. It is evident that
this diagram represents a network in the same sense as the diagram in
Fig. 1. After one has gained enough experience of the type of phase
under consideration, it should be possible to select a mathematical
model for the Gibbs or Helmholtz energy from the beginning and much of
the work described here could probably be carried out simultaneously
just like the network problem of Fig. 1 could be solved by a simultaneous evaluation using computer.
    Admittedly, the separation into two network problems is an over-
simplification. Important information on the properties of the phase
under consideration may come from interaction with other phases, as

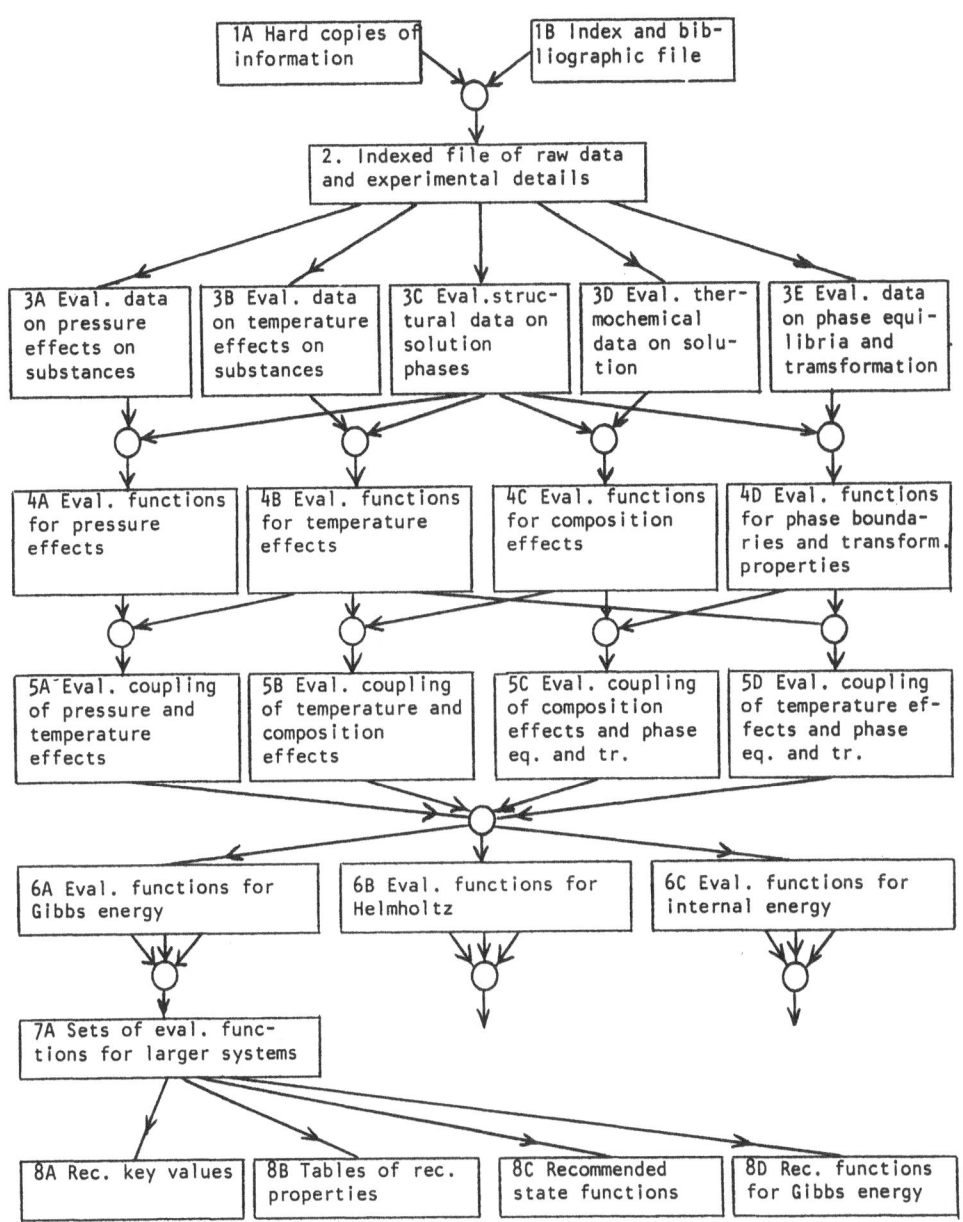

Figure 3. Flow diagram for the preparation of critically evaluated thermodynamic datasets including solution phases.

already illustrated in Fig. 1. When working on that network problem it is thus necessary to realize that each line actually contains a network problem of its own. By solving this more complicated network problem between levels 6 and 7 in Fig. 3 one would finally obtain a set of consistent descriptions of the properties of a group of related phases. It is evident that this work will soon get out of hand when the number of phases is increased and when they require complex models. On a larger scale it will thus be necessary to use the sequential method and even to limit the amount of iterations severely. Naturally, for pure substances one may treat more phases than for solution phases. However, with reference to Fig. 2 it could be emphasized that the liquid is a solution phase which can hardly ever be avoided. It is understandable that in thermochemistry one often leaves the liquid out, a decision which the metallurgist can seldom accept. He will have to consider many solution phases and will have to accept the sequential method for a smaller group of phases than the ordinary thermochemist.

SOLUTION PHASES

As another example of a system with several solution phases we may look at the Fe-O phase diagram which has three solid compounds, all of which have variable composition, see Fig. 4. To approximate them as stoichiometric compounds is not very satisfactory although it is often done. When the variation in composition is small very simple modeling may be sufficient and it has been attempted with some success by Haas (4) in a

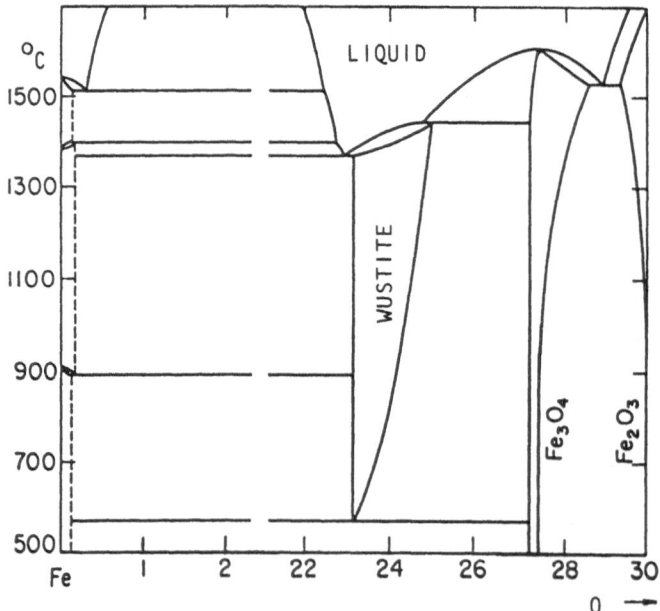

Figure 4. The Fe-O phase diagram

very ambitious network evaluation involving several other Fe compounds. However, the liquid phase was still excluded. When the Fe-0 system was recently evaluated with more ambitious solution models, as many as 40 model parameters had to be assessed (5)

In the alloy field, ambitious modeling of solution phases has been used extensively and for a long time. It may be interesting to examine the reasons for this different attitude. When looking at a collection of alloy phase diagrams, one immediately notices that intermetallic phases very often have a substantial variation in composition. A closer study will reveal the important fact that the same structures often appear in many systems. If we combine this with the fact that metallurgists are often interested in alloys containing many elements it is evident that there is a need to describe metallic phases with models which allow many elements to dissolve. The simple type of modeling which is realistic in a narrow range of composition, only, will not be very useful. It is highly desirable to apply models which cover the whole range of composition theoretically available to a phase. This fact immediately increases the complexity of the network which must be included in an evaluation.

LATTICE STABILITIES

Let us consider the Fe-Cr system, Fig. 5, which shows that Cr can dis-

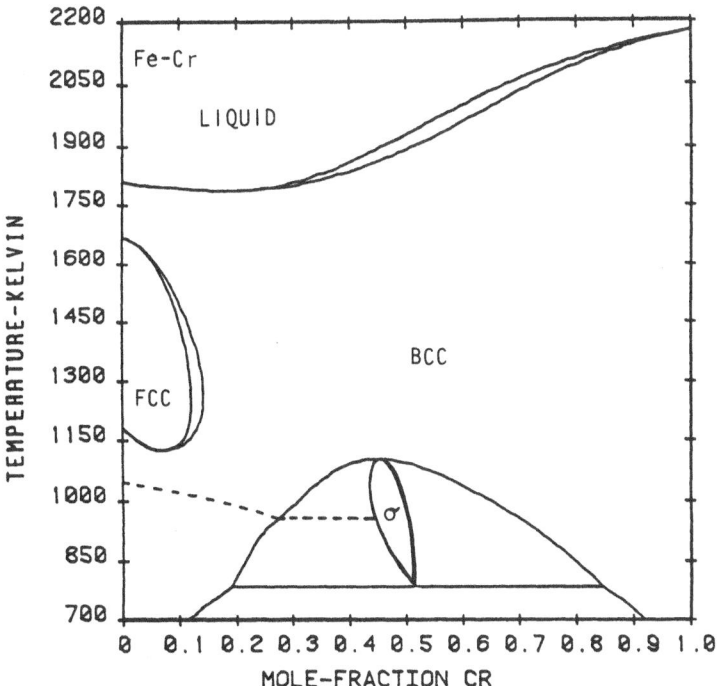

Figure 5. The Fe-Cr phase diagram

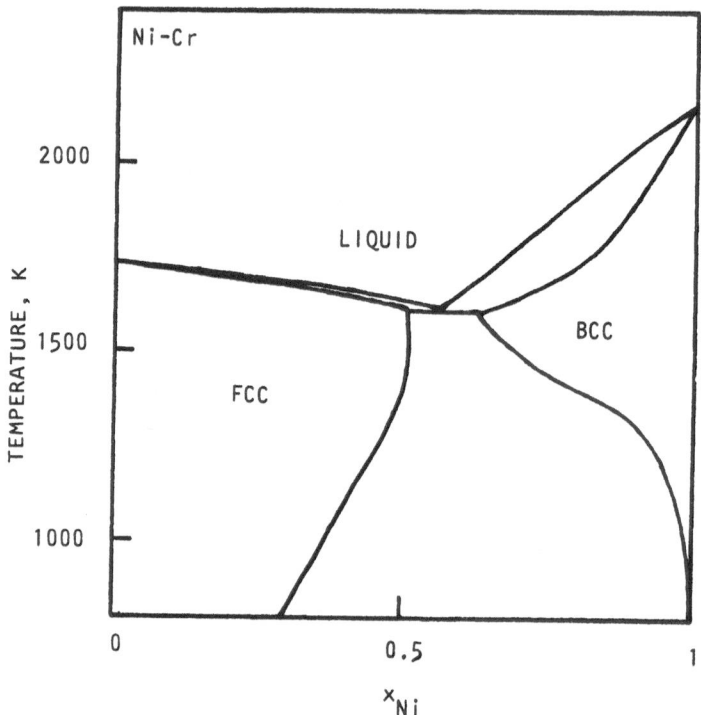

Figure 6. The Ni-Cr phase diagram

solve in the fcc phase but only to about 11%. When modeling this phase
we should consider it all the way to pure Cr and an evaluation will
thus yield some description of pure fcc Cr. Naturally, that description
will not be very reliable and it will to a large extent depend upon the
model we have chosen. In the Ni-Cr system the fcc phase extends to
almost 50%, see Fig. 6, and the extrapolation to pure Cr may be somewhat less uncertain.

These uncertainties do not need to affect the quality of the description in the stable range of the fcc phase. However, in order to
combine the two descriptions in an evaluation of the Fe-Cr-Ni system
it is necessary to use the same description for pure fcc Cr in both
cases. This is a typical network problem. The same situation arises for
each new element added to the alloy. As a basis for evaluation of alloy
systems it is thus necessary to agree on the description of each pure
element in all the common metallic structures. Kaufman introduced the
term lattice stability, defined as a Gibbs energy difference

$$^oG_M^{\alpha \to \beta} = {}^oG_M^{\beta} - {}^oG_M^{\alpha}$$

He produced a very extensive set of lattice stabilities (6). For each
case he considered the most favourable binary systems and usually represented the result in the form of a+bT. When the two structures
appear as stable phases for the element, a and b are immediately ob-

tained from the temperature and heat of transformation. Since the heat capacity, C_p, is obtained from the second derivative of G, it is evident that the expression a+bT is based on the approximation that C_p has the same value for the two phases. Fig. 7 shows C_p data for pure bcc Mo

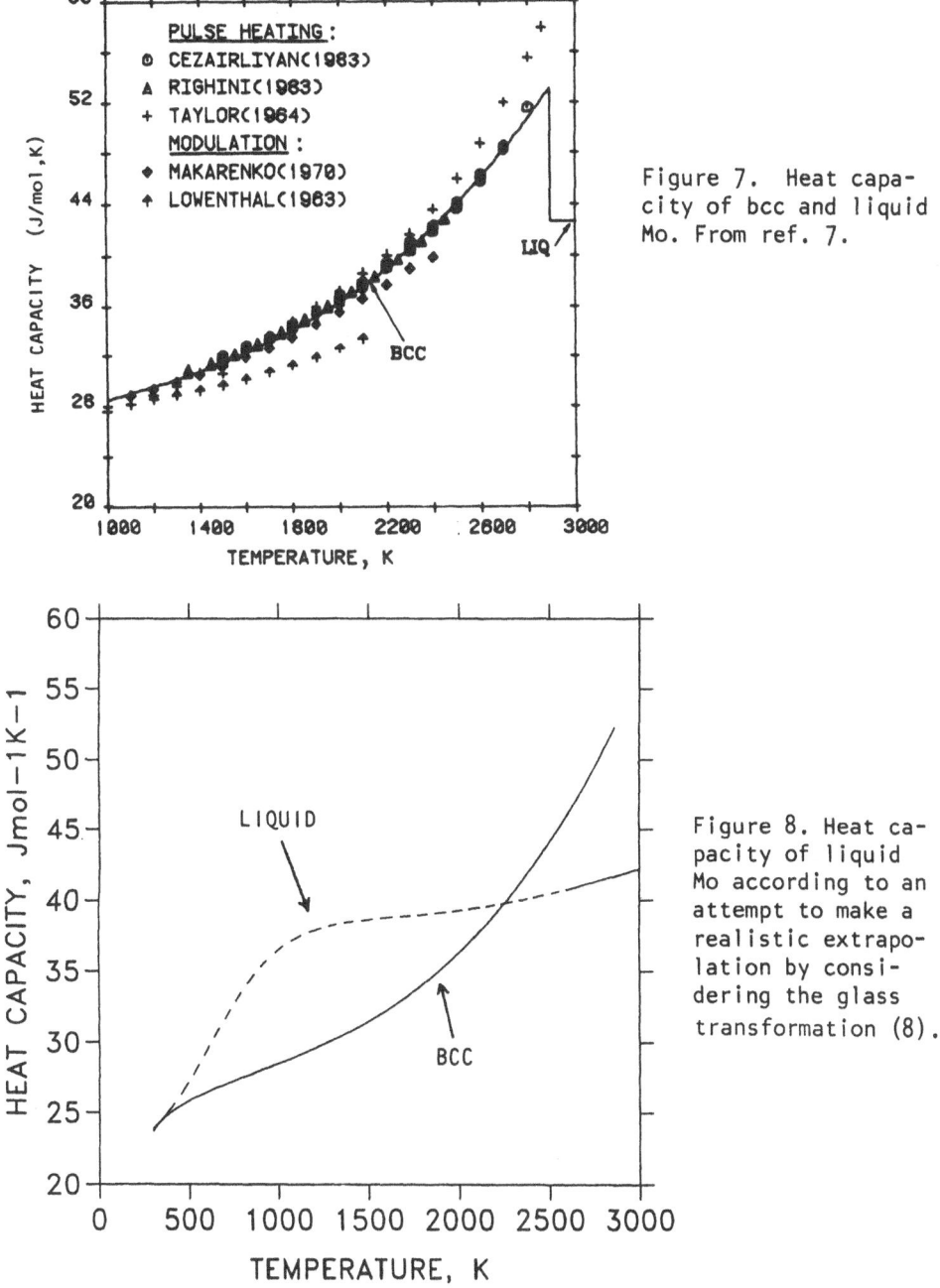

Figure 7. Heat capacity of bcc and liquid Mo. From ref. 7.

Figure 8. Heat capacity of liquid Mo according to an attempt to make a realistic extrapolation by considering the glass transformation (8).

and pure liquid Mo (7). It does not seem very satisfactory to use the full line with its discontinuity at the melting point. The alternative would be to use more realistic extrapolations and thus get separate curves for the phases. Fig. 8 shows the result of such an attempt (8) but the extrapolation of the liquid curve has been dashed in order to indicate that it depends on some very uncertain model. The very drastic decrease at low temperatures represents the so-called glass transition and the relatively large C_p values below the melting point is due to a gradual transformation leading to the glass transition. Unfortunately, the thermodynamics of this gradual transformation is still not understood and today this extrapolation appears very arbitrary. The advantage of Kaufman´s method is that it does not involve any new arbitrary decisions when a new element is evaluated.

THE SGTE CONVENTION

It has recently been suggested by SGTE (Scientific Group Thermodata Europe) that one should at least get rid of the discontinuity at the transformation temperature and this can be done automatically by the application of the following expressions (9)

$$T>T_f : C_P^\alpha = C_P^\ell(T) + [C_P^\alpha(T_f) - C_P^\ell(T_f)](T/T_f)^{-10}$$

$$T<T_f : C_P^\ell = C_P^\alpha(T) + [C_P^\ell(T_f) - C_P^\alpha(T_f)](T/T_f)^{6}$$

The nature of this method is illustrated in Fig. 9. For the Gibbs energy this convention gives almost the same result as the Kaufman method.

SECOND-ORDER TRANSFORMATIONS

There are also important physical phenomena resulting in second-order transformations, e.g. the magnetic transformation. It has a strong thermodynamic effect as illustrated for Fe_3O_4 in Fig. 10. Previously this problem has been handled by applying different sets of parameter values in the C_p expression below and above the Curie temperature (10). The full line in Fig. 10 is obtained with a particular model for the magnetic transformation (11). The fit is not perfect because the evaluation (5) also involved other pieces of information, i.e. a typical result of a network problem.

If the magnetic effect only concerned some pure elements, it would perhaps not be a very serious problem but we shall soon see that it also applies to solutions. It is worth emphasizing that the glass transformation may also be regarded as a kind of second-order transformation and the problems are much the same for all such phenomena.

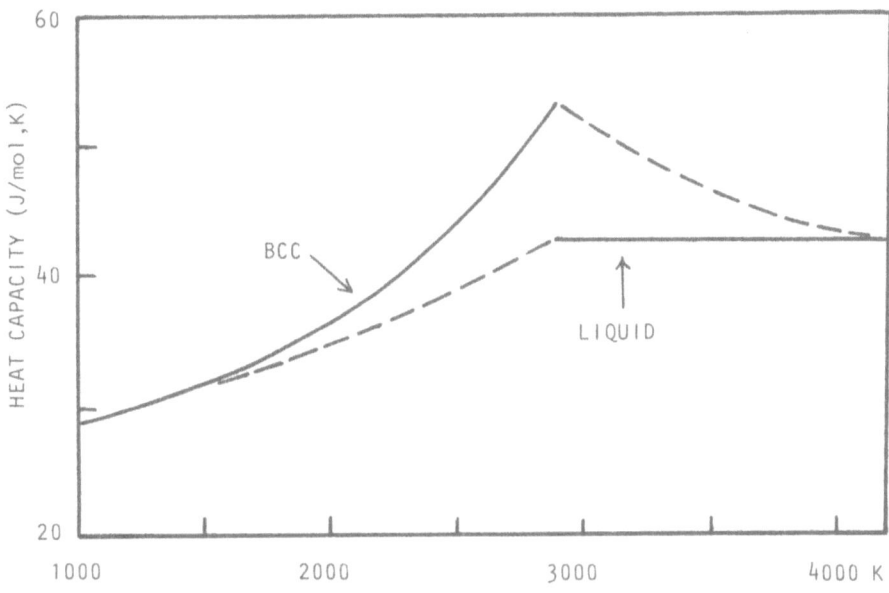

Figure 9.   The nature of the SGTE convention for the extrapolation of heat capacity of bcc and liquid Mo.

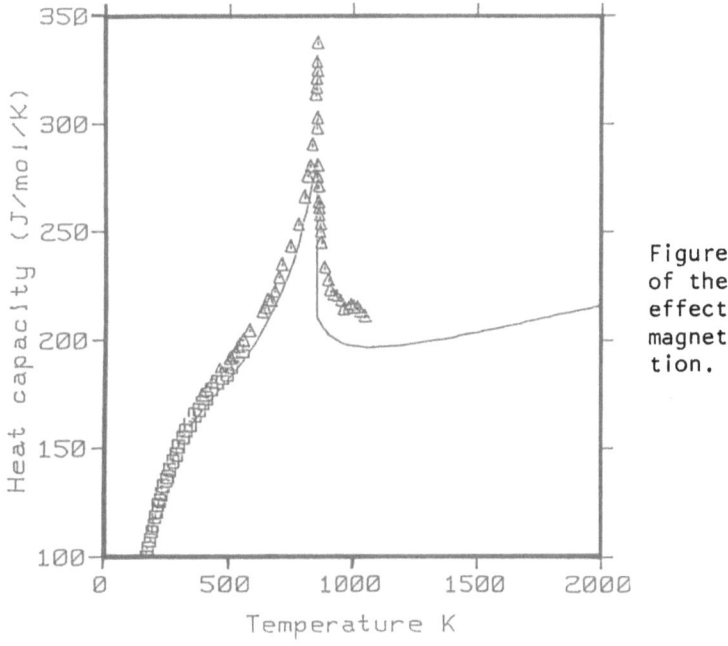

Figure 10. Modeling of the thermodynamic effect of a ferro-magnetic transformation.

SIMPLE SOLUTION BEHAVIOUR

We have seen that it is sometimes necessary to apply complicated models to describe the properties of pure elements. The need of complicated models is exaggerated when one goes to solutions. For the present discussion it is sufficient to consider binary solutions. Usually, the composition dependence of the excess Gibbs energy is represented with a power series and there is rather wide agreement to use a polynomial first proposed by Guggenheim but usually called the Redlich-Kister polynomial

$$^E G_m = x_A x_B \sum_{i=0}^{n} L_i (x_A - x_B)^i$$

This method can be compared to the use of a power serie in T for $C_p(T)$. n=0 gives the regular solution model and n=1 the subregular solution model. However, there are many cases where this method does not give a realistic description of the properties. An example will now be given.

DARKENS TREATMENT

In the spirit of the regular solution model Darken (12) plotted $\ln \gamma_B$ versus x_A where γ_B is the activity coefficient. The regular solution model would yield a horizontal line, the subregular one would yield a straight, sloping line. Fig. 11 shows Darken's result for liquid Mg-Bi.

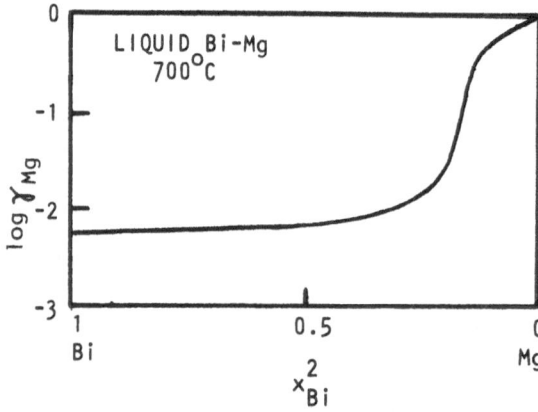

Figure 11. Activity coefficient of Mg in liquid Mg-Bi alloys at 700°C. From ref. 12.

The regular solution model seems to be well satisfied for Mg contents up to about 40% but then there is a drastic change. In order to study this change Darken examined a quantity he called excess stability, defined as

$$\text{Excess stability} = d^2 G_m / dx_B^2 - RT / x_A x_B$$

Fig. 12 shows the striking effect he found in a relatively narrow composition range. He suggested that this indicates a rather sudden transition of the physical properties of the liquid phase. He argued that it should not be reasonable to use the same state of pure Mg as reference in the two regions separated by the sudden change. In the above terminology we could say that one should use two different lattice stabilities for pure liquid Mg and the same conclusion would probably apply to Bi, as well. Darken formulated his suggestion in a mathematical model he called the "quadratic formalism".

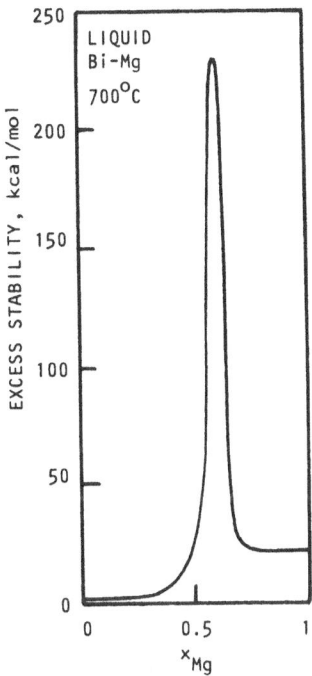

Figure 12. Excess stability of liquid Mg-Bi at 700°C. From ref. 12.

It is interesting that Darken supported his suggestion by emphasizing that the sudden change seems to occur close to the composition of a stoichiometric, solid phase, Bi_2Mg_3, and he showed that a similar form of the excess stability would be obtained according to the ideal gas law if there is compound formation. This model is now often applied to cases similar to Bi-Mg under the name of the associate solution model but usually without reference to Darken.

THE MAGNETIC EFFECT IN SOLUTIONS

It may seem attractive to apply Darken´s suggestion that one should use different reference states in regions separated by a magnetic transition. It would seem that one should use paramagnetic reference states on one side and ferromagnetic reference states on the other. As an example we may examine the fcc phase in the Fe-Co systems, Fig. 13 (13). At 1250 K there is complete solubility between fcc Fe and fcc Co and,

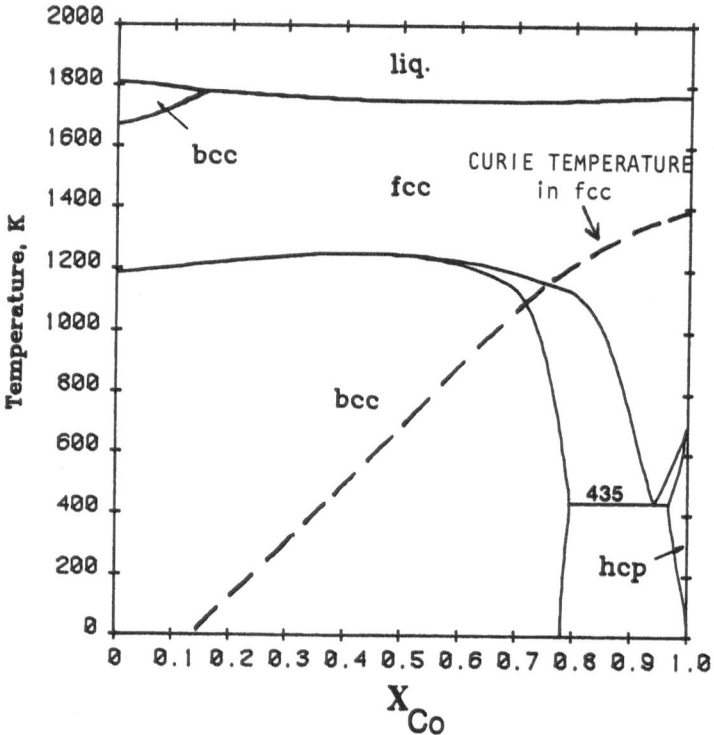

Figure 13. The ferromagnetic transition in the fcc phase of the Fe-Co system (dashed line).

if the magnetic transition were ignored, one would use pure fcc Fe, which is paramagnetic at that temperature, and pure fcc Co which is ferromagnetic. However, following Darken's suggestion it would seem better to use a hypothetical paramagnetic state of pure fcc Co for solutions to the left of the Curie line. In order to find the properties of such a state one could perhaps extrapolate the properties of pure fcc Co from above the Curie temperature. However, the situation is different because the magnetic transformation is very gradual, being of the second order and is spread out over a wide range of temperature or composition. The very complicated thermodynamic alloying effect of this phenomenon is demonstrated by the excess stability plot in Fig. 14. The dashed line shows the behaviour of a subregular solution model. All deviations from that line are due to the magnetic effect. It is evident that one could never hope to describe such a complicated effect except by using a very special model. In fact, the curve in Fig. 14 was calculated from an evaluation of the properties of the Fe-Co system using such a model (11). It is a straight-forward generalization of the magnetic model already introduced for pure elements and compounds.

Inevitably, the results of an evaluation carried out with a magnetic model depends critically upon the properties of the model. In order to make various evaluations of different systems compatible it is

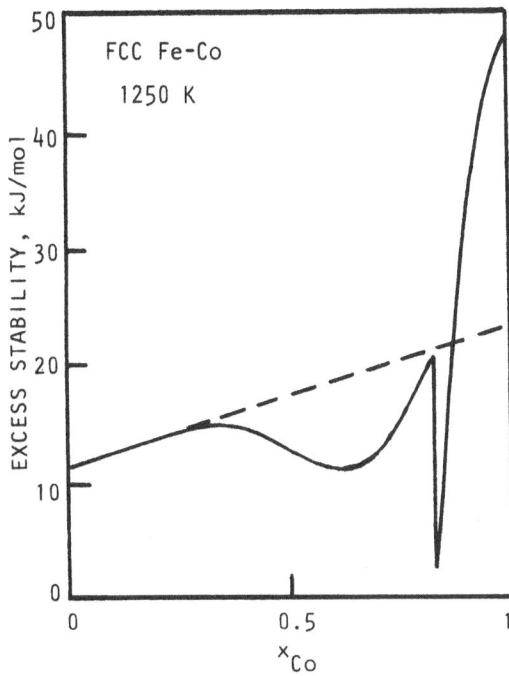

Figure 14. Excess stability of fcc Fe-Co at 1250°K. The dashed line represents the subregular solution model. The difference between the full and dashed lines represents the thermodynamic effect of the ferro-magnetic transformation.

necessary to use the same model. This is just one example of the fact that a crucial point in the establishment of international collaboration is to agree on basic numbers and models.

THE GLASS TRANSFORMATION EFFECT IN SOLUTIONS

The same problem should again arise when one, in the future, would attempt to treat the thermodynamic effect of the glass transformation. Let us for instance discuss the Pb-W system, Fig. 15. The glass temperatures are not known experimentally but have here been estimated by using a straight line between 1/3 of the melting temperature for each element. If one would use a realistic description of the properties of pure liquid W similar to the one sketched for Mo in Fig. 8 but neglect the glass transformation in the solutions, then one would refer the properties of liquid solutions in region 1 to liquid W but in region 2 to glassy W. That does not seem very reasonable since both regions are well above the estimated glass transition of alloys. Due to the very gradual nature of the glass transformation, a satisfactory evaluation of this system cannot be made without a detailed model for the glass transformation, applied to the solutions as well as to the pure elements. For the time being, the best one can do seems to be to neglect the thermodynamic effects of the glass transformation completely and thus to give up all ambitions on a realistic extrapolation of C_p for the supercooled liquid. From this point of view, the SGTE extrapolation, illustrated by Fig. 9, may seem rather attractive.

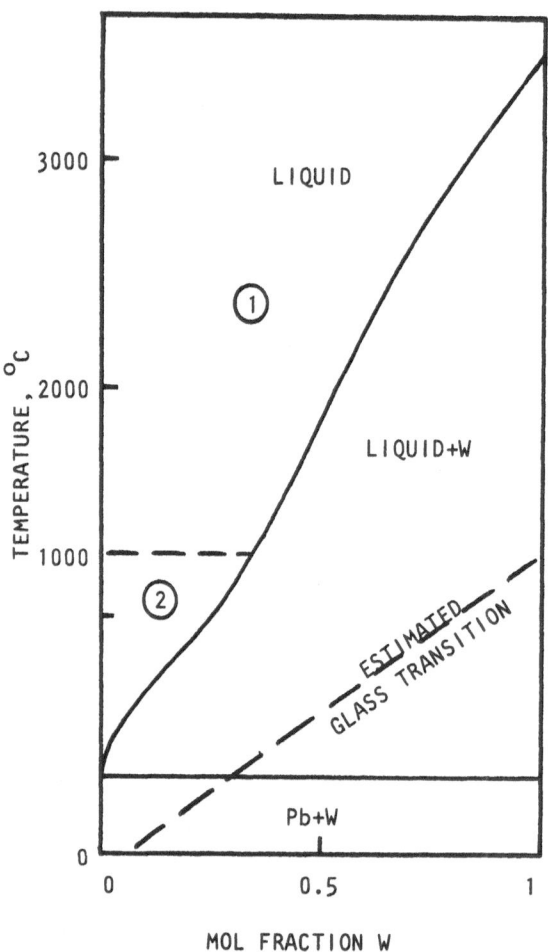

Figure 15. The Pb-W phase diagram with an estimated line for the glass transition.

THE ELECTRONIC SOLUTION EFFECT

When Kaufman (14) evaluated the lattice stabilities of transition metals by the CALPHAD method and plotted the difference in Gibbs energy of fcc and hcp relative to bcc and plotted the results versus the atomic number, then he observed a gradual variation. This has also been found from band-theory calculations (15,16). One may guess that there is a similar variation with composition across a binary system and that effect should be modeled to be included in the description of the solution behaviour of these phases. This is not yet customary and the effect is instead simply included in an expression like the Redlich-Kister polynomial. Naturally, that procedure may result in an error of the evaluated lattice stabilities when the extrapolation in the phase diagram is long. Anyhow, it should be realized that the CALPHAD values of lattice stabilities have been obtained together with parameter values in solution models and together they may yield a satisfactory description of the properties in and close to stable regions. This fact emphasizes the importance of different research groups using the same values and models for lower-order systems as a common basis for the assessment of higher-order systems.

The situation is similar for intermetallic phases when an added element can dissolve in it to some limited extent. It is then necessary to introduce a hypothetical counterphase with the same structure but the maximum alloy content possible theoretically. The Gibbs energy evaluated for such a hypothetical state will of course depend upon the solution model chosen.

In this connection it should also be emphasized that it may be dangerous to combine parameter values from two different datasets even if each one is self-consistent. A method of minimizing the risk would be to store each self-consistent dataset in a databank in such a way that combination of data from two sets is prevented except by a qualified action by the user.

SUMMARY

Even though the computer technique makes it possible to handle large amounts of experimental information, it is not possible to assess a very large network of information in a simultaneous evaluation. The sequential method must be applied sooner or later.

The necessity of using specialized mathematical models for the representation of various physical and chemical effects in substances and solution phases has been demonstrated by some examples. Such models may result in a limited accuracy but allows a larger network of information to be treated in a simultaneous evaluation using some computer-assisted procedure. Furthermore, the values of various model parameters may not always be representative of the quantities they formally refer to. The model parameters obtained from an assessment may still give a reasonable description of the regions covered by the network of information, if treated as a set of parameter values. There is a risk of obtaining incorrect results when values from different sets are combined even if each dataset may have a satisfactory self-consistency. These problems are more severe when many solution phases are involved because they may require more complicated modeling than pure substances.

In order to make collaboration between different assessment groups possible it will be necessary to agree on models and parameter values for simple systems. Otherwise various datasets cannot be combined and each group will be forced to evaluate all the phases and systems it wants to include in the databank.

ACKNOWLEDGEMENT

Generous support from the Swedish Board for Technical Development is gratefully acknowledged.

REFERENCES

1. D. Garvin, V.B. Parker, D.D. Wagman and W.H. Evans, in The Pro-
    ceedings of the Fifth Biennial International CODATA Conference,
    Ed. B. Dreyfus, Pergamon Press, Oxford 1977.
2. O. Kubaschewski and C.B. Alcock, Metallurgical Thermochemistry,
    Pergamon Press, Oxford 1979.
3. CODATA Bulletin 47 (May 1982), 'A systematic Approach to the Pre-
    paration of Thermodynamic Tables' (Report of the CODATA Task Group
    on Internationalization and Systematization of Thermodynamic
    Tables).
4. J. Haas, unpublished research.
5. B. Sundman, TRITA-MAC 344, Royal Inst. of Techn. Stockholm 1987.
6. L. Kaufman, CALPHAD, 1 (1977) 28; 2 (1978) 56,118,119, 296; 3
    (1979) 46.
7. A. Fernández Guillermet, Intern. J. Thermophys. 6 (1985) 395.
8. B. Jönsson and J. Ågren, unpublished research.
9. J.-O. Andersson, A. Fernández Guillermet, P. Gustafson, M. Hillert,
    B. Jansson, B. Jönsson, B. Sundman and J. Ågren, CALPHAD 11, (1987)
    93.
10. L. Kaufman, CALPHAD 1 (1977) 28.
11. M. Hillert and M. Jarl, CALPHAD 2 (1978) 227.
12. L.S. Darken, Trans. AIME 239 (1967) 80.
13. A. Fernández Guillermet, TRITA-MAC 324, Royal Inst. Technology,
    Stockholm, 1986.
14. L. Kaufman, in Metallurgical Chemistry, O. Kubaschewski editor
    (HMSO, London, 1972) p. 373.
15. D.G. Pettifor, in Metallurgical Chemistry, O. Kubaschewski,
    editor (HMSO, London, 1972) p. 191.
16. H.L. Skriver, Phys. Rev. B, 31 (1985) 1909.

MAGNETIC-INDUCED TRICRITICAL POINT IN ALLOYS AND THE LOW-TEMPERATURE Fe-Ni AND Fe-Ni-Cr PHASE DIAGRAMS

Y. Austin Chang
Department of Metallurgical Engineering
University of Wisconsin-Madison
1509 University Avenue
Madison, WI 53706
USA

ABSTRACT.   The phase equilibria at temperatures above and below the tricritical point in binary alloys are discussed in terms of thermodynamic principles. The existence of a second spinodal curve for the disordered phase within the magnetic-induced miscibility gap is rationalized.   The relative interplay of the chemical and magnetic contributions to the stabilities of various types of phase equilibria in binary alloys is also discussed.
   The effect of magnetic contribution to the Fe-Ni and Fe-Ni-Cr phase stabilities and at low temperatures is reviewed. The calculated phase diagrams of the Fe-Ni and Fe-Ni-Cr systems considering the magnetic and non-magnetic terms are compared with experimental results.   These results were obtained either from meteorties or synthesized alloys subjected to irradiation. The meteorties which have cooled slowly in space may have achieved metastable/stable equilibrium conditions. The large amounts of defects introduced by irradiation may have helped to achieve metastable/stable equilibrium conditions.

1.0   Introduction

In a recent paper, Allen and Cahn (1) discussed the phase equilibria in the vicinity of a tricritical point of a binary system. Using a Landau-type expansion of the free energy in terms of temperature, composition and order parameters, Allen and Cahn (1) showed that a spinodal curve for the ordered phase exists from the tricritical point down to lower temperatures as shown in Fig. 1a. But a spinodal curve for the disorder phase is absent in the vicinity of the tricritical point. The unstable region is bounded by the metastable extension of the second-order phase transition temperature and the spinodal curve for the ordered phase. Yet at lower temperatures, the spinodal curve for the disordered phase does exist as also shown in Fig. 1a. This spinodal curve was found for the fcc (Fe,Ni) phase at low temperatures due to magnetic interaction. Subsequently, the appearance of the spinodal curve for the disorderd phase was rationalized thermodynamically (3,4). The phase equilibria shown in Fig. 1a are for alloys such as the fcc (Fe,Ni) alloys which are

85

H. Brodowsky and H.-J. Schaller (eds.), Thermochemistry of Alloys, 85–106.
© 1989 by Kluwer Academic Publishers.

exothermic. On the other hand, when the magnetic alloys are endothermic, we would expect to have two miscibilities at low temperatures, one due to the chemical (non-magnetic) term of the Gibbs energy and the second due to the magnetic interaction as discussed above and shown in Fig. 1a. This type of equilibria occurs for bcc (Fe,Cr) and fcc(Cu,Ni) alloys at low temperatures (5,6). It is evident that the phase equilibria of magnetic alloys at low temperatures are more complex than at higher temperatures where the chemical term dominates. Since experimental determinations of low-temperature phase equilibria under normal conditions are difficult to attain, it is important for us to understand the basic phase relationships which occur at low temperatures. In the present paper, the thermodynamic principles governing the phase equiibria of tri-critical points will be reviewed. The calculated phase equilibria of Fe-Ni and Fe-Ni-Cr at low temperatures involving the tricritical points are compared with the experimentally determined ones. The phase diagrams were determined from alloys subjected to irradiation treatment as well as from meteorites. These meteorites have cooled slowly in space and may have achieved equilibrium condition down to ~150°C (9).

2.0   Thermodynamic Rationalization

The stability of a magnetic phase is governed by two contributions to its total Gibbs energy. These two contributions are the chemical (non-magnetic) and the magnetic terms. Even though the chemical term is exothermic, phase separation may occur at low temperatures. The compositional dependences of the magnetic free energy vary in such a way that addition to the chemical term produces a hump in the free energy curve. This hump then causes phase separation as shown in Fig. 1a. At temperatures higher than the tricritical point T_2 in Fig. 1a, the phase undergoes a second-order phase transformation from a ferromagnetic state to a paramagnetic state. At temperatures lower than the tricritical point, the ferromagnetic phase undergoes a first-order phase transformation to a paramagnetic phase with different composition. The spinodal curve for the ordered phase sp(o) exists at the tricritical point and lower temperatures, but the spinodal curve for the disordered phase sp(d) does not occur at the tricritical point. It exists only at temperatures lower than the tricritical point, i.e., at $T \leq T_4$ shown in Fig. 1a. The appearance of these equilibria has been discussed in terms of the free energy curve, G, the first derivative of the free energy with composition curve, $G' = dG/dx$, and the second derivative of the free energy with composition curve, $G'' = d^2G/dx^2$, for the ordered and disordered phases. In the present papers we will use G'' in terms of the relative stability function to describe the phase equilibria above and below the tricritical point. The relative stability function is defined as

$$re_{St} = G''/{}^{id}G'' \qquad [1]$$

This function is preferred over G'' since it is unity for $x=0$ and $x=1$. Yet, it retains the characteristics of spinodal points, i.e., $re_{St}=0$ at

x(sp).   Figure 1b shows the ^{re}St-vs-composition diagrams for five
selected tempratures from T_1 to T_5. These temperatures are marked in
Fig. 1a. Let us first point out that ^{re}St is discontinuous at T_c for a
second-order phase transition and ^{re}St(d,x_c) > ^{re}St (o,x_c).   It is
positive at temperatures higher than the tricritical point as shown in
Fig. 1b(i). At T_2, the tricritical point, ^{re}St(o,x_c) attains a value of
0 as shown in Fig. 1b(ii). We now have a spinodal point for the ordered
phase.   The value of ^{re}St(d,x_c) is not zero at T_2 and thus we do not
have a spinodal for the disordered phase. At T_3 shown in Fig. 1b(iii),
sp(o) occurs at a value of x > x_c but we still do not yet have a
spinodal for the disordered phase. But at T_4 shown in Fig. 1b(iv),
^{re}St(d,x_c) = 0. We now have the spinodal for the disordered phase sp(d)
at x_c as well as that for the ordered phase at a value of x > x_c. At
temperatures lower than T_4 such as T_5, we have two spinodals, sp(d) at a
value of x < x_c and sp(o) at a value of x > x_c.

It is noteworthy to point that the curvature of the free energy for
magnetic alloys is discontinuous at $T_c(x_c)$ using the phenomenological
equation proposed by Chuang, Schmid and Chang.[8] This equation consists
of two parameters, the Curie parameter T_c and the magnetic moment β. A
knowledge of T_c and β allows us to obtain the magnetic thermodynamic
properties.   The equation has the characteristics of a second-order
phase transition. The free energy, enthalpy and entropy are continuous
at T_c but the specific heat is discontinuous. The difference between
the curvature of the free energy for the paramagnetic phase, G"(para)
and that for the ferromagnetic phase, G"(ferro) at $T_c(x_c)$ is

$$[G''(para) - G''(ferro)]_{T_c(x_c)}$$

$$= 0.633 \; CR \; \ln(\beta + 1) \; (dT_c/dx)^2/T_c \quad \text{for bcc phases} \qquad [2A]$$

$$= 1.967 \; CR \; \ln(\beta + 1) \; (dT_c/dx)^2/T_c \quad \text{for bcc phases} \qquad [2B]$$

where C is an empirical parameter characteristic of an alloy, R is the
gas constant, β is the magnetic moment per atom and T_c is the Curie
temperature as defined earlier.   Since values of C, R, β and T_c are
always positive, the quantity $[G''(para)-G''(ferro)]_{T_c(x_c)}$ must be always
larger than 0 unless dT_c/dx = 0. This means discontinuity occurs in G"
or ^{re}St at $T_c(x_c)$ with ^{re}St(d,x_c) > ^{re}St(o,x_c) as shown in
Figs. 1b (i-iv).

When a magnetic phase is in equilibrium with another phase at high
temperatures, phase separation of this magnetic phase at temperatures
lower than the tricritical point would result in the formation of a
monotectoid isotherm as shown in Fig. 2. The Fe-Ni binary has this type
of equilibrium at low temperatures. At high temperatures, ferrite (α)
coexists with austenite (γ). As the temperature decreases, a monotectoid
reaction γ_1(P) = α + γ_2(F) occurs as shown in Fig. 2. Extension of the
$\gamma_1/(\alpha+\gamma_1)$ phase boundary to the metastable region intercepts $T_c(x_c)$ and
then increases monontonically to a maximum at the spinodal and then
decreases again as shown schematically in Fig. 2. The slope of the
phase boundary exhibits a discontinuity at $T_c(x_c)$ and there is no point

of inflection in the extension of the boundary through this gap as would occur normally in a chemical miscibility gap. This may be demonstrated thermodynamically as given below.

The temperature derivative of the γ_1 phase boundary in equilibrium with α is[10]

$$\frac{dx^{\gamma_1}}{dT} \quad \frac{dS^{\gamma_1}}{dx} - \frac{S^\alpha - S^{\gamma_1}}{x^\alpha - x^{\gamma_1}} \quad / \frac{d^2 G^{\gamma_1}}{dx^2} \qquad [3]$$

where S is the entropy, x is the atom fraction of the second component or Ni in the fcc (Fe,Ni) case, and the superscripts γ_1 and α denote the γ_1-phase and the co-existing α-phase. It is evident from this equation that dx^{γ_1}/dT is ∞ (or dT/dx^{γ_1} is zero) when $d^2G^{\gamma_1}/dx^2$ is zero. In a chemical-induced miscibility gap with two spinodal curves, we would thus have a point of inflection in the extension of the phase boundary through the gap. However, in the present case as shown in Fig. 1 we have only one spinodal curve. Consequently, we do not have a point of inflection in the extension of the $\gamma_1/\alpha+\gamma$ phase boundary through the miscibility gap.

The phase relationships shown in Fig. 1 are for magnetic alloys which have exothermic enthalpies of formation due to chemical contribution. When the chemical term for the enthalpy of formation of magnetic alloys is endothermic, different types of phase equilibria occur as shown schematically in Fig. 3 (a,b,c). Depending on the relative contributions of the chemical and magnetic terms to the free energy of the phase, we may have a stable tricritical point as shown in Fig. 3a or a metastable tricritical point as shown in Fig. 3b. Figure 3c represents a degenerate case when the tricritical point falls on the miscibility phase boundary. Since Lin and Chang (4) have analyzed G, G' and G" as a function of composition at several temperatures, i.e., T_1 to T_2 shown in Fig. 3a for the phase diagram given in Fig. 3, we will not repeat their analysis here. Nevertheless, it is noteworthy to point out that at (T_1, x_1) the alloy undergoes phase transformation from a stable to an unstable region upon cooling as has been discussed in many standard text books on thermodynamics. At (T_4, x_4) the alloy undergoes phase transformation from a metastable to an unstable region upon cooling.

The types of equilibria shown in Figs. 1a and 3 normally occur at low temperatures. When a homogeneous paramagnetic phase is cooled to a two-phase region, phase separation would not occur due to sluggish kinetics, except for meteorites which have undergone slow cooling in space. Alternatively, the alloys may be subjected to electron irradiation resulting in the formation of large amounts of defects. The presence of these defects enhances the diffusional process and permits phase separation. On the other hand, the transformation of an alloy from a paramgnetic state to a ferromagnetic state is not suppressible. It is interesting to discuss the phase transformations of several representative alloys when cooling from the paramagnetic single-phase region to the miscibility gap region for systems with the type of phase equilibria given in Fig. 1. The phase diagram is reproduced in Fig. 4 with five alloys of different cooling paths as given below:

1. $P(x_1^0,T) \rightarrow F(x_1^0,T_1) \xrightarrow{\text{sp}} p(x'',T_1) + F(x',T_1)$

2. $P(x_2^0,T) \rightarrow P(x_2^0,T_1) \xrightarrow{\text{N\&G}} P(x'',T_1) + F(x',T_1)$

3. $P(x_3^0,T) \rightarrow F(x_3^0,T_2) \xrightarrow{\text{sp}} P(x'''',T_2) + F(x''',T_2)$

4. $P(x_4^0,T) \rightarrow P(x_4^0,T_2) \xrightarrow{\text{sp}} P(x'''',T_2) + F(x''',T_2)$

5. $P(x_5^0,T) \rightarrow P(x_5^0,T_2) \xrightarrow{\text{N\&G}} P(x'''',T_2) + F(x''',T_2)$

The symbols P and F denote the paramagnetic and ferromagnetic states, sp denotes spinodal decomposition and N&G denotes decomposition by nucleation and growth. The two phases resulting from alloys 1 and 2 when cooling to a temperature T_1 are the same but the resulting morphologies are rather different. The same holds true for alloys 3, 4, and 5 when cooled to a temperature T_2.

The type of equilibria shown in Fig. 3b and discussed above were found to occur in the bcc phase of (Fe,Cr) alloys (5), fcc(Cu,Ni) (6) and (Fe,Al) alloys with the B2 and DO_3 structures (11). In the last case, the second-order phase transition is not of magnetic origin but of structural disordering.

3.0   The Low-Temperature Fe-Ni and Fe-Ni-Cr Phase Diagrams

3.1   Fe-Ni

Chuang et al. (2) calculated the low-temperature Fe-Ni phase diagram considering both the chemical and magnetic terms of the total free energies of the bcc and fcc phases. A monotectoid isotherm between α, $\gamma_1(P)$ and $\gamma_2(F)$ at 662 K was found. Furthermore, they calculated the metastable miscibility gap of the γ-phase to temperatures below 662 K. The calculated phase boundaries of the metastable miscibility at low temperatures are consistent with the results of Chamberod et al. (12,13). Subsequently, using the data of Reuter et al. (9,14,15) obtained on meteorites, Chuang (17) recalculated the low-temperature Fe-Ni phase diagram including the $Fe_3Ni(L_2)$ and $FeNi(L1_0)$ phases. The phase boundary of $FeNi_3(L1_2)$ was also re-calculated assuming that the homogeneity range does not change much with decreasing temperature. A composite diagram included the various possible metastable equilibria involving α, $Fe_3Ni(L1_2)$, γ_1, γ_2, $L1_0$ and γ' ($FeNi_3,L_2$). We will discuss below the experimental evidence obtained from irradiated (Fe,Ni) aloys and (Fe,Ni) meteorites to support qualitatively the correctness of the calculated diagram.

Figure 5 shows the calculated Fe-Ni phase diagram of at low temperatures. The dashed lines show the calculated metastable equililbria. Chamberod et al. (12,13) studied several (Fe,Ni) alloys at 30, 32, 35 and 40% Ni. These alloys were annealed at 800°C and cooled to room temperature. The same alloys were subsequently subjected to 3 meV electron irradiation with 4×10^{19} electron/cm^2 at 80°C. Figure 6 shows the Mossbauer spectra of the four alloys with and without irradiation treatments. The spectra clearly show that the 30% Ni alloy without irradiation is paramagnetic (top left). On the other hand, the same alloy after irradiation shows the presence of at least one other phase (top right).

The lattice parameters of unirradiated and irradiated alloys at 25°C are shown in Fig. 7 as a function of composition. The unirradiated alloys exhibit the classical Invar behavior, i.e., the lattice parameter attains a pronounced maximum at ~37% Ni. On the other hand, the lattice parameters of the irradiated alloys vary linearly with composition. Since both γ_1 and γ_2 are fcc and have comparable lattice parameters, the x-ray diffraction basically yielded average values of these parameters. All of these results support the idea of phase separation as predicted from thermodynamics. Chamberod et al. also found that the coefficient of thermal expansion of irradiated alloys behaves like a normal substance while that of unirradiated alloys exhibits the classical Invar behavior. In other words, the coefficient of thermal expansion is close to zero at room temperature.

Scorzelli and co-workers (16) studied the Santa Catharina meteorite (~35% Ni) and synthesized (Fe,Ni) alloys, both irradiated and unirradiated, using x-ray diffraction and Mossbauer spectroscopy. Their results for the temperature dependence of the lattice parameter are shown in Fig. 8. Let us first examine the unirradiated (Fe,Ni) alloy. The first derivative of "a" with respect T, i.e., the coefficient of thermal expansion, is close to zero at room temperature. Subsequently, it increases with temperature, a classical Invar behavior. However, after the alloy was subjected to irradiation, "a" increases with temprature from room temperature to 500°C as shown in Fig. 8(b). These results are identical to those reported by Chamberod et al. Let us next examine the results for the Santa Catharina meteorite. The lattice parameter also increases with temperature, similar to that for the irradiated alloys. Since meteorites have been cooled in space over geological time, metastable equilibrium might have been achieved, i.e., the γ-phase has undergone phase separation. Figure 8 (curve C$_2$) also shows the results for the meteorite after subjection to annealing at 800°C for 24 hours. Like the unirradiated (Fe,Ni) alloys, the lattice parameter remains nearly constant at room temperature and then increases with temperature at higher temperatures. Once the meteorites are heated to 800°C, the single-phase austenite is formed. When cooling to room temperature, there is insufficient time for γ-phase separation to occur. These results indicate that the effect of irradiation is to induce large amounts of defects in the lattice and help the alloys to attain metastable/or stable equilibrium states.

.. Scorzelli and Dannon (16) made a detailed analysis of the Mossbauer spectra for the Santa Catharina meteorite which is similar to that shown in Fig. 6 for alloys subjected to electron irradiation. According to their analysis, this meteorite has the following three phases, 38 ± 5% γ_1(P), 12 ± 5% γ_2(F) and 50 ± 5% FeNi(L1$_0$). After heating the meteorite to 450°C (723K) for 15 hours and subjecting it to snock waves with pressures of 100 and 200 kbars, the relative amount of $\gamma_1(P)$ remains essentially the same, that of Ll_0 decreases with corresponding increase in $\gamma_2(F)$. The results are summarized in Table I.

Table I
Results of Scorzelli and Dannon on the Santa Catharina Meteorite

	723 K 15 Hrs.	Shock Wave Pressures	
		100 Kbar	200 Kbar
$\gamma_1(P)$	38 ± 5	36 ± 5	38 ± 5	34 ± 5
$\gamma_2(F)$	12 ± 5	31 ± 5	20 ± 5	42 ± 5
FeNi(Ll_0)	50 ± 5	34 ± 5	42 ± 5	24 ± 5

These results can be rationalized readily in terms of the phase diagram shown in Fig. 5. Heating to 723 K converts some of the Ll_0 phase to $\gamma_2(F)$. It is difficult to rationalize the results obtained from the shock wave experiments except there might have been an abrupt rise in temperature associated with the shock wave. Thus, the shock wave would yield similar results as temperature, i.e., decreasing the amount of Ll_0 with a corresponding increase in $\gamma_2(F)$. These results are again consistent with the stable and metastable phase diagrams presented in Fig. 5.

Let us next explain how the metastable phase equilibria were obtained. According to Reuter et al. (9,14,15) and others (18,19) the ordered FeNi phase (Ll_0) transforms to the fcc phase at 320°C (593K). Their data are shown in Fig. 9. They also determined the compositions of phases present in meteorite samples by means of analytical electron microscopy. According to their calculation, there would be no diffusion below 150°C (423 K) even over geological time. Accordingly, the chemical compositions obtained by them must correspond to this temperature of 423 K as shown in Fig. 8. Chuang took 593K to be the transformation temperature of Ll_0 to fcc and ~ 430 K to be the peritectoid temperature for the formation of $Fe_3Ni(Ll_2)$. The solution models he used for the Ll_0 and Ll_2 phases are those developed by our research group (20-22). As shown in Fig. 9, the calculated phase equilibria involving the $Fe_3Ni(Ll_2)$ and Ll_0 phases are consistent with the results of Reuter et al. Nevertheless, a word of caution is in order. While we are reasonably confident of our calculated miscibility gap below 662 K, the monotectoid temperature, the calculated equlibria involving the $Fe_3Ni(Ll_2)$ and Ll_0 phases are based on two data points mentioned above. Furthermore, the models consider only the nearest interaction and take no other effect into consideration which may be an over-simplification of the real situation. Additional and undoubtedly very difficult experimental investigations on well-controlled irradiated samples and meteorites are needed before a more definitive phase diagram is available at low temperatures.

3.2   Fe-Ni-Cr

Using the same approach as for the binary Fe-Ni system, Chuang and Chang (23) also calculated the Fe-Ni-Cr phase diagram from the melting range to low temperatures. Again, the effect of magnetic term for the fcc phase becomes important in determining the type of phase equilibria for alloys with concentrations near the Fe-Ni binary. Figure 10 shows a schematic diagram at a constant temperature displaying the tricritical point. At Ni concentration higher than the tricritical point, the ferromagnetic phase transforms to the paramagnetic phase by a second-order phase transition. At Ni concentration lower than the tricritical point, the ferromagnetic phase transforms to the paramagnetic phase with a different composition by a first-order phase transition. The spinodal curve for the ordered phase again appears at the tricritical point while that for the disordered phase appears at Ni contents lower than the tricritical point. Figure 11 shows a calculated equilibrium isotherm of Fe-Ni-Cr at 650 K. Since 650 K is below the binary monotectoid reaction: $\gamma_1(P) = \alpha + \gamma_2(F)$, the $\gamma_1(P) + \gamma_2(F)$ two-phase field does not extend to the boundary binary. Also shown in Fig. 11 is the T_0 (α/γ) curve. In order to better use the T_0 (α/γ) curve as given in Fig. 11 several isopleths were calculated. Figure 12 shows such an isopleth with 10 at% Cr. We may use this isopleth much as we do for Fe-Ni binary. Stubbins had studied several (Fe,Ni,Cr) alloys subjected to electron irradiation using a high voltage electron microscope (24). One of the alloys he studied was Fe5Ni10Cr. When this alloy was cooled rapidly to low temperatures, martensite was formed. However, once this alloy was subjected to electron irradiaton at 723 K, the martensite reverted back to austenite. According to Fig. 12, 723 K is higher than the T_0 (α/γ) curve for an alloy with this composition. Any martensite would transform to austenite if there is no kinetic limitation. Under ordinary condition, transformation of martensite to austenite would not occur at this low temperature, but irradiation enhances the transformation. This demonstrates the utility of the calculated phase diagram at low temperatures for practical appilcation. Moreover, the results of Stubbins give qualitative if not quantatitive verification of the calculated Fe-Ni-Cr phase diagram at low temperatures.

4.0   Acknowledgement

The authors wish to thank the National Science Foundation for support through Grant No. NSF-DMR-85-14421 and Dr. J.-C. Lin for his critical review and valuable comments of this manuscript.

5.0   References

1.   S. M. Allen and J. W. Cahn, Bull. Alloy Phase Diagr., 1982, 3, 287; also in Phase Diagrams (ed. by L. H. Bennett, T. B. Massalski and B. C. Giessen) North Holland, New York, 1983, 195.

2.   Y.-Y. Chuang, Y. A. Chang, R. Schmid and J.-C. Lin, Metall. Trans. A, 1986, 17A, 1361.

3. J.-C. Lin, Y.-Y. Chuang and Y. A. Chang, in Rapidly Solidified Alloys and Their Mechanical and Magnetic Properties (ed. by B. C. Giessen, D. E. Polk and A. I. Taub) Materials Res. Soc., Pittsburgh, PA, 1986, 207.

4. J.-C. Lin and Y. A. Chang, Metall. Trans. A, 1987, 18A, in press

5. Y.-Y. Chuang, J.-C. Lin and Y. A. Chang, Calphad, 1987, 11, 1.

6. D. Chakrabarti, D. Laughlin, S.-W. Chen and Y. A. Chang, Bull. Alloy Phase Diagr., 1987, under review.

7. R. Schmid, Y.-Y. Chuang and Y. A. Chang, Calphad, 1986,

8. Y.-Y. Chuang, R. Schmid and Y. A. Chang, Metall. Trans. A, 1985, 16A, 153.

9. K. B. Reuter, PhD Thesis, 1986, Lehigh, University of Bethlehem, PA 18015; K. B. Reuter, D. B. Williams, and J. I. Goldstein, Metall. Trans. 1987, under review.

10. J. L. Meijering, in The Physical Chemistry of Metallic Solutions and Intermetallic Compound Symposium, vol. II, Chem. Pub. Co., Inc., New York, 1960, 124.

11. S. M. Allen and J. W. Cahn, Scripta Met., 1976, 10, 451.

12. A. Chamberod, J. Laughier and J. M. Penisson, J. Magn. Mater., 1979, 10, 139.

13. A. Chamberod, H. Rechenberg and R. de Tournemine, ICM-73, 1974, 3, 170.

14. K. B. Reuter, J. A. Kowalik, D. B. Williams and J. L. Goldstein, Abstract for the XI International Congress on Electron Microscopy, Kyoto, Japan, Aug. 31-Sept. 7, 1986.

15. K. B. Reuter, in Proceedings of the 43rd Annual Meeting of the Electron Microscopy Soc. of America, (Ed: G. W. Bailey) 1985, San Francisco, Press, Inc., Box 6800, San Francisco, CA 94101-6800, 338.

16. R. B. Scorzelli and J. Danon, Physica Scripta, 1985, 32, 143, J; Danon, R. B. Slorzelli, I. Souza-Azevedo, J. Laugier and A. Chamberod, Nature, 1980, 284, 537.

17. Y.-Y. Chuang, unpublished Research, University of Wisconsin-Madison, Madison, WI 1986.

18. J. Pauleve et al., J. de Phys., 1962, 23, 841.

19. J. M. Penisson and A. Bourret, Quatrieme Congress International, Toulowse, 1976, 205.

20. Y. A. Chang and J. P. Neumann, Prog. Solid State Chem., (ed. by G. M. Rosenblatt and W. L. Worrell), 1982, 14, 221.

21. W. W. Liang, Y. A. Chang, S. Lou and I. Gyuk, Acta Met., 1973, 21, 629.

22. I. Gyuk, W. W. Liang and Y. A. Chang, J. Less-Common Met., 1974, 38, 249.

23. Y.-Y. Chuang and Y. A. Chang, Metall. Trans. A, 1987, 18A, 733.

24. J. F. Stubbins, "Phase Stability in Fe-15Ni-10 to 20Cr Under Irradiation", paper presented at the June 23-25, 1986 ASTM Meeting in Seattle, WA.

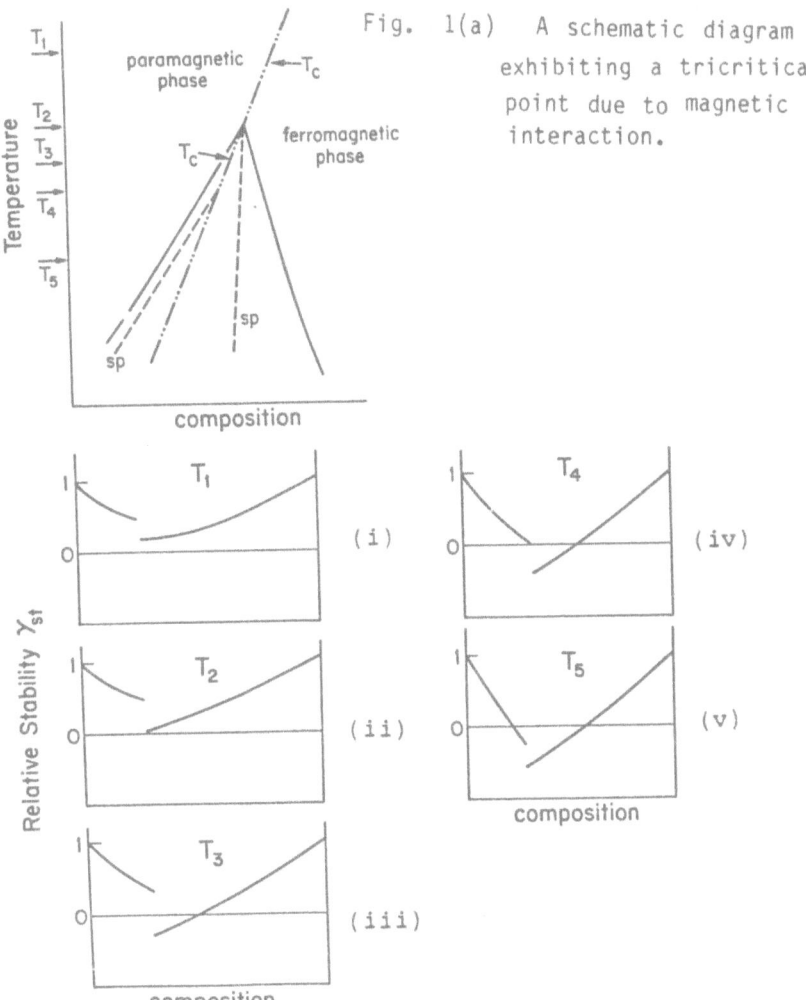

Fig. 1(a) A schematic diagram
exhibiting a tricritical
point due to magnetic
interaction.

Fig. 1(b) The relative stability function as a function of
compositions at temperatures higher and lower than the
tricritical point.

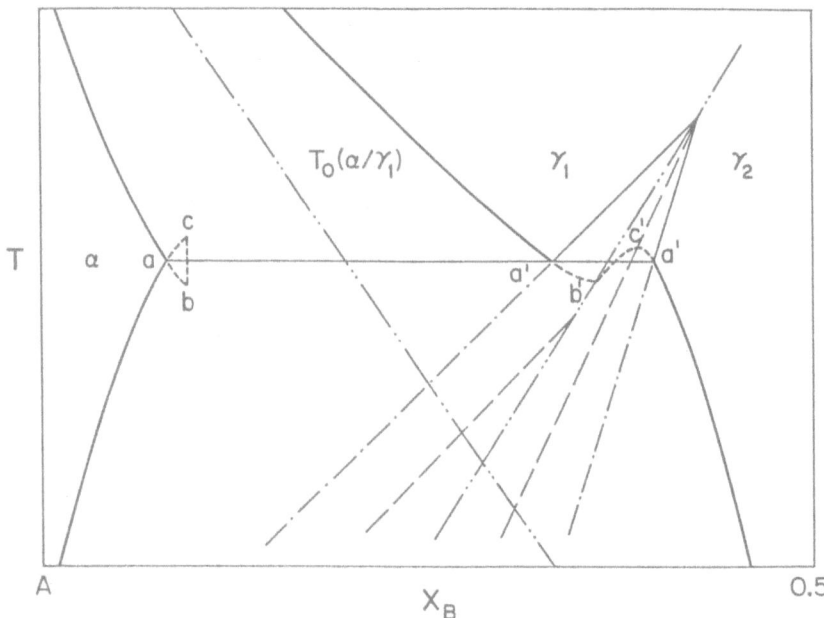

Fig. 2. A schematic phase diagram showing the extension of phase
boundaries through a miscibility gap of the type exhibited
by Fe-Ni binary at low temperatures.

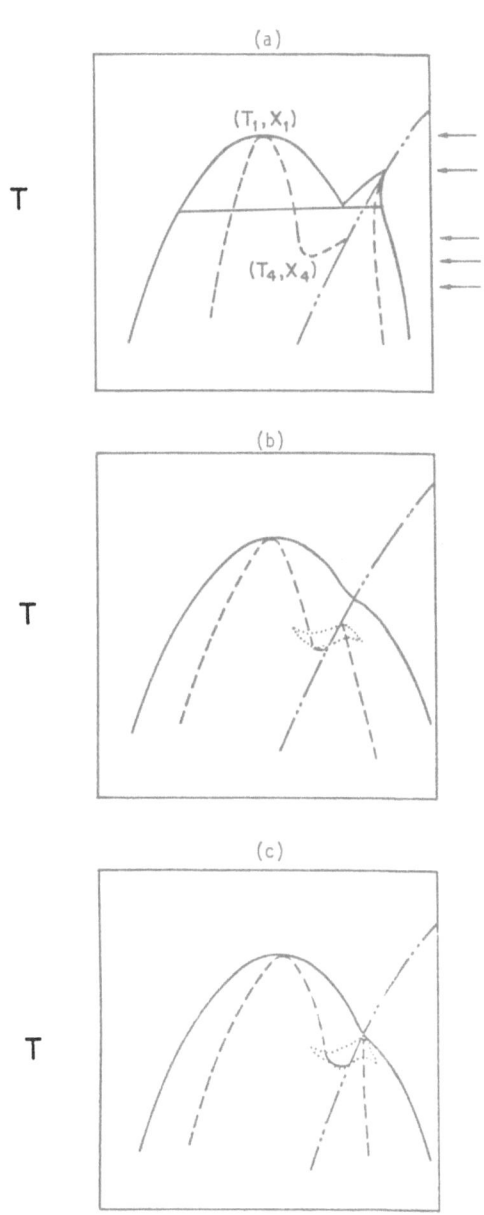

Fig. 3

Appearance of the different types of equilibria when the chemical term is endothermic: (a) the tricritical point is stable, (b) the tricritical point is metastable and (c) the tricritical point coincides with one of the stable miscibility gap phase boundaries. The dotted curves in (b) and (c) show the metastable and unstable phase boundaries.

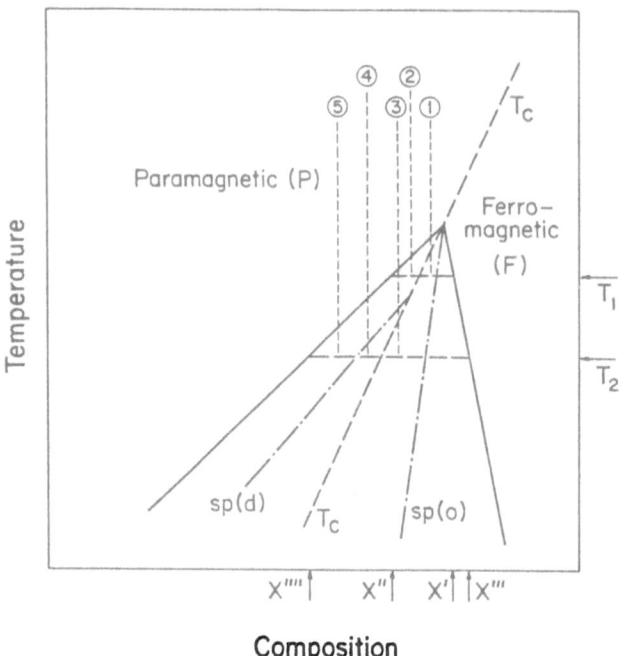

Fig. 4. A schematic phase diagram exhibiting a tricritical point due to magnetic interaction superimposed with transformations of selected alloys from the paramagnetic single phase upon cooling to the miscibility gap.

FE-NI PHASE DIAGRAM

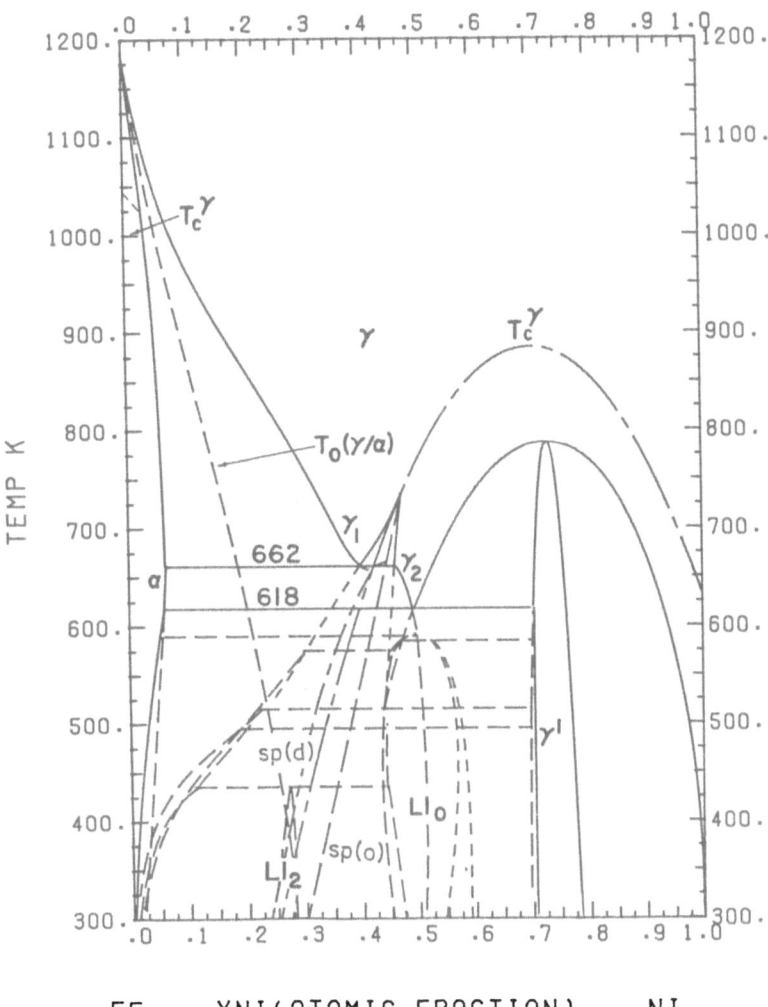

Fig. 5.   A calculated Fe-Ni phase diagram at low temperature by Chuang.[17]

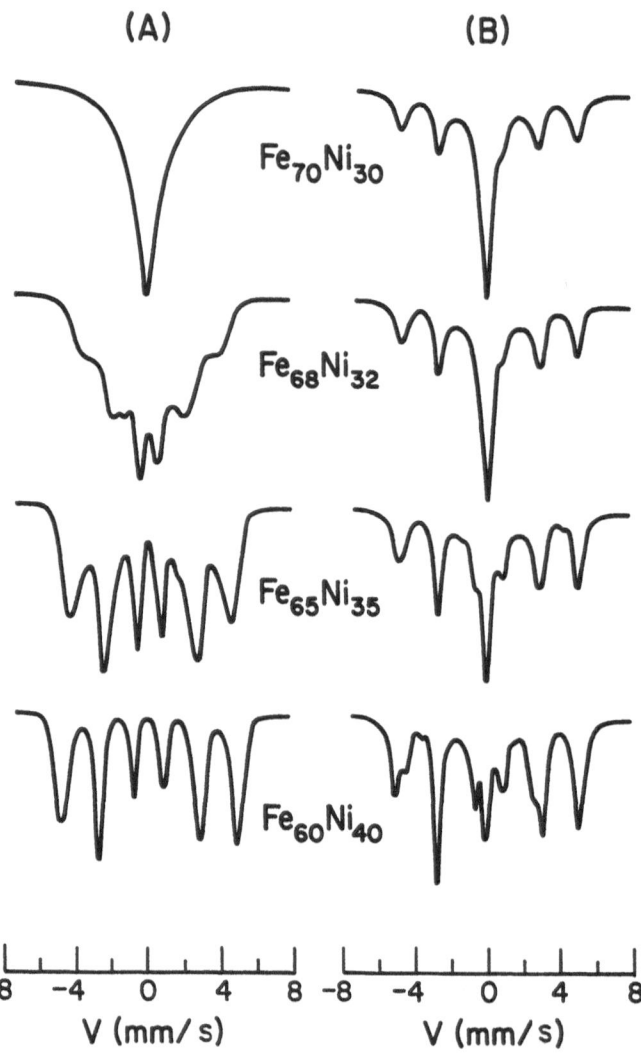

Fig. 6. Mössbauer spectra of (Fe,Ni) alloys at room temperature according to Chamberod et al. The spectra at the left refer to alloys homogenized at 800°C and cooled to room temperature. The spectra at the right refer to those subject to electron irradiation at 80°C.

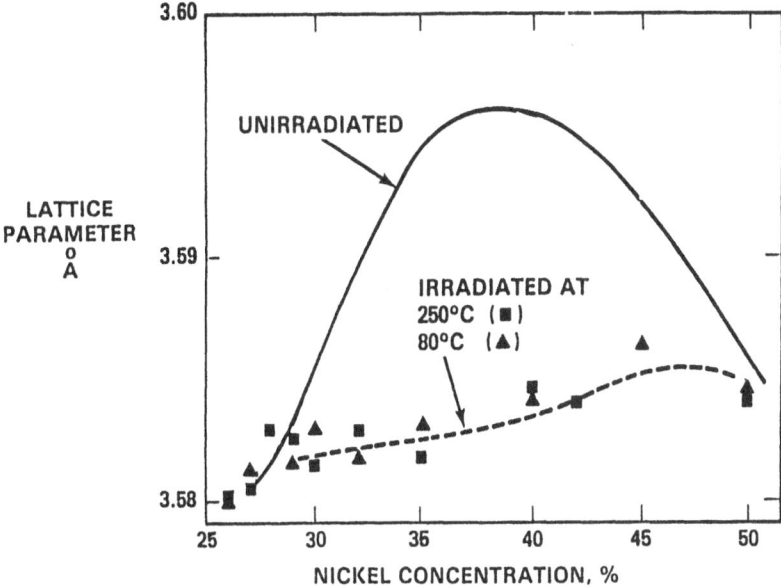

Fig. 7.   Room temperature lattice parameters of (Fe,Ni) alloys from
           25 to 50 at% Ni: comparison between unirradiated and
           irradiated alloys according to Chamberod et al.

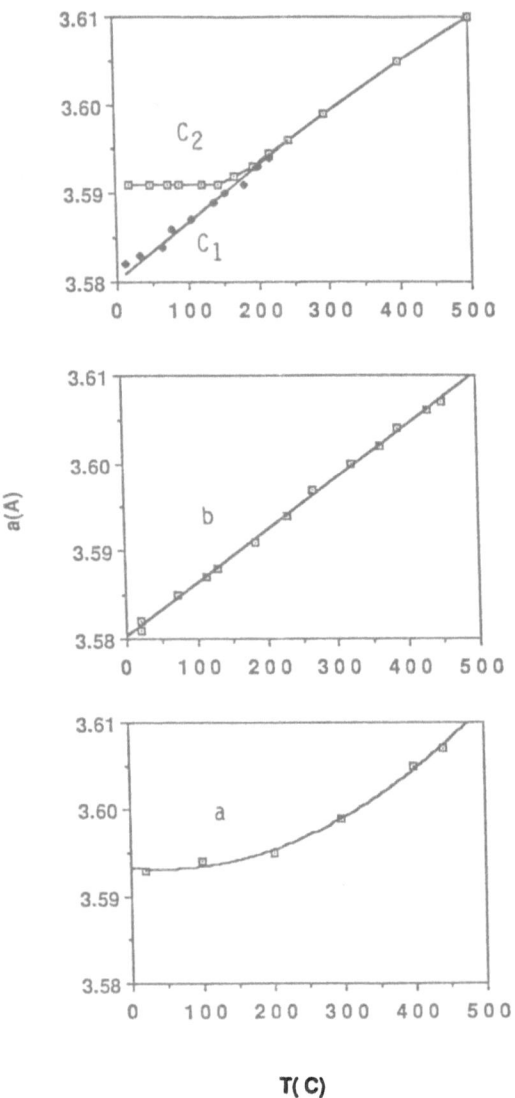

Fig. 8.   Temperature dependence of the lattice parameter of 35% Ni
           alloy according to Scorzelli and Vanon.

C_1: Santa Catharina meteorite, ~35% Ni.
C_2: After the meteoride was annealed at 800°C for 24 hrs.
b: Fe ~35% Ni alloys after irradiation.
a: Fe ~35% Ni alloys before irradiation.

FE-NI PHASE DIAGRAM

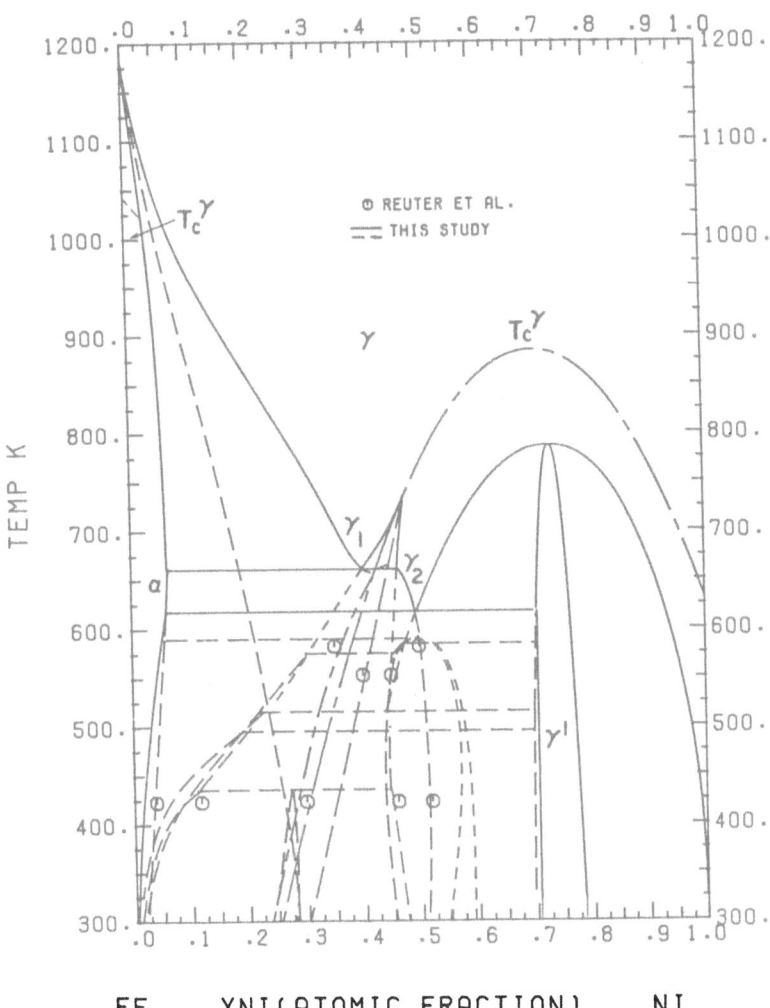

Fig. 9. The Fe-Ni phase diagram at low temperatures comparison between the calculated diagram and experimental data.

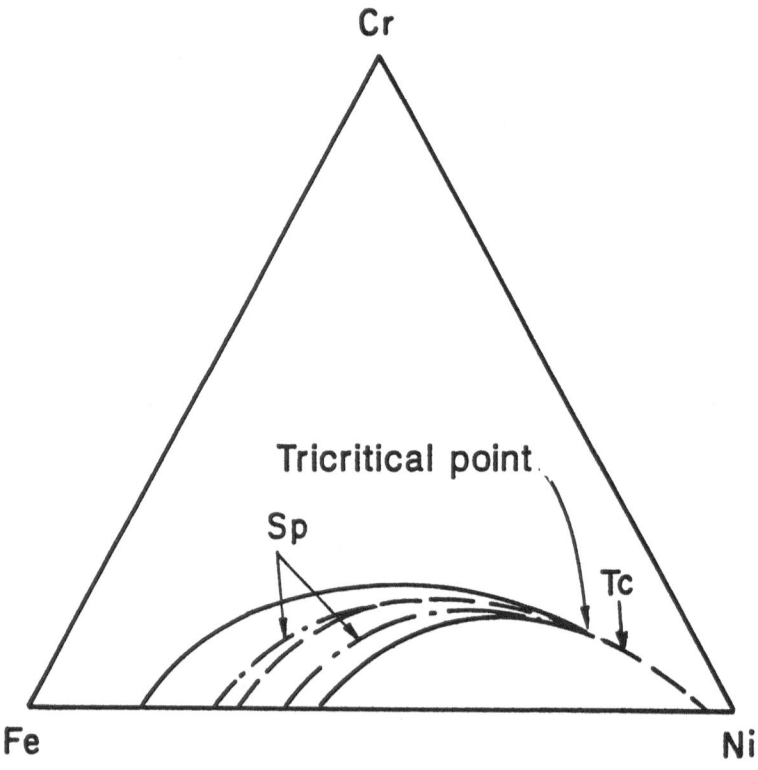

Fig. 10.   A schematic diagram of Fe-Ni-Cr showing the appearance of a
            tricritical point due to magnetic interaction.

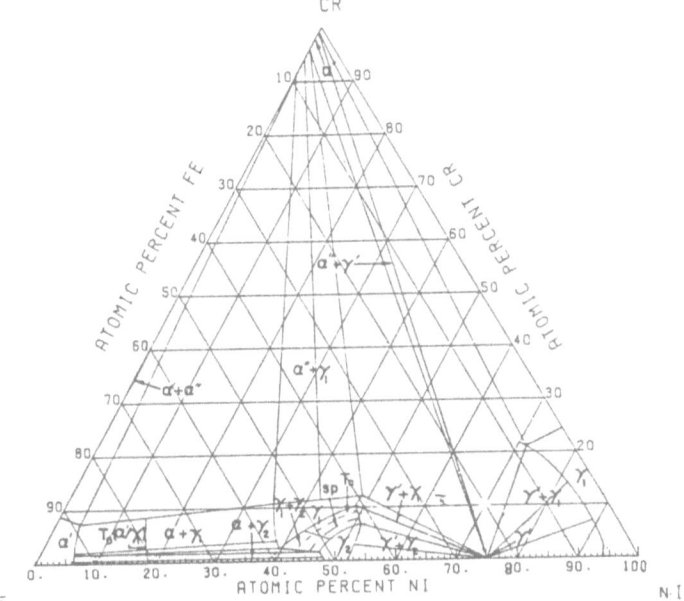

FE-NI-CR PHASE DIAGRAM AT 650 K

Fig. 11. A calculated isotherm of Fe-Ni-Cr at 650 K.

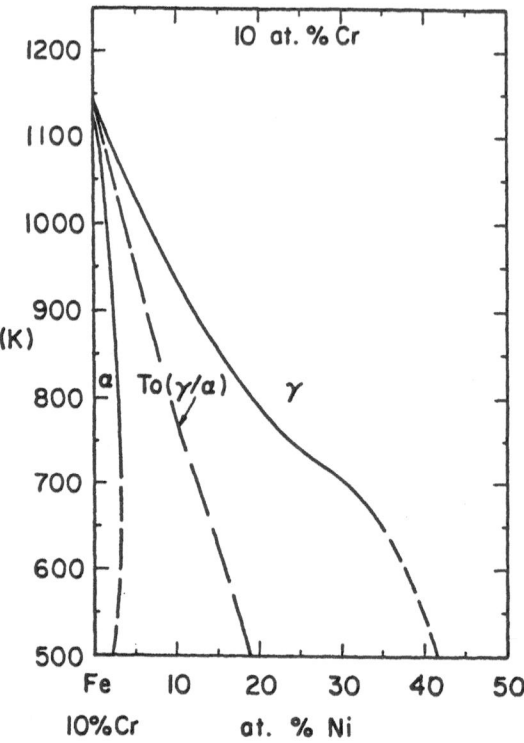

Fig. 12. A calculated isopleth of Fe-Ni-Cr with 10 at% Cr.

APPLICATION OF TERNARY PHASE DIAGRAMS TO GaAs-METAL CONTACTS

Rainer Schmid-Fetzer
Technical University Clausthal
Metallurgical Center
Robert-Koch-Str. 42
D-3392 Clausthal-Zellerfeld
West Germany

ABSTRACT. The usefulness of consistent phase diagram and thermodynamic data as a powerful tool for the development of improved contact materials to compound semiconductors like GaAs is demonstrated. Predictive calculations of ternary Ga-As-M phase diagrams in the solid state using simplifications have been performed as a basis for future work. A simple scheme for representation of thermodynamic driving forces of possible reactions between the various phases is given. The phase diagrams are classified into four different possible types and examples are given for the three types which presumably exist for the metals studied.

1. INTRODUCTION

All microelectronic or optoelectronic semiconductor devices require metallic contacts, either with low resistance (ohmic contacts) or with rectifying behavior (Schottky contacts). The fabrication of reliable contacts is a challenging task, especially for compound semiconductors like GaAs, since interface reactions /1/ are more complex than for elemental semiconductors like Si. These reactions are important for the electrical properties of the contacts /2/ and may be characterized as interdiffusion and/or formation of new phases at the metal/GaAs interface. There is a demand for "better" contacts with resistance to higher temperatures during operation or processing, with long time stability and good dimensional control for higher integrated circuits. This has initiated numerous studies in which typically a metallic thin film (30-150 nm) on GaAs was annealed. The change of microstructure or electrical properties was observed and sometimes related, as summarized by Palmstrøm and Morgan /3/. However, interpretation of the thin film results is often ambigous and difficult due the complexity of gallium-arsenic-metal ternary interactions. Phase diagram data of Ga-As-metal ternary systems would be needed for an understanding of these interactions and the development of improved contact materials /4, 5/, but they are lacking for most systems.

H. Brodowsky and H.-J. Schaller (eds.), Thermochemistry of Alloys, 107-117.
© *1989 by Kluwer Academic Publishers.*

It is the purpose of this study to point out the usefulness of phase diagrams and thermodynamic data for such a development and to present a simple method for a predictive calculation of ternary Ga-As-M phase diagrams under certain assumptions.

2. USEFULNESS OF PHASE DIAGRAMS FOR THE DEVELOPMENT OF IMPROVED CONTACT MATERIALS

The principal problem under consideration is depicted in Fig. 1:

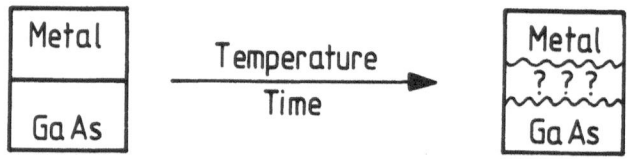

Figure 1. Schematic view of the problem under consideration.

A metallic layer upon a GaAs substrate is heated to a given temperature for a certain time. One would like to know if there are any reactions taking place at the interface. Is interdiffusion of metal into GaAs or of Ga and/or As into the metal of importance? Will completely new phases be formed? What are the thermodynamic driving forces? What is the kinetics, the morphology and the sequence of phase formation? Finally, is the contact system stable or metestable? The sound way of tackling these questions is to start with the ternary Ga-As-M phase diagram and the corresponding thermodynamic data. Then one could calculate the mutual phase stabilities and the final equilibria. Additional information on the reaction kinetics will be much easier interpreted with the phase diagram at hand. This point will be stressed again later.

The problem depicted in Fig. 1 is of technological relevance, since the questions arising from it are just that of the elementary reactions during production or operation of ohmic or Schottky contacts to GaAs. The phase finally formed at the interface next to GaAs is very important for the electrical behavior of the contact. The temperature limit and the reliability of devices is determined by these reactions. This is also true for the degree of miniaturization achievable. For example, the standard "alloyed" ohmic contact is unsuitable, since a liquid phase is formed during processing at the interface. Wetting problems and "balling up" make it difficult to produce finer structure /6/.

The usefulness of phase diagrams is emphasized if the metallizations of Si and GaAs are compared, which is done schematically in Fig. 2.

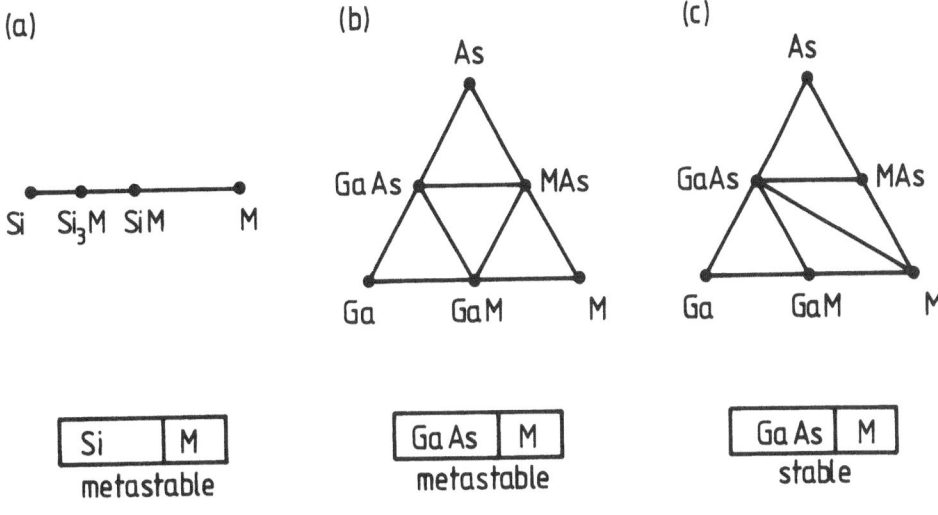

Figure 2. Schematical comparison of phase diagrams and metal contact stabilities for Si and GaAs.

In Fig. 2a an isothermal section at processing or operation temperature of a hypothetical Si-M system with two intermediate phases, Si_3M and SiM, is presented. The contact system Si/M given below is obviously metastable since the silicides will eventually form at the interface. It should be noted that even for this simple looking case of silicon the metallization problem is far from being trivial /7/.

In the case of galliumarsenide there is one additional degree of freedom. Most simple hypothetical Ga-As-M systems are given in Figs. 2b-c with only one compound in each binary. Direct equivalents to the Si_3M and SiM phases of Fig. 2a would be ternary compounds along the GaAs-M cut, which are not given in this simple example. The contact stability will depend on the phase relations among the six phases assumed in both systems. In case of Fig. 2b, which is characteristic for many metals /8/, the contact would be metastable and the reaction

$$GaAs + 2M \rightarrow GaM + MAs \qquad (1)$$

will eventually take place. With a limited amount of metal available in a thin film, the reaction ends in the three phase equilibrum GaAs-GaM-GaAs. The contact would also be metastable in the two other possible cases not shown in Fig. 2, namely if a tie line Ga-MAs or GaM-As exists. The first case is characteristic for the early transition metals while the latter case is not observed in the Ga-As-M systems studied, as discussed later. A tie line GaAs-M, as given in Fig. 2c, represents the only case that results in a stable contact system with the elemental metal. However, if As is evaporated from GaAs during vacuum processing, the GaM phase may form, since the total composition of the condensed system is then shifted into the GaAs-GaM-M triangle of Fig. 2c.

The phase diagram will also be useful to elucidate the kinetics of the interface reaction as given in Fig. 3.

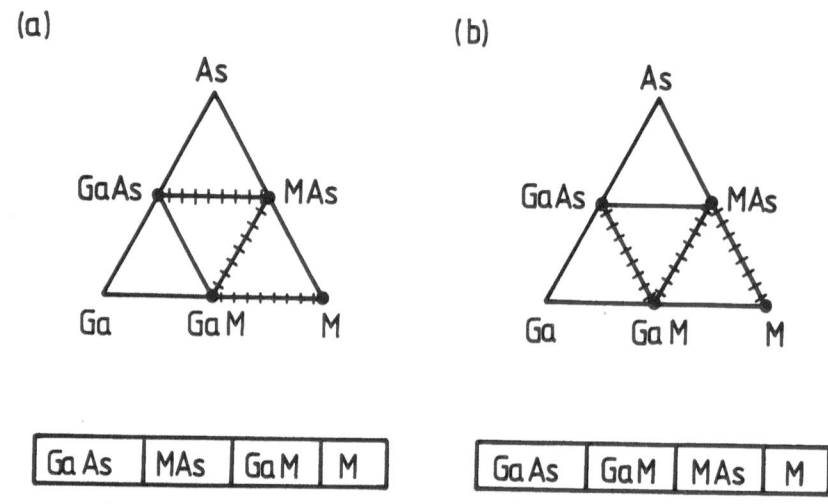

Figure 3. Two possible diffusion paths for the GaAs-M interface reaction in the type of phase diagram assumed in Fig. 2b.

If we consider a diffusion couple M/GaAs for a metal with the simple phase diagram of Fig. 2b, we expect the reaction according to Eq. (1) to occur at the interface. If one of the phases, say MAs, nucleates first at the interface to GaAs we may assume the formation of the other, say GaM, simply from materials balance. Its ease of nucleation at the GaAs/MAs or the MAs/M interface and the relative diffusivities of Ga, As and M through the compound phases or probably along grain boundaries determine the final sequence of phases that eventually forms in the steady state: GaAs/MAs/GaM/M (Fig. 3a) or GaAs/GaM/MAs/M (Fig. 3b).This sequence of phases, the "diffusion path", may be most clearly presented in the phase diagram, as done in Fig. 3. It is obvious that the phase diagram is very useful for this presentation and also for checking the consistency of the diffusion path, since local equilibria would demand that the diffusion path follows tie lines. For given temperature and pressure the diffusion path is fixed in a specific ternary system as pointed out by Kirkaldy and Brown /9/. This concept of diffusion path has first been applied to Pd/GaAs interactions with great success in a most notable study by Lin et al. /10/.

Once the phase diagram and the diffusion path is known, the phase that is stable next to GaAs is identified. Based on that knowledge stable contact interfaces with good morphology may be produced by deposition of the corresponding phase or phase sequence. This route for the development of improved contact materials seems to be more promising than trial and error methods. It has to be complemented, of course, by investigation of the electrical contact properties.

3. CALCULATION OF Ga-As-M PHASE DIAGRAMS

Data on ternary Ga-As-M phase diagrams are not available with very few exceptions /10-14/. It seems therefore reasonable to perform approximate calculations of these diagrams as suggested by Chang /15/. Such calculations are given below and also a simple mapping procedure is described, which is useful in this special case.

3.1 Approximations Used

In order to perform actual predictions of the Ga-As-M phase diagrams without using any experimental information from specific ternary systems, the following four simplifications will be used:

(i) Ternary phases are disregarded.

(ii) Solid solubilities are disregarded.

These restrictions are reasonable since at present information on ternary phases or mutual solubilities is very scarce. Updating of the diagrams by implementation of known ternary phases could be easily done within the scheme used here. Complete mutual solid solubilities, which would alter the diagrams basically, are not expected for the systems in this study.

(iii) The Gibbs energy of formation of the various solid phases ($\Delta G°$) is approximated by the enthalpy of formation ($\Delta H°$)

(iv) The enthalpy of formation is approximated by a value calculated from Miedema's /16/ model ($\Delta H°_{Miedema}$).

The approximation $\Delta G° \approx \Delta H°$ means neglection of the entropy term. This is not a severe assumption, since during the actual phase diagram calculation not the entropy of formation of single phases is neglected but only the difference between those terms for interacting phases. Moreover, for calculations at moderate temperatures the entropy effects become less important. The approximation $\Delta H° \approx \Delta H°_{Miedema}$ is simply done because experimental data of $\Delta H°$ are lacking. Miedema's model was selected since it is considered to be the most general and offers a fair accuracy. This is certainly true if the ratio of accuracy to computational effort is compared to more sophisticated models.

Before starting the calculation, a compilation of all binary Ga-M and M-As compounds that are known to exist was extracted from standard sources /17-23/ and judges have been made in the case of conflicting information. Stoichiometries have been simplified in some cases. For each compound $\Delta H°_{Miedema}$ is calculated. Miedema's model is not applicable for the compound GaAs since no transition metal is involved. Therefore, $\Delta H°(GaAs) = -40.8$ kJ/g-atom has to be taken from the literature /24/.

Under the approximations described above, the relative Gibbs energies of
all pertinent phases are now known and the phase diagram could be calculated using any "black box" computer program. However, it is advantageous to design a special procedure for this case as described below.

3.2 Mapping Procedure

Any two phases in the Ga-As-M system labelled as (I) and (J) may be connected by a line and the question is whether this is a stable tie line
or not. This is depicted in Fig. 4 for a system with the phases (1)
through (6). In this case only 5 possible tie lines, shown dashed, are
crossing the ternary and need further investigation.

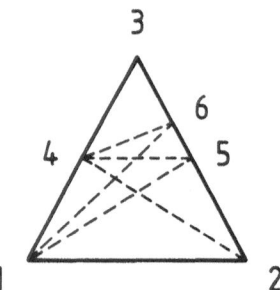

Figure 4.   A ternary with six solid phases and five possible tie lines.
Compare to Ga-As-W systems, Fig. 5a.

If the two lines between (I)-(J) and (K)-(L) are intersecting within the
ternary, the following chemical reaction may be written:

$$(I) + f_1 (J) = f_2 (K) + f_3 (L) \qquad\qquad (2)$$

where f_1, f_2 and f_3 are the proper stoichiometric factors. The Gibbs
energy of reaction of Eq. (2), ΔG°_{rxn}, is readily calculated for all
possible combinations of phases (I) through (L) and can be visualized in
form of a two-dimensional array with rows corresponding to fixed (I)-(J)
and columns to fixed (K)-(L). This is demonstrated in Table I for the
system of Figs. 4 and 5a and the actual data of the Ga-As-W system. Only
those rows where all values of ΔG°_{rxn} are positive correspond to a
stable tie line (I)-(J). This is the case for (2)-(4) and (4)-(5). The
tie line (4)-(6) exists irrespective of the relative phase stabilities,
since there is no other line to compete with. The resulting stable
equilibria for this example are given in Fig. 5a.
       So far all these simple calculations could have been done equally
well with any standard program for calculation of phase diagrams.
However, decisions have to be made if the figures in Table I become
close to zero.

TABLE I

Map of ΔG°_{rxn} of Eq. (2) and the system of Figs. 4 and 5a
in kJ/ 1 mole of atoms on either side of Eq. (2).

** means: tie lines are not intersecting in ternary

(K - L)→	1 - 5	1 - 6	2 - 4	4 - 5	4 - 6
(I - J)					
1 - 5	**	**	-10.1	**	**
1 - 6	**	**	-15.6	-10.7	**
2 - 4	10.1	15.6	**	**	**
4 - 5	**	10.7	**	**	**
4 - 6	**	**	**	**	**

It is the advantage of this mapping procedure compared to "black
box" phase diagram calculation that small positive values of ΔG°_{rxn},
within the order of the approximate error of $\Delta H^\circ_{Miedema}$, are immediately
seen and that alternative tie lines with small negative values of ΔG°_{rxn}
are easily identified from the map. Such alternative tie lines are given
dashed in Fig. 5b for the example of Ga-As-Ti.

A second advantage of calculating such a map is that the thermodynamic driving forces for the reaction between GaAs and metal to form
other phases are given in a comprehensive way. It is very important for
this application that ΔG°_{rxn} is normalized to the amount of 1 mole of
atoms on either side of Eq. (2). This makes ΔG°_{rxn} independent of
factors in the stoichiometry of the compounds and applicable to the
interface reactions as explained below refering to Table I and Fig. 4.

If one considers a sample volume at the M-GaAs or (2)-(4) interface
comprising one mole of atoms and the proper overall composition for
complete reaction of phases (2)-(4) into (1)-(5) we have ΔG°_{rxn}
= 10.1 kJ. For the competing reaction (2)-(4) into (1)-(6) we simply
shift the sample volume so that it comprises the new overall composition
but again one mole of atoms and we have ΔG°_{rxn} = 15.6 kJ. These two
numerical values for the thermodynamic driving force may be compared
directly since they refer to the same "amount". In this example both
reactions will not occur since ΔG°_{rxn} > 0.

If we had chosen the example of titanium, the map would reveal a
number of possible reactions. This map is not given here since there are
52 possible tie lines, however, the possible products of a GaAs-Ti
reaction may be taken from Fig. 5b. During the reaction, if neither Ti
nor GaAs is used up, the transient products could be for example
Ga_2Ti + TiAs or $GaTi_2$ + Ti_4As or many others. The thermodynamic driving
force for each case is given explicitly by simply following the GaAs-Ti
row in the map. This may be helpful in understanding the experimental
interface reactions if kinetic barriers, local composition etc. are also
taken into consideration.

4. DISCUSSION OF Ga-As-M PHASE DIAGRAM TYPES

The phase diagrams calculated for 19 metals ranging from Ti through Au
in the periodic table have been compared to experimental data and
discussed in detail elsewere /8/. As a summary of this discussion, three
different types of phase diagrams may be extracted which are represented
here by the Ga-As-M systems for M = W, Ti and Rh in Figs. 5a-c. The
isothermal sections of the Ga-As-M systems are calculated at 25°C. However, they essentially remain unchanged to about 300°C, in many systems
even higher. The major change is that gallium melts and dissolves certain amounts of metal and arsenic.

   The type of diagram characterized by a GaAs-M equilibrium as given
in Fig. 2c was found for M = W, Re, Ag, Au and is represented by the
Ga-As-W diagram in Fig. 5a. These are the only cases where one would
expect a stable elemental contact, however, some interactions other than
compound formations at the interfaces should be noted /8/.

   The next type of diagram is dominated by very stable metalarsenides
with a characteristic Ga-MAs tie line. These diagrams are found for the
early transition metals Ti, Zr, Hf, Nb, Ta and possibly V and Ga-As-Ti
is given as a typical example in Fig. 5b. The meaning of this diagram
type is that we should find MAs and elemental gallium on top of GaAs
after complete reaction of a thin metal film on GaAs. This has actually
been observed /25/ for the case of tantalum after reaction at 650°C. The
very high oxygen affinity of these metals should also be considered as a
negative feature for device applications.

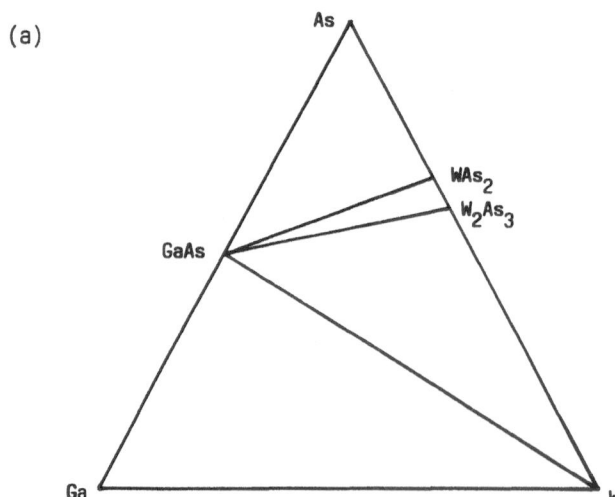

(a)

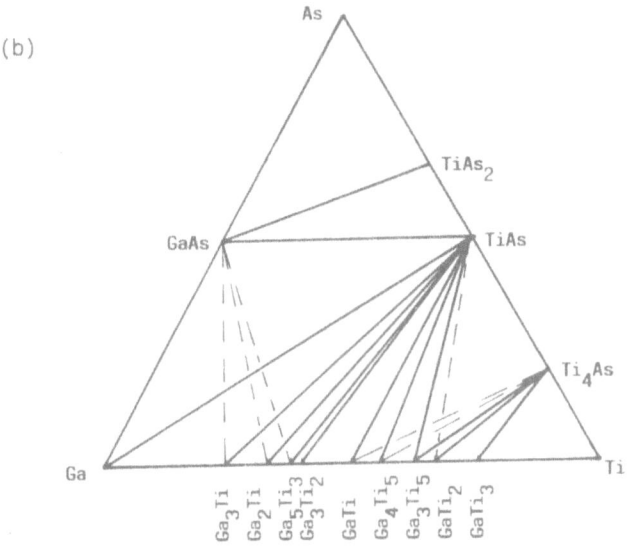

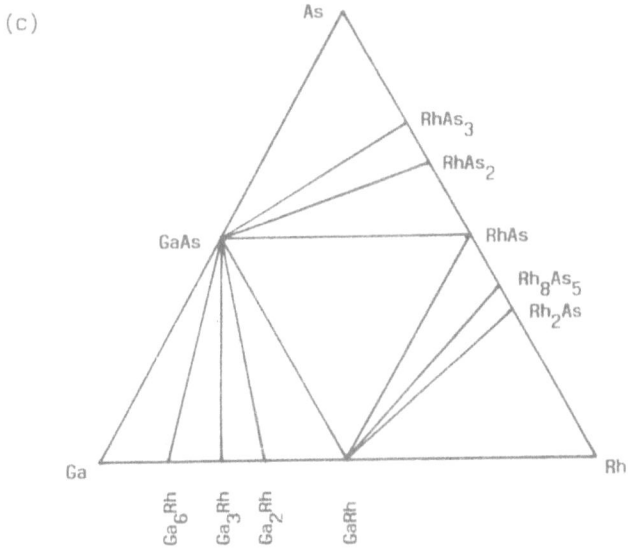

Figure 5.   Isothermal sections of Ga-As-M phase diagrams at 25 °C from approximate calculations for M = W (a), Ti (b) and Rh (c).

The third type of diagram is characterized by not too different stabilities of the binary compounds and is dominated by an equilibrium triangle GaAs-GaM-MAs as depicted in Fig. 2b. This type was found for Cr, Mo, Os, Rh, Ir, Ni, Pd, Pt, Cu and Ga-As-Rh is given as a typical example in Fig. 5c. Compounds are expected to form in equilibrium at the M/GaAs interface. Further studies would be necessary to find out whether these reactions could be controlled to produce suitable contacts.

The last possible type of diagram - if ternary compounds or complete solid solubilities are disregarded - would be dominated by an As-GaM tie line, however, this type was not found among the 19 metals studied. This seems reasonable since one would not expect a GaM compound to be much more stable than GaAs.

5. CONCLUSION

Consistent phase diagram and thermodynamic data are a powerful tool for tackling the materials problems inheret in the metallization of compound semiconductors. Much more work has to be done to improve the availability and accuracy of these data for selected ternary and higher order systems.

6. ACKNOWLEDGEMENT

The author is indebted to both Y. A. Chang and J.-C. Lin for instructive conversations on this topic and to J. Klingbeil for his assistance in programming.

7. REFERENCES

1. A. K. Sinha and J. M. Poate, in: "Thin Films - Interdiffusion and Reactions", J. M. Poate, K. N. Tu and J. W. Mayer (eds.), John Wiley, New York, 407-432 (1978)
2. G. Y. Robinson, in: "Physics and Chemistry of III-V Compound Semiconductor Interfaces", C. W. Wilmsen (ed.), Plenum Press, New York, 73-163 (1985)
3. C. J. Palmstrøm and D. V. Morgan, in: "Gallium Arsenide", M. J. Howes and D. V. Morgan (eds.), John Wiley, 195-261 (1985)
4. T. Sands, J. Metals, 31-33 (Oct. 1986)
5. R. Beyers, K. B. Kim and R. Sinclair, J. Appl. Phys., **61**, 2195-2202 (1987)
6. A. Piotrowska, A. Guivarch and G. Pelous, Solid State Electronics, **26**, 179-197 (1983)
7. B. L. Sharma and S. C. Gupta, Solid State Technology, **23** (5), 97-101 and (6), 90-95 and 110 (1980)
8. R. Schmid-Fetzer, J. Electronic Materials, under review (1987)
9. J. S. Kirkaldy and L. C. Brown, Can. Met. Quart., **2**, 89-115 (1963)
10. J.-C. Lin, K.-C. Hsieh, K. Schulz and Y. A. Chang, J. Mater. Res., under review (1987)

11. M. B. Panish, J. Electrochem. Soc., **114**, 516-521 (1967)
12. M. El-Boragy and K. Schubert, Z. Metallkde., **72**, 279-282 (1981)
13. M. D. Deal, R. A. Gasser and D. A. Stevenson, J. Phys. Chem. Solids, **46**, 859-867 (1985)
14. C. T. Tsai and R. S. Williams, J. Mater. Res., **1**, 352-360 (1986)
15. Y. A. Chang, private communication (1986)
16. A. K. Niessen, F. R. de Boer, R. Boom, P. F. de Chatel, W. C. M. Mattens and A. R. Miedema, Calphad, **7**, 51-70 (1983)
17. M. Hansen and K. Anderko, "Constitution of Binary Alloys" McGraw-Hill, N.Y. (1958)
18. R. P. Elliott, "Constitution of Binary Alloys", First Supplement, McGraw-Hill, N.Y. (1965)
19. F. A. Shunk, "Constitution of Binary Alloys", Second Supplement, McGraw-Hill, N.Y. (1969)
20. W. G. Moffatt, "The Handbook of Binary Phase Diagrams" General Electric Co., Schenectady, N.Y. (1978)
21. T. B. Massalski (ed.), "Binary Alloy Phase Diagrams" Vols. 1 + 2, ASM, Metals Park, Ohio (1986)
22. W. B. Pearson, "Handbook of Lattice Spacings and Structures of Metals and Alloys", Vol. 1, Pergamon Press, N.Y. (1958)
23. W. B. Pearson, "Handbook of Lattice Spacings and Structures of Metals and Alloys", Vol. 2, Pergamon Press, N.Y. (1967)
24. O. Kubaschewski and C. B. Alcock, "Metallurgical Thermochemistry" 5th ed., Pergamon Press, Oxford (1979)
25. A. Lahav and M. Eizenberg, Appl. Phys. Letters, **46**, 430-432 (1985)

ESTIMATED AND MEASURED EXCESS FUNCTIONS AND PHASE DIAGRAM OF TERNARY ALLOYS.

J.P. BROS
Thermodynamique des Systèmes Métalliques
Université de Provence
13331, Marseille cedex 3 , France.

ABSTRACT. Excess functions of formation and equilibrium temperatures of ternary alloys have been determined by calorimetric and potentiometric techniques and differential enthalpic analysis on the whole concentration range. Enthalpies and Gibbs free energies of formation of the binary subsystems have been extrapolated into the ternary system by using empiric relations in which a ternary term has been included. The ternary phase diagrams have been calculated by computer from the ternary Gibbs free energy. Most of the ternary investigated systems exhibit a good agreement between experimental and calculated temperatures . The Al+Bi+Ga and Ag+Au+Ge alloys have been chosen as examples.

RESUME. Les fonctions thermodynamiques d'excès d'alliages ternaires - enthalpie et énergie libre de formation - ont été soit mesurées par calorimétrie et/ou par potentiométrie , soit calculées à partir des grandeurs homologues des binaires limitrophes. Cette approche simultanée par le calcul et l'expérience permet d'obtenir d'une manière satisfaisante et rapide le diagramme d'équilibre du système considéré et conduit à proposer, pour les fonctions d'excès, un ensemble thermodynamiquement cohérent.

1. INTRODUCTION

Since more than fifteen years , the number of experimental and theoritical thermodynamic studies concerning multicomponent alloys increased very rapidly. The main purposes of these researches are on the one hand the knowledge of the liquid and solid metallic state and on the other hand the prediction of the stability of the metallic structures. Accordingly the excess functions of formation of a n-component system (n>2) must be attained because for many alloys thermodynamics provide a fruitful approach of their structure. To obtain this information the major routes are the following :

H. Brodowsky and H.-J. Schaller (eds.), Thermochemistry of Alloys, 119–143.

1) by calculations : excess functions of formation and equilibrium phase diagram can be predicted from the thermodynamic properties of the limiting binary alloys and of the pure components. Consequently models of solutions and special computer programs are necessary.

2) by experiments : this approach consists essentially to measure quantities of mixing and to carry out the deviation of the actual solution under mixing from the ideal solution. The methods of investigation are numerous and the quality of the results depends upon the nature of the components and upon the complexity of the alloy.

But generally the simplicity of models of solution and the difficulties of experimental determinations can lead to a poor agreement between compiled and observed phase diagram and thermodynamic properties. Notwithstanding these discrepancies , the simultaneous use of the two routes provide a substantial assistance in the knowledge of a complexe multicomponent alloy.

To highlight the advantage of this procedure, results obtained for several ternary alloys exhibiting different shapes of phase diagram will be presented in the third part of this report. In the first and second section we have described respectively the experimental techniques applied and the main relations employed to predict the excess functions of formation.

2. EXPERIMENTAL METHODS.

Three techniques - high temperature calorimetry , potentiometry and differential enthalpic analysis - were developed in our laboratory to carry out enthalpies of formation, activities, enthalpies of phase transition, equilibrium temperatures,... of multicomponent alloys.

2.1. High temperature calorimetry.

The choice of the calorimeter depends upon the temperature range investigation . Two types of calorimeter were used : a high temperature Calvet microcalorimeter and a very high temperature calorimeter operable respectively between 500 K-1100 K and 1100 K-1800 K.

2.1.1. On the temperature range 500 K-1100 K .
Method and apparatus : A detailed description of this apparatus has been already published /1,2/. The central alumina block surrounded by two large cylindrical shields in pure alumina are located in a cylindrical furnace heated by four kanthal resistors. Two very similar thermopiles holded in the central block allow to detect thermal effects developed in the cell. These two thermopiles are connected differentially to eliminate the thermal perturbations due to the furnace.

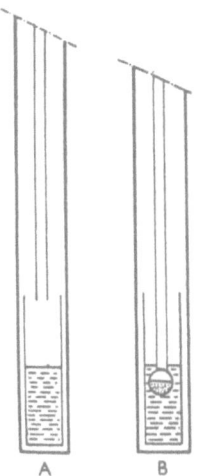

Fig. 1
Experimental cells :
A) direct drop method,
B) break-off ampoule method.

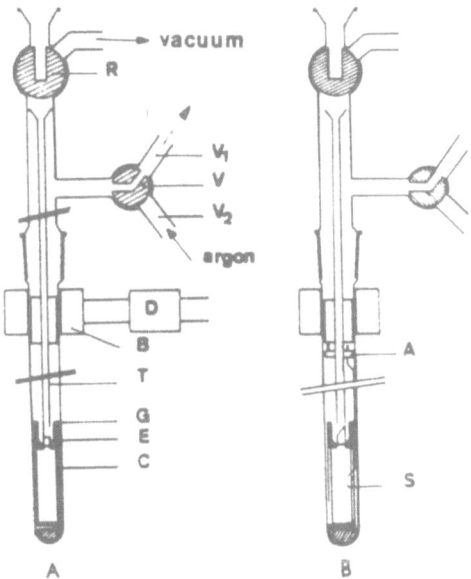

Fig. 2
Experimental cells :
A) indirect drop method
B) with a stirrer.

Each thermopile consists of a thin cylindrical alumina tube (16.9 mm in diameter and 80 mm high) closed at the bottom surrounded by 396 (Pt/Pt + Rh 10 wt %) thermocouples. One of the cell is called "laboratory cell" the other one is " reference cell". The temperature of the calorimeter is regulated or linearly programmed by an electronic device the sensing element of which is a platinum resistance (100 Ω at 0°C).

The experimental techniques worked out in calorimetry for the determination of the enthalpies of mixing are numerous . For all of them , it is essential to eliminate all secondary effects generally arising from metal atmosphere interactions , from solvent or solute-crucible reactions, from difficulties of mixing,..... In order to solve these specific problems various devices have been created. In " Metallurgical thermochemistry"/3/ , the former and also the easier method to obtain the enthalpy of mixing , the so-called " drop method ", has been described by Kubaschewski. This method consists of dropping into the liquid metal (A) - or in the liquid alloy (A+C) - considered as a solvent and maintained at the experimental temperature T_E , a well known amount (m) of the metal B previously stabilized at temperature T_0 (generally the room temperature). This very simple technique (fig. 1- a) has been applied to the Calvet calorimeter. At constant pressure, the recorded thermal effect ($Q_P = \Delta H$) correspond to the following relation :

$$Q_P = m/M_B [\int_{T_0}^{T_{B,fus}} C_{p,Bsol} \cdot dT + \Delta H_{B,fus} + \int_{T_{B,fus}}^{T_E} C_{p,Bliq} \, dT] + Q'_P \quad (1)$$

in which M_B, $C_{p,Bsol}$, $C_{p,Bliq}$, $T_{B,fus}$ and $\Delta H_{B,fus}$ are respectively the atomic mass , the molar heat capacity in solid and liquid state , the temperature and enthalpy of fusion of B. From the experimental value (Q_P), the first , second and third terms must be substracted to obtain Q'_P and accordingly to calculate the enthalpy of formation (ΔH_M). Unfortunately , for some alloys , the uncertainty in these corrective terms is larger than the value of ΔH_M. Of course the accuracy of these evaluations is not great particularly when the enthalpy of mixing becomes very small (or tends towards zero).

In order to palliate these inconveniences /4/ we developed the "indirect drop method" that allows to minimize or to cancel the corrective terms. The added metal (B) is stopped during its drop in the funnel (G) until thermal equilibrium is reached (fig. 2 - A). The mixing is performed by opening the funnel by means of a magnetic device which lifts the tube (T) up and down. If the funnel and the crucible are at the same temperature the recorded thermal effect corresponds directly to the enthalpy of mixing. A small alumina piece (S) bounded to the magnetic device (fig. 2 - B) allows the final liquid alloy to be stirred. This microcalorimetric cell is connected to a vacuum pump (V_1) and to a purified argon circuit (V_2) by a lateral valve (V). For both methods - direct and indirect drop methods- the metal A is introduced in the calorimeter by using a special tap (R) and its drop is directed towards the crucible or the funnel by the tube (T).

The break-off ampoule method previously described by Wittig and Huber /5/ was used to obtain the enthalpy of formation of liquid alloys (for example Hg+Ga) by microcalorimetry /6/. The pyrex - or quartz- ampoule has to be thin enough to break with a single soft press (fig. 1 - b). The thermal effect arising from the ampoule break-off must be very small and reproducible.

Experimental precautions, calibration and accuracy. Careful attention must be paid to obtain reliable results :

- all the experiments are performed under vacuum or under a purified argon atmosphere. Before each experiment cell, tube, crucible are degased during several hours.
- the calorimetric cell is in quartz ; the nature of the crucible (quartz, graphite, alumina,....) depends on the chemical reactivity of the components of alloy and on the experimental temperature.
- the purity of all the metals used during these experiments is about 99,999%.

The calorimeter must be calibrated under experimental conditions either by Joule effect or by drops of alumina or pure metal.

a) by Joule effect : the lower part of the experimental crucible is surrounded by a small resistance that allows to generate a well known thermal effect . This calibration , operable until 900 K, requires precise measurements of time , intensity and resistance.

b) by drops of alumina or pure metal : at the beginning of each experiment some pieces of metal (A) are added to the liquid bath (A) and thermal effects so released allow to know the calibration constant. During and at the end of each experiment the drops of NBS alumina crystals permit a second calibration. A study of the accuracy of calorimetric results /7/ leads to the following conclusion : the standard deviation of calibration experiments is about 1.5 % and the uncertainty in the final value for the enthalpies of mixing is estimated to be ± 3% on the temperature range 100-700°C.

The experimental temperature is measured by using a (Pt / Pt+Rh 10 w.%) thermocouple located in the center of the calorimetric block.

2.1.2. On the temperature range 1100 K - 1800 K.

Method and apparatus. The experimental chamber (fig.3) of the very high temperature calorimeter made of a gas-tight alumina tube (22.5/28.5 mm in diameter and 600 mm high) is contained in a vertical cylindrical furnace heated by a graphite resistor which provides a uniform temperature zone of 140 mm within the reaction tube /8/. The resistor is insulated from the external steel jacket by a layer of graphite felt. This double-walled jacket is water-cooled. The space between the jacket and the experimental chamber can be evacuated and kept under neutral gas atmosphere during the experiment.

The principle of the detector is quite different from that of a Calvet microcalorimeter. In both cases the basic element is always a (Pt /Pt+Rh) thermocouple but now the reference temperature is that of the reference crucible C_t in which no reaction occurs and not that of the calorimetric block. The thermocouples are arranged vertically around the two crucibles (C_t and C_l). The thermocouples are kept vertically by double-bore alumina tubes (B) cemented into a small cylinder. The upper and lower junctions (J_l and J_t) of thermocouples constitute a crown in contact respectively with the laboratory (C_l) and reference (C_t) crucibles. The number, the nature and the arrangement of these thermocouples (fig.4) can be modified in accordance with the experiment ; generally (Pt + 6% wt. Rh/Pt + 30%wt. Rh) are used. The detector is surrounded by an alumina tube (T) acting as a thermal screen. This device is suspended in the center of the reaction chamber by three long double-bore alumina tubes that protect the junction wire of the detector and of the thermocouple (H) measuring the experimental temperature.

With this calorimeter direct - or indirect - drop method can be used. The shape of the alumina crucibles depends on the experiments . Different types of crucibles , schematically reproduced in figure 5 , have been tested. The crucibles b and c allow to avoid the contact and accordingly the reaction of the metal vapors with the platinum wires of the thermocouples. The use of the long crucible (c) modifies the thermal geometry of the calorimeter and makes difficult the preparation of the experiment.

The indirect drop method previously described was adapted to this calorimeter (fig.6) . The metal drops into the calorimeter and remains in the funnel F. The apperture of the funnel is closed by a small alumina sphere (S) of 6 mm. diameter in which is located the junction of a thermocouple. When the sample reaches thermal equilibrium the funnel is opened by lifting the thin alumina rod (M) and the mixing process takes place. The temperature of the funnel and therefore the magnitude of the correction term, depends on its

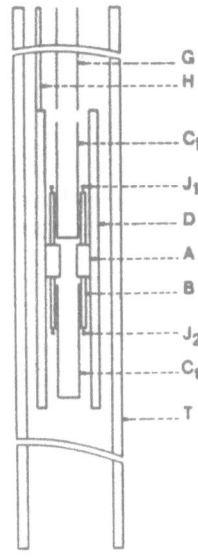

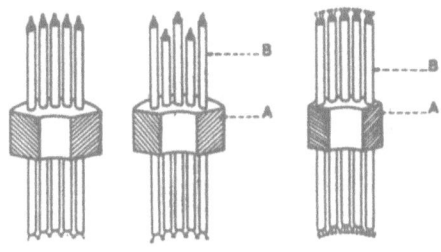

Fig. 4
Vertical sections of the high temperature detector :
a) with 16 solders on the same level,
b) with solders evenly distributed on 2 levels,
c)with 66 solders onthe same level.

Fig. 3
High temperature calorimeter.

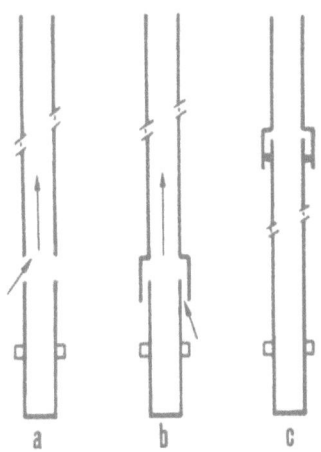

Fig. 5
High temperature calorimetric cells.

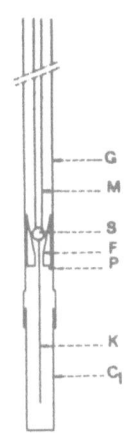

Fig. 6
High temperature calorimetric cells
for indirect drop method.

position in the tube G. The funnel may be attached at distances more or less next of the laboratory crucible i.e. at temperatures equal or smaller than T_E. A thin alumina rod (K) long enough to dip into the liquid bath is added to S and acts as a stirrer ; the vertical shift of K homogenizes the liquid .

Experimental precautions, calibration and accuracy. Drastic precautions must be taken to avoid oxidation of metals and alloys and chemical reactions with tubes and crucibles. To calibrate this calorimeter during experiments , direct drops of pure metal and NBS alumina pieces were used. Enthalpies of formation of binary alloys are obtained with an accuracy varying between 5 and 10%.

2.1.3. <u>Automation.</u> In order to improve the efficiency of these two calorimeters while making the experiments and the exploitation of experimental data easier , a complete automation have been carried out. The main modifications concern the mechanical system to introduce the samples in the crucible and the acquisition and treatment system of the data.

a) The introducer : the manual system enables one to introduce samples at room temperature was replaced by an automatic device the main parts of which are a cylinder holding up several cylindrical "nacelles" containing the samples of metals and an electric motor. These two pieces are enclosed in an gas-tight aluminium jacket. This water cooled jacket allows to keep the metallic samples at constant temperature and under vacuum or under purified argon atmosphere . The rotation of the cylinder brings about the "nacelle" at the top of the drop tube : in this position the "nacelle" turns round automatically and the sample drops in the experimental crucible. The partial rotation of the cylinder is monitored by the microcomputer.

The design of the sample introducer depends upon the calorimeter: with the Calvet calorimeter we used a small introducer the cylinder of which is vertical. This device is directly connected to the quartz experimental cell (9).The high temperature calorimeter introducer (10) presents a double horizontal cylinder with 2x15 "nacelles".

b) The acquistion and treatment system. Generally the determination of the area of the thermogram - corresponding to Q_p - is obtained either by using a manual planimeter (e.g. OTT planimeter) or by the trapezoid method. These methods are timeconsuming and not very accuracy.

The automatic data acquisition /11/ consists of an amplifier , a scanner, a millivoltmeter and a microcomputer ; the thermo-emf values   measured with constant frequency   are fed to the microcomputer (Apple II). From these data the area of the experimental thermogram is computed. The integration procedure includes a correction method for the integral values by means of triangles or rectangles if the thermo emf after the reaction is different from the preperiod base line value. At the end of an experiment , when a correct base line is recorded, a new rotation of the introducer cylinder is ordered.

2.2. Potentiometry.

The formation of a binary - or ternary - alloy, symbolized by a reaction of the type :

$$A + A_aB_bC_c \longrightarrow A_{(a+1)}B_bC_c \qquad \text{at T and P const.}$$

may be investigated by using a concentration cell in which the metal A is the most electropositive. To clarify the presentation an example of measurements with aluminium based ternary alloys - Al+Ga+In alloy - is described below. The cell is represented schematically as following :

$$Al/ \ Al^{3+} \ in \ (\ LiCl+KCl)_{(liq.)}/ \ (Al_aGa_bIn_c)$$

since aluminium is the most electropositive of these three metals : standard electrode potentials for several metals in LiCl+KCl eutectic mixture at 723 K are available from the Laitinen's work /12/ (table I).

Table I

Couple oxydo-reducteur					$E°/V$
Al $^{3+}$	+	3e$^-$	<---->	Al	-1.762
Bi $^{3+}$	+	3e$^-$	<---->	Bi	-0.553
Ga $^{3+}$	+	3e$^-$	<---->	Ga	-0.840
In $^{3+}$	+	3e$^-$	<---->	In	-0.800
Sb $^{3+}$	+	3e$^-$	<---->	Sb	-0.635
Zn $^{2+}$	+	2e$^-$	<---->	Zn	-1.566

Moreover the differences between the standard Gibbs free energies of formation of the chlorides used are sufficiently large to avoid exchange reactions in the experimental temperature range. Some values of the Gibbs free energies of formation of chlorides at 500 K and 1000 K /13/ are reported in the table II.

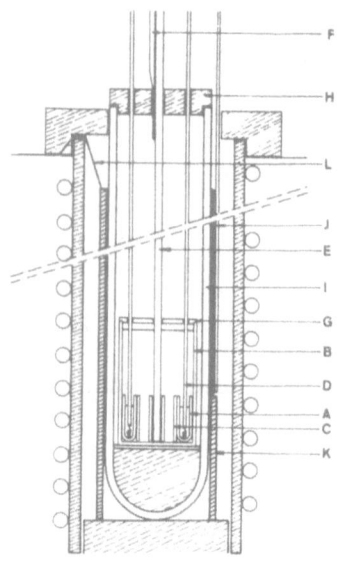

Fig. 7
Schematic diagram
of the electromotive cell.

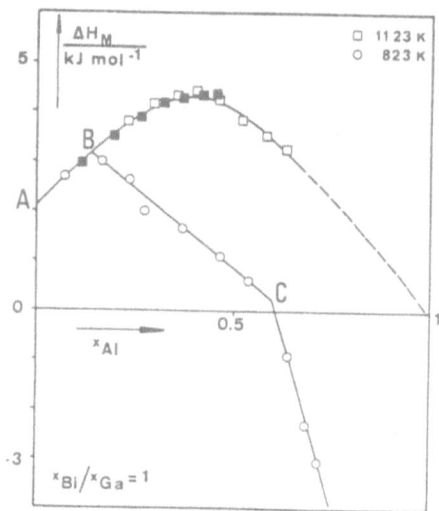

Fig. 8
Enthalpies of formation
of the Al+Bi+Ga liquid alloys.

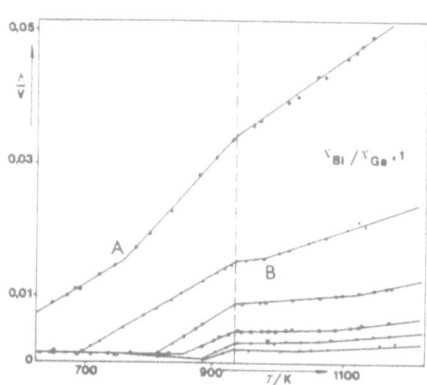

Fig.9
Experimental values of emf
plotted against temperature for the $x_{Bi}/x_{Ga} = 1$ section .

Table II

	Reactions		$\Delta_f G° / kJ$ at	
			500K	1000K
$1/3\,Al +$	$1/2\,Cl_2$	$\longleftrightarrow$	$1/3\,AlCl_3$	- 192	-188
$Al +$	$1/2\,Cl_2$	$\longleftrightarrow$	$AlCl$	- 92	-128
$1/3\,Bi +$	$1/2\,Cl_2$	$\longleftrightarrow$	$1/3\,BiCl_3$	- 93	- 78
$1/3\,Ga +$	$1/2\,Cl_2$	$\longleftrightarrow$	$1/3\,GaCl_3$	- 142	-134
$1/3\,In +$	$1/2\,Cl_2$	$\longleftrightarrow$	$1/3\,InCl_3$	- 138	-125
$K +$	$1/2\,Cl_2$	$\longleftrightarrow$	KCl	- 389	-339
$Li +$	$1/2\,Cl_2$	$\longleftrightarrow$	$LiCl$	- 368	-330
$1/3\,Sb +$	$1/2\,Cl_2$	$\longleftrightarrow$	$1/3\,SbCl_3$	- 98	- 92
$1/2\,Zn +$	$1/2\,Cl_2$	$\longleftrightarrow$	$1/2\,ZnCl_2$	- 170	-142

From the knowledge of the emf values (E) at temperature (T), molar partial Gibbs free energies (ΔG_{Al}) in alloys may be calculated as follows : $-zFE = \Delta G_{Al} = RTLn\,a_{Al}$. The determination of the molar partial entropy (ΔS_{Al}) and molar partial enthalpy (ΔH_{Al}) requires the variation of E versus T since $\Delta S_{Al} = zF(dE/dT)$ and $\Delta H_{Al} = -zF[E - T(dE/dT)]$; F and z correspond respectively to the Faraday constant and to the charge of the cation (z = 3 for Al). All requirements of an emf study concerning the reversibility of the cell, the preparation of electrolyte and alloys and the measurements of E and T must be carrefully observed. Figure 7 shows a schematic view of the cell : the large crucible (B), the lead wire sheath (D), the thermocouple sheath (E) and the small crucibles (A) containing the metal and alloys are in pure alumina. The electrodes crucibles (A) are wedged in place on the bottom of the crucible (B) by several pure alumina pieces(C) . Each cell contains ten electrodes : two with pure aluminium (reference electrodes) and eight with binary or ternary alloys (working electrodes). The tantalum lead wires (D) and (Pt/Pt+10 wt%Rh) thermocouple (E) are aligned by two covers (G and H). Suitable fittings allow purging of the cell during the experiment with a constant purified argon flow. This entire device is heated by a cylindrical vertical furnace with two Kanthal resistances. An electronic regulator keeps the cell temperature constant. A more uniform temperature distribution along the electrodes is obtained by enclosing the cell inside a refractory steel cylinder (K) which also acts as an electric shield.

Emf values are measured using either a digital voltmeter with a large imput impedance ($10^{12}\Omega$) or a manual potentiometer. Carefull attention must be paid to prepare the molten electrolyte. The eutectic mixture of lithium and potassium chlorides (55.8 wt % LiCl and $T_{fus.}$ = 623 K) was dried at 523 K during 24 hours under vacuum and melted under purified argon. After addition of aluminium chloride (5 wt %) the liquid electrolyte mixture was decanted in a special pyrex apparatus and quickly transferred to the cell , an argon flow being applied at the same time. Then this cell was heated above the melting temperature of alloys and reference metal. Emf measurements were carried out at intervals of 30 K in the experimental temperature range. After each temperature change , thermal equilibrium must be reached to perform emf measurements. The purity of the metals in these experiments is about 99.999%. The error in cell temperature measurements is ±1 K.

It is difficult to evaluate correctly the accuracy of emf measurements but several criteria allowed us to test our experimental method :

a) the emf between two reference electrodes (pure aluminium) was always lower than 0.0002 V at moderate temperature (623 K < T< 1100 K) and lower than 0.0005 V at upper temperature (1100 K < T < 1300 K);

b) all measurements were repeated on the next day to check that the differences between first and second values were not larger than 0.0005 V ;

c) for each composition , at least three independent experiments were carried out to test the reproducibility ;

d) Emf measurements performed for well-known binary alloys exhibit a linear variation with temperature : from these two straight lines $\Delta_{mix}S_{Al} = nF(dE/dT)$ and $\Delta_{mix}H_{Al} = nF(TdE/dT - E)$ values have been obtained and compared with previously published data ;

e) The enthalpy of the reference metal (ΔH_{fus}) can be calculated from the slope difference of the E =f(T) curves at $T_{fus.}$. When aluminium is the reference metal the (ΔH_{fus}) values so obtained are very near 10750 J.mol^{-1} , value published by Hultgren et al./14/.

2.3. Differential thermal analysis.

The shapes of the $\Delta H_M = f(x_A)$ and E = f(T) curves obtained for many ternary alloys at several temperatures allow to determine some points of the equilibrium phase diagram ; for instance the break points A and B observed in figures 8 and 9 correspond to the appearance of a new phase (for example : liq <---> sol. + liq.).But generally the number of information so deduced is very scarce and the majority of equilibrium phase diagrams is obtained by using thermal methods. Several books and papers give detailled descriptions of these methods. Ternary systems described in this

reports have been studied with the following apparatus :
- between 273 K and 773 K we used a differential thermal analyser (from Arion company)   with two cylindrical cells $(1.7 \ cm^3)$ surrounded by chromel-constantan thermocouples ,
- between 200 K and 1000 K with a scanning calorimeter we have obtained simultaneously   on the one hand temperatures and enthalpies of transition and on the other hand molar heat capacities ,
- between 400 K and 1200 K a linear temperature programmer was added to the regulator of the Calvet high temperature calorimeter ; so modified, this apparatus becomes a very sensitive differential enthalpic analyser but its heating rate never exceeds 12 $K.h^{-1}$. This very slow heating rate   permits a clear separation of the thermal effects   and a good calculation of enthalpies of transition ,
- between 500 K and 1800 K , in the graphite resistor furnace , the very high temperature detector is replaced by a special DTA device constituted by two alumina crucibles the bottom of which are in contact with the junction of the thermocouples.

3. CALCULATION OF EXCESS FUNCTIONS AND EQUILIBRIUM PHASE DIAGRAM.

The analytical representation of the thermodynamic properties of a multicomponent system and the calculation of its phase diagram are two indissociable problems.

3.1. Calculation of excess functions of formation.

Recently, for binary alloys , new physical models (15,16) allowing "a priori" calculations of excess functions of formation have been published. From these results, it seems likely that tentative calculations can be carried out in the near future to obtain phase diagrams. But , in general, these theoritical models are not available for multicomponents alloys.
     For a better inderstanding of new multicomponent materials , the need of relations to evaluate their thermodynamic properties have been emphasized by several researchers (17,18). The methods available to calculate thermodynamic properties of multicomponents alloys are derived either from physical models (quasi-chemical model, surrounded atom model (19),....) or from empirical interpolations of limiting binary data (Redlich-Kister (20), Kohler (21), Colinet (22), Muggianu(23), Toop(24), Bonnier(25), Hillert(26), Hoch-Arpshofen (27,28)...

3.2. Phase diagram calculation.

The calculation of phase diagram from Gibbs free energy expressions takes the two following steps :
a) the Gibbs free energy of every phase must be expressed as an analytical function of temperature and molar fraction ;
b) at several temperatures , the phase diagram will be compiled by equating the chemical potentials of all components or by minimizing the Gibbs free energy of all possible competing phases.
With the advent of computers many groups of researchers (Lukas(29), Kaufman (34), Oonk (30), Jansson (31),..) have proposed programs to calculate ternary - or higher order - phase diagrams from the data of the binary subsystems. Some of these methods have been used in the following examples.

4. RESULTS

To illustrate the use of these experimental methods and computer calculations we have gathered in the following part the results obtain for two ternary alloys exhibiting typical equilibrium phase diagrams , viz.

Al+ Bi+Ga system with a large ternary liquid miscibility gap ,
Ag+Au+Ge system with a eutectic valley.

4.1. Al+Bi+Ga system.

For the three limiting binary systems, we gathered the information refering to the equilibrium phase diagram on the one hand and to the experimental enthalpies and Gibbs free energies of formation on the other hand.

<u>Al+Bi system</u> : The phase diagram compiled by Hultgren/14/ from Bonnier /32/ and Predel /33/ works shows a large miscibility gap in the liquid state . The coordinates of the eutectic and critical points are respectively $x_{Bi} = 0.9944$ with $T_E = 543$ K and $x_{Bi} = 0.175$ with $T_C = 1323$ K. At the monotectic temperature $T_M = 930$ K the limiting molar fraction of the gap are $x_{Bi} = 0.0055$ and 0. 84.

Molar enthalpy and excess molar entropy of mixing /34/ can be represented by the following relations :

$$\Delta_{mix}H_m \text{ (Al+Bi,l)} = x_{Bi}x_{Al} (- 16.17\ x_{Bi} + 43.37) \text{ in kJ.mol}^{-1}$$

$$\Delta_{mix}S^E_m \text{ (Al+Bi,l)} = x_{Bi}x_{Al}\ 14.21 \quad \text{in J.mol}^{-1}.K^{-1}$$

<u>Al+Ga system</u> : The phase diagram exhibts an eutectic point ($T_E = 299.6$ K and $x_{Ga} = 0.973$) very near pure gallium /35-39/. At the eutectic temperature the solid solubility of gallium in aluminium is about 8.8 at. %. On the other hand the solublity of aluminium in solid gallium is not exactly known. From Predel's/38/ and Girard's/4/ results the molar enthalpy of formation (in kJ.mol^{-1}) is :

$$\Delta_{mix}H_m \text{ (Al+Ga,l)} = x_{Ga}x_{Al}\ (\ 0.700\ x^2_{Ga} - 1.909\ x_{Ga} + 3.428)$$

Among emf measurements , those of Predel /40/, Gaune/41/ and Eslami/42/ are particularly coherent. From these results the excess molar entropy of mixing (in J.mol^{-1}.K^{-1}) is deduced :

$$\Delta_{mix}S^E_m \text{ (Al+Ga,l)} = x_{Ga}x_{Al}\ (0.8390x^2_{Ga} - 2.44x_{Ga} + 2.192)$$

<u>Bi+Ga system</u> :
Electrical /43/ and calorimetric /44/ measurements confirm the phase diagram published by Predel/45/ from D.T.A. experiments. At the monotectic temperature $T_M = 495$ K the limiting molar fractions of the miscibility gap are $x_{Bi} = 0.615$ and 0. 085 . The coordinates of critical and eutectic points are respectively $x_{Bi} = 0.30$ with $T_C = 535$ K and $x_{Bi} = 0.0022$ with $T_E = 302.5$ K.

Predel /46/ and Gambino/44/ found very close enthalpy values by calorimetry. Molar enthalpy of formation(in kJ.mol^{-1}) is given by :
$$\Delta_{mix}H_m \text{ (Bi+Ga,l)} =$$
$$x_{Bi}x_{Al}\ (\ -9.30\ x^3_{Bi} + 16.97\ x^2_{Bi} - 11.12x_{Bi} - 10.88)$$
On the other hand mixing entropy and Gibbs free energy proposed by Yatsenko /47/ and by Predel/46/ are noticeably different. From Gambino's work /44/ we retained for the excess molar entropy of mixing the following equation :
$$\Delta_{mix}S^E_m \text{ (Bi+Ga,l)} =$$
$$x_{Bi}x_{Ga}\ (\ 2.731x^2_{Bi} + 2.137x_{Bi} - 2.747) \quad \text{in J.mol}^{-1}.K^{-1}$$

<u>Al+Bi+Ga system</u>.

Calorimetry : Experimental determinations of the enthalpy of formation were performed at several temperatures : 725, 817, 823,

948, 972, 981, 985, 1073, 1123, 1136, 1142 and 1170 K. Six ternary sections were investigated : $x_{Al}/x_{Ga} = 1/3$, 1/1, 3/1 and $x_{Bi}/x_{Ga} = 1/3$, 1/1 and 3/1. The figure 8 exhibits calorimetric results obtained at 1123 K, 823 K and 725 K with $x_{Bi}/x_{Al} = 1/1$. The curve without discontinuity corresponds to a monophase liquid alloy; the appearance of new phases in the liquid entails break points (B and C) and linear parts on $\Delta_{mix}H = f(x_{Al})$ graph. These discontinuities allow to detect some points of the equilibrium phase diagram.

Potentiometry : Eighteen alloys belonging to the following sections : $x_{Bi}/x_{Ga} = 1/7$, 1/3, 1/1 and 3/1 with $0.10 < x_{Al} < 0.90$ were studied on the temperature range 623 K - 1223 K. In figure 9 are gathered the emf values measured for the following alloys : $x_{Bi}/x_{Ga} = 1/1$ and $x_{Al} = 0.1$, 0.3, 0.5, 0.7, 0.8 and 0.9. The break points corresponding to the appearance of new phases are gathered in the table III :

Table III

x_{Bi}/x_{Ga}	x_{Al}	T/K	
3/1	0.10	785	497
	0.30	1093	815
	0.50	1149	839
	0.70		889
	0.80	1133	913
1/1	0.30	987	689
	0.50	1119	799
	0.70	1109	853
	0.80	1079	883
	0.90	1005	903
1/3	0.10	678	
	0.30	813	(597)
	0.50		720
	0.70		815
1/7	0.30	781	(563)
	0.50		685
	0.70		767

Differential thermal analysis : We studied sisteen alloys located

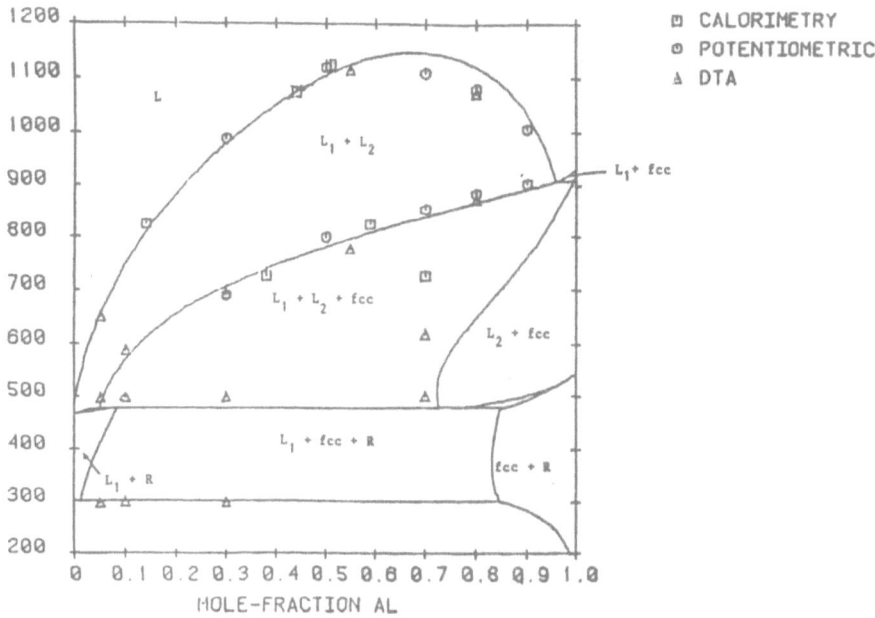

Fig.10
Calculated and experimental section
(x_{Bi}/x_{Ga} =1) of the Al+Bi+Ga equilibrium phase diagram.

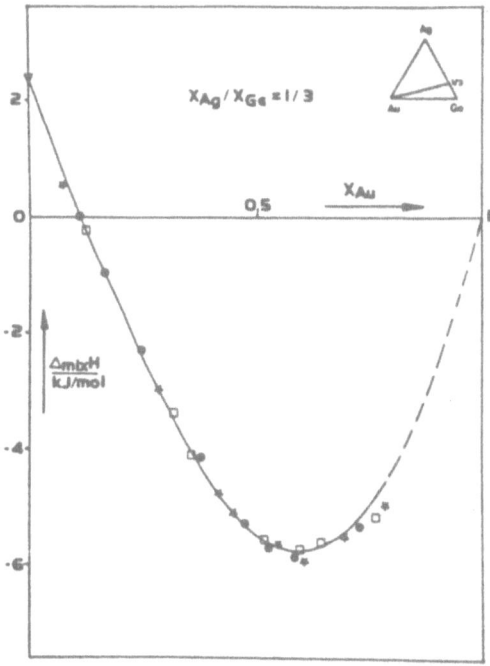

Fig.11
Enthalpies of formation
of the Ag+Au+Ge liquid alloys.

on the ternary sections (x_{Bi}/x_{Ga} = 1/3, 1/1 and 3/1) previously explored by potentiometry. When temperature was increasing two endothermic peaks were recorded : the first one between 295 and 298 K and the second one , sometimes split in two parts between 494 and 508 K. All these temperatures are obtained with a mean accuracy of ±3 K.

Results collected in the following table show the presence of two invariant horizontal planes cooresponding to the coexisting phases :
- liquid + $Bi_{(sol.)}$ + $Ga_{(sol.)}$ + Aluminium-rich solid solution at about 297 K.
- liquid 1 + liquid 2 + Bi(sol.) + Alumunium-rich solid solution at about 497 K.

Table IV

x_{Bi}/x_{Ga}	x_{Al}	T/K		
3/1	0.10	296	497	
	0.20	297	505	
	0.43	501	830	
1/1	0.05	295	497	649
	0.10	298	496	587
	0.30	298	498	692
	0.55	777	1115	
	0.70	500	617	
	0.80	871	1073	
1/3	0.05	296	498	617
	0.10	297	497	682
	0.20	297	498	549
	0.285	297	498	597
	0.50	296	498	724

The excess thermodynamic functions have been described by means of the Muggianu's equation. The vertical sections (x_{Bi}/x_{Ga} = 3/1, 1/1 and 1/3) were calculated and plotted by means of a computer program developed by Jansson (31). Experimental and calculated temperatures are in good agreement (see fig. 10) excepted in the solid rich aluminium region.

For the Al+Ga+X ternary alloys with X = Bi, Ge, In, Sb, Sn,... equilibrium phase diagrams have been predicted from thermodynamically consistent sets of data.

4.2. Ag+Au+Ge system

Ag+Au system : Gold and silver which have many close physico-chemical features can be mixed in any proportions . The Ag+Au phase diagram is very simple ; liquidus and solidus lines are

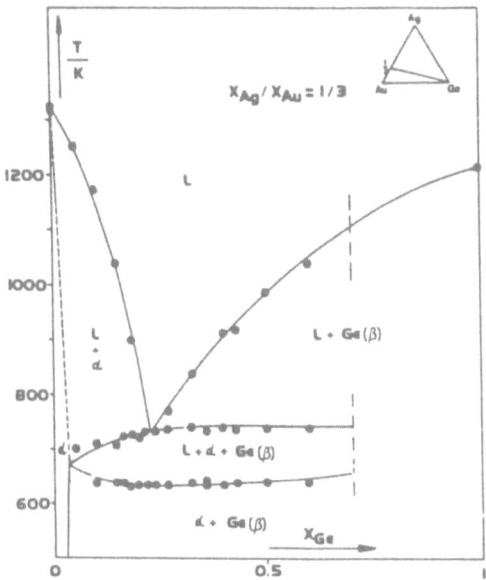

Fig.12
Experimental section (x_{Ag}/x_{Au} = 1/3)
of the Ag+Au+Ge phase diagram
obtained by D.T.A.

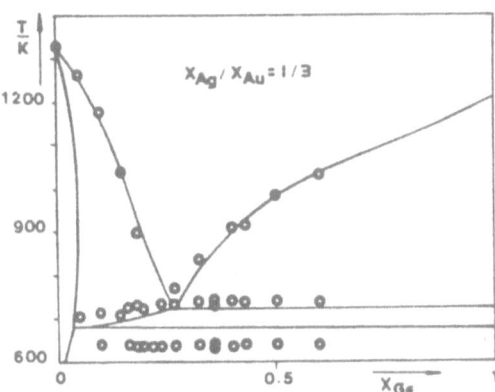

Fig.13
Calculated section (x_{Ag}/x_{Au} = 1/3)
of the Ag+Au+Ge phase diagram.

very closed altogether (at $x_{Au} = 0.50$ the gap between the liquidus and the solidus lines is about 1.7 K). This alloy have been subject of several thermodynamic investigations /48-54/ . Using a high temperature calorimeter , Miane /54/ determined the enthalpies of formation of Ag+Au liquid alloys at 1700 K on the entire concentration range. Miane's results can be represented by the following relation (in kJ.mol^{-1}) :

$\Delta_{mix}H_m(Ag+Au,l) =$

$x_{Ag}x_{Au}(-33.12 \; x^3_{Ag} + 55.99 \; x^2_{Ag} - 23.49 \; x_{Ag} - 14.73)$

Ag+Ge system : This system has an eutectic point shifted towards silver ($x_{Ge} = 0.24$ with $T_E = 924$ K). The solid solubility of germanium in silver is about 9.6 mol% at the eutectic temperature whereas the solubility of silver in germanium is almost negligible.Using calorimetric techniques Itagaki /55/, Wittig /56/ and Castanet /57/ have measured the enthalpies of formation of Ag+Ge liquid alloys. These results are in good agreement and can be expressed by the following equation (in kJ.mol^{-1}) :

$\Delta_{mix}H_m(Ag+Ge,l) =$

$x_{Ag} \; x_{Ge} (21.813 \; x^3_{Ge} - 65.624 \; x^2_{Ge} + 76.034 \; x_{Ge} - 16.08)$

Au+Ge system : The equilibrium phase diagram exhibits a simple eutectic point ($x_{Ge} = 0.27$ with $T_E = 629$ K) shifted towards pure gold and a slight solubility of germanium into gold ($x_{Ge} = 0.032$ at the eutectic temperature)/58,59/. Predel/60/, Hager/61/ and Itagaki /62/ have measured the enthalpy of formation of this system. These different values , always negative and asymmetrical exhibit a important discrepancy. Some measurements of the enthalpy of formation of this alloy were performed at 1373 K by calorimetry. Results lead to the following equation :

$\Delta_{mix}H(Au+Ge,l) = x_{Au} \; x_{Ge} (- 17.11 \; x_{Au} - 10.904)$ in kJ.mol^{-1}

Ag+Au+Ge system : Calorimetric measurements (fig.11) were carried out at 1373 K on the following sections :

$x_{Au}/x_{Ge} = 1/3$ with $0 < x_{Ag} < 0.670$

$x_{Au}/x_{Ge} = 1/1$ with $0 < x_{Ag} < 0.726$

$x_{Au}/x_{Ge} = 3/1$ with $0 < x_{Ag} < 0.565$

and

$x_{Ag}/x_{Ge} = 1/3$ with $0 < x_{Au} < 0.78$

$x_{Ag}/x_{Ge} = 1/1$ with $0 < x_{Au} < 0.60$

$x_{Ag}/x_{Ge} = 3/1$ with $0 < x_{Au} < 0.70$

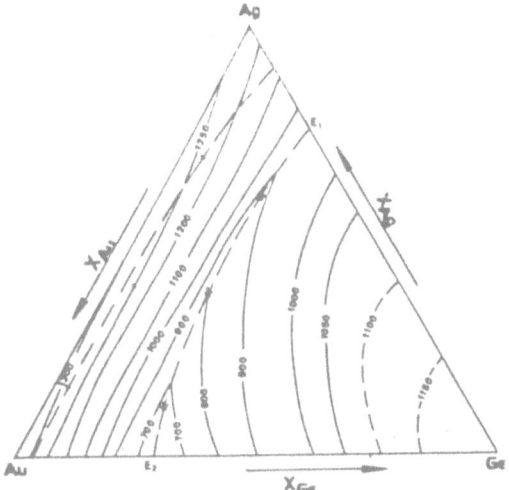

Fig. 14
Isothermal sections of the liquidus
of Ag+Au+Ge phase diagram.

Our experimental results were represented with a modified Muggianu equation in which a ternary term has been included :
$$x_{Ag}x_{Au}x_{Ge}(-41.535\ x_{Ag} - 98.815\ x_{Au} + 47.209\ x_{Ge})$$
Temperatures of phase transition were determined by differential thermal analysis between 500 K and 1300 K for 41 ternary alloys located on the sections $x_{Au}/x_{Ag} = 1/3$ (see fig.12),1/1 and 3/1.

All experimental values of the enthalpies of formation were fed into the Jansson program /31/ and some ternary sections were calculated . The agreement (fig. 13) between calculated and measured equilibrium temperatures is satisfactory on a large temperature and molar fraction range. The surface of the liquidus of this system is reported in the figure 14. Similar studies have been carried out for the Ag+Au+Si system and Ag+AU+Sn and Ag+Au+Pb alloys are under investigation.

5. CONCLUSION.
The knowledge of the thermodynamic properties of a n-component alloy needs careful experimental and calculated approaches. The combination of these two routes allows i) to choose the best experiments and to limit drastically their number, ii) to calculate the equilibrium phase diagram, iii) to provide a thermodynamically-consistent set of the excess fonctions of formation.

ACKNOWLEDGEMENTS.
The experimental parts of this work were performed at the Université de Provence (Laboratories : "Systèmes Energétiques et Transferts Thermiques" and "Thermodynamique des Systèmes Métalliques". The author is gratefully indebted to Dr. M. Gambino, G. Girard and S. Hassam for their unvaluable assistance.

REFERENCES

1/ E. Calvet et H. Prat ,
    Microcalorimétrie, Masson Ed., Paris 1955
2/ M. Gaune-Escard,
    Thèse Doct. Etat, Univ. de Provence, Marseille , 1972.
3/ O. Kubaschewski and C.B. Alcock ,
    Metallurgical Thermochemistry ,
    5th Edition, 1979, Pergamon Press, New York.
4/ C. Girard ,
    Thèse Doct. Etat , Univ. de Provence, Marseille , 1985.
5/ F. E. Wittig , F. Huber ,
    Z. Electrochem.,1956, <u>60</u>, 1181.

6/ M. Gaune-Escard , J.P. Bros ,
   Thermochimica Acta , 1979, <u>31</u>, 323.
7/ G. Hatem ,
   J. Chim. Phys. , 1986, <u>83</u>, 197.
8/ Setaram Company, Lyon, France.
9/ P. Rebouillon ,
   Rapport DEA, Univ. de Provence, Marseille, 1987 .
10/ E. Hayer, to be published.
11/ G. Hatem, P. Gaune, J.P. Bros, F. Gehringer, E. Hayer ,
   Rev. Sci. Instrum. ,1981, <u>52</u>(4), 585 .
12/ H.A. Laitinen, C.H. Liu ,
   J. Amer. Chem. Soc., 1958, <u>80</u>,1015
13/ T.B. Reed ,
   Free energy of formation of binary compounds, M.I.T. press,
   1971.
14/ R. Hultgren, P.D. Desai, D.T.Hawkins, M. Gleiser, K.K. Kelley
   *Selected values of thermodynamic properties of metals and
   alloys metals*, American Soc. for Metals, Metals Park, Ohio, USA,
   1973.
15/ A.R. Miedema ,
   Philips Techn. Rev., 1976, <u>36</u>(8), 217 .
16/ A. Pasturel ,
   Thèse Doct. Etat Sci. , Grenoble, 1985 .
17/ I. Ansara , Comparison of methods of thermodynamic calculation
   of   phase   diagrams . *International Metals Reviews* , 1979 ,
   vol.1,pp 20-53 .
18/ I. Ansara , Prediction of Thermodynamic Properties of Mixing
   and Phase Diagrams in Multicomponent Systems. *Metallurgical
   Chemistry Proceeding of a symposium held at Brunel University
   and N.P.L.* , 1971, pp. 403-430, London HMSO 1972 SBN 11
   480026 X.
19/ Brion B., Mathieu J.C., Hicter P., Desré P.,
   J. Chim. Phys. , 1969, <u>66</u>, 1238
20/ O. Redlich and A.T. Kister ,
   Ind. Chem. Eng. , 1948, <u>40</u>, 345.
21/Kohler F.
   Monatsh. Chem. , 1960, 91,738
22/ C. Chatillon-Colinet, I. Ansara, P. Desré, E. Bonnier
   Rev. Int. Htes Temp. et Refract. , 1969, <u>6</u>, 227.
23/ Y.M. Muggianu, M. Gambino, J.P. Bros
   J. Chim. Phys., 1975, <u>72</u>, 83.
24/ G.W. Toop
   Trans. A.I.M.E. , 1965, <u>233</u>, 850.
25/ E. Bonnier, R. Caboz,
   C.R. Acad. Sci. , 1960, <u>250</u>, 527.

26/ M. Hillert ,
   Calphad, 1980, 4(1), 1 .
27/ M. Hoch and I. Arpshofen ,
   Z. Metallkde , 1984, 75, 23.
28/ M. Hoch,
   Calphad ,1987, 11, 219.
29/ H.L. Lukas, E.Th. Emig and B. Zimmermann ,
   Calphad , 1977, 1, 225 .
30/ H.A.J.Oonk , *Phase Theory*, Elsevier Scientific Publishing
   Company New York, 1981.
31/ B. Jansson , *Internal Report D 19*, Royal Inst. Tech. S-10014,
   Stockholm, Sweden, 1979.
32/ R. Martin-Garin , G. Massart , P. Desré , E. Bonnier.
   C.R. Acad. Sci. ,1966, C 262, 335.
33/ B. Predel , H. Sandig ,
   Mater. Sci. Eng. ,1969, 4, 49.
34/ L.Kaufman , J. Nell , K. Taylor , F. Hayes .
   Calphad ,1985, 5, 185.
35/ E. Jenkel .
    Z. Metallk. , 1934, 26, 249.
36/ N.A. Puschin , O.P. Micic ,
    Z. Anorg. Chem. ,1937, 234, 233.
37/ V.N. Danilin , S.P. Yatsenko ,
    Zh. Prikl. Khim. ,1968, 41 , 1463.
38/ B. Predel , A.W. Stein,
    J. Less Comm. Metals , 1969, 17, 377.
39/ALCOA Res. Lab. in E.H. Wright , *L.A. Willey Alcoa Res. Lab. Techn.*
    Paper 16, 1, 1960.
40/ B. Predel , V. Schallner ,
    Z. Metallkde., 1969, 60, 869.
41/ J.L. Gaune,
    Thèse Doct. Spc., Univ. de Provence, Marseille , France, (1974).
42/ H. Eslami ,
    Thèse Doct. Sci. , Univ. de Provence, Marseille, France, (1979).
43/ S.P. Yatsenko, V.I. Kononenko ,
    Inorg. Mater. ,1967, 3, 1367
44/ M. Gambino , J.P. Bros , F. Ajersch, I. Ansara ,
    Thermochimica Acta ,1976, 14, 305 .
45/ B. Predel,
    Z. Phys. Chem. ,1960 , 24, 206 .
46/ B. Predel , M. Frebel , W. Gust,
    J. Less Common Metals ,1969, 17, 391 .
47/ S.P. Yatsenko, V.N. Danilin ,
    Izv. Akad. Nauk. SSSR Neorg. Mat. ,1968, 4, 863 .
48/ E. Jaenecke ,

Metallurgie, 1911, <u>8A</u>, 597 .

49/ U. Raydt ,
   Z. Anorg. Chem.,1912, <u>75</u>, 58 .

50/ C. Wagner
   Acta Met. , 1954, <u>2</u>, 242 .

51/ J.L. White ,
   Trans. AIME , 1959, <u>215</u>, 178 .

52/ J. Markali, P. Thorensen,
   Acta Chem. Scand. , 1961, <u>15</u>, 31 .

53/ C.J. Cooke , W. Hume-Rothery ,
   Acta Met. ,1961, <u>9</u>, 982 .

54/ J.M. Miane ,
   Thèse Doct. Spéc. , Univ. de Provence, Marseille, 1979 .

55/ K. Itagaki , A. Yazawa ,
   Nippon Kinzohu Gakkaishi , 1968, <u>32</u>, 1294.

56/ F.E. Wittig ,
   Private communication reported by Castanet R. (see ref. 57)

57/ R. Castanet, Y. Claire , M. Laffitte ,
   J. Chem. Phys. ,1969, <u>66</u>, 1276 .

58/ B. Predel , H. Bankstahl ,
   Less Common Met. , 1975, <u>43</u>, 191 .

59/ B. Legendre , C. Souleau ,
   J. Chem. Research (S) 1977, 306 ; (M) 1977, 3701 .

60/ B. Predel , W. Stein ,
   Z. Naturforsch ,1971 , <u>26</u>, 4, 722.

61/ J.P. Hager , S.M. Howard , J.H. Jones
   Met. Trans. ,1973, <u>4</u>, 2383 .

62/ K. Itagaki , A. Yazawa
   Trans. JIM ,1975, <u>16</u>, 679 .

CALORIMETRIC METHODS IN METALLURGY

R. CASTANET
Centre de Thermodynamique et Microcalorimétrie du CNRS,
26, rue du 141e R.I.A.,
F-13003 Marseille
France

ABSTRACT. The calorimetric methods used for the study of metallic systems are described and discussed. Special attention is paid to calorimetry of reacting systems, i.e. direct reaction calorimetry, dissolution calorimetry, combustion calorimetry, ... and non-reacting systems are considered as additional methods from this point of view. Emphasis is placed on their particular fields of application. Some recent applications developed during last ten years are given.

During the last ten years several papers reviewing experimental thermodynamics have been written, mainly by Komarek (1-6), Kubaschewski (7,10), Spencer (8), Navrotsky (9), Predel (11,12), Bruzzone (13) and Hertz (14). However, those especially devoted to calorimetric methods are scarce or deal with a restricted area of interest as, for example, geological problems (9). Then, my object is to try to emphazise particularly the experimental procedures used for calorimetric measurements according to the different aims of experimentators (measurements of integral or partial enthalpies of formation of liquid or solid alloys, determination of phase boundaries ...) and according to experimental conditions (high vapor pressure of components or alloys, high oxydability, high temperature, ...).

1. INTRODUCTION

The main purpose of the thermodynamic investigation on a system is to determine the Gibbs energy of formation with respect to composition, temperature and pressure with such an accuracy that it is possible to derive integral and partial quantities (ΔH, ΔS, Δc_p, ...) for all the phases. Then we can describe the system (for example calculating the phase diagram in agreement with the experimental one) and, theoretically, resolve any problem including of course those associated with industrial processes. The thermodynamic functions of formation can be fitted according to the following equations :

$$\Delta G^{XS}(x,T) = a + bT + cT \, Ln \, T + dT^2 + \ldots$$

145

H. Brodowsky and H.-J. Schaller (eds.), Thermochemistry of Alloys, 145–168.
© 1989 by Kluwer Academic Publishers.

$$\Delta H\,(x,T) \;=\; a \;-\; cT \;-\; dT^2 \;+\; \ldots$$

$$\Delta S^{XS}(x,T) \;=\; -\,b \;-\; c \;-\; c\,\mathrm{Ln}\,T \;-\; 2\,dT \;+\; \ldots$$

$$\Delta C_p^{XS}(x,T) \;=\; -\,c \;-\; 2\,dT \;+\; \ldots$$

where a, b, c, d, ... are functions of the type : $a = x(1-x)\,\sum_i a_i x^i$
in the case of a binary system. Obviously, these coefficients have no physical meaning and can be used only in order to reproduce experimental data as close as possible.

In my opinion, the best way for obtaining ΔG with respect to temperature is to measure the Gibbs energy of formation at a given temperature, $\Delta G^f(T_0)$, and the enthalpy of formation vs. T, $\Delta H^f(T)$, in the temperature range $T_1 < T < T_h$ where informations are needed :

$$\Delta C_p^f(x,T) \;=\; \frac{d\Delta H^f(T)}{dT}$$

(1) $\Delta H^f(T) = f(x,T) \longrightarrow \Delta H^f(T_0)$

(2) $\Delta G^f(T_0) = f(x)$

$$\Delta S^f(T_0)$$

$$\Delta G^f(x,T) \;=\; \Delta H^f(T_0) + \int_{T_0}^T \Delta C_p^f \, dT \;-\; T\left|\Delta S^f(T_0) + \int_{T_0}^T \frac{\Delta C_p^f}{T}\,dT\right|$$

$$=\; \Delta G^f(x,T_0) + \int_{T_0}^T \Delta C_p^f(x,T)\,dT \;-\; T\int_{T_0}^T \frac{\Delta C_p^f(x,T)}{T}\,dT$$

As you can see the calorimetric part is very important. The $\Delta C_p^f(x,T)$ determination can be performed according to two ways :

reaction calorimetry : $\Delta C_p^f = \dfrac{d\Delta H^f(T)}{dT}$

enthalpimetry : $\Delta C_p^f = C_p - \sum_i x_i C_{p,i}^o$

where $C_{p,i}^o$ is the heat capacity of the pure i component.

The choice between these two possibilities depends on different experimental parameters and will be developed later.

This lecture will be divided in two parts : the first one concerning calorimetry of reacting systems, the second one related to calorimetry of non-reacting systems. Yet, this distinction raises some difficulties. To what part belongs, for example, the peritectic formation of an intermediate compound ? That is called "peritectic reaction" and one can actually deduce the enthalpy of reaction from the overall DTA heat effect, for instance. Since we have to restrict reacting systems to those for which we start from - or we arrive to - a non-equilibrium state, it is easy in this case to check that measuring the enthalpy of a peritectic reaction does not correspond to reaction calorimetry. But how should be the recrystallization of a glass considered ?

In any case the problem of classification is a difficult one and we have to keep in mind that the boundary between reacting and non-reacting systems is not well-defined and that in some occasions enthalpimetry can lead to enthalpy of reaction and vice-versa.

2. REACTION CALORIMETRY

2.1. Direct reaction calorimetry (DRC)

By this method, the two components (or several components in the case of multicomponent systems) react together to give the alloy in the desired equilibrium state. It is probably the most commonly used method in the world. Indeed :
   - it can be employed with many kinds of calorimeters, even, with some limitations, with adiabatic apparatus ;
   - it leads to enthalpy of formation of solid and liquid phases ;
   - in many cases it allows to obtain the partial enthalpies of formation of components ;
   - it can be used to determine phase boundaries

As an example, here is developed what it can be realized in the case of an hypothetical A-B binary system whose phase diagram is shown on Figure 1.

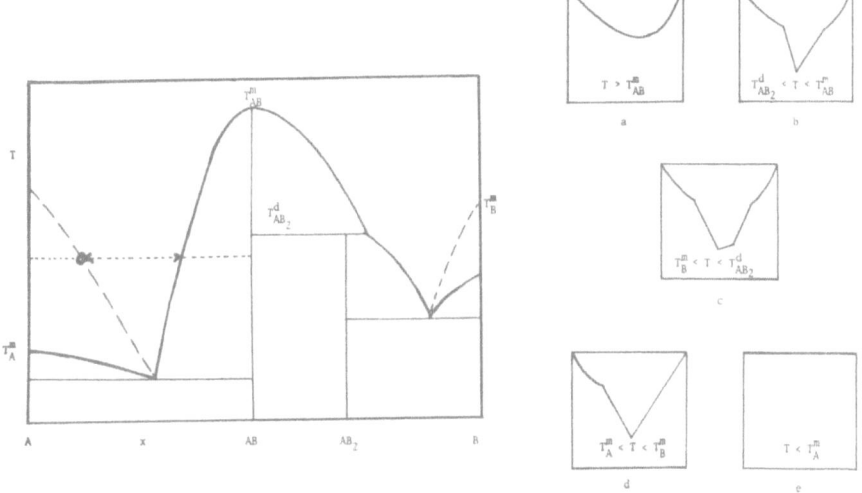

Figure 1. Hypothetical binary phase diagram and enthalpy of formation at different temperatures (see text).

a) at $T < T_{AB}^m$, we obtain the integral enthalpy of formation of the liquid phase in the whole range of concentration from the heat effects corresponding to successive additions of B into the melt which consists of pure A before the first addition of B. If the quantities of B added to the bath are small enough (i.e. if the variations of the molar fraction of B,

148

δx_B, due to the additions of B are small), $\delta\Delta H^f / \delta x_B$ can be considered as the slope of ΔH, so partial enthalpy of B, ΔH_B, can be directly obtained. Obviously, the values of δx_B depends on the curvature of $\Delta H(x_A)$. The measurements can be performed by adding A instead of B. The agreement between the two sets of data is a good test of quality. When measurements are carried out at different temperatures, the excess heat capacity of liquid alloys $\Delta C_p = d\Delta H^f/dT$, can be deduced. Some results concerning Mn-Ge (15) and In-Te (16) liquid alloys obtained with a high-temperature Calvet calorimeter are shown on Figures 2 to 4.

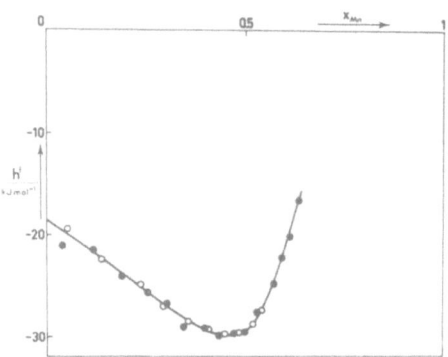

Figure 2. Partial molar enthalpy of Mn at 1278 K in liquid Mn-Ge alloys with reference to solid γ-Mn (15)

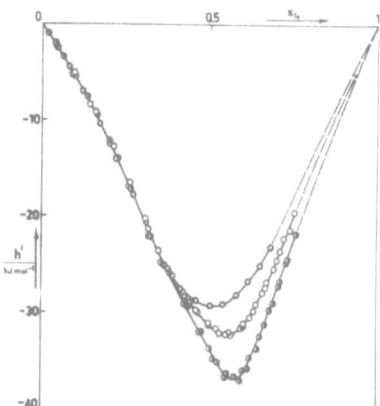

Figure 3. Molar integral enthalpy of formation of In-Te liquid alloys at 1340 K (o), 1123 K (o) and 987 K (●) with reference to both pure liquid components (16)

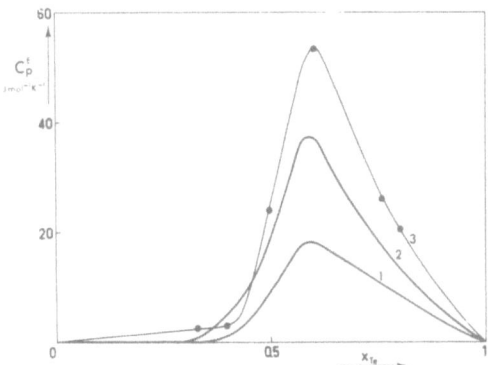

Figure 4. Heat capacity of liquid In-Te alloys determined by DRC (16) in 1123-1340 K (curve 1) and the 987-1123 K (curve 2) temperature ranges. Curve 3 : from C_p measurements at 1023 K by adiabatic calorimetry (17).

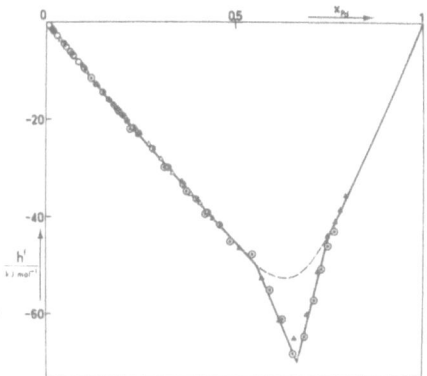

Figure 5. Molar integral enthalpy of formation of Pd-Ge alloys at 1270 K with reference to both pure liquid components (18).

b) At $T_{AB_2}^d < T < T_{AB}^m$, the same types of measurements lead to the enthalpy of formation of the liquid phase on the rich A and B-sides, but in the two-phase regions, the AB solid compounds precipitates and we obtain its enthalpy of formation. Δc_p^f of the solid is also available if measurements are performed at different temperatures. From the break of the integral enthalpy or from the jump of the partial one, the phase boundary can be deduced at the given temperature. Results concerning the Pd-Ge system (18) are shown on Figures 5 and 6.

c) At $T_B^m < T < T_{AB_2}^d$, the $\Delta H^f(x)$ curve shows two peaks, correspondding to the formation of the AB and AB_2 compounds.

d) At $T_A^m < T < T_B^m$, the results are similar to c) for $x_B \leqslant 0.5$ but for $x_B > 0.5$ further additions of B on solid AB does not yield equilibrium states since solid B does not react with solid AB.

e) Finally, no measurement can be performed below T_A^m since there is no liquid to allow the reaction from pure components. Then, in the case of the

150

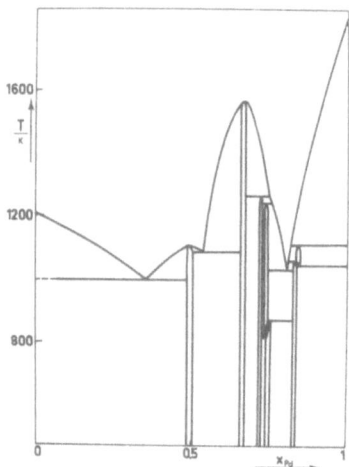

Figure 6. Pd-Ge phase diagram. The solid points were deduced from the breaks of ΔH against x_{Pd} curve obtained by DRC (1270 K).

dotted line phase diagram (Figure 1), the enthalpy of formation of AB_2 cannot be obtained, nor that of the liquid phase below the lowest melting pure component. However, starting from a master alloy, prepared outside of the calorimeter with a composition corresponding to the liquid state at the temperature of measurements (circle on Fig. 1) it is possible to determine the partial enthalpies of formation of B and A adding successively B then A, into the same bath. Then, the integral enthalpy of formation of the liquid is obtained :

$$\Delta H^f = x_A \, \Delta H^f_A + x_B \, \Delta H^f_B$$

and by further additions of B into 2 (see Fig. 1) the enthalpy of formation of AB can be measured. The same series of measurements can be carried out on the B-rich side, yielding the enthalpy of formation of AB_2. Figures 7 and 8 show the results obtained at 1380 K for the Pd-Si liquid alloys (19). ΔH^f is calculated in this way from two separate sets of measurements and the accuracy is lowered because there is no reason for the Gibbs-Duhem relation :

$$\Delta H^f_B = - \int_1^{x_B} \frac{x_A}{x_B} \, d(\Delta H^f_A)$$

to be followed.

As it can be seen on Figure 8, the partial enthalpy of formation of Pd, ΔH^f_{Pd}, calculated from ΔH^f_{Si} (full line) does not completely agree with the experimental determinations (open circles).

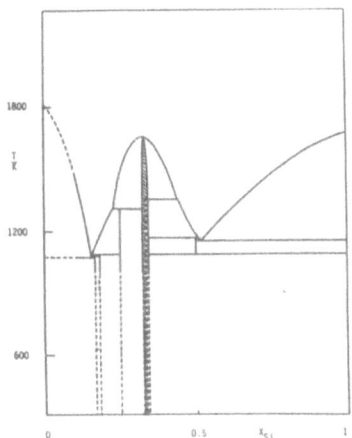

Figure 7. Pd-Si phase diagram.

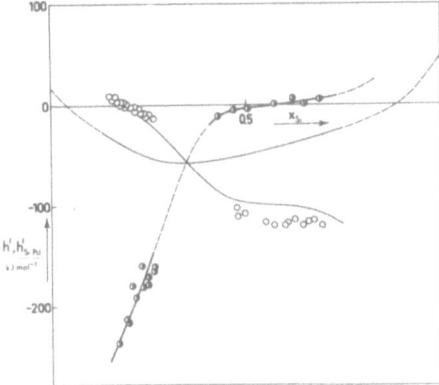

Figure 8. Molar partial enthalpy ofSi (●) and Pd (o) at 1380 K in liquid Pd-Si alloys, measured separately from DRC (see text) and referred to both solid components (19).

Experimental procedures

The most simple (and in many cases the best) procedure is the direct drop method (20) : pure solid B at T_O (near room temperature, in general) is dropped into the bath inside of the calorimetric cell at T_C (Figure 9). Then, we measure the enthalpy of the reaction :

$$n_B B(s, T_O) + n_A A(l \text{ or } s, T_C) \rightarrow (n_A + n_B) AB(l \text{ or } s, T_C)$$

152

where A or B have to be liquid at T_c. The enthalpy variation of B from T_0 to T_c has to be known to obtain the enthalpy of formation of AB with reference to both pure liquid components at T_c.

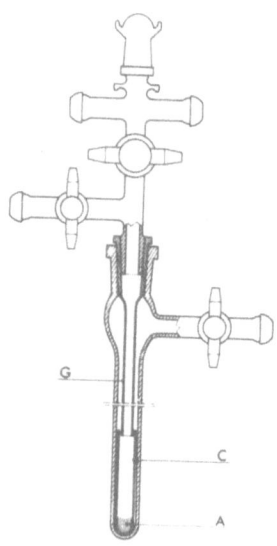

Figure 9. Calorimetric cell used in DRC without preheating (direct drop method) (20).

Such an enthalpy variation can be easily measured with the same apparatus, by dropping pure B into the empty calorimetric cell. Besides, one has to consider the fact that this enthalpic term is not necessarily a source of error. Indeed, when the enthalpy of formation of AB referred to liquid components is negative, the positive heat content of B (including heat of melting) reduces the overall measured heat effect and then results in a zero method. The experimental error is reduced when the heat content of B is well known as it is the case for many pure metals. For the Cu-Si liquid alloys (21), we found $\Delta H^f_{min} = -14.35$ kJ mol^{-1} refered to pure liquid Cu and supercooled Si at T = 1370 K and $x_{Si} = 0.25$. According to Hultgren et al. (22) the heat content variation from 298 to 1370 K and the enthalpy of melting of Si are respectively 27.11 and 50.55 kJ mol^{-1}. The overall heat effect measured is then only + 5.07 kJ mol^{-1}. DDM can be used, not only for alloys but also for metal-gas systems. Gerdanian and col. (23,24) determined partial enthalpies of oxygen in metal-oxygen systems with respect to composition for non-stoichiometric phases and Kleppa et alii (25-29) applied the same method to metal-hydrogen systems.
When the enthalpy of formation referred to the selected reference state is positive, the accuracy is improved by preheating. Such an indirect drop method can be realized either by stoping the sample in the calorimetric cell just above the reacting vessel until its temperature reaches T_c (Figure 10), or by a furnace placed above the calorimeter. The use of a furnace

allows the experimentator to stabilize the sample at a temperature different of T_C.

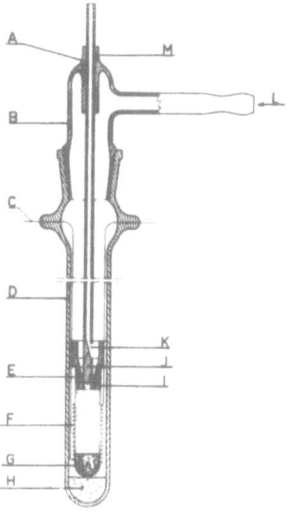

Figure 10. Calorimetric cell used in DRC with preheating (indirect drop method) (30).

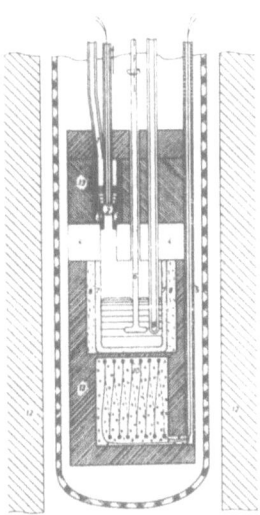

Figure 11. High-temperature calorimeter for determination of mixing and solution enthalpies (36).

A cryostat in place of a furnace can be employed when the sample is liquid at room temperature. The first procedure was used mainly by Bros with a high-temperature calorimeter (up to 1750 K) for Au-Ag-Ge (30) and Au-Ag-Si

(31) liquid alloys. The metal added is solid. Another device used by Wittig and Huber (32, 37), Ticknor and Bever (33), Misra et al. (34) operates with liquid samples. Sommer et al. (35) used a similar procedure to measure the enthalpy of mixing of alkali and rare earth metals. Oehme and Predel (36) developed a calorimeter operating up to 1300 K for the study of liquid mixtures (Fig. 11). In all instances several samples can be introduced successively into the crucible leading to ΔH^f with respect to concentration and to partial values in a single series of measurements.

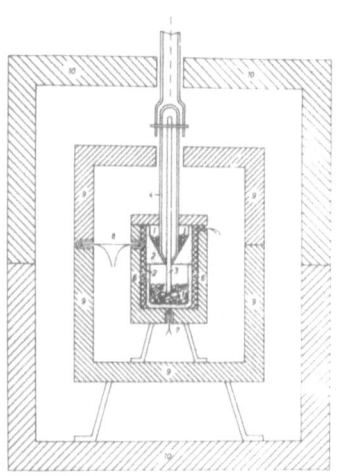

Figure 12. High-temperature adiabatic calorimeter for determination of mixing enthalpies of alloys (38).

For such a goal, adiabatic calorimetry can be used. As an example, Itagaki and Yazawa (38) determined the heat of mixing of Pb and Bi (Fig. 12). When the vapour pressure of the bath is too high at the temperature of measurement a low vapour pressure protecting liquid is placed on the metallic bath (Fig. 13) to avoid evaporation and attack of sample during the obtention of thermal equilibrium. This method was used by Claire et al. (39) for measuring the mixing enthalpy of Tl and Hg above the melting point of Tl. Another method avoiding evaporation of the second component before and after mixing has been used by Bros (40) and by Castanet et al (41) : the metal to be added is placed in a glass vacuum sealed ampoule and broken by a piston which close the calorimetric cell (Fig. 14).

Besides, in the case of liquid alloys presenting strongly negative heats of mixing there is no loss of high vapour pressure component out of the bath even at high temperature, because the activity of this component is strongly lowered by heteroassociations in the melt. In 1979, we measured by DRC (54) for $0 < x_{Te} < 1/3$ the heat of mixing of the Ag-Te liquid alloys adding pure Te into liquid Ag at T = 1378 K i.e. above the boiling point of Te (Fig. 15). Actually for $x_{Te} < 1/3$ (Ag_2Te) there are no free Te atoms in the bath (Fig. 16). Moreover, in order to avoid evaporation during dissolution it is useful to add a compound of the system instead of the pure high vapour pressure component. This procedure was used by Kotchi et al (56) to

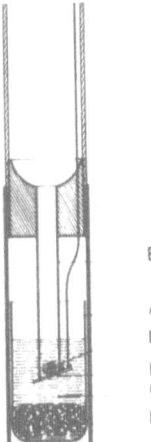

Figure 13. Calorimetric cell used in DRC with preheating and high vapor pressure of the melt (indirect drop method) (39).

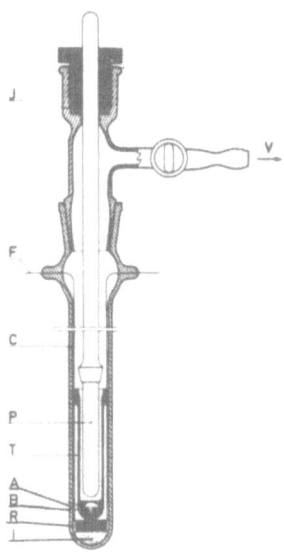

Figure 14. Calorimetric cell used in DRC (break of ampoules) (41).

determine the enthalpy of formation of the Pb-PbSe melts adding solid PbSe into liquid Pb.

Finally, the more sophisticated and yet very elegant levitation method can be used which eliminates all possibilities of side reactions since the mixing process occurs without any contact. It was used by Frohberg and Betz (42) who measured the heat of mixing of Fe-Cu and Nb-Si alloys, and by Arpaci and Frohberg (43) who determined the heat of mixing of Mo-Si liquid

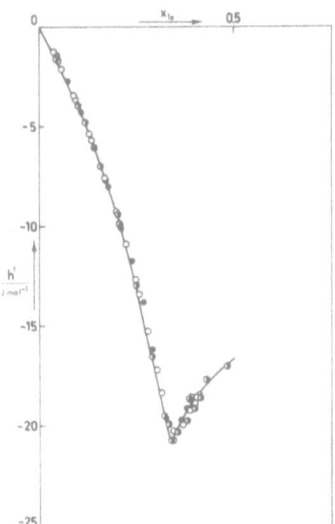

Figure 15. Molar integral enthalpy of formation of liquid Ag-Te alloys referred to both liquid components (55). ▼1378 K, ◐ 1302 K, o 1255 K, ● 1235 K

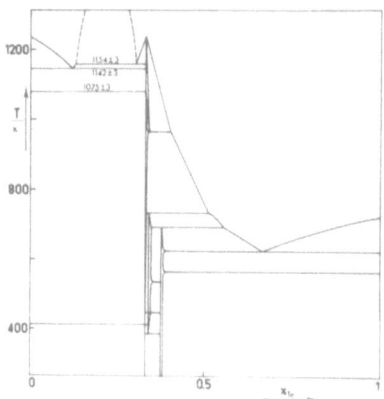

Figure 16. Ag-Te phase diagram.

alloys at 3087 K.

The previous methods are especially well-suited for measuring the enthalpy of liquid alloys. However, measuring the enthalpy of solidification of the corresponding alloys, they lead to the enthalpy of formation of the solids. The melting enthalpy of solid alloys can be obtainec by D.T.A. or from enthalpy measurements as seen later. The Gathers' et al. (44) method allows by this way to determine the heat of formation of solids up to 8000 K. Yet, there are some methods yielding the formation enthalpy of solids from direct reaction in the solid state. The first one was developed by Kubaschewski and Dench (45,46) whose adiabatic calorimeter was further improved by Kubaschewski and Grundman (47). The calorimeter works up to 1900 K. The two pure powdered components are intimately mixed and heated from a T_O temperature, where no reaction takes place up to a temperature T, where they react to lead to the alloy. Then, the sample is heated again from T_O to T and the enthalpy of formation is obtained as the difference between the two thermal effects. This method was used by some other workers, as for example by Ferro et al (48,49). According to another procedure used by Robins and Jenkins (132) for intermetallic compounds, the reaction between mixed powdered components is initiated by the ignition of a tablet of thermite and then goes on by itself if its enthalpy is strongly negative. The third method was developed by Gachon et al. (50-53). Small pellets of powdered components mixed in suitable proportions are introduced into the calorimeter, at a temperature just below the melting point of the compound, so that the reaction occurs. Gachon (54) obtained in this way the enthalpy of formation of several (Fe, Co, Ni)-(Ti, Zr) and (Pd,Pt)-(Ti, Zr, Hf) solid compounds. However, the use of such methods requires special attention to the final state of the sample, the main source of error being incomplete reaction.

DRC has been also employed to study oxide systems. The method is unfortunately not applicable to silicates since reactions between silica and metallic oxides are generally to slow. DRC can be used only when the two components differ strongly in acidity as, for example, in the formation of lead vanadates from liquid V_2O_5 and solid PbO (57).

2.2. Dissolution calorimetry (DC)

DRC can be used for measuring the enthalpy of formation of solid compounds or alloys, but we saw that this method presents some limitations for such a purpose. For solids, DC has been very largely used, measuring separately the heat of dissolution of the compound A_xB_{1-x} and that of a mechanical mixture xA + (1-x)B into the same solvent :

(1) $\qquad A_xB_{1-x} + S \rightarrow [xA, (1-x)B]_S \qquad \Delta H_1$

(2) $\qquad xA + (1-x)B + S \rightarrow [xA, (1-x)B]_S \qquad \Delta H_2$

When the final concentrations of the bath are the same for (1) and (2), the enthalpy of formation of A_xB_{1-x} can be deduced as :

$$\Delta H^f = \Delta H_2 - \Delta H_1$$

From a practical point of view, preparing the mechanical mixture with the same x/1-x ratio as for A_xB_{1-x} and obtaining the same final bath concentrations with respect to the solvent is not so easy. Then, the more generally used procedure consists in measuring the heats of dissolution at high dilution of A_xB_{1-x}, A and B into S vs. concentration and to extrapolate at infinite dilution :

(3) $\qquad n_A A(T_O) + NS(T_C) \rightarrow [A]_S \ (x_A = \dfrac{n_A}{n_A + N}, \ T_C)$

$$(4) \qquad n_B B(T_O) + NS(T_C) \rightarrow [B]_S \ (x_B = \frac{n_B}{n_B + N}, \ T_C)$$

$$(5) \qquad n_C A_x B_{1-x}(T_O) + NS(T_C) \rightarrow [xA,(1-x)B]_S \ (x_C = \frac{n_C}{n_C + N}, \ T_C)$$

For $x_A = x_B = x_C = 0$, the final states of (3) and (4) are the sum of those of (5) since there is no interaction between A and B. Then :

$$\Delta H^f(T_O) = x \ \Delta H^\infty_{diss.A} + (1-x) \ \Delta H^\infty_{diss.B} - \Delta H^\infty_{diss.A_x B_{1-x}}$$

The experimental procedures used are obviously the same as in the case of DRC. In general, the temperature of solutes, T_O, and the bath, T_C, are different. Then, ΔH^f is obtained at T_O and the method can supply data at room temperature or (with a cryostat above the calorimeter) at low temperature, where DRC cannot work. Generaly speaking, the accuracy of DC is inferior to that of DRC since enthalpies of formation are obtained as sum of three terms, and since the result is often small as compared to dissolution enthalpy values. However, the accuracy can be improved by using, when possible, one of the two components as solvent.

Many years ago, the solvents used for dissolution were mainly aqueous inorganic solvents, such as hydrochloric or hydrofluoridric acids, even for alloys. The results were obtained as small differences of very large enthalpies of dissolution, so accuracy was often very poor. Later on, metallic solvents were used. A good solvent has to fullfill some requirements :

- to dissolve solutes in a reasonnable time, typically twenty minutes in the case of a heat flow calorimeter ;
- to have a low vapour pressure enough in order not to yield losses by evaporation ;
- to give moderate heats of dissolution of solutes to minimize errors.

The first and the third conditions are unfortunately contradictory in some aspects. The choice of the solvent is the main problem in dissolution calorimetry, such as that of liquid electrolytes in the case of E.M.F.. There is no universal solvent. Liquid tin is probably the most frequently used because it can dissolve many metals and alloys at low temperature. It was mainly used at the time when high temperature calorimetry was not developed as now. In 1952 Ticknor and Bever (58) used it for measuring the heat of formation of Ag-Au alloys. It was also very much used by Kleppa for binary (59) and ternary (60) alloys. With some transition metals, the times of dissolution into Sn can become very long and the temperature has to be increased. For example, Meyer-Liautaud et al. (61-62), must perform dissolution experiments in Sn at 1173 K in order to determine the enthalpies of formation of some Sm-Co, Sm-Co-Cu, Y-Co-Cu and Ce-Co-Cu solid phases.

Many other liquid metals were used for the same goal : aluminium by Mathieu et al. (63), copper and copper alloys (64,65), nickel and iron (66), gallium (72) and lead (77). As examples of specific solvents used for specific materials we have to mention Kleppa who determined the enthalpies of formation of Mn_5C_2 into $Mn_{0.6}Ni_{0.4}$ (67), of B_4C into $Mn_{0.6}Ni_{0.4}B_{0.02}$

(68) of Mn_2B, MnB and MnB_2 into Cu (69), of LaB_6 into Pt_2B (70) and CrB_2 into Pd_2B (71). The use of a furnace placed above the calorimeter is particularly useful when the compound to be dissolved is stable only at high temperature. M. Baier et al. (73) determined in this way at $T_O = 956$ K the enthalpy of formation of the β-phase of the Ag-Al system ($\sim Ag_3Al$) which is stable only between ~ 870 and 1053 K.

In order to get the enthalpy of formation of some transition metal silicides, Kleppa used a novel method (74). By a drop method he measured separately the heat effects corresponding to the following reactions (where Me $\equiv$ Ti and V) :

(6) 0.05 $MeSi_2$(s, 298 K) + 0.85 Pd(s, 298 K) $\rightarrow$

$Pd_{0.85}Si_{0.10}Me_{0.05}$(l, 1400 K)

(7) 0.85 Pd(s, 298 K) + 0.10 Si(S, 298 K) + 0.05 Me(s, 298 K) $\rightarrow$

$Pd_{0.85}Si_{0.10}Me_{0.05}$(l, 1400 K)

combining equations (6) and (7), we have :

(8) 0.05 Me(s, 298 K) + 0.10 Si(s, 298 K) $\rightarrow$ 0.05 $MeSi_2$(s, 298 K)

and then :

$$\Delta H^f(MeSi_2, s, 298 \text{ K}) = \Delta H_7 - \Delta H_6.$$

The method is very close to DC using liquid Pd-Si alloy as solvent, to DRC and to enthalpimetry. The same procedure was applied recently by Kleppa (75) to the determination of the heat of formation of PdTi using Cu-Si instead of Pd-Si. In each case, the "solvent", i.e. Pd-Si or Cu-Si, had a composition corresponding to low melting point for which the dissolution process is complete.

Obviously, DC can be used not only for the determination of heats of formation of alloys from their components but also in order to get heats of transformation in the solid state. For example, Hertz (76) obtained, many years ago, the heat of the order-disorder transition in $AuCu_3$ with respect to the order parameter from dissolution experiments into Sn.

DC has also been used for providing data of interest in geology (9). Many years ago, hydrofluoric acid solutions were used as solvent and still now HF dissolution calorimetry can be useful in some special cases, for example for hydrous phases and, more generally speaking, when the enthalpy of dissolution is needed at or near room temperature (Table I). Torgeson and Sahama (78) obtained in this way the heats of formation of Mg_2SiO_4, $MgSiO_3$ and $CaSiO_3$. Oxide melt calorimetry has many advantages : ease of dissolution of compounds containing Al_2O_3 and MgO, small enthalpies of dissolution, ability to work with small amounts of samples (9). Another very important point is the possibility to work with buffered solvents. Then, the dependence on concentration of heats of dissolution is very low and can be neglected. It is no longer necessary to extrapolate the results at infinite dilution. Many solvents have been used (9, 79). Among them one can note $2PbO-B_2O_3$, $3Na_2O-4MoO_3$, $9PbO-3CdO-4B_2O_3$, $0.35\ Na_2O-0.35\ B_2O_3-0.30$ SiO_2 and some mixture of $Na_2O-P_2O_5$. A series of chromite spinels (80), some silicates and germanates (81) and tungstates (82) of Mg, Co, Ni, Cu and Zn were studied as well as the systems $MgO-Al_2O_3-SiO_2$ (83), $NaAlSi_3O_8-KAlSi_3O_8$ (84), $Na_2O-K_2O-SiO_2$ (85) and sodium borosilicates glasses (86).

TABLE I

Solvents for high temperature molten salt calorimetry (from ref. 9).

Solvent	Temp. range °C	Uses	Limitations	Comments
$2PbO-B_2O_3$	650-900	Excellent for SiO_2, Al_2O_3, GeO_2, Fe_2O_3, Cr_2O_3, Ga_2O_3, MgO, CoO, NiO, CuO, ZnO, CdO, CaO, Cu_2O.	forms insoluble precipitates with TiO_2 SnO_2, Mn_2O_3. Canno bet used under reducing atm.. Can be used for Mn^{2+} (and Fe^{2+}) containing oxides under inert atm. when oxides can be maintained.	Solvent of choice when applicable
$9PbO-3CdO-4B_2O_3$	650-900	same as above.	same as above.	above generally preferable.
$3Na_2O-4MoO_3$	650-750	good for TiO_2, SnO_2, Mn_2O_3, MgO, CoO, NiO, ZnO, CdO.	SiO_2 and other acid oxides dissolve slowly. Not used under reducing atm., vaporizes significantly. at higher temp.	does not form a glass upon cooling
$0.35Na_2O-0.35B_2O_3-0.35SiO_2$	800-900	Good for SiO_2, Al_2O_3, MgO, CaO, and FeO. MnO under reducing atmospheres.		hygroscopic ; caref. prepared and stored.

2.3. Precipitation calorimetry (PC)

The precipitation method relies on the ability to induce precipitation of the phase under investigation by addition of one of its components to a dilute solution of the other component in a liquid metal solvent. It was used for the first time by Bryant and Pratt (87,88) for the determination of the enthalpy of formation of NiSi, PdSi and some III-V compounds, as InAs, InSb and GaSb, for which dissolution calorimetry experiments by Schottky and Bever (89) in liquid tin were unsuccessful. The PC procedure is the following : one of the component is dissolved into the solvent. By addition of the second component, a solid compound AB is formed and precipitates if its solubility is small enough. For such a procedure, solubility of AB in the solvent has to be smaller than that of the first component. This condition is one explanation to the failure of dissolution calorimetry. The distinction between the two techniques is clearly demonstrated by the following equations :

Precipitation calorimetry

$$(9) \quad B(s) + \frac{x}{1-x}[A]_S \rightarrow \frac{1}{1-x}A_xB_{1-x}(s) \qquad \Delta H_1$$

$$(10) \qquad\qquad\qquad A(s) \rightarrow [A]_S \qquad \Delta H_2$$

$$(11) \quad (1-x)B(s) + xA(s) \rightarrow A_xB_{1-x}(s) \qquad \Delta H^f$$

$$\Delta H^f = (1-x)\,\Delta H_1 + x\,\Delta H_2$$

Dissolution calorimetry

$$(12) \qquad\qquad\qquad A(s) \rightarrow [A]_S \qquad \Delta H_1$$

$$(13) \qquad\qquad\qquad B(s) \rightarrow [B]_S \qquad \Delta H_2$$

$$(14) \qquad A_xB_{1-x}(s) \rightarrow x[A]_S + (1-x)[B]_S \quad \Delta H_3$$

$$(15) \quad xA(s) + (1-x)B(s) \rightarrow A_xB_{1-x}(s) \qquad \Delta H^f$$

$$\Delta H^f = x\,\Delta H_1 + (1-x)\,\Delta H_2 - \Delta H_3$$

PC calorimetry is based on the synthesis of the phase under investigation whereas DC involve a de-alloying process. In both cases, final states need to be carefully checked in order to avoid any incomplete precipitation (PC) or incomplete dissolution (DC). However, the field of application of PC is somewhat restricted.

2.4. Combustion calorimetry (CC)

In combustion calorimetry, still more than in dissolution calorimetry résults are obtained as small differences of large numbers and the accuracy is often poor. Then if a direct calorimetric measurement can be done it may be more precise. However, oxygen bomb combustion calorimetry is a well developed technique which can be used when no other method is available (90). Obviously, these restrictions are valid only for metallic systems, and do not apply to oxides (or halogenides) for which CC is in fact DRC as used by Charlu et al. (91, 92) for the study of Ti, V and W oxides with a high temperature calorimeter.

Some other gases can be used in CC. Very recently, Kuwata et al. (93) used an oxygen-nitrogen mixture in order to determine the enthalpy of formation of the TiS_x solid solution (1.95 < x < 2.00). Actually, not only temperature but also oxygen partial pressure are important factors for complete combustion. In the case of TiS_x, unreacted sulfide increased with increasing oxygen pressure. Halogens, mainly fluorine, were used in CC (94, 95). Generally speaking, fluorine is better than oxygen because metal fluorides deviates less from the stoichiometry than oxides and moreover, it is often possible to get complete reaction by forming a single fluoride instead of incomplete combustion and multiple oxidation states as with another oxidant (95). The main problem in CC is, once again, the final states of the combustion products which have to be carefully defined (96).

3. NON-REACTION CALORIMETRY (NRC)

We will discuss here non-reacting calorimetry methods, i.e. determination of heat contents, only from the point of view of their relations with reaction calorimetry. Actually, since high temperature heat content measurements are easier than enthalpy of formation determinations, heats of reaction are mostly deduced from reaction calorimetry at $T < 1800$ K and heat content experiments carried out at higher temperature, according to the scheme :

$$xA(T_1) + (1-x)B(T_1) \rightarrow A_xB_{1-x}(T_1) \qquad \Delta H^f(T_1)$$

$$xA(T_1) + (1-x)B(T_1) \rightarrow xA(T_2) + (1-x)B(T_2) \quad \Delta H_2$$

$$A_xB_{1-x}(T_1) \rightarrow A_xB_{1-x}(T_2) \qquad \Delta H_3$$

$$xA(T_2) + (1-x)B(T_2) \rightarrow A_xB_{1-x}(T_2) \qquad \Delta H^f(T_2)$$

$$\Delta H^f(T_2) = \Delta H^f(T_1) + \Delta H_3 - \Delta H_2$$

The method can obviously be employed to determine $\Delta H^f(T_2)$ at low temperature or at the 298.15 K standard temperature. Usually heat content measurements below room temperature are performed by adiabatic calorimetry whereas drop calorimetry or adiabatic scanning calorimetry are used above room temperature. Methods of measuring heat capacities of solids at high pressure and low-temperature were reviewed by Loriers-Susse (122). At very high temperature pulse-heating techniques are employed.

The most accurate measurements are obtained by adiabatic calorimetry (7,8), which is used up to 1800 K : Andrews et al. (97) from 300 to 550 K, Cash et al. (98) from 300 to 1200 K, Rogez et al. (99) from 800 to 1800 K and Kubaschewski and Grundman (100) from 700 to 1700 K. The adiabatic drop calorimeter of Conard et al. (101) works from 400 to 1400 K.

Levitation calorimetry (42,43) can be used up to 3000 K without any side reactions but above this limit, losses due to thermal radiation increase drastically. Then, experiments of very short durations have to be carried out. It is the case of all the methods where a wire-shaped metallic sample is electrically heated during a very short time (108-111).

Shpilrain et al built a pulse-differential apparatus working up to 1800 K and 300 bar (102), then up to 3000 K (103). Petrova and Chekhovskoi (104, 105) determined the heat capacity of TiC and NbC from 1600 to 2400 K with an automated pulse-method and that of Nb was measured up to 5000 K by Mozharov and Savatimskii (106). Finally, Seydel et al. (107) developed a submicrosecond-pulse-heating method for obtaining simultaneously specific heats, electric resistivities, specific volumes and thermal expansion coefficients of liquid metals up to 12000 K.

In a more restricted temperature range but including probably 80 % of needs, drop calorimetry still works mainly with heat flow calorimeters. Heat content measurements can be carried out from high-temperature to room-temperature or to 273K using an ice calorimeter as in the case of Denielou et al. (112-114). The method developped by Oelsen et al. (115) can also be used for a room temperature calorimeter. Special attention has to be given to non-equilibrium final states which can be obtained on cooling. In order to avoid such problems drop calorimetry from room to high tempera-tures (the so-called "reverse drop calorimetry") is used with high temperature calorimeters. For example, we measured by this method the heat content of the $Au_{0.75}Si_{0.25}$ liquid alloys between 625 and 1350 K (116) and the heat capacity of the tellurium-rich Ge-Te eutectic (117, 118)(Figure 17). The high ability of these melts to give glasses prevented the use of the high-to low-temperature method.

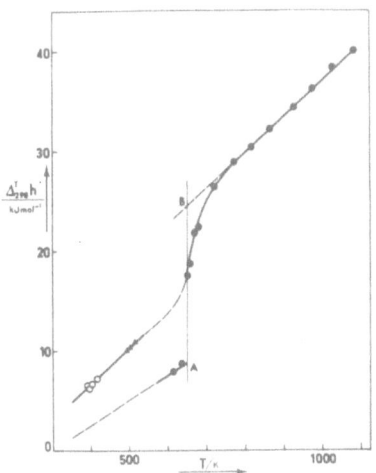

Figure 17. Molar enthalpy variation, $\Delta_{298}h$, from 298 to T/K of the Te-rich eutectic of Ge-Te alloys in the crystalline, supercooled and stable liquid states (117,118). ● stables states (drop method), o supercooled liquid (drop method),▲ supercooled liquid (DSC).

When heat content measurements cover a temperature range which includes the melting point (or any first order transition), the enthalpy of melting can be deduced as a function of temperature by difference between the heat of the solid and of the liquid phase. $\Delta_H{}^{fus}$ (T) is an hypothetical value, often needed in direct reaction calorimetry for reference state changes. Holm and Gronvold (119) and Blachnik and Igel (120) performed many such measurements on cryolites and on chalcogenides.

4. CONCLUSION

In this paper, it was impossible to cover all fields of application of calorimetry, even from the particular point of view of metals and oxides. Heat content measurements vs. temperature can be used to investigate many particular phenomena such as order-disorder transformation (for example, C_p measurements made by Charrin et al. (121) on Cd-Mg alloys), structural relaxation and crystallization of amorphous materials (123) or glass transitions (124). Deneuville et al (131) developed for the study of order-disorder transitions a novel method using a heat flow twin calorimeter at a temperature where the sample is disordered (double drop method) : The difference between the heat effects produced by simultaneous heatings of ordered and disordered alloys into the two thermoelectric cells in opposition yield directly the enthalpy of order.

Important areas, such as determination of heats of adsorption or use of calorimetry for corrosion studies, reviewed respectively by Gravelle (125, 126) and by Kuhn and Sham El Din (127), had to be discarded. The aging of supersatured solid solutions, the annealing of deformed, irradiated or quenched metals or alloys can be investigated by isothermal calorimetry, except for the beginning of the phenomena as shown by Zahra et al. (128).Many C_p and enthalpies of state changes are nowadays performed by differential scanning calorimetry (using mainly commercial apparatus) and such techniques, although dynamic then non-equilibrium methods, produce more and more data. However, they were disregarded in this paper. It was also not possible to speak of calibration which is a main problem in calorimetry, or of automatization of apparatus, as done by Regelsberg et al. (129) or by Hatem et al. (130) for example. Calorimetry is the oldest part of the thermodyn amics. Its fields of application and its different branches are really tremendously wide.

REFERENCES

(1) K.L. Komarek, in Metallurgical Chemistry, ed. O. Kubaschewski, HMSO, London, 16 (1972) 75.
(2) K.L. Komarek, Z. Metallkde, 64 (1973) 325 and 406.
(3) K.L. Komarek, Ber. Bunsenges. Phys. Chem., 81 (1977) 936.
(4) K.L. Komarek, Ber. Bunsenges. Phys. Chem., 87 (1983) 709.
(5) H. Ipser and K.L. Komarek, Z. Metallkde, 75 (1984) 11.
(6) K.L. Komarek and H. Ipser, Pure and Appl. Chem., 56 (1984) 1511.
(7) O. Kubaschewski, P.J. Spencer, W.A. Dench, in Chemical Thermodyamics, vol. 1, pp. 317, The Chem. Soc., 1973.
(8) P.J. Spencer, Pure and Appl. Chem., 47 (1976) 267.
(9) A. Navrotsky, Phys. Chem. Min., 2 (1977) 89.
(10) O. Kubaschewski, Thermochim. Acta, 22 (1978) 199.
(11) B. Predel and I. Arpshofen, Thermochim. Acta, 22 (1978) 211.
(12) B. Predel, Calphad, 6 (1982) 199.
(13) G. Bruzzone, Thermochim. Acta, 96 (1985) 239.
(14) J. Hertz, J. Thermal. Anal., 30 (1985) 1227.
(15) R. Castanet, Z. Metallkde, 77 (1986) 26.
(16) H. Saîd and R. Castanet, High Temp. High Pressures, 10 (1978) 681.
(17) S. Takeda, H. Okazaki and S. Tamaki, J. Phys. C : Solid State Phys., 15 (1982) 5203.
(18) R. Castanet and R. Chastel, Z. Metallkde, 78 (1987) 97.
(19) C. Bergman, R. Chastel, M. Gilbert, R. Castanet and J.-C. Mathieu, J. Phys., 41 (1980) C8-591.
(20) T. Kang and R. Castanet, J. Less-Comm. Met., 51 (1977) 125.
(21) R. Castanet, J. Chem. Thermodyn., 11 (1979) 787.
(22) R. Hultgren, P. Desai, D. Hawkins, M. Gleiser, K. Kelley and D. Wagmans. 'Selected values of the thermodynamic properties of the elements', Am. Soc. Met., Ohio, USA, 1973.
(23) C. Picard and P. Gerdanian, J. Solid State Chem., 14 (1975) 66.
(24) G. Boureau and P. Gerdanian, Acta Metall., 24 (1976) 717.

(25) G. Boureau, O. J. Kleppa and P. Dantzer, J. Chem. Phys., **64** (1976) 5247.

(26) G. Boureau and O.J. Kleppa, J. Chem. Phys., **65** (1976) 3915.

(27) P. Dantzer and O.J. Kleppa, J. solid State Chem., **24** (1978) 1.

(28) P. Dantzer and O.J. Kleppa, J. Solid State Chem., **35** (1980) 34.

(29) O.J. Kleppa, Ber. Bunsenges. Phys. Chem., **87** (1983) 741.

(30) S. Hassam, M. Gaune-Escard and J.-P. Bros, High Temp. Sci., **16** (1983) 131.

(31) S. Hassam, M. Gaune-Escard and J.-P. Bros, Actes J.C.A.T., XIV (1983).

(32) F.E. Wittig and F. Huber, Z. Elektrochem., **60** (1956) 1181.

(33) L.B. Ticknor and M.B. Bever, J. Metals, **4** (1952) 941.

(34) S. Misra, H.P. Singh and P.U. Nayak, Indian J. Technol., **6** (1968) 254.

(35) F. Sommer, J.J. Lee and B. Predel, Z. Metallkde, **71** (1980) 818.

(36) G. Oehme and B. Predel, Thermochim. Acta, **22** (1978) 267.

(37) F.E. Wittig and G. Keil, Z. Phys. Chem. Neue Folge, **144** (1985) 253.

(38) K Itagaki and A. Yazawa, Trans. J. Inst. Met., **10** (1969) 259.

(39) Y. Claire, R. Castanet and M. Laffitte, J. Less-Comm. Met., **31** (1973) 83.

(40) J.-P. Bros, Thesis, Aix-Marseille (France), 1968.

(41) R. Castanet, J.-P. Bros and M. Laffitte, J. Chim. Phys., **65** (1968) 1536.

(42) M.G. Frohberg and G. Betz, Ber. Bunsenges. Phys. Chem., **87** (1983) 782.

(43) E. Arpaci and M.G. Frohberg, Z. Metallkde, **76** (1985) 440.

(44) G.R. Gathers, J.W. Shammer and R.L. Brier, Res. Sci. Instrum., **47** (1976) 471.

(45) O. Kubaschewski and W.A. Dench, Acta Metall., **3** (1955) 339.

(46) W.A. Dench, Trans. Faraday Soc., **59** (1963) 1279.

(47) O. Kubaschewski and J. Grundmann, Ber. Bunsenges. Phys. Chem., **81** (1977) 1239.

(48) R. Ferro and R. Capelli, Atti Accad. Naz. Lincei, **34** (1963) 659.

(49) R. Ferro, R. Capelli and A. Borsese, Thermochim. Acta, **10** (1974) 13.

(50) J.C. Gachon, J. Giner and J. Hertz, Scripta Met., **15** (1981) 981.

(51) J.-C. Gachon, M. Dirand and J. Hertz, J. Less-Comm. Met., **85** (1982)1.

(52) J.-C. Gachon and J. Hertz, Calphad, **7** (1983) 1.

(53) J.-C. Gachon,M. Dirand and J. Hertz, J. Less-Comm. Met., **92** (1983)307

(54) J.-C. Gachon, Doct. Sci. Phys., Nancy I (France), 1986.

(55) R. Castanet and C. Bergman, J. Chem. Thermodyn., **11** (1979) 83.

(56) K.P. Kotchi, R. Castanet and J.-C. Mathieu, J. Thermal Anal, in press.

(57) T. Yokokawa and O.J. Kleppa, Inorg. Chem., **3** (1964) 954.

(58) L.B. Ticknor and M.B. Bever, J. Met., **4** (1952) 941.

(59) O.J. Kleppa, J. Phys. Chem., **59** (1955) 354.

(60) O.J. Kleppa, J. Phys. Chem., **60** (1956) 842.

(61) F. Meyer-Liautaud, C.H. Allibert and R. Castanet, J. Less-Comm. Met., **127** (1987) 243.

(62) F. Meyer-Liautaud, S. Derkaoui, C.H. Allibert and R. Castanet, J. Less-Comm. Met., **127** (1987) 231.

(63) J.-C. Mathieu, B. Jounel, P. Desré and E. Bonnier, Proc. IAEA Symp : Thermodyn. Nucl. Mater., Vienna, 1967.

166

(64) M.G. Benz and J.F. Elliott, Trans. Metall. Soc. AIME, 230 (1964) 706.
(65) R.D. Dokken and J.F. Elliott, Trans. Metall. Soc. AIME, 233 (1965)
      1351.
(66) F. Wooley and J.F. Elliott, Trans. Metall. Soc. AIME, 239 (1967)
      1872.
(67) O.J. Kleppa and K.C. Hong, J. Chem. Thermodyn., 10 (1978) 243.
(68) K.C. Hong and O.J. Kleppa, J. Chem. Thermodyn., 10 (1978) 797.
(69) O.J. Kleppa and S. Sato, J. Chem. Thermodyn., 14 (1982) 133.
(70) L. Topor and O.J. Kleppa, J. Chem. Thermodyn., 16 (1984) 993.
(71) L. Topor and O.J. Kleppa, J. Chem. Thermodyn., 17 (1985) 109.
(72) H. Saïd and R. Castanet, J. Less-Comm. Met., 68 (1979) 213.
(73) M. Baier, C. Chatillon-Colinet and J.-C. Mathieu, Ann. Chim., Sci.
      Mater., 6 (1981) 291.
(74) L. Topor and O.J. Kleppa, Met. Trans. A, 17A (1986) 1217.
(75) L. Topor and O.J. Kleppa, Z. Metallkde, 77 (1986) 633.
(76) J. Hertz, Mem. Sci. Rev. Metall., 63 (1966) 781.
(77) W. Vogelbein, M. Ellner and B. Predel, Thermochim. Acta, 44 (1981)
      141.
(78) O. Torgeson and T. Sahama, J. Am. Chem. Soc., 70 (1948) 2156.
(79) O.J. Kleppa, CNRS Intern. Coll., n° 201 Thermochimie, 1972, pp. 119.
(80) F. Müller and O.J. Kleppa, Z. anorg. Allgem. Chem., 397 (1973) 171.
(81) A. Navrotski, J. Inorg. Nucl. Chem., 33 (1971) 4035.
(82) A. Navrotski and O.J. Kleppa, Inorg. Chem., 8 (1969) 756.
(83) T.V. Charlu, R.C. Newton and O.J. Kleppa, Geochim. Cosmochim. Acta,
      39 (1975) 1487.
(84) J. Rogez, R. Chastel, C. Bergman, C. Brousse, R. Castanet and J.-C.
      Mathieu, Bull. Minéral., 106 (1983) 119.
(85) J. Rogez and J.-C. Mathieu, Phys. Chem. Liq., 14 (1985) 259.
(86) A. Navrotsky, R.L. Hervig, B.N. Roy and M. Huffman, High Temp. Sci.,
      19 (1985) 133.
(87) S. Martosudirdjo and J.N. Pratt, Thermochim. Acta, 10 (1974) 23.
(88)   A.W. Bryant and J.N. Pratt, Metallurgical Chemistry Symp., Brunel
      Univ.- NPL, UK, Paper 1.2, Ed. O. Kubaschewski, London HMSO, 1972.
(89) N.F.Schottky and M.B. Bever, Acta Metall., 6 (1958) 320.
(90) C.E. Holley Jr. and E.J. Huber Jr., 'Combustion calorimetry of metals
      and simple metallic compounds', in Experimental Thermodynamics,
      Combustion calorimetry, Ed. by S. Sunner and M. Mansson, Pergamon,
      Oxford, 1 (1979) 211.
(91) T.V. Charlu and O.J. Kleppa, J. Chem. Thermodyn., 5 (1973) 325.
(92) T.V. Charlu, O.J. Kleppa and T.B. Reed, J. Chem. Thermodyn., 6 (1974)
      1065..
(93) N. Kuwata, M. Nishio, H. Himode and M. Wakihara, Thermochim. Acta,
      109 (1986) 181.
(94) P. Gross, C. Hayman and M.C. Stuart, Proc. Brit. Ceram. Soc., 8
      (1967) 39.
(95) W.N. Hubbard, G.K. Johnson and V.Y. Leonidov, 'Combustion in
      fluorine and other halogens', in Experimental Thermodynamics,
      Combustion calorimetry, Ed. by S. Sunner and M. Mansson, IUPAC,
      Pergamon, Oxford, 1 (1979) 255.
(96) R. Sabbah and M. Coten, Thermochim. Acta, 49 (1981) 307.
(97) J.T.S. Andrews, P.A. Norton and E.F. Westrum Jr., J. Chem.
      Thermodyn., 10 (1978)949.

(98) W.M.Cash, E.E. Stansbury, C.F. Moore and C.R. Brooks, Rev. Sci. Instrum., **52** (1981) 895.

(99) J. Rogez and J. Le Coze, Rev. Phys. Appl., **15** (1980) 341.

(100) O.Kubaschewski and J. Grundmann, Ber. Bunsenges. Phys. Chem., **81** (1977) 1239.

(101) B.R. Conard, R. Sridhar and J.S. Warner, J. Chem. Thermodyn., **12** (1980) 817.

(102) E.E. Shpilrain, D.N. Kagan and S.N. Ulyanov, Teplofiz. Vys. Temp., **18** (1980) 1184.

(103) E.E. Shpilrain, D.N. Kagan and S.N. Ulyanov, Teplofiz. Vys. Temp., **19** (1981) 1040.

(104) I.I. Petrova and V.Y. Chekhovskoi, Teplofiz. Vys. Temp., **16** (1978) 1226.

(105) I.I. Petrova and V.Y. Chekhovskoi, Teplofiz. Vys. Temp., **19** (1981) 603.

(106) G.I. Mozharov and A.I. Sovatimskii, Teplofiz. Vys. Temp., **19** (1981) 954.

(107) U. Seydel, H. Bauhof, W. Fucke and H. Wadle, High Temp. High Pressures, **11** (1979) 35.

(108) A. Cezairliyan, M.R. Morse, H.A. Beckman and C.W. Beckett, J. Res. Nat. Bur. Stand., **A74** (1970) 65.

(109) A. Cezairliyan, J.L. McClure and C.W. Beckett, J. Res. Nat. Bur. Stand., **A75** (1971) 1.

(110) K.W.Henry, D.R. Stephens, D.J. Steinberg and E.B. Royce, Rev. Sci. Instrum., **43** (1972) 1777.

(111) G.R. Gathers, J.W. Shanner and D.A. Young, Phys. Rev. Letters, **33** (1974) 70.

(112) L. Denielou, Y. Fournie, J.P. Petitet and C. Tequi, Rev. Int. Hautes Temp. Refract., **8** (1971) 119.

(113) L. Denielou, J.P. Petitet and C. Tequi, J. Chem. Eng. Data, **27** (1982) 129.

(114) L. Denielou, J.P. Petitet and C. Tequi, J. Chem. Eng. Data, **29** (1984) 116.

(115) W. Oelsen, K.H. Rieskamp and O. Oelsen, Arch. Eisenhüttenwes, **26** (1955) 253.

(116) R. Castanet and C. Bergman, Ann. Chim. Sci. Mater., **4** (1979) 419.

(117) R. Castanet and C. Bergman, J. Less-Comm. Met., **68** (1979) 119.

(118) R. Castanet and C. Bergman, Phys. Chem. Liquids, **14** (1985) 219.

(119) B.J. Holm and F. Gronvold, Acta Chem. Scand., **27** (1973) 2043.

(120) R. Blachnik and R. Igel, Z. Naturforsch., **29b** (1974) 625.

(121) L. Charrin, R. Castanet and M. Laffitte, C.R. Acad. Sci., **282C** (1976) 19.

(122) C. Loriers-Susse, High Temp. High Pressures, **12** (1980) 119.

(123) C. Bergman, I. Avramov, C.Y. Zahra and J.-C. Mathieu, J. Non-Cryst. Solids, **70** (1985) 367.

(124) P. Verdier and R. Castanet, J. Thermal Anal., **31** (1986) 865.

(125) P.C. Gravelle, J. Thermal. Anal., **14** (1978) 53.

(126) P.C. Gravelle, Catalysis Rev., **16** (1977) 37.

(127) A.T. Kuhn and A.M. Sham El Din, Surface Tech., **20** (1983) 55.

(128) A.-M. Zahra and M. Laffitte, Analusis, **6** (1978) 133.

(129) M. Regelsberger, R. Wernhardt and M. Rosenberg, J. Phys. E : Sci. Instrum., **19** (1986) 525.

168

(130) G. Hatem, P. Gaune, J.-P. Bros, F. Gehringer and E. Hayer, <u>Rev. Sci. Instrum.</u>, **52** (1984) 585.

(131) J.L. Deneuville, D. Gratias, C. Chatillon-Colinet and J.-C. Mathieu, <u>High Temp. High Pressures</u>, **9** (1977) 445.

(132) D.A. Robins and J. Jenkins, <u>Acta Met.</u>, 3 (1955) 598.

Results of Enthalpy Measurements on Refractory Metals by Levitation Calorimetry

Rongshan Lin and Martin G. Frohberg
Institut fuer Metallurgie - Allgemeine Metallurgie -
Technische Universitaet Berlin
Joachimstaler Strasse 31/32
D-1000 Berlin 15

ABSTRACT. Levitation calorimetry technique was used to measure the enthalpy of solid and liquid refractory metals (Molybdenum, Niobium, Tantalum and Tungsten) in the temperature range 2282 to 4020 K. Particular attention was paid to the accuracy of optical temperature measurement. Convection and radiation heat losses during the fall of the sample were evaluated and considered in the enthalpy. For the solid and liquid state of the metals enthalpy-temperature functions are given. From the experimental data the enthalpies and entropies of fusion were calculated.

1.   Introduction

The main difficulties in the traditional calorimetric measurement arise first from obtaining and maintaining high temperature of the sample above 2000 K and second finding a suited crucible material that does not react with the sample. Levitation calorimetry consisting generally of a levitation heating unit and a conventional drop calorimeter solves these problems. In a high-frequency induction heating coil the sample is levitated and heated contactlessly. Thus contamination of the sample is eliminated. The temperature of the sample can easily and rapidly reach 3000 K and more.

In the past years measurements of enthalpy increment of refractory metals were started in some laboratories by means of the levitation calorimetry[1-3]. The results of the experiments proved great reliability of thermodynamic data particularly in the solid state. In our laboratory a levitation heating apparatus in combination with an isoperibol calorimeter was used to measure the enthalpy of Mo, Nb, Ta and W both in the solid and chiefly in the liquid state [4-6]. Only from these experimental results the exact values of the enthalpy and the entropy of fusion can be calculated.

2.   Experimental Equipment

Fig.1 illustrates the experimental apparatus consisting of a levitation chamber and a block-type isoperibol calorimeter. A high-frquency generator (60 kW, 900/450kHz) is employed to obtain the high temperatures. The high frequency current flowing through an induction coil produces an alternating electromagnetic field. The sample is levitated in this electromagnetic field and heated to the expected temperature. Three sight glas-

169

H. Brodowsky and H.-J. Schaller (eds.), Thermochemistry of Alloys, 169–176.

ses are used for observation and temperature measurement of the levitated sample. A video camera is installed to record the behaviour of the sample in course of levitation and fall. After the temperature of the sample is determined by a disappearing filament optical pyrometer in two wavelengthes (650 nm and 547 nm respectively), the generator is switched off and the sample drops into the calorimeter. Now the calorimetric measurement starts.

The calorimeter consists of a nickel-plated copper block resting on teflon edges. The block has a conical hole with a copper cup inside for receiving the dropped sample. The cup is covered with a foil of same material as the sample to avoid alloying of the sample with the block. A quartz thermometer positioned in the block registrates the temperature increment of the calorimeter. The block is placed in a stainless steel container with a constant outer jacket temperature of 25 ± 0.001 ° C. Before each experiment the whole apparatus inculding the levitation chamber and the calorimeter is evacuated by a turbomolecular pump to a final pressure of 2×10^{-5} - 5×10^{-6} mbar. Then it is filled with pure Argon or Helium (99.9999%). A gate valve separates the calorimeter from the levitating unit to protect radiation.

3. Method of evaluation

The enthalpy of the specimen is calculated from the following equation:

$$H(T) - H(298) = C_k \, \Delta T_k \, \frac{M}{m} + Q_t \, \frac{M}{m} + \left[H(T_f) - H(298) \right] , \quad (1)$$

C_k - calibrated calorimeter constant,
ΔT_k - corrected temperature rise of the calorimeter,
M - molecular mass,
m - sample mass,
Q_t - heat losses during drop of the sample,
T - temperature of the levitated sample,
T_f - temperature of calorimeter at the end of main period,
σ - radiation constant.

The first term is the quantity of the heat measured in the calorimeter. the third term is the enthalpy increment calculated from the temperature at the end of main period to the room temperature. The second term is a correction term which takes into account the heat losses by radiation and convection during the fall of the sample from the levitating coil into the calorimeter. These heat losses are evaluated by the expression:

$$Q_t = A \, \Delta T \int_0^{tf} \alpha \, dt + A \, \varepsilon_h \sigma \, (T^4 - T_a^4) t_f \qquad (2)$$

A is the sample surface and is assumed to be constant in the course of dropping. This assumption has been checked for liquid samples by continuous photographic recording. ΔT is the difference between the sample temperature and the ambient temperature T_a. During the fall through a gas the sample is treated as a stationary sphere in a flowing gas stream. In this case the heat transfer coefficient α can be calculated by means of the equation proposed by Ranz and Marshall [7]. The second term is the Stefan-Bolzmann's formula for radiation. The hemispherical total emissivity ε_h was taken from literature. The amount of the heat losses depends on the size of the sample and the kind of inert gas used. For example, the heat losses of molybdenum vary in the range of 1.5-4.0 kJ/mol (= 1.8-4%) which have to be corrected in the enthalpy determined.

The possible largest source of error in levitation calorimetry is surely the temperature determination of the levitated sample. Thus particular attention is paid to this topic. For each specimen the brightness temperatures are measured optically by the disappearing filament pyrometer in two wavelengthes (650 nm and 547 nm). The true temperature of the sample was calculated by means of the spectral emissivity ε_{650} or ε_{547} of the refractory metals. The deviation of two values of the true temperature was not allowed to be greater than 5 K. The spectral emissivities ε_{650} and ε_{547} were determined by us at the melting points of the refractory metals and assumed to be constant in the liquid range

4. Results and Discusion

All experimental results were treated by the least squares methode and represented by a type of the equation proposed by Hoch[8].

The determined values of solid and liquid molybdenum are plotted in Fig.2. Our results are shown by the dark dots. Here we have very good agreement with former works. In the liquid state we got a linear function. Comparision with the results of Berezin et al[2]. shows differences of about 5%. Those authors did not consider the convective heat losses during the fall of the sample. Values reported by Trerton and Margrave [1] lie about 10% below the ours. This is due to the different experimental conditions. They did not give imformation concerning the heat losses of the sample in their conditions. Besides they did not measure the enthalpy in the solid state. Therefor, a further comparision is not possible.

The following enthalpy-temperature functions are valid for the solid state:

$$H(T)-H(298) = -8159 + 24.943\ T + 1.3101 \times 10^{-3} T^2 + \tag{3}$$

$$+\ 2.0015 \times 10^{-10} T^4 + 1.8009 \times 10^5 T^{-1}\ (\ J\ /\ mol\ )\ ,$$

and for liquid state:

$$H(T) - H(298) = 128112 + 40.353 \left[\ T - T_{mp}\ \right]\qquad (\ J\ /mol\ )\ . \tag{4}$$

The results of the experiment on solid and liquid niobium are shown in Fig.3. There is a similar deviation in the liquid state as mentioned before. But the agreement of the values in the solid state is remarkable[9]. This reduces to the strongly dependance of the heat losses of the sample on the temperature.

The enthalpy increment of the solid state is described by the function:

$$H(T) - H(298) = -7899 + 24.943\ T + 9.4887\ 10^{-4} T^2 +$$

$$+\ 1.7490 \times 10^{-10}\ T^4 + 1.1224 \times 10^5 T^{-1}\qquad (J/mol)\ , \tag{5}$$

and that of the liquid state by:

$$H(T) - H(298) = 107967 + 41.781 \left[T - T_{mp} \right] \qquad (J/mol) \quad . \quad (6)$$

Fig.4 illustrates the experimental results of solid and liquid tantalum. For this metal we were the first to measure the enthalpy in the liquid state. In the solid state there is again a very good conformity to the results of other investigators[10-12].

The dependence of the enthalpy on the Temperature is given for solid state by the following equation:

$$H(T) - H(298) = -7750 + 24.9430T + 7.3111 \times 10^{-4}T^2 +$$
$$+ 9.6954 \times 10^{-11}T^4 + 7.4860 \times 10^4 T^{-1} \quad (J/mol) , \qquad (7)$$

and for liquid state:

$$H(T) - H(298) = 127239 + 44.6348 \left[T - T_{mp} \right] \qquad (J/mol) \quad . \quad (8)$$

The results of solid and liquid tungsten are shown in Fig.5. One can recognize by the good agreement of our results with the previous values in the low temperature range[13-15]. We refer the deviation near the melting point to the fact the authors took into account only 3.5% heat losses during the fall of the sample which exclude only the radiation heat losses. Our own measurements showed for the same falling time heat losses of about 7% .

By curve fitting we get following function to calculate the enthalpy increment of solid tungsten:

$$H(T) - H(298) = -7873 + 24.9430 T + 3.7474 \times 10^{-4}T^2 +$$
$$+ 1.5716 \times 10^{-10}T^4 + 11.9851 \times 10^4 T^{-1} \qquad (9)$$

and of liquid tungsten:

$$H(T) - H(298) = 168364 + 53.7628 \left[T - T_{mp} \right] \qquad (J/mol) \quad . \quad (10)$$

With the enthalpy values of the solid and liquid metals at their melting points we have calculated the enthalpy and the entropy of the fusion. They are summarized in Table 1. A detailed and careful error analysis was performed and shows a systematic error of max. 2% for the solid state. We suggest that this magnitude of error is not far exceeded for the liquid state. At the time we are engaged in measuring the enthalpy of chromium in the solid and liquid state. But there are a lot of difficulties which have to overcome, viz. the preparation of the sample and the evaporation problem.

5. Acknowlegdement

We wish to thank the Deusche Forschungsgemeinschaft for the financial support.

6. References

1) Treverton, J.A. and J.L. Margrave
Proc. of the Fifth Symp. Thermophysical Properties, Newton, Mass. 30. Sept.-
20. Okt. 1970
2) Berezin, B.Y., V.Y. Chekhovskoi and A.E. Sheindlin
High Temp. - High Press. 3 (1971), 287/297
3) Stretz, L.A. and R.G. Bautista
Metall. Trans. 5 (1974), 4, 921/928
4) Frohberg, M.G. and G. Betz
Thermodynamics and Kinetics of Metall. Proce. (ICMS-81)
The Indian Institute of Metals, Calcutta (1985) 81/90
5) Arpaci, E. and M.G. Frohberg
Z. Metallkunde 73 (1982), 9, 548/551
6) Arpaci, E. and M. G. Frohberg
Z. Metallkunde 75 (1984), 8, 614/618
7) Ranz, W.E. and W.R. Marshall
Chem. Eng. Progr. 48 (1952) 173/180
8) Hoch, M.
High Temp. - High Press. 1 (1969) 531/542
9) Sheindlin, A.E., B.Y. Berezin and V.Y. Chekhovskoi
High Temp. - High Press. 4 (1972) 611/619
10) Berezin, B.Y. and V.Y. Chekhovskoi
Izvest. Akad. Nauk. SSSR Met. 3 (1977) 51/57
11) Hoch, M. and Johnston
J. Phys. Chem. &% (1961) 855/860
12) Conway, J.B. and R.A. Hein
High Temp. - High Press. 1 (1969) 535.
13) Martynyuk, M.M. and V.I. Tsapkov
Izvest. Akad. Nauk. SSSR. Met. 6 (1974) 63/67
14) Shpil'rain, E.E., D.N. Kagan and L.S. Barkhatow
High Temp. - High Press. 4 (1972) 605/609
15) Leibowitz, L., M.G. Chasanov and L.W. Mishler
Trans. of the Metall. Soc. of AIME 245 (1969) 981/984

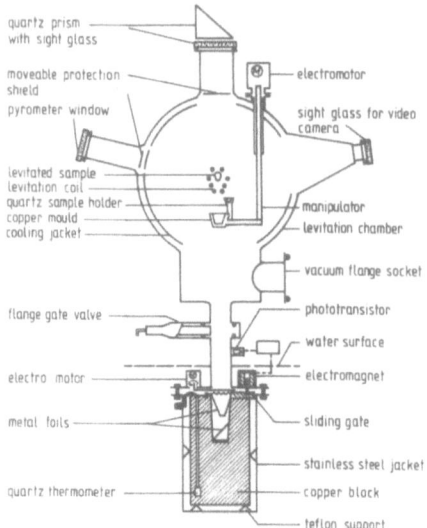

quartz prism
with sight glass

moveable protection
shield

pyrometer window

levitated sample
levitation coil
quartz sample holder
copper mould
cooling jacket

flange gate valve

electro motor

metal foils

quartz thermometer

electromotor

sight glass for video
camera

manipulator

levitation chamber

vacuum flange socket

phototransistor

water surface

electromagnet

sliding gate

stainless steel jacket

copper block

teflon support

Figure 1. Schematic drawing of the levitation calorimeter

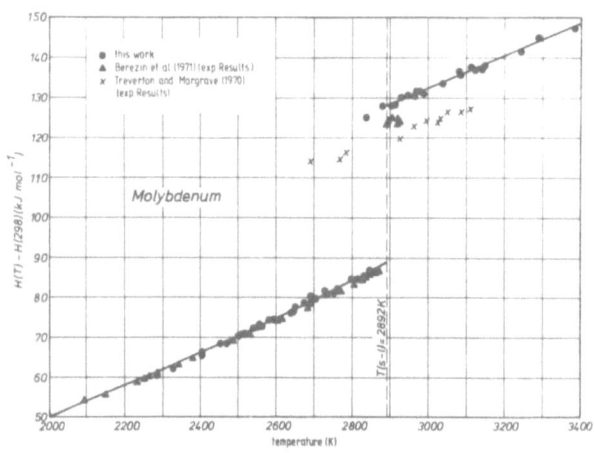

Figure 2. Enthalpy-temperature diagram for molybdenum

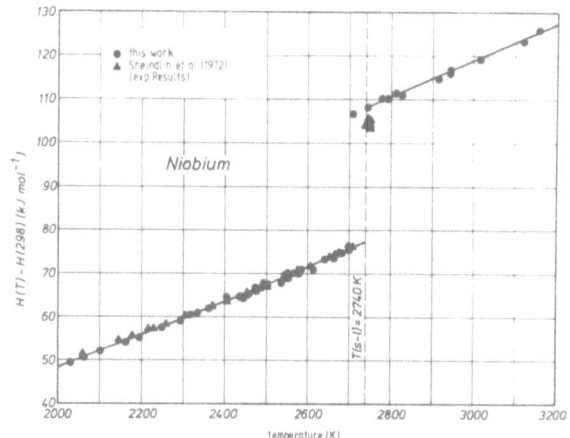

Figure 3. Enthalpy-temperature diagram for niobium

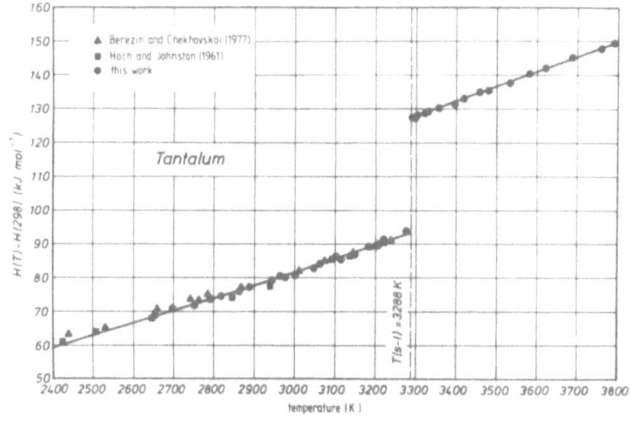

Figure 4 .   Enthalpy-temperature diagram for tantalum

176

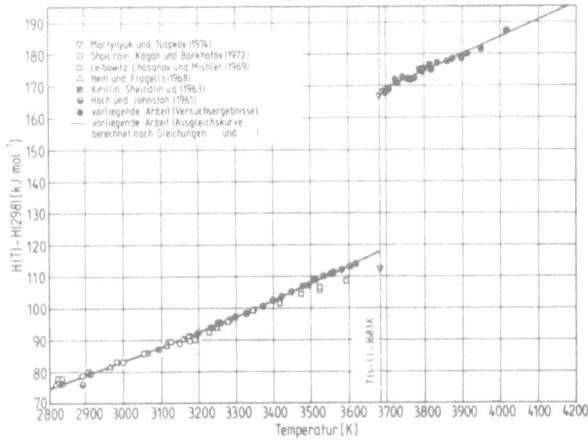

Figure 5. Enthalpy-temperature diagram for tungsten

Nb (2740 K)	Mo (2892 K)	
41,77	40,35	← C_p (J mol^{-1}K^{-1})
30,5	39,1	← ΔH_f (kJ mol^{-1})
11,1	13,5	← ΔS_f (J mol^{-1}K^{-1})
Ta (3288 K)	**W (3683 K)**	
44,63	53,7	
33,7	54,5	
10,25	14,8	

Melting data of refractory metals

Table 1. Melting data of refractory metals

STUDY OF THE Hg-In LIQUID ALLOY BY USING THE PIEZOTHERMAL METHOD (1 bar < P < 3 kbar).

M. Gambino*, P. Rebouillon*, J.P. Bros* et L. Ter Minassian**.

* Thermodynamique des Systèmes Métalliques
Université de Provence , 3 Place V. Hugo,
13331 Marseille cedex 3 , France.

** Laboratoire de Chimie Physique,
Université Pierre et Marie Curie,
11 rue P. et M. Curie, 75231 Paris Cedex 05, France.

ABSTRACT.
        Under isothermal conditions, the determination of the amount of heat induced to flow by a change in pressure allows to determine the expansivity α. Using a Calvet calorimeter and the piezothermal method the expansivity of Hg+In alloys has been measured in the pressure range 1- 3000 bars. Two liquid alloys ($x_{In} = 0.075$ and $x_{In} = 0.143$) have been studied at 262.5 K and 264.4 K, respectively. From these measurements we have deduced the variation of the liquidus temperature against pressure on the one hand and the melting enthalpy of the stoichiometric compound (Hg_6In) on the other hand .

1. INTRODUCTION.

        Temperature and pressure , two intensives properties, play a fundamental role in thermodynamics. Of course direct measurements of pressure and temperature effects on a material have a great practical interest. Because their determinations involve experimental difficulties, the pressure effects on condensed systems are not so well known than the effects of temperature variations. This lack of information is strongly harmful to understand several phenomena in various fields of research (geology, synthesis of new materials,...).
        The main purpose of this contribution is to show that the variation of some thermodynamic properties of condensed phases can be studied in a large pressure range by using the piezothermal method. As an example the Hg+In system has been chosen ; the calorimetric measurements were performed near 265 K between 1 bar and 3000 bars. The variations of expansivity and melting temperature have been determined. In the same way the melting enthalpy of the stoichiometric compound (Hg_6In) has been obtained and this result has been compared to the corresponding value measured under atmospheric pressure with a scanning calorimeter.

H. Brodowsky and H.-J. Schaller (eds.), Thermochemistry of Alloys, 177–188.
© 1989 by Kluwer Academic Publishers.

2. EXPERIMENTAL.

2.1. High pressure measurements.

a) Method.
The piezothermal method, developed recently (1-7) is based on the Maxwell equations :

$$(\partial S/\partial P)_T = - (\partial V/\partial T)_P = - \alpha V \quad \text{and} \quad (\partial S/\partial V)_T = (\partial P/\partial T)_P = \alpha/\beta$$

in which $\quad \alpha = (1/V)(\partial V/\partial T)_P$ (1) $\quad$ and $\quad \beta = - (1/V)(\partial V/\partial P)_T$ (2)

are the expansivity at constant pressure and the isothermal compressibility, respectively .
Under isothermal and reversible conditions , equation (1) takes the form :

$$dQ = - \alpha V T \, dP$$

Therefore the thermal effect ΔQ arising from a pressure variation ΔP generated in the calorimetric cell containing the liquid sample under investigation is related to the thermal expansivity of the cell and of the liquid phase by the following equation :

$$\alpha_s - \alpha_c = - (1/V_c T).(\Delta Q/\Delta P)$$

where V_c is the constant internal volume of the high pressure cell and α_s and α_c are the expansivity of the sample and of the cell, respectively .

b) Apparatus.
The main parts of this set-up are the calorimeter, the high pressure cell, and the pressure generator.
- *the calorimeter* : The Calvet calorimeter used in this work has a single thermopile, the cell of which (18 mm internal diameter and 80 mm height) is surrounded by 400 chromel-alumel thermocouples. The electromotive force of this thermopile is recorded against time (E = f(t)) and the area between this curve and the base line is proportional to the heat effect developed in the cell. The thermostat consists of a large metallic block the temperature of which is maintained constant by an electronic regulator. This apparatus works up to 500 K.
- *the high pressure cell* : The cell (Fig. 1) is made of a standard (4 kbars) high pressure tube (stainless steel Z 6 CN 18-09); its internal and external diameter are 3.2 and 9.52 mm., respectively. The lower part of the high pressure tube is closed by a special plug. Thermal coupling between the high pressure tube and the calorimetric cell ((2)

in Fig. 1) is obtained by means of a sliding copper cylinder (80 mm height). Two copper shunts ((1) and (3) in Fig. 1) disposed on both sides of the calorimetric cell allow two thermal couplings between the high pressure tube and the calorimetric block((A) in Fig. 1) maintained at the constant experimental temperature. With this set-up and assuming that heat is liberated uniformely inside the high pressure tube between the two thermal shunts , the volume V_C is well defined. This value as well as the thermal relaxation time between the high pressure cell and the thermostat depends slightly on the mutual arrangement of the copper junctions. The estimation of the heat liberated between the thermal junctions and propagated along the thermocouples is the main problem. Assuming the establishment of a quasi-stationary state, the volume V_C can be calculated : $V_C = 0.66$ cm^3.

- *the high pressure generator* : The cell is connected to the hydraulic pressure generator. It is a screw-captan type from Nova-Swiss. Pressure measurements are performed by means of a strain gauge from Sedene Co. The hydraulic fluid used in this present work was heptane.

c) Calibration and accuracy.

Pressure measurements are performed with an error of about ± 0.5 % until 3 kbar. The experimental temperature is determined with an estimated accuracy of ± 0.5°C from the thermoelectric power of a chromel-alumel thermocouple located in the calorimetric block.

The calibration of this apparatus , based on the isothermal compression of a perfect gas, allows to check the estimation of V_C . The expansivity of a perfect gas is defined to be $\alpha = 1/T$; so the heat liberated in a volume V_C when compressed is given by the following relation :

$$Q = -V_C.\Delta P$$

where α_c has been neglected ($\alpha_c << \alpha$). Previously, the thermal fluxmeter of the calorimeter was calibrated by Joule effect.

2.2. Atmospheric pressure measurements.

a) Method.

To obtain both temperature and enthalpy of phase transition , the conventional method - the differential enthalpic analysis - has been employed. Simultaneously , with the same sample the molar heat capacity against temperature ($Cp = f(T)$) has been measured.

b) Apparatus.

The scanning calorimetric method has been described by several authors. The apparatus used in this work (DSC 111 purchased from Setaram Co.) is designed as a Calvet calorimeter : two cylindrical cells - laboratory and reference cells - surrounded by a thermal fluxmeter are located into a metallic block the temperature of which can be linearly programmed. These cells (6 mm. in diameter and 12 mm. in length) in stainless steel are made gastight by using a special cover and a metallic O-ring. The two fluxmeters are connected in opposition . Such an apparatus is suitable for measurements between - 100°C and 750°C.

c) Calibration and accuracy.

By differential enthalpic analysis (D.E.A.), enthalpies and temperatures of melting have been measured. This calorimeter was calibrated by Joule effect ; during D.E.A. runs, thermal effects generated by the fusion of pure metals (Hg, Ga, In, Sn,...) allow to check this calibration. To control Cp determination, the molar heat content of pure alumina, provided by N.B.S., has been determined in the same temperature range (223 K<T<303 K) . The heating rate is about 2 K.mn^{-1}. Temperatures of phase transitions were obtained with an error of about ± 0.3 K . For the enthalpies of fusion the accuracy is about 2%. The molar heat content of these alloys was carried out with an accuracy less than 2%.

3. THERMODYNAMIC PROPERTIES of THE Hg+In SYSTEM.

To perform the first experiments under pressure the Hg+In system was chosen for the following reasons :
- the equilibrium phase diagram exhibits several congruently melting temperatures compounds, the melting temperatures of which are between - 40°C and 0°C. In the temperature range (-80°C, + 100°C) , high pressure experiments on these alloys do not need special modifications of the experimental cell,
- the excess functions of formation of this binary system have been determined by several authors,
- the excess volume of formation versus composition is well known ,
- the introduction of these liquid alloys in the experimental cell is quite easy.

Equilibrium phase diagram.

Many results concerning this phase diagram have been published but all are not in good agreement. Compiled by Shunk (8), this phase diagram has been considerably modified by Hultgren (9) to take into account the last published data. The main features of the Hg+In phase diagram (fig.2) are :
- a peritectic point (Tp = 382 K) corresponding to the $HgIn_{11}$ phase,
- two congruently melting intermediate phases (HgIn and Hg_6In),
- two incongruently melting intermediate phases ($HgIn_2$(hypothetic) and Hg_4In),
- two eutectic mixtures (-37.2 °C at x_{In} = 0.347 and -31.0°C at x_{In} = 0.615)

Excess functions of formation.

The molar enthalpies of formation of the liquid Hg+In alloys carried out by Kleppa (10) and Bros (11) are in good agreement. The $\Delta_{mix}H_m = f(x_{In})$ is negative and exhibits a slight asymmetry ; the coordinates of the extremum point are : - 2.25 kJ.mol^{-1} at $x_{In} = 0.50$. No temperature variation was found indicating $\Delta C_p = 0$.

From potentiometric and vapor pressure measurements the molar partial Gibbs energies have been obtained (9).
At 298 K, $\Delta_{mix}G_m = - 3.81$ kJ.mol^{-1} at $x_{In} = 0.50$ (Fig.3).

The molar volume of several Hg+In mixtures has been measured : $\Delta V_m = f(x_{In})$ is negative in the whole concentration range (12).

Up to now no critical assessment of these data has been performed and so, no set of thermodynamic data self consistent with the equilibrium phase diagram has been proposed.

4. RESULTS.

4.1. Influence of pressure on the liquidus temperature.

All the liquid alloys were prepared in a glove box under pure argon. Mercury and indium were of high purity (99.9999 mole percent Hg and 99.99 mole percent In). The experimental molar fraction is known with an accuracy of about ± 0.001.
The liquid alloys were poured into the high pressure cell ; care was taken to keep the level above their of the first thermal shunt. The remaining empty volume was filled with heptane chosen as hydraulic fluid.
Under isothermal condition, the pressure was increased , step by step , from 1 bar to 3000 bars then decreased in the same way. Far from the melting region, 200 bars steps (ΔP) were used and the equilibrium was reached in 20 minutes. Near the melting point ΔP was reduced (ΔP = 50). Several runs were performed with the following alloys :
a) with $x_{In} = 0.075$: at 262.5 K two series of experiments were performed and all the results are gathered in the figure 4 and table 1. The crystallization takes place at 2.42 kbars (first run) and 2.44 kbars (second run). The important thermal effect corresponding to the solidification of the Hg$_6$In phase has been measured. Under atmospheric pressure , at $x_{In} = 0.075$, the temperature of the liquidus is : 255 K. So, the variation of the melting point of this alloy is about 3.09 K/kbar.
b) with $x_{In} = 0.143$: using the same route, the expansivity of this alloy was determined at 264.4 K ; the results are gathered in table 1 and figure 5. The crystallization, observed during three experimental runs, begins at 1.820, 1.830, and 1.825 kbars, respectively. Under atmospheric pressure, 258.8 K is the temperature of fusion of this alloy. Consequently, the variation of the melting temperature against pressure is : 3.08 K/kbar. These two results are in good agreement and must be compared with the variation of the temperature of fusion of mercury : 5K/kbar proposed by Bridgman .

4.2. Enthalpy of fusion.

From the heat effects recorded during the crystallization , the molar melting enthalpy of the compound (Hg_6In) was deduced :

$$\Delta H_{(fus,Hg6In)} = 2.7 \pm 0.2 \ kJ.mol^{-1} \ \ at \ 1.8 \ kbar.$$

Using a differential scanning calorimeter the variation of the molar heat capacity of of Hg_6In and $HgIn_{(0.075)}$ have been measured between 223 K and 303 K. From these experimental results, reported in table 2 and figures 6 and 7, the melting enthalpy of the stoichiometric compound has been obtained under atmospheric pressure :

$$\Delta H_{(fus,Hg6In)} = 2.62 \pm 0.2 \ kJ.mol^{-1} \ \ at \ 1 \ bar.$$

These two results are quite similar and the pressure effect on the melting enthalpy of this stoichiometric compound is small.

CONCLUSION.

These results show that the expansivity, the temperature and the enthalpy of solidification of molten alloys can be obtained with accuracy by using the piezothermal method in a large pressure range. The next experiments performed with other Hg+In liquid alloys will allow to determine the pressure effect on the equilibrium phase diagram and on the thermodynamic properties.

BIBLIOGRAPHY.

1/ Ter-Minassian L., Petit J.C., Nguyen Van Kiet , Brunard L.
     J. Chim. Phys. <u>67</u>, 265,(1970)

2/ Petit J.C. , Ter-Minassian L. ,
     J. Chem. Thermod. <u>6</u>, 1139, (1974).

3/ Ter-Minassian L. , Pruzan Ph.
     J. Chem. Thermod. <u>9</u>, 375, (1977).

4/ Fuchs A.H., Pruzan Ph., Ter-Minassian L.,
     J. Phys. Chem. Solids <u>40</u>, 369, (1979).

5/ Ter-Minassian L. , Pruzan Ph.
     J. Chem. Phys. <u>75</u>, 3064,(1981).

6/ Alba C., Ter-Minassian L., Denis A., Soulard A.
     J. Chem. Phys. <u>75</u>, 3064,(1981).

7/ Randzio S.L., Ter-Minassian L.
     Thermochimica Acta <u>113</u>, 67, (1987).

8/ Shunk F.A.
     Constitution of Binary Alloys, Second Suppl.,
     Mc Graw Hill Ed. (1969).

9/ Hultgren R., Desai P.D., Hawkins D.T., Gleiser M., Kelley K.K.
     Selected Values of the Thermodynamic Properties of Binary
     Alloys. Amer. Soc. for Metals, Metals Park, Ohio, USA (1973).

10/ Kleppa O.J.
     Acta Met. <u>8</u>,435,(1960).

11/ Bros J.P. , Lefèvre M.,
     Bull. Soc Chim. Fr. , <u>8</u>, 2582,(1966).

12/ Kleppa O.J., Kaplan M.
     J. Phys. Chem., <u>61</u>, 1120, &1957).

184

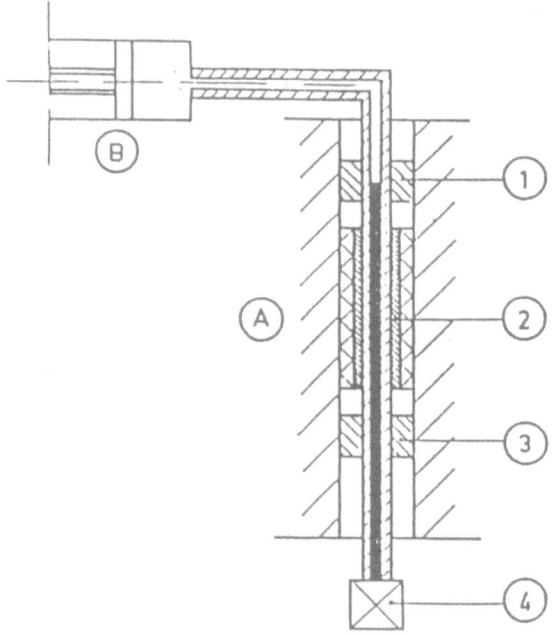

Fig. 1 : Experimental arrangement of the high pressure cell
(A : calorimetric block, B : high pressure generator, 1 and 3 : copper
shunts, 2 : calorimetric cell, 4 : Bridgman tap).

262.5 K	Xln = 0.075	T=264.4 K	Xln=0.143
P/bar	alpha	P/bar	alpha
175	1.607	106	1.514
215	1.588	134	1.537
514	1.543	200	1.518
630	1.541	337	1.495
844	1.525	434	1.525
982	1.525	590	1.472
1168	1.517	601	1.477
1312	1.524	846	1.490
1480	1.505	848	1.469
1621	1.495	1003	1.444
1792	1.497	1092	1.458
2015	1.497	1288	1.464
2066	1.478	1363	1.424
2215	1.474	1502	1.415
2299	1.437	1606	1.440
		1777	1.385
		1779	1.406

Table 1 : Experimental values of α (in 10 E-4/K-1) versus P.

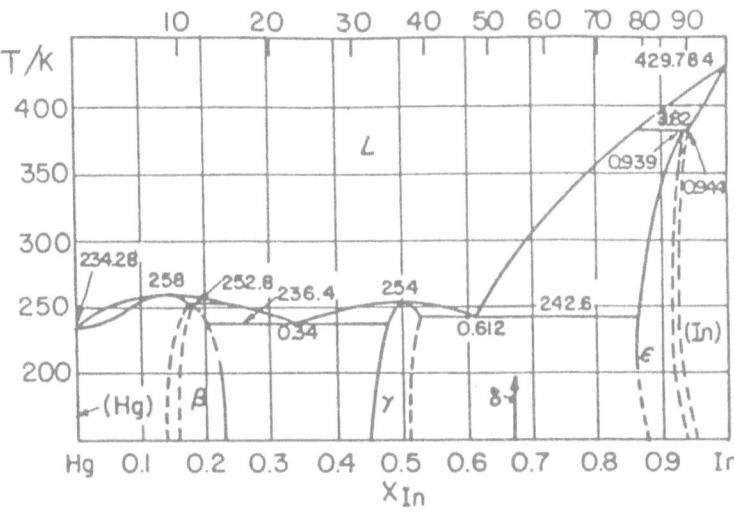

Fig. 2 : The Hg-In equilibrium phase diagram from Hultgren et al. (ref.9).

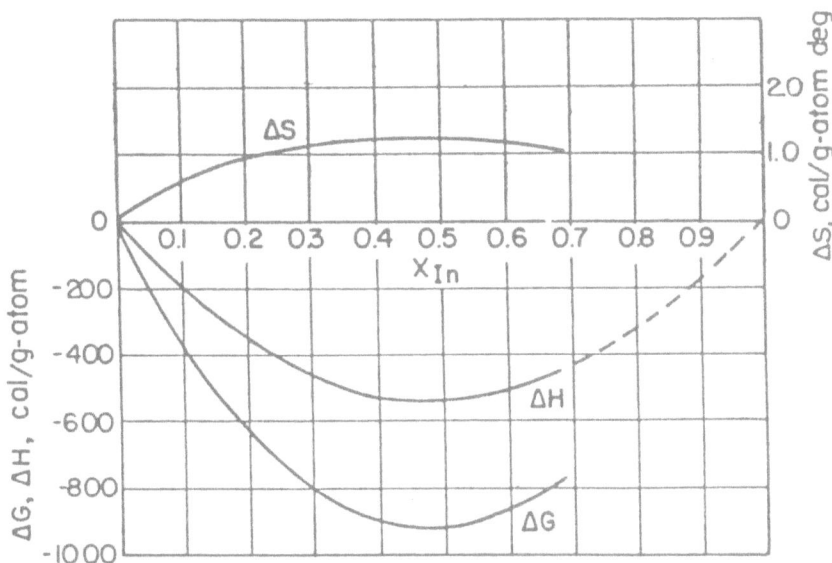

Fig. 3 : The thermodynamic properties (entropy, enthalpy and free energy) of the Hg-In system (ref. 9).

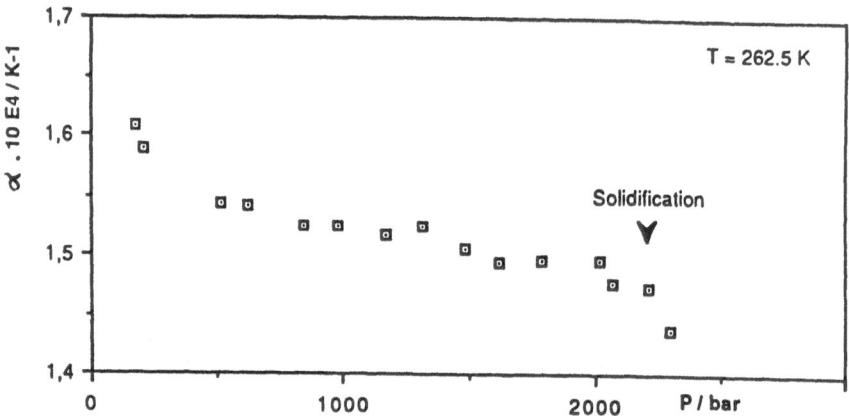

Fig.4 : Variation of the expansivity (a) of the Hg $_{(0,925)}$ In$_{(0,075)}$ alloy versus pressure at 262.5 K.

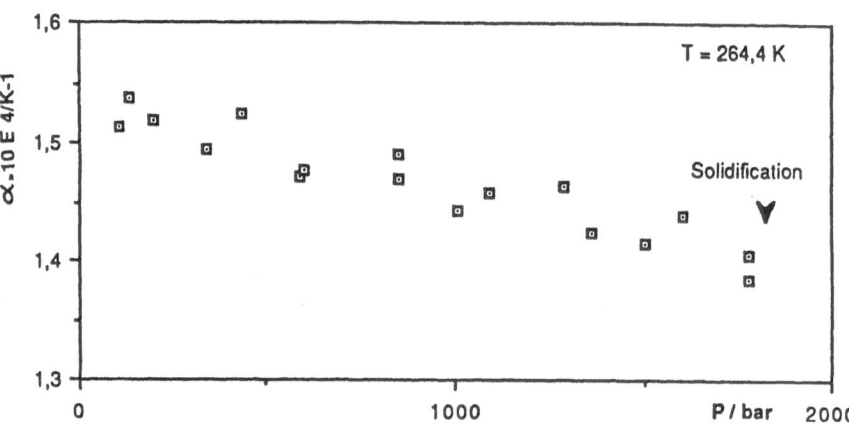

Fig.5 : Variation of the expansivity (a) of the Hg $_{(0,857)}$ In$_{(0,143)}$ alloy versus pressure at 264.5 K.

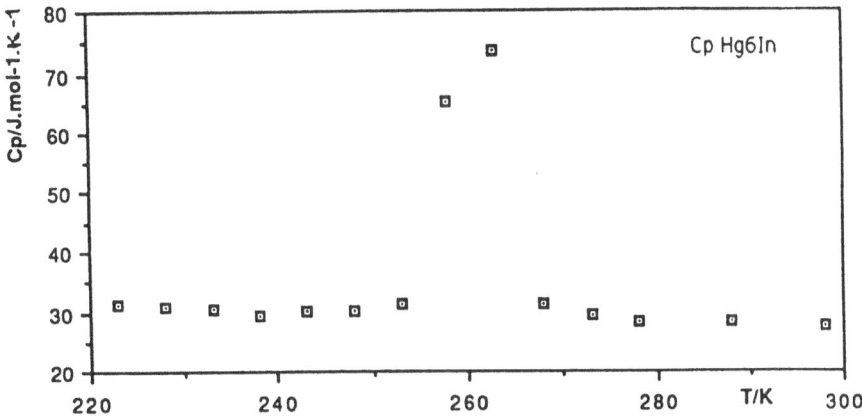

Fig.6 : Variation of the molar heat capacity of the Hg $_{(0,857)}$ In$_{(0,143)}$ alloy versus temperature at 1 bar.

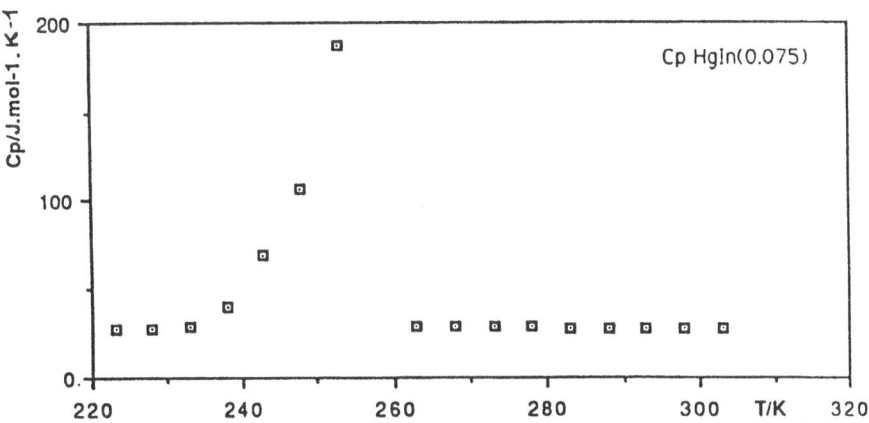

Fig.7 : Variation of the molar heat capacity of the Hg $_{(0,925)}$ In$_{(0,075)}$ alloy versus temperature at 1 bar

T/K	HgIn(0.075) Cp(J/mol/K)	Hg6In Cp(J/mol/K)
223	27.2	31.4
228	28.0	30.8
233	28.7	30.6
238	40.7	29.6
243	68.9	30.3
248	107.4	30.3
253	186.8	31.2
258	.....	65.1
263	29.5	73.5
268	28.4	31.4
273	28.6	29.3
278	28.4	28.3
283	28.3	.....
288	28.2	28.3
293	28.0	.....
298	27.8	27.6
303	27.6	.....

Table 2 : Experimental values of the molar heat capacity of the Hg_6In and $Hg_{(0.925)} In_{(0.075)}$ alloys versus temperature at 1 bar.

EXPERIMENTAL DETERMINATION OF THE ENTHALPIES OF FORMATION OF BINARY
TRANSITION - RARE EARTH METAL ALLOYS

C. Colinet and A. Pasturel
Laboratoire de Thermodynamique
et Physico-Chimie Métallurgiques,
ENSEEG, BP 75
38402 Saint Martin d'Hères Cédex
France

ABSTRACT. A review is presented of enthalpies of formation obtained
using aluminium solution calorimetry for solid alloys of transition
metals with rare earth metals. The results are compared with literature
data. Even large discrepancies between the different values are
observed, all the results allow to discern the trends of the enthalpies
of formation along the rare earth series and along the transition metal
series, say the end of the first row and the Ni column. The results are
compared with predictions derived from semi empirical model of Miedema
and coworkers and from band theory models.

1. INTRODUCTION

During the last decade there has been greatly increased interest in the
thermodynamic of alloys, in particular with respect to alloys formed by
and among the transition metals. Miedema and coworkers (1,2) developed
an extremely simple scheme for predicting the enthalpies of formation
of metallic alloys. Several workers applied a tight-binding approach to
calculate the enthalpies of formation of transition metal alloys (for
more details, see Colinet et al. in this volume). Often the
calculations of the enthalpies of formation were limited to typical
cases. Only Watson and Bennett (3) and Colinet and al. (4) presented
predicted values of enthalpies of formation for a large number of
binary transition metal or rare earth metal alloys, the calculations of
Watson and Bennett (3) being restricted to the equiatomic composition.
    Often the lack of detailed thermodynamic information in transition
metal based alloys made difficult realistic comparisons between the
theoretical calculations and experimental values and it has happened
more than once that theorists have compared their results not with
experimental data but with the semi empirical estimates of Miedema et
al (1).
    Therefore the alloys formed by early transition metals with late
transition metals have received a particular attention and a large
number of experimental informations about the enthalpies of formation
in these systems have been recently published by Gachon et al (5),

H. Brodowsky and H.-J. Schaller (eds.), Thermochemistry of Alloys, 189–202.

Kleppa and coworkers (6-9), Ansara et al. (10), Henaff et al. (11). Gachon et al. (5) used a direct high temperature reaction calorimetric method while Topor and Kleppa (19) focused on a very original and powerful high temperature mixing calorimetric method.

The same holds for alloys between rare earth elements and late transition metals which up to some years have rarely been the object of systematic studies dealing with their thermodynamic properties. Let us recall the earlier systematic investigations performed in these systems : Deodhar and Ficalora (12) have deduced the enthalpies of formation of the (Gd, Dy, Er) $(Fe,Co,Ni)_2$ compounds from differential thermal analysis of the formation reaction of the compounds ; Palenzona et Cirafici (13) used a dynamic differential calorimetric method to obtain the enthalpy of formation of various equiatomic compounds in the LnPd and LnPt series.

From a couple of years several authors have independently performed experimental investigations of thermodynamic properties in rare earth-transition metal binary alloys. Let us mention their works : Watanabe and Kleppa (14) determined the enthalpies of mixing and the enthalpies of formation of compounds in the La-Ni and (Sc,Y,La,Lu)-Cu systems using a mixing calorimetric method. Subramanian and Smith used solid electrolyte electromotive force cells to determine the thermodynamic data of formation of the compounds in the Y-(Fe,Co,Ni) systems (15), Sommer and coworkers determined the enthalpies of formation of intermetallic compounds of cobalt, nickel and copper with dysprosium, erbium and gadolinium and of copper with lanthanum (16,17) by tin solution calorimetry. F. Meyer-Liautaud et al. used also tin solution calorimetry to determine the enthalpies of formation of compounds in the Sm-Co systems and in the $R(Co_{1-x}Cu_x)_5$ series (R = Sm, Y and Ce). (18).

Since some years, the authors of the present communication have determined the enthalpies of formation of compounds in the series $LnNi_5$(Ln = La,Ce,Pr,Sm,Gd,Y) (19-21) and in substituted series of the same $CaCu_5$ structure (22-26). More recently the enthalpies of formation of intermetallic compounds of gadolinium with iron, cobalt, nickel, palladium and platinum have been investigated (27).

In the present paper we shall review some of the results of our calorimetric studies. The experimental results will be compared with recent literature data in the purpose to discern trends in the evolution of the enthalpies of formation along the rare earth series as well as along the transition metal series. Finally our experimental results and the literature data shall be compared with Miedema's estimates and band theory model predictions.

2. EXPERIMENTAL

A dissolution calorimetry method has been used to determine the enthalpies of formation of the rare earth-transition metal compounds. The enthalpy of formation of a compound is deduced from the heat of dissolution of this compound and its components in a solvent bath. The major advantage of this method is due to the fact that the compound is prepared and analyzed before the calorimetric measurements. If the compound becomes easily oxydized, the samples can be prepared and weighed in a gloves box and handled under inert atmosphere.

The choice of the solvent bath is crucial, as a matter of fact it must allow a good dissolution of the compound and of the constituents elements. From a point of view of uncertainties a good choice for the solvent bath is a parent element of the studied compound, but this cannot often been realized due to the high temperatures of fusion of the transition and rare earth elements and to their high reactivity.

HCl(aq) has been used by Semenenk (28) and by Hubbard et al. (29) to determine the enthalpy of formation of the $LaNi_5$ compound. This solvent presents the advantage that the dissolution experiments are performed at room temperature avoiding the problems connected with reactive elements at high temperature. Metallic solvent bath such as liquid aluminium or tin are now well used, because they generally allow good dissolution of all the rare earth metals as well as several transition metals. Sommer and coworkers used a tin solvent bath at T = 1100 K (16,17), F. Meyer-Liautaud et al. also used a tin solvent bath but at T = 1173 K (18). In our studies we have used an aluminium solvent bath at a temperature about 1000 K except for the dissolutions of platinum and platinum based alloys for which a temperature of 1090 K was required to obtain a rapid dissolution of the samples.

2.1. Experimental procedure

The enthalpy of formation of the $Ln_m Mt_n$ compound at the temperature T_o, $\Delta fH(T_o)$ corresponds to the enthalpy of the reaction

$$m \ Ln \ (cr, T_o) + n \ Mt(cr, T_o) \rightarrow Ln_m Mt_n(cr, T_o)$$

Its value is obtained indirectly by measuring the enthalpies of dissolution of the pure components Ln and Mt and of the intermetallic compound in a same metallic bath, molten aluminium at a temperature T, in the present case, according to the following reactions :

$$m \ Ln(cr, T_o) + N \ \{ \ mx \ Ln + (1-mx) \ Al \} \ (1, T)$$
$$\rightarrow \quad (N+m) \ \ m(x+ \ x) \ Ln + \{1-m(x+ \ x)\} \ Al \quad (1, T)$$

$$n \ Mt(cr, T_o) + N \ \ nx \ Mt + (1-nx)Al \quad (1, T)$$
$$\rightarrow \quad (N+n) \ \ n(x+ \ x)Mt + \{ \ 1-n(x+ \ x)\} \ Al \quad (1, T)$$

$$Ln_m Mt_n(cr, T_o) + N \ \ mx \ Ln + nx \ Mt + (1-mx-nx) \ Al \quad (1, T)$$
$$\rightarrow \quad (N+m+n) \ \ m(x+ \ x) \ Ln+n(x+ \ x) \ Mt + \{ \ 1-(m+n)(x+ \ x)\} \ Al \quad (1, T)$$

At infinite dilution of Ln and Mt in liquid aluminium one may write :

$$\Delta fH(Ln_m Mt_n, cr, T_o) = m \ Q^\infty \ (Ln) + n \ Q^\infty \ (Mt) - Q^\infty \ (Ln_m Mt_n)$$

where $Q^\infty \ (Ln)$, $Q^\infty \ (Mt)$, $Q^\infty \ (Ln_m Mt_n)$ are the values at infinite dilution of the thermal effects determined for the successive dissolution reactions. If no dependence of Q on composition x is observed, one may regard the Q^∞ value as a mean value. Otherwise Q^∞ is obtained by a linear regression.

$$Q = Q^\infty + a(x+ \delta x/2)$$

The values of $Q^\infty(Ln)$ and $Q^\infty(Mt)$ must be determined for a same temperature T of the solvent bath as for the $Ln_m Mt_n$ dissolutions. In order to correct for possible differences of temperature of the elements in the solution experiments we have assumed that the partial molar enthalpies of Ln or Mt in liquid aluminium do not vary with temperature within an interval of 100 K and used the enthalpy results compiled by Hultgren et al (30) for the elements.

The formulation (I) of the enthalpy of formation of the compound shows clearly the major disadvantage of the solution calorimetry method. The enthalpy of formation is deduced from three independent measurements. If the solvent bath is a parent element of the compound this number of measurements is reduced to two. If the absolute value of the enthalpy of formation is small the relative uncertainty can be rather large even if the uncertainties connected to the individual measurements are small.

2.2. Experimental details

An isoperibol calorimeter previously described (31) has been used for the majority of the experiments needed for the results proposed in this communication. The calorimetric cell is made in graphite in which is enclosed an alumina crucible containing the solvent bath. The dissolution experiments are performed under vacuum ($P=10^{-4}$ Pa). Stainless steel foils which are around the calorimetric cell prevent oxydation of the bath.

Due to the fact that the highest temperature reached in the isoperibol calorimeter is around 1000 K, a Calvet type twin calorimeter (1200 K) was used to perform the dissolutions of Pt, Pt based compounds and for some Fe and Co based compounds. The Calvet type twin calorimeter has been described previously (32,33). The cells of the calorimeter consist of quartz tubes in which are inserted alumina crucibles. The solution experiments are carried out in a flowing atmosphere of argon. To prevent oxydation of the bath, stainless steel or titanium tubes are maintained inside the quartz liner just above the crucible.

The experimental procedure is the same for the two calorimeters and has been previously described (34,35). The samples are dropped into the bath from a thermostatted block at a temperature T° about 300 K. The calibration runs are carried out by successive additions of the solvent element (Al 99.999%). The individual additions of the solute elements consist of small pieces of about 50mg (isoperibol calorimeter) and 20mg (Calvet microcalorimeter). The rare earth elements and the compounds which become easily oxydized are prepared and weighed in a gloves box and handled under inert atmosphere before the calorimetric measurements.

3. EXPERIMENTAL RESULTS AND COMPARISON WITH LITERATURE DATA

3.1. Enthalpies of formation of gadolinium based alloys

The enthalpies of formation of the compounds of gadolinium with Fe,Co,Ni,Pd and Pt and of the compounds of yttrium with Ni have been determined (27). In figure 1 our experimental data are plotted as function of the transition metal composition. In the two-phase regions the enthalpy of formation is indicated by means of straight lines connecting the data points. The uncertainties connected with all these results are very different. In the case of Pd and Pt based alloys the relative uncertainty is less than 3 and 5% respectively. For the Gd-Ni system the uncertainty can reach 10% in the more unfavorable case. For the Gd-Fe and Gd-Co systems the maximum of the relative uncertainty is rather large respectively 39% and 26% (As mentionned above the enthalpy of formation is relatively small compared with the measured heat effects). Thermodynamic data of some compounds considered here have been reported in literature. Deodhar and Ficalora (12) have deduced the enthalpies of formation of Gd $(Fe,Co,Ni)_2$ compounds these values are less negative than the values found by us. Schott and Sommer (16) have determined the enthalpies of formation of some compounds in the Gd-Ni and Gd-Co system. The values are slightly less negative than our obtained values, but as can be shown in figure 1, the shape of the enthalpy of formation versus composition presents the same asymmetry as found by us, the maximum of the absolute value of the enthalpy of formation being obtained for the composition 0.66 of the transition metal. Subramanian and Smith (15) determined the Gibbs energy of formation of various Y-Ni compounds. The enthalpy of formation was deduced from the temperature variation of the Gibbs energy of formation. The values of ΔfH proposed by Subramanian and Smith are reported in Figure 1, they are of the same order of magnitude as ours, but the composition dependence has a different shape. Paasch and Schaller (37) used electromotive force measurements to obtain thermodynamic properties of Gd-Pd alloys of various compositions. The enthalpy of formation given by these authors for the compound $GdPd_3$ is in rather good agreement with the value found by us. Palenzona and Cirafici (13) obtained the enthalpy of formation of various compounds LnPd and LnPt. For GdPd they report a value which deviates considerably from ours, for GdPt their value is very close to the value found by us for the compound Gd_3Pt_4 which has a slightly higher Pt composition (see Figure 1).

Inspection of Figure 1 shows that the enthalpies of formation become more exothermic going from Fe, Co to Ni and also going from Ni, Pd to Pt. The asymmetric shape of the enthalpy of formation versus composition is less pronounced for Pt based alloys. The behaviour of Y with Ni appears to be very similar than the one of Gd with Ni.

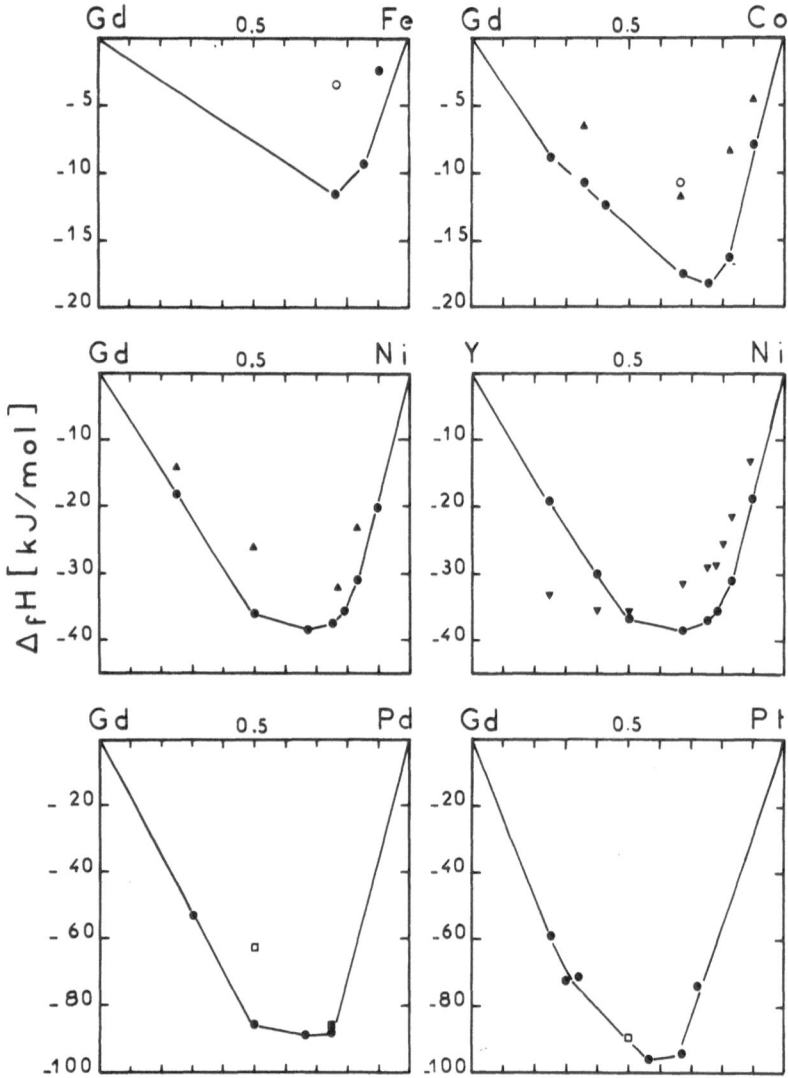

Figure 1 : Enthalpies of formation of intermetallic compounds in the Gd-Fe, Gd-Co, Gd-Ni, Y-Ni, Gd-Pd and Gd-Pt systems. Experimental data are from Ref. 26 (●) ; Schott and Sommer, Ref. 16 (▲) ; Subramanian and Smith, Ref. 15 (▽) ; Palenzona and Cirafici, Ref. 13 (□), Deodhar and Ficalora, Ref. 12 (○) and Paasch and Schaller, Ref. 37 (■)

3.2. Enthalpies of formation in the LnNi₅ series

Some years ago we have reported on calorimetric data for the enthalpies of formation of the (La,Ce,Sm,Gd)Ni₅ compounds (20-21). These values are plotted as function of the lanthanide atomic number in Figure 2. Let us recall that the enthalpy of formation of the LaNi₅ compound has been the object of several determinations, the interest to this compound being due to the fact that LaNi₅ compound is able to absorb reversibly large amounts of hydrogen using convenient conditions. The first determinations (19,28,36) led to values in the range -19.9 to -23.6 kJ/at.g as the more recent ones obtained by Watanabe and Kleppa (14), Pasturel et al. (20), Hubbard et al (29) and Colinet et al. (26) are more exothermic and in very good agreement together (see Figure 2). Let us point out that the two last values have been determined by different methods, respectively acid solution calorimetry and alumininium solution calorimetry using the same LaNi₅ sample. For the GdNi₅ compound there is another determination performed by Schott and Sommer (15). As pointed out above the value quoted by these authors is less exothermic than the value we have obtained. These authors have also determined the enthalpies of formation of DyNi₅ and ErNi₅ compounds by tin solution calorimetry.

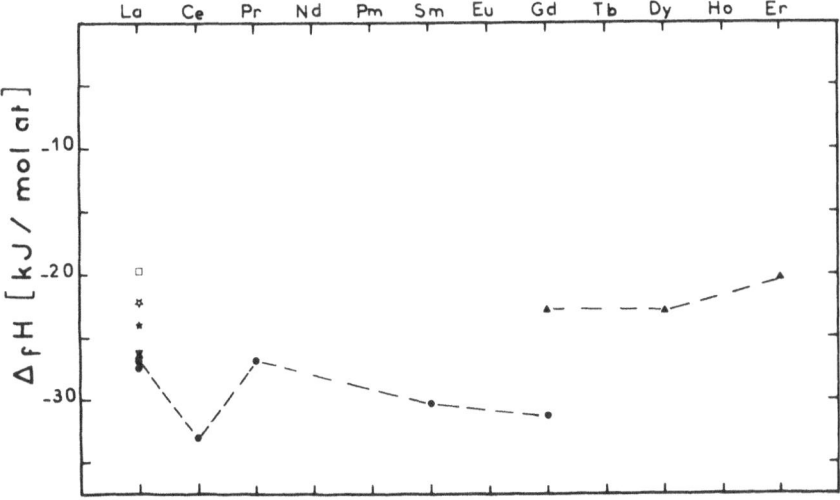

Figure 2 : Enthalpies of formation of LnNi₅ intermetallic compounds. Experimental data are from Ref. 21 and 25 (●) ; Schott and Sommer, Ref. 16 (▲) ; Watanabe and Kleppa, Ref. 14 (▼) ; Hubbard et al, Ref. 29 (△) ; Semenenko et al, Ref. 28 (□) ; Rezukhina and Kutsev, Ref. 36 (✬) and Chatillon et al, Ref. 19 (✸).

Considering all these values reported in Figure 2, one observe a slight decrease of the enthalpy of formation in the series from lanthanum to gadolinium. The values quoted by Schott and Sommer (16) suggest that the enthalpies of formation become quite constant for the heavy lanthanide elements. A same behaviour for the light lanthanide elements when alloyed with cobalt and copper is observed for the same stoechiometry (18).

3.3. Enthalpies fo formation in the Ln(Ni, Co, Cu)$_5$ series

The enthalpies of formation of LaCo$_5$ and LaCu$_5$ compounds have been determined (26,23). These values are plotted as function of the transition metal atomic number in Figure 3. The minimum of the enthalpy of formation is obtained for the Ni based alloy. It is interesting to compare these results with those obtained for the compounds LaNi$_4$Mt(Mt = Fe, Co, Cu) (22) where one atom of nickel is substituted by another transition element. The same evolution of the enthalpies of formation is observed.

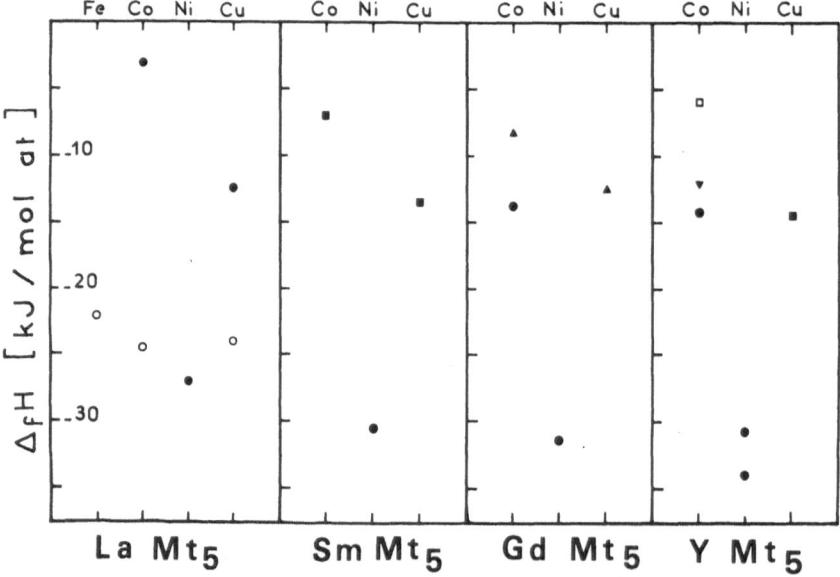

Figure 3 : Enthalpies of formation of LaMt$_5$, LaNi$_4$Mt and (Sm,Gd,Y)Mt$_5$ intermetallic compounds (Mt = Fe, Co, Ni, Cu). Experimental data are from Ref. 21, 23 and 26 (●) ; Ref. 22 (○) for the substituted LaNi$_4$Mt compounds ; Sommer et al, Ref. 16 and 17 (▲) ; Meyer et al, Ref. 18 (■).

In Figure 3 we have also reported the enthalpies of formation of the parent compounds substituting lanthanum by samarium or gadolinium or yttrium. The enthalpies of formation of $SmCo_5$, $SmCu_5$ YCo_5 and YCu_5 have been determined by Meyer-Liautaud et al. (17) using tin solution calorimetry. The value proposed by these authors for the YCo_5 compound is less exothermic than the value we have obtained. Values reported for the $GdCu_5$ and $GdCo_5$ compounds have been obtained by Sommer et al. (16,17).In spite of the discrepancies between the different values, one observe the same trends when Sm, Gd and Y are alloyed with Co, Ni, Cu as for the corresponding La based alloys. The similar behaviour, characterized by quite identical values of the enthalpies of formation, of Y and Gd when alloyed with Co, Ni and Cu is again observed.

4. COMPARISON WITH MODEL PREDICTIONS

As regards systematic predictions of alloy enthalpies of formation, Miedema, Niessen and coworkers (1,2), Watson and Bennett (3) and Colinet and Pasturel (4) have proposed estimates for a large number of binary transition metal alloys involving a rare earth element.

In the previous publications dealing with our experimental determinations of the enthalpies of formation of Gd based alloys we have compared our results with the estimates quoted above and we refer to these publications for more details (26). These comparisons can be summarized by plotting the experimental formation enthalpies $\Delta_f H(exp)$ versus the corresponding calculated values $\Delta fH(calc)$. Such plots comparing experimental and calculated values using the Miedema's estimates (1,2) and Colinet and Pasturel predictions (4) are shown in Figures 4 and 5 respectively. The comparison is limited to Gd based alloys to simplify the figures.

Inspection of the results given in Figure 4 leads to the following remarks. The deviations from Miedema's model predictions increase with increasing absolute values of $\Delta_f H$. Considering the calculated values, Niessen et al. (2) note that the uncertainties may be as much as 20%, owing to the small uncertainties in the parameters listed by these authors. As mentionned above the experimental values present relative uncertainties which depend on the studied system but in the whole the relative uncertainties are more important for low absolute $\Delta_f H$ values. The scattering of the data points with respect to the straight line in Figure 4 may reflect these uncertainties for low absolue absolute $\Delta_f H$ values but certainly not for large absolute $\Delta_f H$ values, say for the Pt and Pd based alloys. These systematic deviations from model calculations were earlier considered by Niessen et al (2). If one considere the scattering of the data points corresponding to Ni based alloys, one observe a very good agreement however inspection of the corresponding compositions shows that for low transition metal composition the data points tend to be in the lower side of the straight line while for high transition metal composition the data points tend to lie above it, this confirm the previous observations concluding that the experimental composition dependence is more asymmetric than the calculated one.

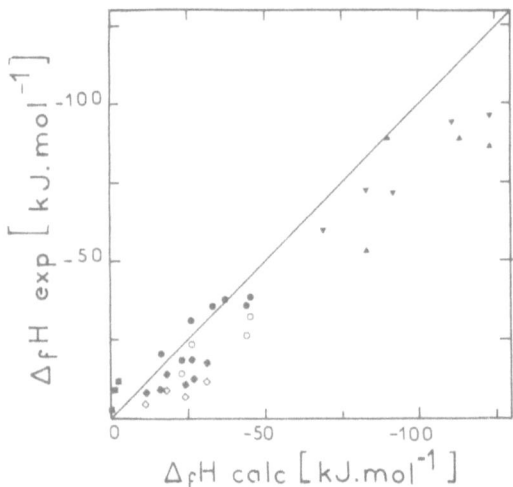

Figure 4 : Comparison between experimental values of the formation enthalpies, ΔfH exp, and calculated values, ΔfH cal, using the Miedema's model (1,2), for compounds of the systems Gd–Fe (■), Gd–Co (◇, ◆), Gd–Ni (○, ●), Gd–Pd (▲) and Gd–Pt (▼). The empty and full symbols refer to the results obtained by Schott and Sommer (16) and Colinet et al. (27) respectively.

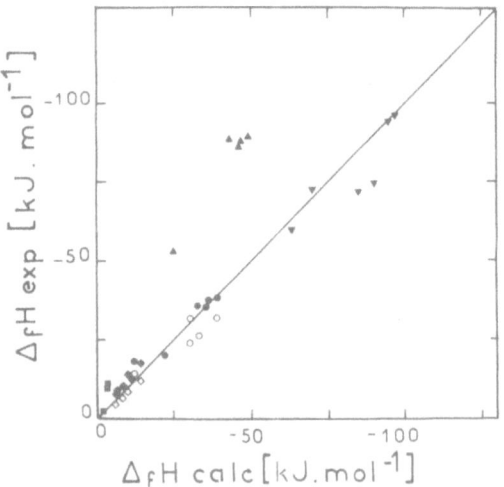

Figure 5 : Comparison between experimental values of the formation enthalpies, ΔfH exp, and predicted values, ΔfH cal, by Colinet and Pasturel (4) for compounds of the systems Gd(Fe,Co,Ni,Pd,Pt). For additionnal keys, see Figure 4 caption.

Inspection of the results given in Figure 5 shows the following. The agreement between the experimental values of the enthalpies of formation and the values predicted by Colinet and Pasturel (4) is quite satisfactory except in the case of the Gd-Pd alloys where the experimental values are largely more exothermic than the calculated one, the quite small relative uncertertainty (less than 3%) of the experimental values cannot explain the discrepancy. Compared to the Miedema's predictions the asymmetrical shape of the enthalpy of formation versus composition is very well represented.

As conclusion from a couple of years a large number of experimental determinations of enthalpies of formation in transition-rare earth metal alloys have been performed. Even if quite large discrepancies between different sets of determinations are observed, the general trends of these enthalpies of formation along the rare earth series as well as along the transition metal series, say the end of the first row and the Ni colomm, are now well known. These data have allowed to discuss the validity of different models and are basic informations which can be used in phase diagram synthesis.

REFERENCES

1   A.R. Miedema, P.F. de Châtel and F.R. de Boer, 'Cohesion in alloys Fondamentals of a semi-empirical model', Physica, 100B (1980) pp.1-28 and references herein.

2   A.K. Niessen, F.R. de Boer, P.F. de Châtel, W.C.M. Mattens and A.R. Miedema, 'Model predictions for the enthalpy of formation of transition metal alloys II', Calphad, 7 (1983) pp.51-70.

3   R.E. Watson and L.H. Bennett, 'Optimized prediction for heats of formation of transition metal alloys', Calphad, 5 (1981) pp.25-40.
    R.E. Watson and L.H. Bennett, 'Optimized predictions for heats of formation of transition metal alloys II', Calphad, 8 (1984) pp.307-321.

4   C. Colinet, A. Pasturel and P. Hicter, 'Trends in cohesive energy of transition metal alloys', Calphad, 9 (1985) pp.71-99.
    C. Colinet and A. Pasturel, 'Trends in cohesive energy in transition rare earth metal alloys', Calphad, to be published.

5   J.C. Gachon, M. Dirand and J. Hertz, 'Enthalpic and structural studies of the NiZr system', J. Less-Common Met., 92 (1983) pp.307-315.
    J.C. Gachon, J. Charles and J. Hertz, 'Different ways to find the thermodynamic functions describing the formation of binary alloys', Calphad, 9 (1985) pp.29-34.
    J.C. Gachon and J. Hertz, 'Enthalpies of formation of binary phases in the systems FeTi, FeZr, CoTi, CoZr, NiTi and NiZr by direct reaction calorimetry', Calphad, 7 (1983) pp.1-12.
    J.C. Gachon, M. Dirand and J. Hertz, 'The enthalpies of formation of the intermediate phases $Co_{0.33}Zr_{0.67}$, $Co_{0.50}Zr_{0.50}$, $Co_{0.67}Zr_{0.33}$ and $Co_{0.80}Zr_{0.20}$ by direct reaction calorimetry at high temperature' J. Less-Common Met., 85 (1982) pp.1-9.
    J.C. Gachon, J.L. Jorda, J. Charles and J. Hertz, 'Thermodynamic study of the ZrRh system', presented at Calphad Fulmer Grange UK, July 9-11 1986.

6 H. Yokokawa and O.J. Kleppa, 'Thermochemistry of liquid alloys of transition metals. II (Copper + Titanium) at 1372 K', J. Chem. Thermodyn., 13 (1981) pp.703-715.

7 O.J. Kleppa and S. Watanabe, 'Thermochemistry of alloys of transition metals. Part III Copper-Silver, -Titanium, -Zirconium, and -Hafnium at 1373K', Metall. Trans., 13B (1982) pp.391-401.

8 O.J. Kleppa and L. Topor, 'Thermochemistry of binary liquid gold alloys : the systems (Au+Cr), (Au+V), (Au+Ti), and (Au+Sc) at 1379K', Metall. Trans., 16A (1985) pp.93-99.

9 L. Topor and O.J. Kleppa, 'Enthalpy of formation of PdTi by high temperature mixing calorimetry', Z. Metallkde., 77 (1986) pp.633-636.
L. Topor and O.J. Kleppa, 'Standard enthalpies of formation of PdZr and PdHf', Metall. Trans, in press.
L. Topor and O.J. Kleppa, 'Standard enthalpies of formation of RhTi, RhZr and RhHf', J. Less-Common Met., 135 (1987) pp.67-75.
L. Topor and O.J. Kleppa, 'Thermochemistry of the intermetallic compounds RuTi, RuZr and RuHf', Metall. Trans., in press.

10 I. Ansara, A. Pasturel and K.H.J. Buschow, 'Enthalpy effects in amorphous alloys and intermetallic compounds in the system Zr-Cu', Phys. Stat. Sol. (a), 69 (1982) pp.447-453.

11 M.P. Henaff, C. Colinet, A. Pasturel and K.H.J. Buschow, 'Study of the enthalpies of formation and crystallization in the system Zr-Ni', J. Appl. Phys., 56 (1984) pp.307-310.

12 S.S. Deodhar and P.J. Ficalora, 'A study of the reaction kinetics for the formation of rare earth - transition metal laves compounds', Metall. Trans., 6A (1975) pp.1909-1914.

13 A. Palenzona and S. Cirafici, 'Thermodynamic and crystallographic properties of RePd intermetallic compounds', Thermochim. Acta, 12 (1975) pp.267-275.
A. Palenzona and S. Cirafici, 'Thermodynamic and crystallographic properties of RPt intermetallic compounds', Thermochim. Acta, 25 (1978) pp.252-256.

14 S. Watanabe and O.J. Kleppa, 'A thermochemical study of liquid and solid alloys {(1-x) La+x Ni} at 1376 K', J. Chem. Thermodynamics, 15 (1983) 633-644.
S. Watanabe and O.J. Kleppa, 'Thermochemistry of alloys of transition metals : Part IV. Alloys of copper with scandium, yttrium, lanthanum and lutetium', Metall. Trans., 15B, (1984) pp.357-368.

15 P.R. Subramanian and J.F Smith, 'Thermodynamics of formation of Y-Co alloys', Metall. Trans., 16A (1985) pp.1195-1201.
P.R. Subramanian and J.F. Smith, 'Thermodynamics of formation of Y-Fe alloys', Calphad, 8 (1984) pp.295-305.
P.R. Subramanian and J.F. Smith, 'Thermodynamics of formation of Y-Ni alloys, Met. Trans., 16B (1985) pp.577-584.

16 J. Schott and F. Sommer, 'Determination of the enthalpies of formation at intermetallic compounds of cobalt and nickel with dysposium, erbium and gadolinium', J. Less-common Met., 119 (1986) pp.307-317.

17 F. Sommer, J. Schott and B. Predel, 'Thermodynamic investigations of Cu-Dy, Cu-Er, Cu-Gd and Cu-La alloys', J. Less-Common Met., 125 (1986) pp.175-181.

18 F. Meyer-Liautaud, S. Derkaoui, C.H. Allibert and R. Castanet, 'Structural and thermodynamic data on the Pseudobinary phases $R(Co_{(1-x)}Cu_x)_5$ with R=Sm, Y, Ce', J. Less-Common Met., 127 (1987) pp.231-242.
F. Meyer-Liautaud, C.H. Allibert and R. Castanet, 'Enthalpies of formation of Sm-Co alloys in the composition range 10-22at%Sm', J. Less-Common Met., 127 (1987) pp.243-250.

19 C. Chatillon-Colinet, H. Diaz, J.C. Mathieu, A. Percheron-Guegan and J.C. Achard, 'Determination of the enthalpies of formation of $LaNi_5$ and $LaNi_4Al$ by a calorimetric method', Am. Chim. Fr., 8 (1979) pp.657-663.

20 C. Colinet and A. Pasturel, 'A thermodynamic study of cerium behaviour in hexagonal $CeNi_5$ compound', Phys. Stat. Sol., (a) 80 (1983) K75-79.

21 A. Pasturel, C. Colinet, C. Allibert, P. Hicter, A. Percheron-Guegan and J.C. Achard, 'A theoretical and experimental study of the enthalpies of formation of $LaNi_5$- Type compounds, Phys. Stat. Sol., (b) 125 (1984) pp.101-106.

22 A. Pasturel, C. Chatillon-Colinet, A. Percheron-Guegan and J.C. Achard, 'Thermodynamic properties of $LaNi_4M$ compounds and their related hydrides', J. Less-Common Met., 84 (1982) pp.73-78.

23 A. Pasturel, F. Liautaud, C. Colinet, C. Allibert, A. Percheron-Guegan and J.C. Achard, 'Thermodynamic study of the $LaNi_{5-x}Cu_x$ system', J. Less-Common Met., 96 (1984) pp.93-97.

24 F. Meyer-Liautaud, A. Pasturel, C.H. Allibert and C. Colinet, 'Enthalpies of formation of the La-Cu intermetallic phases', J. Less-Common Met., 110 (1985) pp.75-80.

25 F. Meyer-Liautaud, A. Pasturel, C.H. Allibert and C. Colinet, 'Thermodynamic study of the valence state of cerium and hydrogen storage in $Ce(Ni_{1-x}Cu_x)_5$ compounds', J. Less-Common Met., 110 (1985) pp.119-126.

26 C. Colinet, A. Pasturel, A. Percheron-Guegan and J.C. Achard, 'Enthalpies of formation and hydrogenation of $La(Ni_{(1-x)}Co_x)_5$ compounds', J. Less-Common Met., 134 (1987) pp.109-122.

27 C. Colinet, A. Pasturel and K.H.J. Buschow, 'Study of the enthalpies of formation in the Gd-Ni system', Met. Trans., 17A (1986) pp.777-780.
C. Colinet, A. Pasturel and K.H.J. Buschow, 'Study of the enthalpies of formation in the Gd(Fe,Co,Pd,Pt) systems', Met. Trans., 18A (1987) pp.903-907.
C. Colinet, A. Pasturel and K.H.J. Buschow, 'Short-Range order and stability in Gd-Ni and Y-Ni systems', J. Applied. Phys., in press.

28 K.N. Semenenko, R.A. Sirotina and A.P. Savchenkova, 'Thermochemical study of intermetallic compounds in the lanthanum-nickel system', Russ. J. Phys. Chem., 53 (1979) p.1356.

29 W.N. Hubbard, P.L. Rawlins, P.A. Connick, R.E. Stedwell and P.A.G. O'Hare, 'The standard enthalpy of formation of $LaNi_5$. The enthalpies of hydriding of $LaNi_{5-x}Al_x$', J. Chem. Thermodynamics, 15 (1983) pp.785-798.

30 R. Hultgren, P.D. Desai, D.T. Hawkins, M. Gleiser, K.K. Kelley and D.D. Wagman, Selected values of the thermodynamic properties of the elements, American Society for Metals, Metals Park, Ohio, (1973).

31 J.C. Mathieu, F. Durand and E. Bonnier, 'Use of a vacuum calorimeter to measure the heats of dissolution of Ge, Al and Ag in Sn at 700°C, Enthalpy measurements on Sn, ZrB_2, TiB_2 BN and B_4C', Thermodynamics, Vol.I, International Atomic Energy Agency, Vienna, 1966.

32 E. Calvet and H. Prat, Microcalorimetrie, applications physico-chimiques et biologiques, Masson, Paris 1956.

33 C. Colinet, A. Bessoud and A. Pasturel, 'Thermodynamic investigation of (Ni, Pd, Pt)-(Al,In) alloys', Z. Metallkde., 77 (1986) pp.798-804.

34 M. Jeymond, D. Landaud, M. Legardeur and A. Pasturel, 'Microcomputer controlled experimentation in calorimetry. Application to the determination of the partial enthalpies of 3d transition metals in liquid aluminium', Thermochim. Act. 55 (1982) pp.301-306.

35 C. Chatillon-Colinet et J.C. Mathieu, 'Mesures des enthalpies partielles des constituants d'une phase liquide et de l'enthalpie de formation d'une phase intermédiaire par calorimétrie de dissolution', Rapport LTPCM (1979) TM-01, Saint Martin d'Hères France.

36 T.N. Rezukhina and S.V. Kutsev, 'Thermodynamic properties of intermetallic compounds in tha La-Ni system', Russ. J. Phys. Chem., 56 (1982) pp.1-6.

37 S.S. Paasch and H.J. Schaller, 'Thermodynamic properties of Pd-X -alloys, with X=Gd,Y,Ce', Ber. Bunsenges. Phys. Chem., 87 (1983) pp.812-814.

CALORIMETRIC STUDIES OF GUINIER-PRESTON ZONES IN ALUMINIUM-BASE ALLOYS

A.-M. ZAHRA
Centre de Thermodynamique et Microcalorimétrie du CNRS
26, rue du 141e R.I.A.
13003 Marseille
France

ABSTRACT. Isothermal microcalorimetric measurements as well as DSC-runs at various heating rates were applied to the study of the aging phenomena in Al-Cu, Al-Zn and Al-Zn-Mg alloys. The precipitation of Guinier-Preston zones was examined at different temperatures as function of homogenisation temperature and quench rate which determine the concentration of lattice defects ; its temporal evolution shows scaling behaviour and agrees with computer simulations based on a statistical model. The conditions of formation, evolution and dissolution of the metastable phases were established and the corresponding enthalpies determined. The binding enthalpies between solute atoms and Guinier-Preston zones were derived in dependence of their mean size. Indications of precipitate stability, size distribution and decomposition mechanism were obtained.

1. INTRODUCTION

When Al-rich solid solutions are cooled to temperatures where unmixing occurs, metastable phases appear and provoke a time and temperature dependent evolution of all physical and chemical properties. These aging phenomena permit the technological application of light metal alloys, as the presence of very small particles within the matrix improves the mechanical behaviour. The most typical example of coherent phases are the Guinier-Preston (G.P.) zones observable in nearly all age-hardenable Al-alloys. The present contribution summarises the different informations obtained on G.P. zones in binary and ternary alloys by means of calorimetric measurements. These are applied in conjunction with other experimental techniques, mostly electron microscopy, in order to ascertain a correct interpretation and gain structural insight.

2. EXPERIMENTAL

Over the past 20 years, the aging behaviour of various binary and ternary alloys which belong to the Al-Cu, Al-Zn and Al-Zn-Mg systems, has been examined. In each case, solid solutions were first produced by homogenising

H. Brodowsky and H.-J. Schaller (eds.), Thermochemistry of Alloys, 203–212.

1 mm thick specimens in the single phase region ; they were then quenched into iced water or cooled in air, thus initiating the formation of metastable phases from the supersaturated solid solutions. For a given solute concentration, the unmixing kinetics is determined by the concentration of quenched-in point defects, as vacancies transport the solute atoms and assemble them at low teperatures into bi- or tridimensional G.P. zones. The excess free vacancy concentration is in general the greater, the higher the temperature difference during the quench as well as the cooling rate have been.

The precipitation of metastable phases may be followed within a calorimeter isothermed at different temperatures ; alternatively, it may be carried out in thermostats for different periods of time and the aged sample subjected to heating in a DSC-apparatus during which the phases redissolve.

The *isothermal* phase separation is accompanied by rather low heat outputs -dH/dt which last for long times. Only one apparatus is capable of measuring them in an open system, the differential heat conduction microcalorimeter of the Tian-Calvet type which incorporates two thermoelectric piles consisting of a great number of thermocouples. It allows to follow the time evolution of the process, so that thermodynamic and kinetic data are obtained simultaneously. The later aging process proceeds slowly, so that a correction for the thermal inertia of the apparatus becomes unnecessary. As reference material in the second thermoelectric pile, pure Al may be used. Calibration is carried out by means of the Joule effect. Care must be taken to imitate the experimental conditions as closely as possible.

The high sensitivity of the microcalorimeter limits its performance, if the experiment starts even before introducing the sample into the detector, like in aging studies. Then the heat effects characterising the reaction under study are superimposed by the ones due to the heating or cooling of the specimen to the working temperature. The latter effects may never be determined accurately, but fall off exponentially and become eventually negligible. Hence results for the first half an hour are subject to a great uncertainty, and it is advisable to study mainly processes which start slowly, so that the initial error may be neglected.

Thermal studies at constant *heating or cooling* rates may be carried out with any commercial DSC-apparatus which offers reproducible base lines. At very low rates, the heat effects may become too small to be measurable, and a microcalorimeter must be used, as the latter examines about 100 times greater sample masses. By convention, endothermal (+ dH/dt) heat effects should be drawn downward and exothermal (- dH/dt) ones upward.

The DuPont thermal analyser used in the present study possesses one thermoelectric disc which supports the alloy and reference samples. Its sensitivity varies with temperature and must be assessed for each experimental set-up. The calibration has to be carried out under identical conditions and checked, if it is rate independant. It is performed either by melting small beads of metals the heats of fusion of which are well established (like In, Sn, Zn) or by using materials with well known heat capacity values (Cu, alumina, f.i.). The correction for the thermal inertia of the apparatus becomes important for rates above 10 °C/min.

Enthalpy values are invariably obtained by integrating the heat flow curves over time. Data acquisition and treatment are performed by microcomputers.

3. EXAMPLES OF CALORIMETRIC RESULTS

3.1. Precipitation kinetics of G.P. zones

Quite generally, the rapid decomposition of a supersaturated solid solution
gives rise to an initially strong heat output which decreases with time.
Simultaneous nucleation and growth lead to heat effects which increase with
time up to a peak value the position of which depends on temperature ; beyond
this maximum, fairly no new precipitates appear.
    Figure 1 shows the isothermal precipitation kinetics of G.P. zones from
an Al-15 wt% Zn solid solution homogenised at 400 °C and transferred to
different temperatures.

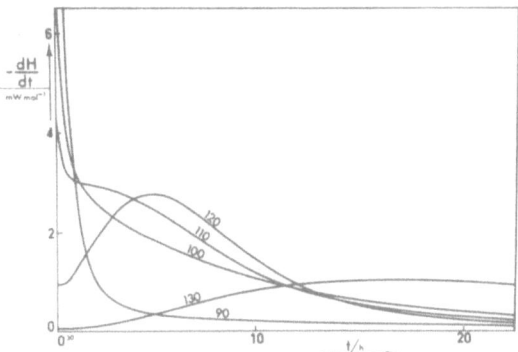

Figure 1. Exothermal heat flows accompanying the G.P. zone formation in Al-15
% Zn alloys at the temperatures indicated.

    At 130 °C, the nucleation rate of G.P. zones is rather constant over the
first couple of hours. The transition G.P. zones→α'_R-phase which is known to
set in after approximately 10 h, is very gradual and does not manifest itself
in the thermal curve. Its maximum shifts to shorter times with falling
temperature.
    Nucleation difficulties are also observed in Al-4 % Cu solid solutions
after direct cooling from 525 °C to temperatures between 120 and 225 °C
(figure 2). The G.P. zones appear the most rapidly at 140 °C. Above 175 °C,
another metastable phase, θ', precipitates. θ'' never appears as first phase.
    Figure 3 confirms the strong influence of homogenisation temperature
and quench rate from the single phase region on the kinetics of G.P. zone
precipitation around room temperature, in the case of an Al-21 % Zn alloy.
After water quenches, the rate of G.P. zone formation and growth is seen to
rise first of all with homogenisation temperature to about 400 °C, then
decreases. This inversion effect is related to the appearance of dislocation
loops and slip bands which reduce the free vacancy concentration. If the
solid solution is cooled in air, G.P. zones appear much more slowly, but
continue to grow for long times, as less vacancy sinks are present. After
prolonged aging, the G.P. zone dimensions will thus be the greater, the lower
the homogenisation temperature has been, as indeed confirmed by electron
microscopy.

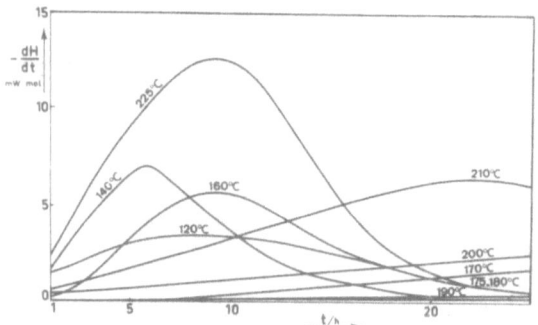

Figure 2. Exothermal heat flows characterising the precipitation of G.P. zones (below 175 °C) and of θ'-phase (above 175 °C) in Al-4 % Cu alloys.

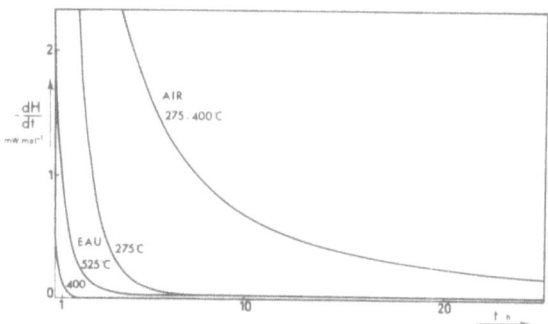

Figure 3. G.P. Zone formation in Al-21 % Zn alloys at 30 °C, after water (EAU) quenches or air cooling from the given temperatures.

Calorimetric studies performed at constant or linearly rising temperatures show that G.P. zones are formed as first precipitate in an Al-5 % Zn - 1 % Mg alloy at all temperatures below 120 °C. In contrast to binary Al-Zn alloys, the aging process lasts for very long times due to the strong interaction between Mg atoms and vacancies. Increasing the homogenisation temperature leads to a reinforced precipitation during an initial period (below 30 min), whereas the later evolution is independent of the vacancy concentration at lower aging temperatures.

3.2. Enthalpy values

The heat flows integrated over time represent the enthalpies of G.P. zone formation (studied within an isothermal microcalorimeter) or dissolution (within a DSC-apparatus). Figure 4 illustrates their evolution in the case of Al-4 % Cu alloys aged at the temperatures indicated after water quenches (T) or short reversion treatments at 230 °C (R). The latter bring about complete dissolution of preexisting G.P. zones and reduce the free vacancy concentration to its equilibrium value at 230 °C, so that only a low vacancy supersaturation is available for the subsequent aging at temperatures below 230 °C. Few experimental methods allow to detect the slow evolution of the reverted state at room temperature or 70 °C. Typical enthalpy values are of the order of 200 J/mol.

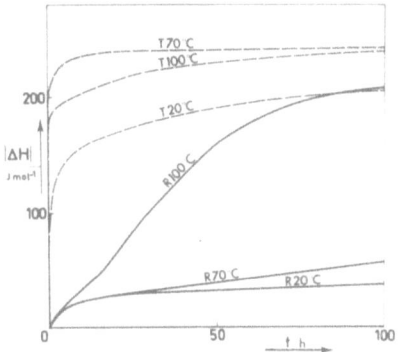

Figure 4. Evolution of enthalpy during G.P. zone formation in quenched (T) or reverted (R) Al-4 % Cu solid solutions.

Combined with the metastable solubility value of Cu in Al, the enthalpies characterising the fully aged state permit to deduce the binding enthalpies between the Cu-atoms present in G.P. zones. The latter vary strongly with the mean precipitate size estimated from the electron diffraction patterns. They attain 0.24 eV for G.P. zones greater than 10 nm, for which the elastic and interfacial energy contributions become stable.

At the contrary, if the same sort of calculation is applied to Al-10 and 15 % Zn alloys, values of 0.07 to 0.08 eV are obtained for the binding enthalpy between Zn-atoms present either in G.P. zone or in the α'_R-phase. As it does not depend notably on the nature and dimension of the precipitate, it may be concluded that G.P. zones and α'_R represent the same phase from the thermodynamic point of view, and that the elastic and interfacial energy contributions owing to their presence in the matrix are relatively small.

3.3 Reversion conditions

The best redissolution conditions for G.P. zones may easily be verified by thermal analysis, as shown in figure 5 for an Al-4 % Cu alloy containing initially G.P. zones. If the latter are incompletely dissolved, they cause an endothermal effect during a DSC-scan, as in case of A (reversion temperature 200 °C), B (210 °C) and C (220 °C). Only a 3 min-hold at 230 °C brings about the fully reverted state characterised by a low vacancy concentration.

This figure also demonstrates the influence of the heating rate on the further evolution. At rates above 10 °C/min, θ'-précipitation sets in after complete reversion of G.P. zones, whereas at lower rates, the G.P. zones transform into the θ"-phase.

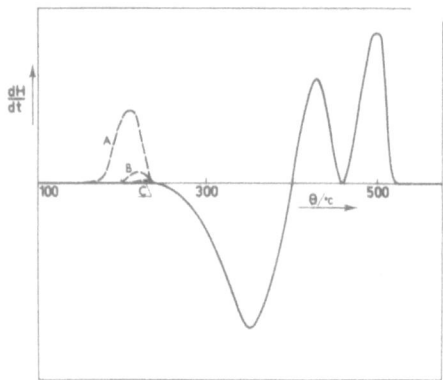

Figure 5. DSC-curves obtained at 20 °C/min on Al-4 % Cu alloys. The first endotherm corresponds to the dissolution of small amounts of G.P. zones which subsist after incomplete reversion treatments at 200 °C (A), 210 °C (B) or 220 °C (C).

3.4. Verification of phase diagrams

The phase diagram proposed for the Al-Zn system does not foresee the existence of G.P. zones in an Al-3 % Zn alloy aged above 0 °C. Very small endothermal effects of the order of 1 J/mol are, however, observed, when this alloy is first aged at temperatures $\leq$ 80 °C and submitted to reversion in a DSC-apparatus. High resolution electron microscopy confirms the presence of Zn-rich clusters of irregular shape in room temperature aged alloys : their sizes are about 1 to 1.5 nm. The transition from the single to the two phase region seems to be a very gradual one.

3.5. Histogram of precipitates

The size distribution of metastable precipitates in Al-base alloys cannot be determined by conventional electron microscopy, as G.P. zones smaller than 2 nm are barely detected. A rapid indication may, however, be obtained indirectly from DSC-runs on aged alloys, as the G.P. zone size determines its stability, hence its reversion temperature. Thus the abscissa of the DSC-curve represents the particle dimension, $\emptyset$, whereas the ordinate is proportional to $\emptyset^3$ times the number of particles of a given size. This correspondence is not generally valid, as certain conditions have be fulfilled : there must be negligible phase evolution during the heating in the DSC-apparatus, hence the interaction between vacancies and solute atoms must be small ; the reversion temperature must depend on the dimensions of the particles the dissolution of which must involve approximately the same heat per solute atom going into solid solution. This last condition is obeyed in the case of Al-Zn alloys. Low vacancy concentrations are ensured by examining alloys in the aged state.

The DSC-curves recorded during the heating of Al-10 % Zn solid solutions quenched from 275, 400 or 525 °C and aged for 4 d at room temperature are given in figure 6. The shapes of these endotherms exhibit two

(400 and 525 °C) or three (275 °C) maxima ; the main peaks appear in the order 525, 400, 275 °C , as is also true after much longer aging times.

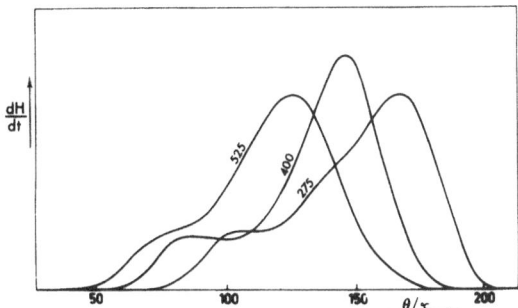

Figure 6. DSC-curves obtained at 20 °C/min on Al-10 % Zn solid solutions quenched from the temperature indicated and aged for 4d at room temperature. The peaks indicate the dissolution of spherical, ellipsoidal and rhombohedral (α'_R) G.P. zones.

Electron microscopic observations confirm that the peaks observed correspond to the reversion of spherical (below 4 nm), ellipsoidal (below 7 nm) G.P. zones and the α'_R-phase ; the latter appears only after quenching from low temperatures which enables zinc diffusion over much longer times than water quenches, and corresponds to large G.P. zones. It is again confirmed that the mean precipitate diameter decreases when the homogenisation temperature is raised.

Once the general pattern is known, DCS-curves may be used to deduce the phases present in aged Al-Zn alloys, as well as their relative amounts.

3.6. Decomposition mechanism

The microcalorimetric studies permit to calculate the excess energy $\Delta U = U_t - U_\infty$ stored at any moment in the decomposing solid solution, by integration of the heat flux liberated between a given time t and ∞. Statistical approaches describe the later decomposition stages by simple hyperbolic laws ; this may be checked by a double-logarithmic representation of the experimental points. Figure 7 gives some examples ; binary and ternary Al-alloys were homogenised at different temperatures and directly introduced into a microcalorimeter isothermed at different temperatures. The curves show that the evolution of the excess energy during G.P. zone formation may not be described by a single law of the type $\Delta U \propto t^{-b}$, as the exponent increases continuously toward infinity.

Computer simulations based on the kinetic Ising model were carried out on alloys containing 10 atom % of solute, using the scaling factors drawn from the microcalorimetric runs. These simulations are in excellent agreement with the experimental data despite all simplifications introduced by the Ising model. They fit to a linear combination of the two laws which represent the elementary mechanisms, namely cluster diffusion and coagulation to which corresponds $\Delta U \propto t^{-1/6}$, and monoatomic diffusion represented by

$\Delta U \propto t^{-1/3}$, such that the reduced excess energy becomes :

$$\Delta U^* = \gamma_1 t^{*-1/3} + \gamma_2 t^{*-1/6}$$

Figure 8 shows the values of $\Delta U^* t^{*1/3}$ taken from the experiments on various Al-Zn alloys and from computer simulations, as function of $t^{*1/6}$. A small difference exists between the values obtained on solid solutions quenched deeply into the miscibility gap and those for shallow quenches near the solvus. This difference has also been observed in measurements of the structure function.

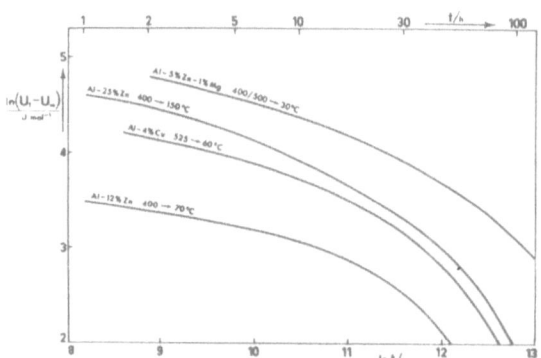

Figure 7. Evolution of excess energy during G.P. zone formation, as calculated from the calorimetric curves, for the alloys and heat treatments stated.

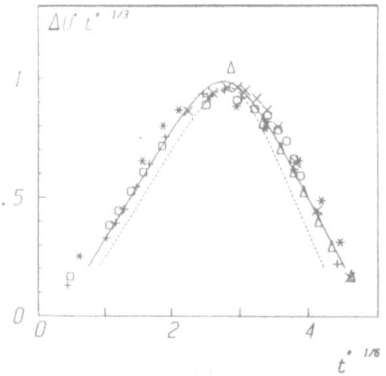

Figure 8. Plot of $\Delta U^* t^{*1/3}$ vs. $t^{*1/6}$ for the case of deep (full curve) and shallow (dashed curve) quenches into the miscibility gap.

According to this figure, the system evolves in two stages after an initial period ; the first one is characterised by a predominance of the $t^{-1/6}$-law ; finally, the $t^{-1/3}$-law overrules. The cross-over between the two regimes occurs rather sharply.

The time evolution of the excess energies during phase separation is seen to display like the structure function a universal behaviour, as scaling of the data corresponding to different compositions, temperatures and even systems is observed.

4. SUMMARY

The aging phenomena in Al-base alloys are due to the formation of metastable phases including G.P. zones which are difficult to detect in their initial stages owing to their small sizes. This article demonstrates that the calorimetric technique counts amoung the most sensitive ones, as it allows to measure extremely slow evolutions. The very beginning of isothermal studies suffers from poor reproducibility , but the experiment may be extended over weeks thanks to the high stability of the microcalorimeter. The number of experimental results is strongly increased by simultaneous DSC-work which excels in rapidity and ease of application, but yields curves which are in general not easy to interprete. The correct calibration may be checked by studying a phase precipitation by isothermal microcalorimetry and its dissolution by DSC.

The major advantage of isothermal microcalorimetry lies in the fact that it allows to obtain kinetic as well as thermodynamic data. Nucleation and growth processes may be recognised unambiguously, if they proceed slowly; the temporal evolution of phase separation is shown to be universal. No abrupt changes occur in the vicinity of coherent phase boundaries.

DSC-studies allow in particular to examine the stability of metastable phases during subsequent heating, which is of practical importance. The curves reflect their nature and size distribution in aged Al-Zn alloys. They also yield valuable enthalpy values.

REFERENCES

A.-M. Zahra and M. Laffitte, *Analusis* 6 (1978) 133.
  'La microcalorimétrie à flux appliquée à des problèmes métallurgiques.'
A.-M. Zahra and M. Laffitte, *Scr. Metallurg.* 8 (1974) 165.
  'Etude microcalorimétrique de la précipitation après réversion dans un alliage Al-Cu 4 %.'
A.-M. Zahra, C.Y. Zahra and M. Laffitte, *Scr. Metallurg.* 9 (1975) 879.
  'Contribution à l'étude du diagramme des phases métastables dans le système Al-Cu du côté riche en Al.'
A.-M. Zahra, M. Laffitte, P. Vigier and M. Wintenberger, *Aluminium* 52 (1976) 357.
  'Bildung der Guinier-Preston-Zonen und der θ"-Phase bei kontinuierlicher Erhitzung einer Legierung vom Typ AlCu4.'
A.-M. Zahra, M. Laffitte, P. Vigier and M. Wintenberger, *Mém. Sci. Rev. Mét.* 74 (1977) 561.
  'Corrélations entre enthalpie, limite d'élasticité et taille des zones de Guinier-Preston dans un alliage Al-4 % Cu.'
A.-M. Zahra, C.Y. Zahra, M. Laffitte, W. Lacom and H.P. Degischer, *Z. Metallkde.* 70 (1979) 172.
  'Entmischungsvorgänge in einer Al-5 % Zn-1 % Mg-Legierung. Teil I : Kalorimetrische Untersuchungen.'
A.-M. Zahra, C.Y. Zahra and M. Laffitte, *Z. Metallkde.* 70 (1979) 669.
  'Calorimetric study of the aging behaviour of an Al-10 % Zn alloy.'

A.-M. Zahra, C.Y. Zahra and J.-C. Mathieu, *Z. Metallkde.* **71** (1980) 54.
'Calorimetric aging study of an Al- 15 wt% Zn alloy .'

H.P. Degischer, C.Y. Zahra and A.-M. Zahra, *Z. Metallkde.* **73** (1982) 635.
'Size distribution of metastable phases in Al-10 and -15 wt% Zn alloys.'

W. Lacom, H.P. Degischer, A.-M. Zahra and C.Y. Zahra, *Z. Metallkde.* **73** (1982) 781.
'Decomposition processes in an Al-5 % Zn-1 % Mg alloy. Part III : Reversion studies.'

A. Charai, C. Boulesteix, C.Y. Zahra and A.-M. Zahra, *Scr. Metallurg.* **18** (1984) 759.
'Metastable phases and clusters in an Al-6 % Zn alloy.'

A.-M. Zahra, C.Y. Zahra, A. Charai and C. Boulesteix, *J. Therm. Anal.* **30** (1985) 671.
'Evidence for GP-zone formation in an Al-3 % Zn alloy.'

A.-M. Zahra, R. Toral and J. Marro, *Cal. Anal. Therm.* **15** (1984) 52.
'Apport de la microcalorimétrie aux théories de décomposition de solutions solides sursaturées.'

J. Marro, R. Toral and A.-M. Zahra, *J. Phys.* **C18** (1985) 1377.
'Time evolution of the excess energy in supersaturated solid solutions : microcalorimetric experiments, computer simulations and theory.'

THERMODYNAMIC PROPERTIES OF THE CuS - Cu$_2$S SYSTEM

Svein Stølen, Fredrik Grønvold
Department of Chemistry
University of Oslo
Blindern, N-0315 Oslo 3
Norway

ABSTRACT. The thermodynamic properties of the CuS - Cu$_2$S system have been studied by means of adiabatic-shield calorimetry in the temperature range 5 to 1000 K. The heat-capacity curves show the presence of a large number of transitions which have been characterized structurally by powder X-ray and neutron diffraction. A revised phase diagram is presented. The heat-capacity data serve as a starting point for evaluation and discussion of the thermodynamic properties of the CuS - Cu$_2$S system. The calorimeter used for the high-temperature measurements and the experimental procedure are described in some detail.

INTRODUCTION

In a series of papers on the thermodynamics of copper sulfides [1-4] the heat capacities of nine copper-sulfide samples, Cu$_x$S with x = 1.00, 1.75, 1.80, 1.85, 1.90, 1.95, 1.98, 1.995 and 2.00, are reported. These data serve as a starting point for the present evaluation and discussion of the formation properties of the Cu - S system. The compositional dependence of the Gibbs energy of formation and the enthalpy and entropy of formation in the non-stoichiometric high-digenite field, have been interpreted in terms of a statistical-mechanical model.

For a long time the copper sulfur system was assumed to contain only three intermediate solid phases, covellite (CuS), digenite (Cu$_{1.80}$S) and chalcocite (Cu$_2$S). More recent studies have, however, revealed a large number of new phases, and the phase relations at low temperatures are not yet unambiguously determined. In the present series of papers three new major findings are reported: a low temperature structural distortion of CuS [2,5], a new decomposition temperature for Cu$_{1.75}$S, anilite [3,6], and the existence of a low-temperature phase with composition between Cu$_{1.95}$S, (djurleite), and Cu$_2$S, (chalcocite). This low-temperature phase undergoes a first order transition to one with tetragonal structure at 225 K [4].

A revised version of the equilibrium diagram for the region CuS to Cu$_2$S is proposed in figure 1. Here the tetragonal phase is assumed to

H. Brodowsky and H.-J. Schaller (eds.), Thermochemistry of Alloys, 213–220.
© *1989 by Kluwer Academic Publishers.*

Table I. <u>Enthalpy and entropy of [1/(3-x)]Cu$_{2-x}$S at 298.15 and 750 K.</u>

x in Cu$_x$S	1.00	1.75	1.80	1.85	1.90	1.95	1.98	1.995	2.00
[S(T) - S(0)] / (J/K mol)									
298.15 K	33.68	39.09	39.14	38.66	38.72	38.90	(38.80)	(38.80)	38.76
750 K	57.91	69.45	(69.62)	69.23	69.21	69.55	70.04	70.41	70.55
[H(T) - H(0)] / (J/mol)									
298.15 K	4725	5398	5406	5336	5311	5314	(5300)	(5300)	5267
750 K	16781	19701	(19723)	19727	19802	20052	20345	20544	20640

Figures in parenthesis are extrapolated

respectively. In Cu$_{1.75}$S, figure 2b, three effects are observed. At 314 K (arrow 1) anilite decomposes into CuS and low digenite, whereas low digenite transforms into high digenite around 355 K (arrow 3). The intermediate effect (arrow 2) reflects changes in the phase limits of low digenite.

The experimental heat capacities in the low- and high-temperature regions were fitted to polynomials in temperature by the method of least squares. The fitting and especially the joining of the fitted segments is thouroughly checked, in the low-temperature range by plotting dC$_p$/dT versus T. The polynomials are integrated by Simpsons rule to yield values of the thermodynamic functions. Within the transition regions the heat- capacity values are read from large scale plots and the thermodynamic functions are integrated manually. At the lowest temperatures the heat capacity data are smoothed with the aid of plots of C$_p$/T versus T^2 and functions evaluated by extrapolation of this function. The enthalpy and entropy at 298.15 and 750 K are given in table I. The enthalpy at 750 K shows a smooth variation with composition, whereas the entropy of Cu$_{1.80}$S seems to be slightly high. In the following evaluation of the formation properties the entropy of Cu$_{1.80}$S at 298.15 K is adjusted to 39.00 J/K mol.

In order to calculate the formation data, the Gibbs energy or the enthalpy of formation at one temperature must be known. For this purpose the Gibbs energy of formation of Cu$_2$S at 298.15 K is used as a starting point, and the mean value of those reported by Mills [10] and by Potter [11], -28550 J/mol selected. The thermodynamic properties of formation of Cu S at selected temperatures are given in table II. For the calculations, the thermodynamic properties of copper is taken from Pavel and Stansbury [12], whereas the data for solid and liquid sulfur is taken from West [13]. For gaseous sulfur the data compilation by Grønvold et al. [14] is used. The present data approximate the mean value of the earlier studies and are in good agreement with those by e.g. Arndt and Kordes [15] and Ferrante et al. [16].

The Gibbs energy for the other copper sulfides, except CuS, are evaluated by Gibbs Duhem integration of activity data at 790 K. The consistency of the reported activity data [17-19] is good. From these data, the enthalpy and entropy of formation are evaluated for all copper sulfides and given in table II. The present thermodynamic results were also compared with Gibbs-Duhem integrations of activity data at different temperatures by Rau [17], by Nagamori [18] and by Peronne and Balesdent [19]. Representative examples of the general good agreement is shown in figure 3, which shows the results for Cu$_{1.75}$and Cu$_{1.98}$S. For Cu$_{1.95}$S the

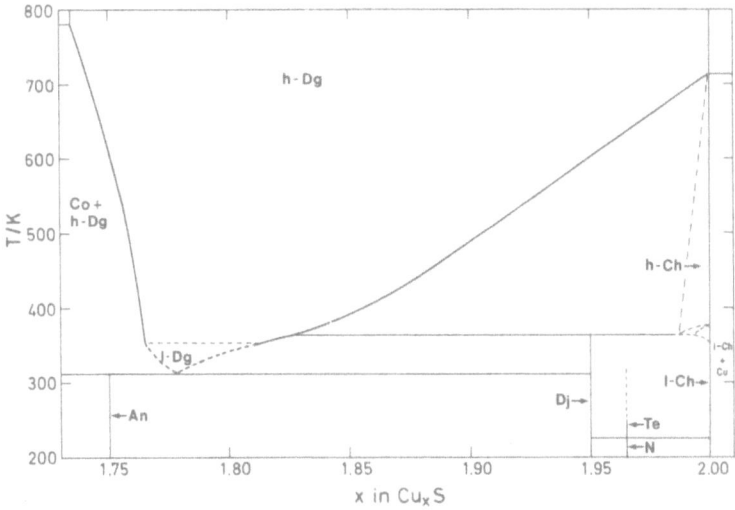

Figure 1. Actual part of the Cu - S phase diagram.

be metastable at room temperature. Djurleite is schematically represented
as a line phase since the extension of its homogeneity range still remains
unknown. The new low temperature phase is provisionally assumed to have the
composition $Cu_{1.965}S$.

EXPERIMENTAL

The high-temperature measurements were made in an intermittently
heated adiabatic-shield calorimeter described in detail in [7] (for
other experimental details, see [1]). The sample is enclosed in an
evacuated and sealed silica glass tube of about 50 cm^3 volume, tightly
fitted into the silver calorimeter. The sample containers have a central
well for the heater and the platinum resistance thermometer, axially
located in the cylindrical silver calorimeter. The calorimeter is
suspended inside a double-walled silver shield system with enclosed
heaters. Outside the shield is a heated guard system, also of silver.
The whole assembly is placed in a vertical tube furnace. A resistance
thermometer made of ~ 10 m 0.2 mm Pt wire wound on 1.0 mm alumina
tubes is used for temperature measurments, whereas the temperature
differences between the corresponding parts of the calorimeter and
shield are measured by means of Pt/Pt10%Rh thermocouples. The
amplified signals are recorded and also used for automatic control
of the shield heaters to maintain quasi-adiabatic conditions during
input and drift periods. The temperature of the guard body is kept
automatically 0.4 K below that of the shield, while the temperature

of the furnace core is kept 10 K lower to secure satisfactory
operation of the control units.

The resistance thermometer has been calibrated locally at the
ice, steam, tin and zinc points, and the temperatures are measured
with a high precision resistance bridge (ASL MODEL F-18). The derived
temperatures are judged to correspond to IPTS 1968 within 0.08 K,
whereas the precision is within 0.00002 K. The energy inputs from
a constant- current supply are measured with a Hewlett-Packard 3455A
Digital Voltmeter with an accuracy of 0.025%.

The computer-operated experiments are started after obtaining a
low and steady temperature drift, preferably less than 10 mK over a 10
min interval. This is obtained by shifting the control zero on all three
shield units. The energy input is then given by means of a suitable
constant current through the heater windings, usually 0.3 A for about
10 min - equivalent to a temperature increase of about 10 K -. The heater
potential is measured every s, whereas the current is measured every
10 s. The energy input is integrated every 10 s as the product of
current, time and the average value of the potential, (neglecting
the highest as well as the lowest value). Under ordinary conditions,
i.e. when no phase transitions or phase reactions takes place in the
sample, the calorimeter temperature reaches its new equilibrium
value after about 30 - 40 m in the room temperature range, after
about 20 m at 600 K and within 10 m at 1000 K. The temperature drift is
recorded and normally used for adjustment of the temperature increment
in the calculation of the heat capacity. In transition regions, smaller
energy inputs and/or longer equilibrium intervals are used.

Heat capacity measurements of the calorimeter with empty sample
container were carried out in separate series of experiments, and
usually represent about 40 - 55 % of the total heat capacity - outside
the transition regions - small corrections are applied for differences
in mass of the silica glass containers and for temperature excursions
of the shields from the calorimeter temperature.

For a description of the low temperature adiabatic measurements
see [8-9].

RESULTS AND DISCUSSION

Representative examples of the obtained heat capacity results are
given in figure 2, where the heat-capacity curves for $Cu_{1.95}S$ and $Cu_{1.75}S$
are shown. Twice the standard deviation in the measured heat capacity is
about 1% from 5 to 30 K, 0.1% from 30 to 300 K and 0.3% in the higher
temperature region. In all samples pronounced endothermic effects,
originating from phase transitions, phase reactions/decompositions, are
observed. Figure 2a shows the four main effects present in $Cu_{1.95}S$.
At 223 K (arrow 1) a not properly characterized low-temperature structure
transforms into a tetragonal one. The transition is of first order and shows
appreciable temperature hysteresis. At 358 K (arrow 2) the low-temperature
(= djurleite) phase decomposes into high digenite and high chalcocite, see
figure 1. Effects represented by arrows 3 and 4 originate from phase reactions,
i.e. the changing boundaries of the high-chalcocite and high-digenite fields,

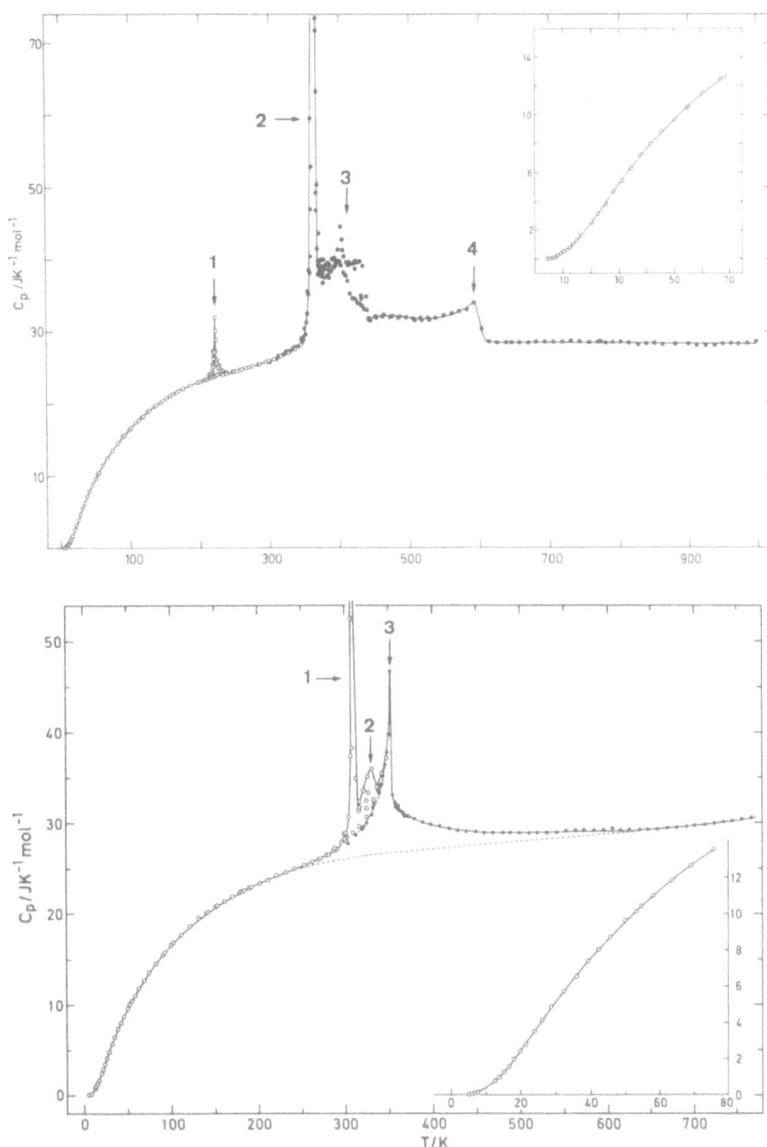

Figure 2. Heat capacity (a), of $(1/2.95)Cu_{1.95}S$ and (b), of $(1/2.75)Cu_{1.75}S$.

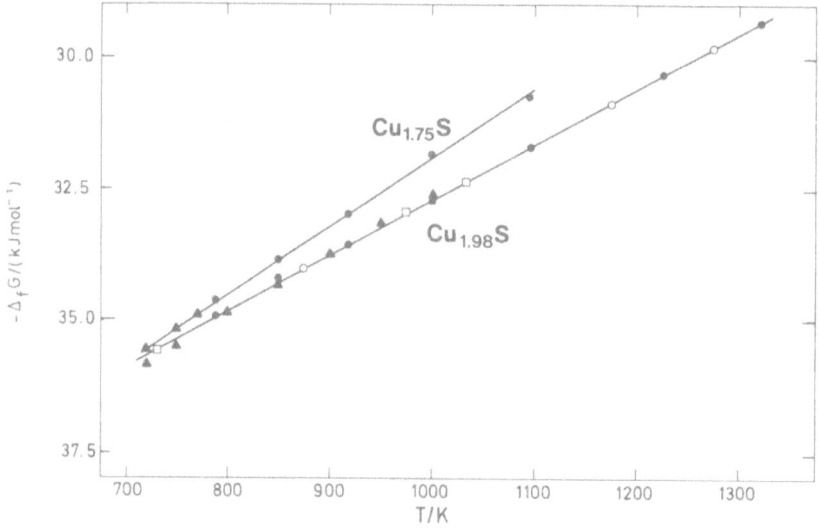

Figure 3. Gibbs energies of formation of (1/2.75)Cu$_{1.75}$S and (1/2.98)Cu$_{1.98}$S. ▲, present values; ◻, Nagamori (18); o, Rau (17); ●, Peronne and Balesdent (19).

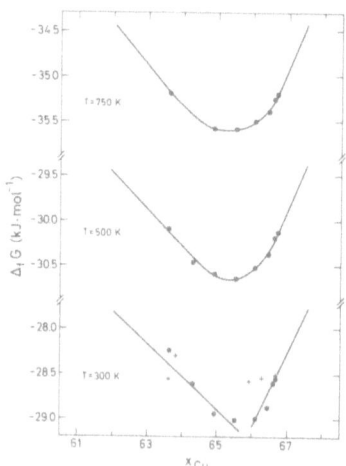

Figure 4. Compositional dependence of the Gibbs energy of formation. ●, present values; +, Potter (11).

agreement is somewhat poorer. This is due to the presence of the metastable tetragonal phase. Although the total enthalpy at high temperatures may be approximately correct, the entropy of $Cu_{1.95}S$, obtained from the heat capacity measurements, appears to give rise to an erroneous temperature dependence of the Gibbs energy of formation. The deficiency in the entropy probably originates from the fact that the transition to the phase mixture stable above ~ 300 K occurs at too high temperature. The formation properties for CuS, see table II, are obtained by selecting the value $\Delta_f G(298.15 \text{ K}) = -24800$ J/mol for 1/2 CuS. This value is much lower than the recommended value [20], -26800 J/mol. The present value does, however, give consistent values with regard to the sulfur-rich stability limit of high digenite. The very good consistency of the obtained results is seen in figure 4, where the obtained Gibbs energies of formation are plotted versus composition for T = 300, 500 and 750 K. At 300 K the stable phases are CuS, $Cu_{1.75}S$, djurleite and Cu_2S. The points marked with +, are taken from Potter [11]. The straight lines for $x_{Cu} < 63.63$ and > 66.66 represent tangents to the neighbouring phases, CuS and Cu, respectively (negligible solid solubility of S in Cu is assumed)

The presently obtained temperature dependence of the Gibbs energy of formation for Cu_2S and CuS are in reasonable agreement with results obtained by Potter [11] (aquous EMF technique). The change in the Gibbs energy of formation of CuS from 298.15 to 717.82 K, 1055 J/mol is 55 J/mol higher than obtained by Potter. For Cu S the increase from 298.15 to 710 K is 3525 J/mol, or about 200 J/mol lower than that obtained by Potter. For $Cu_{1.75}S$ the data derived by us lead to a much steeper rise in the Gibbs energy than proposed by Potter [11]. A similar discrepancy is observed between our data and the EMF measurements by Mathieu and Rickert [21] for $Cu_{1.80}S$. According to Nowak et al. [22] the application of the cell used

$$Cu(s)|Cu^{2+}(aq)|Cu_{2-x}S(s)$$

for evaluation of thermodynamic properties is doubtful as the cell can hardly ever be at thermodynamic equilibrium. As the potential of the $Cu_{2-x}S$

Table II. Standard enthalpy, entropy and Gibbs energy of formation for $(1/3-x)Cu_{2-x}S$ at 300, 500 and 750 K (standard states: Cu(s), and S(rh), S(l) and 1/2S (g) at the respective temperatures).

x in Cu_xS	2.00	1.995	1.98	1.95	1.90	1.85	1.80	1.75	1.00
$-\Delta_f H$ (J/mol)									
300	26764	26795	27063	27152	27216	27067	26730	26343	24478
500	25655	25692	26073	26260	26213	26193	26037	25763	26244
750	43341	43536	44144	44714	45377	45691		45826	54427
$\Delta_f S$ (J/K mol)									
300	5.989	6.026	6.028	6.134	5.981	5.918	6.285	6.362	1.075
500	8.955	8.955	8.637	8.510	8.830	8.786	8.875	8.664	-3.266
750	-10.87	-11.04	-11.66	-12.29	-13.05	-13.48		-14.18	-35.67
$-\Delta_f G$ (J/mol)									
300	28567	28603	28872	28993	29010	28842	28615	28252	24800
500	30133	30169	30391	30515	30628	30586	30474	30095	24611
750	35193	35255	35400	35495	35590	35585		35191	27677

electrode in solution of cupric salts is nearly always more positive than the Cu electrode under the same condition, the placement of both electrodes in the same cell containing Cu^{2+} ions should give the following reactions:

$$Cu(s) + Cu^{2+}(aq) \rightarrow 2\ Cu^+$$

$$Cu_{2-x}S(s) + 2y\ Cu^+ \rightarrow Cu_{2-x+y}S(s) + y\ Cu^{2+}(aq)$$

The sulfide electrode should, hence, be continuously reduced and change composition during experiment.

REFERENCES

1. F. Grønvold and E.F. Westrum, Jr. J. Chem. Thermodyn. 1987, 19, xxx.
2. E.F. Westrum, Jr., S. Stølen and F. Grønvold J. Chem. Thermodyn. 1987, 19, xxx.
3. F. Grønvold, S. Stølen, E.F. Westrum, Jr. and C. . Galeas J. Chem. Thermodyn. 1987, 19, xxx.
4. S. Stølen, F. Grønvold and E.F. Westrum, Jr. to be published in J. Chem. Thermodyn.
5. H. Fjellvåg, F. Grønvold, S. Stølen, A.F. Andresen, R. Müller-Käfer and A. Simon to be published in Z. Kristallogr.
6. F. Grønvold and E.F. Westrum, Jr. Am. Mineral. 1980, 65, 574.
7. F. Grønvold Acta Chem. Scand. 1967, 21, 1695.
8. E.F. Westrum, Jr., G.T. Furukawa and J.P. McCullough in "Experimental Thermodynamics, Vol. I." J.P. McCullough and D.W. Scott : Editors. Butterworths: London, 1968, p. 133.
9. E.F. Westrum, Jr. in "Proceedings NATO Advanced Study Institute on Thermochemistry." Da Silva, R.; Editor. Reidel: New York, 1984, p. 745.
10 K.C. Mills Thermodynamic Data for Inorganic Sulfides, Selenides and Tellurides. Butterworths: London, 1974.
11. R.W. Potter, Jr. Econ. Geol. 1977, 72, 1524.
12. R.E. Pavel and E.E. Stansbury J. Phys. Chem. Solids 1965, 26, 607.
13. E.D. West J. Am. Chem. Soc. 1959, 81, 29.
14. F. Grønvold, P. Drowart and E.F. Westrum, Jr. In "The Chemical Thermodynamics of Actinide Elements and Compounds, Part 4". IAEA: Wienna 1984, pp. 249 - 251.
15. D. Arndt and E. Kordes Z. Anorg. Allg. Chem. 1968, 359, 1.
16. M.J. Ferrante, J.M. Stuve, G.E. Daut and L.B. Pankratz U.S. Bureau of Mines Rep. Invest. 8305, 1978.
17. H. Rau J. Phys. Chem. Solids 1974, 35, 1415.
18. M. Nagamori Met. Trans. 1976, 7B, 67.
19. R. Peronne and D. Balesdent J. Chem. Thermodyn. 1983, 15, 295.
20. D.D. Wagman, W.H. Evans, V.B. Parker, R.H. Schumm, I. Halow, S.M. Bailey, K.L. Churney and R.L. Nuttall J. Phys. Chem. Ref. Data 1982, 11, suppl. 2.
21. H.J. Mathieu and H. Rickert Z. Phys. Chem. NF 1972, 79, 315.
22. P. Nowak, W. Barzyk and A. Pomianowski J. Electroanal. Chem. 1984, 171, 355.

MASS-SPECTROMETRIC DETERMINATION OF THERMODYNAMIC MIXING EFFECTS OF
ALLOYS

A. Neckel
University of Vienna
Institute for Physical Chemistry
Währingerstraße 42
A-1090 Vienna
Austria

ABSTRACT. The combination of a Knudsen cell with a mass spectrometer is
particularly suitable for the determination of thermodynamic activities
and thermodynamic mixing functions of liquid and solid alloys. A
significant problem in the application of this technique is created by
small changes in the instrumental sensitivity when the sample is
changed. This effect precludes the accurate determination of thermodynamic activities by measuring the intensities of the ion currents
produced by the alloy and the pure component in two different
experiments, one with the alloy and the other with the pure component.
In the present article an overview is given on the various methods
which have been proposed to overcome this difficulty. One group of
methods is based upon the use of dual, triple or multiple cells.
However, most widely applied are intensity ratio methods, whereby the
ratio of the intensities of the ion currents of two species present in
the vapour phase above an alloy is determined in a single experiment.
Particularly, the intensity ratio-integration methods are thoroughly
described and discussed.

1. INTRODUCTION

The combination of a Knudsen cell with a mass spectrometer has proved
to be a very valuable tool for the determination of thermodynamic
activities and thermodynamic mixing function of condensed phases at
high temperatures. This technique offers the particular advantage of
identifying all species present in the vapour phase in equilibrium with
a condensed phase. The high sensitivity of a mass spectrometer permits
the measurement of very low partial pressures. The accessible range of
vapour pressures is about 10^2 to 10^{-5} Pa. Extension of the range to
higher pressure in the order of 10^5 Pa can be achieved by combining the
transpiration method with mass spectrometry [1-3].
    Several reviews of high temperature mass spectrometry can be found
in literature. Drowart and Goldfinger [4] described the method and
discussed some results. Grimley [5] delivered a general account on this
technique. Fabian [6] presented an overview on the application of mass

H. Brodowsky and H.-J. Schaller (eds.), Thermochemistry of Alloys, 221–246.

spectrometry in the field of metallurgy and materials science. The investigations of thermodynamic properties of alloys have been reviewed by Raychaudhuri and Stafford [7]. Stafford [8] also discussed the limitations in applying mass spectrometry to high-temperature equilibrium studies. Comprehensive reviews of molecular species observed in high temperature vapours and their bond energies have been provided by Gingerich [9,10]. Sidorov et al. [11] discussed the appli-cantion of mass spectrometry to studies of ion-ion, ion-molecule, and molecule-molecule equilibria in high temperature vapours. A general review of the determination of thermodynamic properties of condensed phases has been presented by Chatillon et al. [12]. A survey of recent mass spectrometric investigations on gaseous species and on the deter-mination of thermodynamic properties of condensed phases in equilibrium with the gas phase has been given by Drowart [13].

2. INSTRUMENTATION

To demonstrate the principle of the experimental set-up with the Knudsen cell-mass spectrometry, the Knudsen cell together with the ion source, as designed by Tomiska [14] and used in the group of the author, is displayed in Fig.1. The Knudsen cell consists of an outer cell top, produced from Ta, Mo or W, and a sample liner top, made from a suitable material which is stable against the alloy to be investi-gated. The sample liner is closed by a lid, in which the effusion orifice is drilled. The Knudsen cell is mounted on a supporting system which can be operated from outside the UHV housing. It allows the proper positioning of the effusion orifice with respect to the entrance slit of the ion source. The Knudsen cell is heated by electron bombardement to the desired temperature. Temperatures up to 2200 K can be reached. Heating system and Knudsen cell are surrounded by radiation shields. The ion source of the mass spectrometer has to be protected by a cooling jacket against the heat evolved by the heating system. A molecular beam escapes through the effusion orifice of the Knudsen cell. The diameter of the effusion orifice amounts to between 0.5 and 1.5 mm, to guarantee Knudsen effusion conditions. The molecular beam traverses an adjustable aperture and a shutter system, which can be operated mechanically or automatically. It can be closed to protect the ionisation chamber against contamination. The closing and opening of the shutter enables the distinction between ions originating from the background gases and ions of the molecular beam. The shutter is provided with a small rectangular slit, which can be positioned using a micrometer system. This beam defining slit enables the variation of the intensity and the shape of the molecular beam, thus giving the possibility to identify background gases, creeping of the sample etc. The ion source, shown in the upper part of Fig.1, was constructed to meet the specific demands of high temperature investigations employing a Knudsen cell [14]. It differs from a conventional ion source as used, e.g., in organic chemistry. The atoms or molecules of the molecular beam entering the ion source are ionized by electron impact. The electrons are emitted from a heated filament. Molecular beam, ionizing

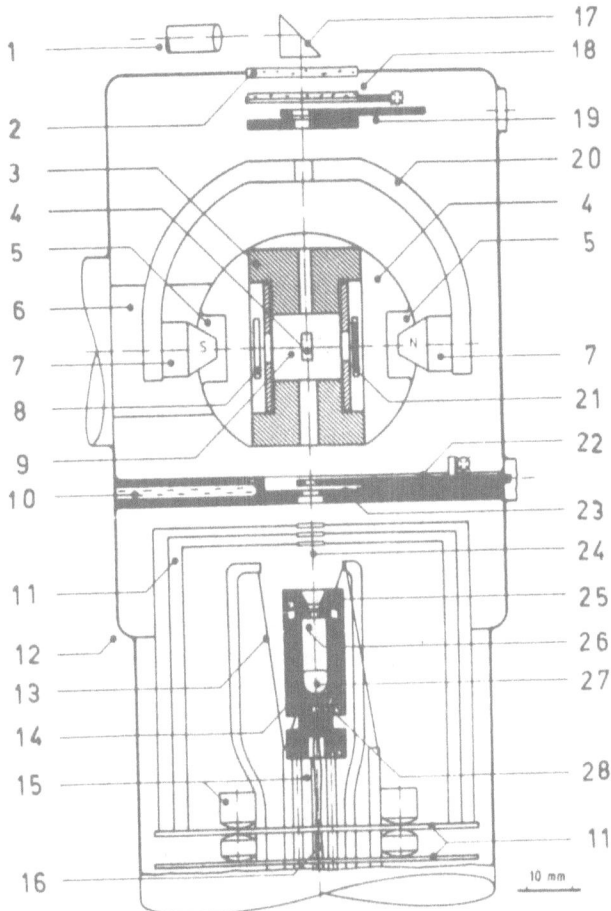

Figure 1. Configuration of Knudsen cell and ion source. 1, Optical pyrometer; 2, UHV window; 3, ionisation chamber; 4, source slit; 5, lens electrodes; 6, adjustable exit slit; 7, collimator magnets; 8, electron target; 9, electron beam (from 21 to 8); 10, cooling jacket; 11, radiation shields; 12, UHV housing (schematic; upper part: rectangular block, lower part: cylindrical); 13, electron bombardment heating; 14, outer Knudsen cell (Mo,Ta,W); 15, supporting system; 16, thermocouple; 17, prism; 18, movable cover slide; 19, shutter system II (electromagnetic); 20, magnet support; 21, filament and grid; 22, shutter system I (electromagnetic and mechanical); 23, adjustable aperture (silver); 24, molecular beam; 25, outer cell top (Ta,Mo,W); 26, sample liner top (Al_2O_3,...) and effusion orifice; 27, sample; 28, sample liner (Al_2O_3,...). According to Tomiska [14].

electron beam and the resulting ion beam are positioned at right angles
to each other. Collimators force the electrons onto spiral lines to
increase the probability of collisions with the neutral particles of
the molecular beam. It is of great importance that the conditions in
the ion source remain constant, also during long-time experiments. The
positive ions formed are accelerated and channeled into the analyser
tube of the mass spectrometer, where they are separated according to
their mass to charge ratio, detected using a secondary electron
multiplier and recorded. The atoms or molecules of the molecular beam
which are not ionized leave the ion source on the opposite side and are
condensed on the shutter system II. The UHV window is protected by a
movable cover slide against condensing particles. The cleanliness of
the UHV window is of great importance for the temperature measurement.
The effusion orifice can be aimed at by means of a prism through the
window and through the ionization chamber with an optical pyrometer.
The temperature, however, is not only measured by an optical pyrometer,
but also with a thermocouple. The electron bombardement heating
prevents direct contact of the thermocouple with the outer Knudsen cell.
By the use of a special shaped sample liner, designed for the
application in thermal analysis, the thermocouple can be brought into
direct contact with the sample liner.

Different types of mass spectrometers, which are based on different principles for the mass separating system, are applied in high
temperature chemistry. Most commonly used are single focusing instruments. All single-focusing spectrometers are equipped with a magnetic
sector field with an angle of deflection of 60°, 90° or 180°. The
magnetic sector field has a direction focusing action.

Double-focusing instruments, which contain in addition to a
conventional magnetic sector field an electrostatic sector field, use
both direction and velocity focusing. By this principle, a substantial
gain in the resolution has been achieved.

Time-of-flight instruments, which are based on the measurement of
velocity of ions, have also been applied for the study of vaporization
processes because of their rapid mass scan, although they may have a
lower resolution and less sensitivity than a magnetic sector instrument.

With a quadrupole mass filter, the mass separation is achieved by
the excitation of ion oscillations generated by an electric radio-
frequency field. A quadrupole mass filter is sensitive. However, its
sensitivity is strongly mass dependent. The main advantage of a
quadrupole mass filter is its compact form and the possibility of
bringing the analyser very close to the vaporizing surface. An
increasing number of applications of quadrupole mass filters for the
investigation of high temperature systems has been reported in the
last years.

3. VAPOUR PRESSURE MEASUREMENT

The partial pressure P_i of a species i in the vapour phase of the
Knudsen cell is related to the intensity $I^{(\ell)}_i$ of the measured ion
current produced by the isotope ℓ of the species i by the equation:

$$P_i = \frac{I_i^{(\ell)} T}{C\, \sigma_i(E)\, \gamma_i\, N_i^{(\ell)}} = \frac{I_i^{(\ell)} T}{S_i} \tag{1}$$

[4], where T is the absolute temperature, $\sigma_i(E)$ the electron impact cross section at the electron energy E, γ_i the multiplier gain, and $N_i^{(\ell)}$ the relative abundance of the isotope ℓ of the species i. The sensitivity factor C includes geometric and electrostatic factors of the Knudsen cell and the ion source and is independent of the species i; thus C contains also the area of the effusion orifice and the correction factor for non-ideality of the Knudsen cell and the effusion-orifice. An explicit expression for the sensitivity factor for an electron impact ion source has been derived by Tomiska [14].

Ionization cross sections σ_i can be determined experimentally or can be calculated [2,15-22]. However, the accuracy of the avaible ionization cross sections is not very high, they may be in error by 50%. The ionization cross sections depend upon the energy of the ionizing electrons. The electron accelerating energy can be varied by varying the voltage on the trap electrode in the ion source with respect to the filament from which the electrons are emitted. Fig.2 shows the typical

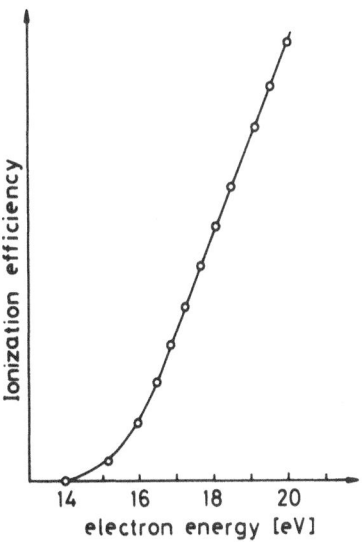

Figure 2. Dependence of the ionization efficiency on the electron accelerating energy.

dependence of the ionization efficiency, which is proportional to the ionization cross section, on the accelerating electron energy. The

ionization efficiency rises above zero at a distinct electron energy.
The minimum energy required for the incident electrons to form
positive ions is called appearance potential. The rise of the ionization efficiency is exponential in character for a range of 1 or 2 eV,
then becoming linear. At higher electron energies, a maximum usually
appears in the curve, which is not shown in Fig.2.

The secondary electron multiplier gain γ_i can be measured by means
of a Faraday collector or by pulse counting.

The factors in the denominator of equation (1) are sometimes
combined in a factor S_i, which can be determined by calibration. The
most commonly used calibration methods have been discussed by Drowart
and Goldfinger [4] and Grimley [5]. A completely new technique of
calibration has recently been introduced by Kematick et al. [21], who
combined the measurement of mass loss using a microbalance, with the
simultaneous measurement of the intensity of the ion current.

4. METHODS FOR THE DETERMINATION OF THERMODYNAMIC ACTIVITIES AND THERMODYNAMIC EXCESS FUNCTIONS

The main task in alloy thermodynamics is the determination of the
thermodynamic activity a_i of a component i in a condensed phase. If, at
low pressures, the fugacity of the component i in the vapour phase can
be regarded as equal to its partial pressure P_i, the activity a_i is
given by

$$a_i = x_i f_i = \frac{P_i}{P_i^o} , \tag{2}$$

where x_i is the mole fraction of component i in the alloy phase, and f_i
its activity coefficient. P_i^o is the vapour pressure of the component i
in a standard state, usually the state of the pure component i at the
same temperature T.

It might be expected that the activity a_i could be determined,
according to equation (2), by the mass spectrometric measurement of P_i
and P_i^o, in two successive experiments, one with the alloy and the other
with the pure component i. By forming the ratio of the intensities of
the ion currents obtained in both these experiments the activity a_i
should be obtained, since all other factors in equation (3) should
cancel, when the same experimental conditions are applied:

$$a_i = \frac{P_i}{P_i^o} = \frac{(I_i^{(\ell)})_a}{(I_i^{(\ell)})^o} \cdot \frac{(C\,\sigma_i(E)\,\gamma_i\,N_i^{(\ell)})^o}{(C\,\sigma_i(E)\,\gamma_i\,N_i^{(\ell)})_a} = \frac{(I_i^{(\ell)})_a}{(I_i^{(\ell)})^o} \tag{3}$$

(In equation (3), the index a refers to the alloy and the superscript ^{o}
to the pure component.) However, small changes in the experimental

sensitivity between different experiments seem to be unavoidable, thus preventing the accurate determination of thermodynamic activities by simply forming the ratio of the intensities of the ion currents obtained in successive experiments. These changes, which affect the sensitivity factor C in (1), are probably caused by small shifts in the position of the Knudsen cell which lead to large changes in the intensity of the ion current, or by changes in the ion optics due to surface coatings. In order to avoid this undesired effect, the intensities of the ion currents produced by the alloy and the pure component have to be measured in a single experiment, without breaking the vacuum to make a sample change.

Various approaches have been developed to overcome the described difficulty.

4.1. Methods based on the use of dual, triple or multiple cells.

A multiple cell consists of different compartments. Usually, one or two compartments contain the pure element, or elements, respectively, the other compartments alloys. The molecular beam effusing from the effusion hole of each compartment is successively analysed by means of a mass spectrometer. By forming the ratio of the intensities of the ion currents produced by the alloy and the pure element and correcting for

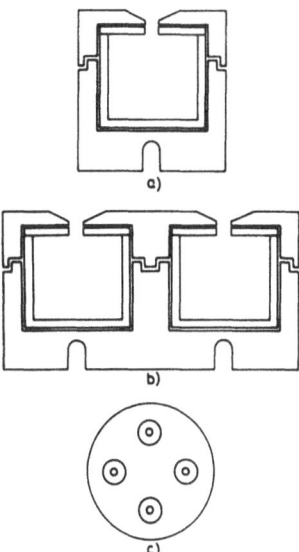

Figure 3.   Schematic diagrams of various Knudsen cells used in mass spectrometry. (a): "standard" cell; (b): dual cell; (c): multiple cell (top view).

possible differences in the effusion holes, the thermodynamic activity is obtained. Various, very ingenious systems of dual, triple, and multiple cell arrangements have been developed [22-33]. Fig.3 shows schematically various types of Knudsen cells: a standard cell (a), a dual cell (b), and a multiple cell (c).

To demonstrate the principle of the multiple cell method, the quad-cell sample system used by Johnston and Palmer [32], shown in Fig.4, will be briefly discussed. The quad-cell is suspended on a

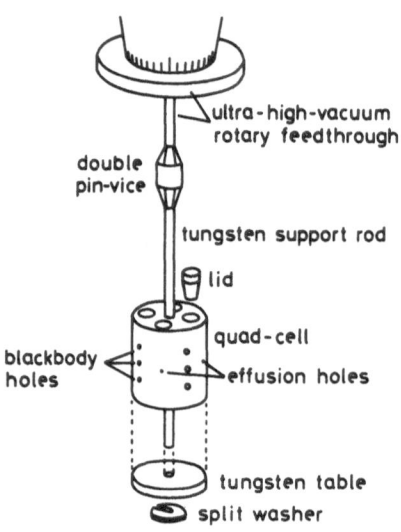

Figure 4.   The quad-cell of Johnston and Palmer [32].

rotatable suspension. The effusion holes of the four compartments are situated on the jacket side of the cylinder forming the quad-cell. Two compartments can be filled with the pure elements, the other compartments contain alloys. By rotating the quad-cell, the effusion orifices of the different compartments can be successively positioned precisely opposite to the entrance slit of the ion source. As the effusion holes can be slightly different the relative sensitivity factors for the individual effusion holes are determined by filling all compartments with the same element.

Alternatively, DeMaria and Piacente [24] used a multiple cell with the effusion holes of the four compartments placed on the top side of a cylinder, which is rotated about its cylindrical axis. (Compare Fig.3c) Furthermore, the positioning of the effusion holes of the individual cells can be achieved by angular displacement of the multiple cell, a technique which is used by Chatillon and coworkers [29]. For other designs the cited literature should be consulted.

4.2. Triple cell isotope method

Hoch and coworkers [34,35] devised a triple cell which is shown schematically in Fig.5. The alloy is placed in one of the lower compartments. The alloy contains the component i with the relative

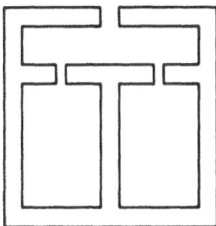

Figure 5. Schematic diagram of a triple cell, used with the "triple cell isotope method" by Hoch et al. [34,35]

isotopic abundances $N_i^{(k)}$ and $N_i^{(k')}$ of the isotopes k and k', respectively. The pure component i with a different isotopic composition $(N_i^{(k)})^o$ and $(N_i^{(k')})^o$ is placed in the other compartment. In the upper compartment the vapour effusing from the two lower compartments is mixed and a single molecular beam formed. The activity of the component i in the alloy can be calculated from the measured intensities of the ion currents of the different isotopes and the known abundances. The activity is given by [7,34,35]

$$
a_i = \frac{(I_i^{(k)}/I_i^{(k')})(N_i^{(k')})^o - (N_i^{(k)})^o}{(h_a/h_r)\, N_i^{(k)} - (I_i^{(k)}/I_i^{(k')})N_i^{(k')}} \ . \tag{4}
$$

The ratio of the "hole factors" (h_a/h_r) of the effusion orifices can be determined by placing chemically pure component i of each of the respective isotopic compositions in the alloy and reference compartment.

4.3. Intensity ratio-integration methods

4.3.1. Intensity ratio-integration method, involving only atomic species

4.3.1.1. Binary systems. No additional experimental equipment is needed if the intensities of the ion currents due to the partial pressures of two monatomic species in the vapour phase above a binary alloy are measured in the same experiment. By forming the ratio P_2/P_1, the sensitivity factor C in equation (1) cancels:

$$\frac{P_2}{P_1} = \frac{I_2^{(m)}}{I_1^{(\ell)}} \cdot \frac{\sigma_1(E)\gamma_1 \, N_1^{(\ell)}}{\sigma_2(E)\gamma_2 \, N_2^{(m)}} = \frac{I_2}{I_1} \cdot \frac{\sigma_1(E)\gamma_1}{\sigma_2(E)\gamma_2} \; . \tag{5}$$

P_2 and P_1 are the partial pressures of the components 2 and 1 in the vapour phase above the alloy. I_2 and I_1 are the intensities of the ion currents, corrected for isotopic distribution:

$$\frac{I_2^{(m)} N_1^{(\ell)}}{I_1^{(\ell)} N_2^{(m)}} = \frac{I_2}{I_1} \; . \tag{6}$$

This method requires that both components of the alloy have measureable and comparable vapour pressures over a wide range of compositions and temperatures. A special advantage of the method is the fact that only a standard Knudsen cell is needed. In order to evaluate the thermodynamic mixing functions of an alloy, relations have to be found which connect a thermodynamic function with the ratio of the partial pressures P_2/P_1.

4.3.1.1.1. <u>Evaluation procedure of Neckel and Wagner</u>. Neckel and Wagner (1969) suggested the following procedure [36]: The excess potential μ_i^E of a component i is given by

$$\mu_i^E = RT \ln f_i = RT \ln \frac{P_i}{P_i^o x_i} \; . \tag{7}$$

By forming the difference $(\mu_2^E - \mu_1^E)$ and noting that this difference is equal to the partial derivative of the excess Gibbs energy with respect to x_2, we obtain

$$\mu_2^E - \mu_1^E = \frac{\partial G^E}{\partial x_2} = RT \ln \frac{P_2 x_1}{P_1 x_2} - RT \ln \frac{P_2^o}{P_1^o} \; . \tag{8}$$

Replacing the ratio of the partial pressures P_2/P_1, in (8), by (5) gives

$$\mu_2^E - \mu_1^E = \frac{\partial G^E}{\partial x_2} = RT \ln \frac{I_2 x_1}{I_1 x_2} - RT \ln \frac{\sigma_2(E)\gamma_2 P_2^o}{\sigma_1(E)\gamma_1 P_1^o} \; . \tag{9}$$

The second term in (9) depends only on temperature, but not on concentration. It will be abbreviated by $C^G(T)$

$$C_G(T) = RT \ln \frac{\sigma_2(E)\gamma_2 P_2^o}{\sigma_1(E)\gamma_1 P_1^o} \quad . \tag{10}$$

The resulting equation (11) forms the basis for the evaluation of thermodynamic mixing functions from mass spectrometric measurements:

$$\frac{\partial G^E}{\partial x_2} = RT \ln \frac{I_2 x_1}{I_1 x_2} - C^G(T) \quad . \tag{11}$$

The function $RT \ln (I_2 x_1 / I_1 x_2)$ will be called "intensity ratio function". In principle, $C^G(T)$ could be determined from the ratio of the cross sections, multiplier gains, and the vapour pressures of the pure components. However, these quantities are generally not precisely known. For determining the constant $C_G(T)$, the thermodynamic condition

$$\int_{x_2 = 0}^{x_2 = 1} \left(\frac{\partial G^E}{\partial x_2} \right) dx_2 = G^E(x_2 = 1) - G^E(x_2 = 0) = 0 \tag{12}$$

can be used, which leads to

$$C^G(T) = RT \int_{x_2 = 0}^{x_2 = 1} \ln \frac{I_2(1-x_2)}{I_1 x_2} dx_2 \quad . \tag{13}$$

Since the ionization cross sections $\sigma_i(E)$ depend on the energy of the ionizing electrons, the same energy has to be used for all experiments in order to keep $\sigma_i(E)$ constant.

Having determined $C^G(T)$, $(\partial G^E / \partial x_2)$ can be obtained according to (11) by subtracting $C^G(T)$ from the intensity ratio function. Integration of the partial derivative of the excess Gibbs energy leads to the excess Gibbs energy (14) and, by well-known thermodynamic relations, the activity coefficients (15a,b) can be calculated:

$$G^E(x_2) = RT \int_{x_2 = 0}^{x_2 = x_2} \ln \frac{I_2(1-x_2)}{I_1 x_2} dx_2 - C^G(T) x_2 \quad , \tag{14}$$

$$\mu_1^E = RT \ln f_1 = G^E - x_2 \left(\frac{\partial G^E}{\partial x_2} \right) , \tag{15a}$$

$$\mu_2^E = RT \ln f_2 = G^E + (1-x_2) \left(\frac{\partial G^E}{\partial x_2} \right) . \tag{15b}$$

Differentation of μ_i^E with respect to x_j leads to

$$\frac{\partial \mu_i^E}{\partial x_j} = - x_j \frac{\partial}{\partial x_j} \left(\frac{\partial G^E}{\partial x_j} \right) . \tag{16}$$

The described procedure presupposes that the investigated phase extends over the entire concentration range. Wagner and St.Pierre [37] were the first who adapted the intensity ratio method to treat a system (Fe-Sn) with a miscibility gap.

The heats of mixing H^E can be determined in a completely analogous manner from measurements of the temperature dependence of the ratio of the intensities of the ion currents. Starting from the thermodynamic expression (17), which relates the partial derivative of the heat of mixing to the partial derivative of the excess Gibbs energy, and replacing $(\partial G^E/\partial x_2)$ by (11), we obtain the basic equation (19) for the determination of the heat of mixing H^E from the ratio of the intensities of the ion currents:

$$\frac{\partial H^E}{\partial x_2} = \frac{\partial \left(\frac{1}{T} \frac{\partial G^E}{\partial x_2} \right)}{\partial (1/T)} , \tag{17}$$

$$\frac{\partial H^E}{\partial x_2} = \frac{\partial}{\partial (1/T)} \left[R \ln \frac{I_2 x_1}{I_1 x_2} - \frac{c^G(T)}{T} \right] , \tag{18}$$

$$\frac{\partial H^E}{\partial x_2} = R \frac{\partial \ln(I_2/I_1)}{\partial (1/T)} - c^H . \tag{19}$$

The constant C^H is related to $C^G(T)$ by (20):

$$C^H = \frac{\partial(C^G(T)/T)}{\partial(1/T)} = L_1^o - L_2^o \quad . \tag{20}$$

Assuming that the ionization cross sections σ_i and multiplier gains γ_i are temperature independent, it can be shown that C^H is equal to the difference of the heats of vaporization (sublimation) L_i^o of the pure components [38,39].

Analogously to the evaluation of $C^G(T)$, equ.(13), C^H can be determined by integration, according to

$$C^H = R \int_{x_2 = 0}^{x_2 = 1} \frac{\partial \ln(I_2/I_1)}{\partial(1/T)} \, dx_2 \quad . \tag{21}$$

Having determined C^H, the partial derivative of the heat of mixing with respect to x_2, the heat of mixing, and the partial molar heats of mixing can be determined.

4.3.1.1.2. Evaluation procedure of Belton and Fruehan. Another, however equivalent formalism was proposed by Belton and Fruehan (1967) [40,41]. These authors start from the Gibbs-Duhem equation and arrive at the following relations for the chemical excess potential

$$\mu_i^E(x_j) = RT \ln f_i = -RT \int_{x_j = 0}^{x_j = x_j} x_j \, d \left[\ln \frac{I_j(1-x_j)}{I_i x_j} \right] \tag{22}$$

and for the partial molar heat of mixing

$$H_i^E(x_j) = -R \int_{x_j = 0}^{x_j = x_j} x_j \, d \left[\frac{\partial \ln(I_2/I_1)}{\partial(1/T)} \right] . \tag{23}$$

These formulae can also be obtained starting from (16) and taking into account (11).

4.3.1.1.3. Algebraic intensity ratio method. Previously, the integrations involved in the determination of $C^G(T)$ and C^H were performed numerically after the experimental values for the functions $RT \ln(I_2 x_1/I_1 x_2)$ and $R \, \partial \ln(I_2/I_1)/\partial(1/T)$, respectively, were fitted graphically by a curve.

In order to avoid the cumbersome numerical integration, an

analytical procedure has recently been developed [42] which is based on the representation of an excess function Z ($Z = G^E$, H^E, S^E) by a polynomial. This may be the Margules equation [43], the Redlich-Kister equation [44,45] or another suitable polynomial. It turned out that the "Thermodynamically Adapted Power-series" (T.A.P.-series) proposed by Tomiska [46], has special advantages. Furthermore, Tomiska [47] has proved that all the various expansions which use polynomials of integer powers of the mole fraction are necessarily equivalent. The result of an algebraic best fit of given experimental data depends only on the highest power of the mole fraction x in the applied polynomial, but not on an arbitrary splitting of the coefficients of any power of x into various terms. The set of parameters for one type of polynomial can be unambiguously transformed into the set of parameters of another type of polynomial.

The T.A.P.-series has to form

$$Z = x(1-x) \sum_{n=1}^{N} B_n^Z x^{n-1}$$

$$= x(1-x) \left[B_1^Z + B_2^Z x + B_3^Z x^2 + \ldots \right] , \tag{24}$$

where B_n^Z are adjustable parameters and N is their total number. ($x = x_2$). The partial derivative of Z with respect to x and the partial molar excess functions Z_i are given by

$$\frac{\partial Z}{\partial x} = \sum_{n=1}^{N} B_n^Z \left[nx^{n-1} - (n+1)x^n \right] , \tag{25}$$

$$Z_1 = x^2 \sum_{n=1}^{N} B_n^Z \left[x^{n-2}(1-n) + nx^{n-1} \right] , \tag{26a}$$

$$Z_2 = (1-x)^2 \sum_{n=1}^{N} B_n^Z nx^{n-1} . \tag{26b}$$

Using (25) for Z equal G^E and inserting the obtained expression for ($\partial G^E/\partial x$) into equation (11) we arrive at

$$RT \ln \frac{I_2(1-x)}{I_1 x} = C^G(T) + \sum_{n=1}^{N} B_n^G \left[nx^{n-1} - (n+1)x^n \right] . \tag{27}$$

$c^G(T)$ can be regarded as an additional adjustable parameter. The constant $c^G(T)$ and the parameters B_n^G are adjusted by means of a least square method, to fit the experimental values of the intensity ratio function $RT \ln(I_2 x_1 / I_1 x_2)$ as closely as possible.

The enthalpies of mixing H^E and the excess entropies S^E can be determined from the temperature dependence of the excess Gibbs energies. Assuming that the enthalpies of mixing and the excess entropies are temperature independent in the investigated temperature interval, the parameters of the T.A.P.-series for G^E, H^E, and S^E will be connected by the relation

$$B_n^G(T) = B_n^H - T B_n^S , \qquad (28)$$

whereby the parameters B_n^H and B_n^S are regarded as temperature independent.

The algebraic regression of all experimental values for the intensity ratio function, measured at various concentrations and temperatures, yields the parameters of the T.A.P.-series for G^E, H^E, and S^E. This approach has been termed "Algebraic Intensity Ratio" method.

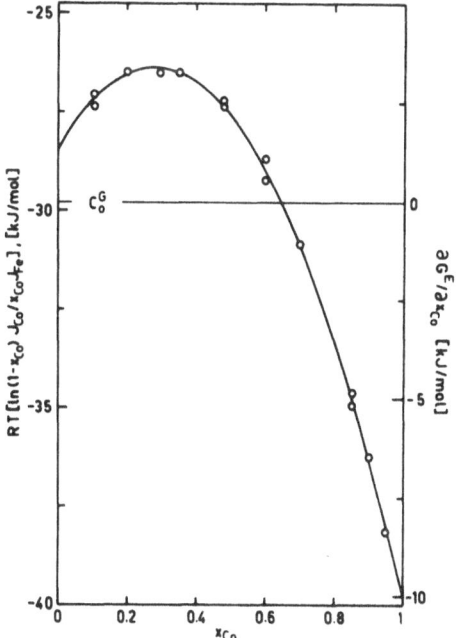

Figure 6. The intensity ratio function $RT \ln \left[(1-x_{Co}) I_{Co} / x_{Co} I_{Fe} \right]$ for the system Fe-Co at 1873 K (left scale). $\partial G^E / \partial x_{Co}$ for the system Fe-Co at 1873 K (right scale). According to Tomiska and Neckel [48].

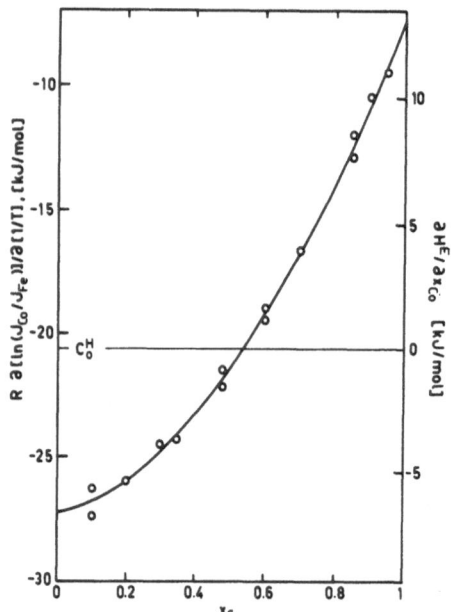

Figure 7. The function $R \ \partial \ln(I_{Co}/I_{Fe})/\partial(1/T)$ for the system Fe–Co at 1873 K (left scale). $\partial H^E/\partial x_{Co}$ for the system Fe–Co at 1873 K (right scale). According to Tomiska and Neckel [48].

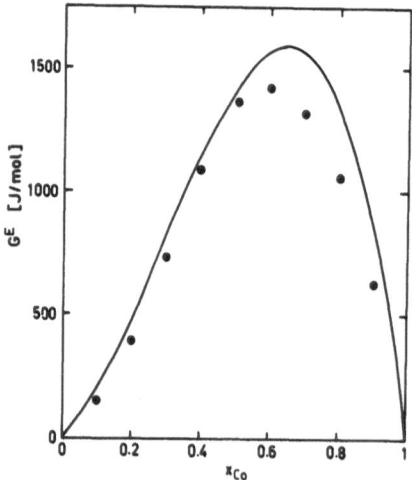

Figure 8. G^E for the system Fe–Co at 1873 K. Tomiska and Neckel [48]: ———— ; Belton and Fruehan [40]: ● (1863 K). According to [48].

To demonstrate the capability of the "Algebraic Intensity Ratio" method, the determination of the thermodynamic excess functions of liquid Fe-Co alloys will be discussed [48]. The ratios of the intensities of the ion currents of ^{56}Fe and ^{59}Co were determined in 14 runs for 10 different compositions in temperature intervals of 80° to 160°C around a temperature of 1600°C. The logarithms of the measured intensity ratios for a given composition x were fitted by an algebraic linear regression according to

$$\ln (I_{Co}/I_{Fe}) = d_o(x) + d_1(x) \frac{1}{T} \quad . \tag{29}$$

From the parameters $d_o(x)$ and $d_1(x)$, the values of the intensity ratio function RT $\ln(I_{Co}(1-x_{Co})/I_{Fe}x_{Co})$ for T = 1873 and the investigated compositions x were calculated. Fig.6 shows the intensity ratio function plotted versus the mole fraction of Co. The experimental points were fitted by a two-parameter T.A.P.-series. The regression yields also the integration constant C^G. Subtraction of C^G from the intensity ratio function gives $\partial G^E/\partial x_{Co}$. The subtraction of C^G means a shift of

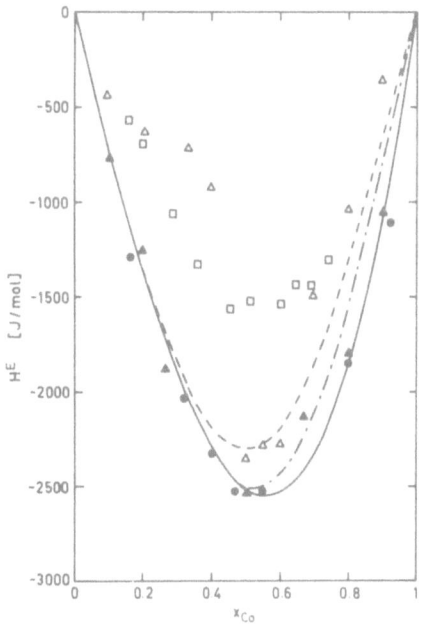

Figure 9. H^E for the system Fe-Co at 1873 K. Tomiska and Neckel [48]: ——— ; Sirota and Breslav-Maslennikov [49]: $\triangle$ (1873 K); Murayama and S.Banya [50]: -- (1873 K); Predel and Mohs [51]: $\bullet$; Tozaki et al. [52]: $\blacktriangle$ (1823-63 K); Batalin et al. [53]: $\square$ (1873 K); Iguchi et al. [54]: -·- (1844 K). According to [48].

the origin of the ordinate axis. The abscissa has to be shifted in such a way that the area between the abscissa and the curve is zero, to satisfy (12). That means that the positive area above the abscissa must be equal to the negative area below the abscissa. For this reason this method has sometimes been termed "equal area method". Fig.7 shows a plot of the function $R(\partial \ln I_{Co}/I_{Fe})/\partial(1/T)$ versus x_{Co}. This function differs by the constant C_H from $\partial H^E/\partial x_{Co}$. The algebraic regression yields the parameters of T.A.P.-series for G^E, H^E, and S^E and the integration constants $C^G(T)$ and C^H. Fig.8 shows the excess Gibbs energy of this system represented by the solid curve. Comparison is made with the G^E-values obtained by Belton and Fruehan [40].

The heat of mixing obtained mass spectrometrically is shown in Fig.9 and compared with the values of other authors. Good agreement is found with the calorimetrically determined values by Predel and Mohs [51]. Tomiska [54] has investigated the solid Fe-Co alloys in the range of the f.c.c. phase by the same technique. He used the data obtained by both these investigations to calculate the phase diagram to check the internal consistency.

4.3.1.2. <u>Ternary systems</u>. Ternary systems can be treated in an analogous manner. The necessary formulae were derived by Neckel and Wagner (1969) [56] and by Belton and Fruehan (1970) [57,58].

The evaluation of the thermodynamic mixing functions from the mass spectrometrically determined ratios of the intensities of the ion currents can be based upon the equations (30) and (31), which are of the same form as the basic equation (11) for binary systems:

$$\left(\frac{\partial G^E}{\partial x_2}\right)_{x_3} = RT \ln \frac{I_2 x_1}{I_1 x_2} - C^G_{2,1}(T) \quad , \tag{30}$$

$$\left(\frac{\partial G^E}{\partial x_3}\right)_{x_2} = RT \ln \frac{I_3 x_1}{I_1 x_3} - C^G_{3,1}(T) \quad , \tag{31}$$

where $C^G_{j,i}(T)$ is given by

$$C^G_{j,i}(T) = RT \ln \frac{\sigma_j(E)\gamma_j \, P^o_j}{\sigma_i(E)\gamma_i \, P^o_i} \quad . \tag{32}$$

The application of the intensity-ratio method requires that at least two components are volatile. Also, for ternary systems, an improvement in the evaluation procedure can be expected from the application of an algebraic evaluation technique [59].

4.3.2. <u>Intensity ratio-integration method, involving polyatomic species</u>. The methods which have been discussed until now were based on the measurement of the intensities of the ion currents of monatomic species in the vapour phase above an alloy. However, there exist numerous alloys systems in whose vapour phase predominantly polyatomic species are found. In such cases, the intensities of the ion currents of poly-atomic species can be used to determine the thermodynamic properties of the condensed phase in equilibrium with the vapour phase. Such a procedural approach presupposes not only that the equilibrium between the vapour phase and the condensed phase is established, but also that thermodynamic equilibrium between the various species in the vapour phase exists. Furthermore, it will be assumed that no fragmentation of the polyatomic species occurs. The complications caused by fragmenta-tion will be discussed later.

Let us consider, as an example, a system whose vapour phase consists mainly of A_2 and AB molecules. From the equilibrium in the vapour phase

$$A_2 + B = AB + A \tag{33}$$

follows

$$K = \frac{P_{AB}P_A}{P_{A_2}P_B} \quad (34) \; , \qquad \frac{P_B}{P_A} = \frac{1}{K} \cdot \frac{P_{AB}}{P_{A_2}} \; , \tag{35}$$

where P_{A_2} and P_{B_2} are the partial pressures of the polyatomic species in the vapour phase and K is the equilibrium constant.
Using (8)

$$\mu_B^E - \mu_A^E = \frac{\partial G^E}{\partial x_B} = RT \ln \frac{P_B x_A}{P_A x_B} - RT \ln \frac{P_B^o}{P_A^o} \tag{36}$$

and replacing the ratio P_B/P_A by (35), and the ratio P_{AB}/P_{A_2} by the ratio of the intensities of the ion currents, (37) is obtained

$$\frac{\partial G^E}{\partial x_B} = RT \ln \frac{I_{AB} x_A}{I_{A_2} x_B} - C_{AB,A_2}^G (T) \; , \tag{37}$$

where $C_{AB,A_2}^G (T)$ contains mass spectrometric and thermodynamic quanti-ties which are independent of concentration. The evaluation of the thermodynamic excess functions of the binary system A-B based upon equation (37) is completely analogous to the procedure for binary systems, which has already been discussed in detail in paragraph 4.3.1.1. For other systems with other vapour phase equilibria, modified in-tensity ratio functions can be readily derived.

The example, which has been considered above, corresponds to the situation in the system Bi-Te, which was studied by Belton and Fruehan [58]. The vapour above Bi-Te alloys consists predominantly of Bi_2, Te_2 and BiTe molecules. Belton and Fruehan based their evaluation of the activity coefficients of Bi upon the formula

$$\ln f_{Bi} = - \int_{x_{Bi}=1}^{x_{Bi}=x_{Bi}} x_{Te} \, d \ln \left[\frac{I_{BiTe} x_{Bi}}{I_{Bi_2} x_{Te}} \right] , \tag{38}$$

which follows from the approach used by these authors. The activity coefficients obtained for Bi agree very closely with values from galvanic cell studies [60]. Other examples were presented by Fruehan [61].

A somewhat different treatment is required if the vapour phase above an A-B system contains A, B, and AB species in measurable quantities. From the equilibrium in the vapour phase

$$A + B = AB \tag{39}$$

follows

$$P_B = \frac{P_{AB}}{P_A} \cdot \frac{1}{K} , \tag{40}$$

where P_{AB} is the partial pressure of the AB molecules and K the equilibrium constant. The chemical excess potential μ_B^E is then given by

$$\mu_B^E = RT \ln \frac{P_B}{P_B^o x_B} = RT \ln \frac{P_{AB}}{P_A x_B} - RT \ln P_B^o K . \tag{41}$$

Replacing the ratio P_{AB}/P_A by the ratio of the intensities of the ion currents leads to

$$\mu_B^E = RT \ln \frac{I_{AB}}{I_A x_B} - D(T) , \tag{42}$$

where D(T) contains mass spectrometric and thermodynamic quantities which are independent of concentration. D(T) could be determined by extrapolating the measured values of RT $\ln(I_{AB}/I_A x_B)$ to $x_B = 1$. However, more convenient would it be to represent μ_B^E according to (26) by a polynomial and to determine D(T) and the parameters of the polynomial by a least square fit.

The described situation is encountered in the case of the Au-Cu system, whose vapour phase contains Au and Cu atoms as well as AuCu molecules. The approach presented above was used to determine the activity coefficients of Au and Cu [36].

4.4. The "monomer-dimer" method

Frequently, oligomeric species A_n and B_m, particularly dimeric species, are present in the vapour phase above an A-B alloy. In this case, the monomer-dimer method, introduced by Berkowitz and Chupka [62], can be applied. These authors studied salt systems whose vapour phase contained monomers and dimers. Their approach can be generalized to cases where a monomer A and a oligomer A_m or two different oligomers (e.g. Sb_2 and Sb_4) are present in measurable concentrations. With this method, the ratio of the intensities of the ion currents of the oligomer and the monomer (I_{A_m}/I_A) for the alloy and for the pure element (reference) are measured in separate experiments. The activity a_A is then given by [7];

$$a_A = \left[\left(\frac{I_{A_m}}{I_A} \right)_{alloy} \middle/ \left(\frac{I_{A_m}}{I_A} \right)_{ref} \right]^{1/(m-1)} . \tag{43}$$

By forming the ratios, contained in (43), not only the sensitivity factor C cancels, but also the mass spectrometric quantities σ_i and γ_i. The corrections for fragmentation which have to be applied with this method were derived by Belton and Fruehan [58]. The determination of the activities in liquid Bi-Sn alloys by Rieckert et al. [63] serves as an example for the application of this technique.

5. FRAGMENTATION

Fragmentation can be a major source of error in systems where the vapour phase contains higher concentrations of polyatomic molecules. Fragmentation influences not only the partial pressure of the polyatomic molecules but also the partial pressures of the monatomic species. A thorough discussion of the problem has been given by Stafford [8]. Some rules have been proposed in literature [8,64] which allow estimating a possible fragmentation. Let us consider, as an example, the formation of parent and fragment ions from a dimer A_2:

$$A_2 \longrightarrow A_2^+ : AP(A_2^+/A_2) = IP(A_2) . \tag{44}$$

The appearance potential $AP(A_2^+/A_2)$ of the parent ion A_2^+ is equal to the ionization potential $IP(A_2)$ of the dimer A_2.

$$A_2 \longrightarrow A^+ + A : AP(A^+/A_2) = IP(A) + D(A-A) . \tag{45}$$

The appearance potential $AP(A^+/A_2)$ of the fragment ion A^+ is equal to the sum of the ionization potential $IP(A)$ of the atom A and the dissociation energy $D(A-A)$ of the A-A bond.

If we want to measure the ion current produced by the neutral monatomic species in the presence of a dimer, we should use an electron energy which does not significantly exceed $AP(A^+/A_2)$. Even so, the electron energy has to be high enough to ionize the atom A. These two energies $(IP(A)$ and $AP(A^+/A_2))$ differ by the dissociation energy $D(A-A)$, which is typically 2 or 3 eV. Consequently, an electron energy between these two values has to be applied. However, a low electron energy reduces the intensity of the ion current I_A. The ion current produced by the dimer A_2 can be measured, unaffected by fragmentation, if $AP(A_2^+/A_2)$ is lower than $AP(A^+/A_2)$. Similar rules can also be established for other polyatomic molecules.

The presence of fragmentation contributions can be recognized by breaks in the ionization efficiency curves.

For the intensity ratio method involving only monatomic species (paragraph 4.3.1.1.) a technique was proposed by Howard and Hager [65] which allows testing the presence of fragment ions due to dissociative ionization of molecules and to correct the intensities of the ion currents of the atomic species for dissociative ionization.

6. DISCUSSION AND COMPARISON OF THE VARIOUS MASS SPECTROMETRIC METHODS

The advantage of the dual- and multiple cell methods is their ability to permit the determination of the activity at a single composition. Alloys with significant differences in the partial pressures of the components or with only one volatily component can be investigated. The data treatment is simple. The problems arising with dual and multiple cells are: (i) the difficult reproducibility of the successive cell positionings, (ii) the occurrence of temperature gradients in the bigger multiple cells, and (iii) possible concentration changes of the alloy during lenghtened measurement, because element and alloy has to be measured in the same experiment.

The triple cell isotope method requires distinct, isotypically enriched samples. Diffusion from the reference cell into the alloy cell has been considered as a possible source of error.

The intensity ratio-integration methods are limited to systems whose components have comparable partial pressures. The use of numerical integration techniques requires measurements over the entire composition range or at least in the concentration range between $x_i = 1$ and x_i, the concentration of interest. Miscibility gaps can be included. If, however, the algebraic intensity ratio method is applied, the concentration range for the measurements can be limited, provided that the polynomials used for the representation of the excess functions describe their concentration dependences properly. The special advantage of the intensity ratio-integration method is that only a "standard" Knudsen cell is needed.

The application of the intensity ratio-integration method involving polyatomic species presupposes relatively high concentrations of polyatomic molecules in the vapour phase. Possible fragmentative ionization has to be considered.

The "monomer-dimer" method offers the advantage of determining the activity at a single composition. It requires, however, the presence of a monomer and an oligomer or of two oligomers in the vapour phase. Fragmentation may be crucial and a major source of error.

7. CONCLUSION

There are many other problems; for example the temperature measurement, temperature gradients in the cell, and possible reactions of the sample with the cell liner, etc, which could not be discussed in this article.

In conclusion, it has been attempted in this overview to demonstrate that mass spectrometry offers an efficient method for the determination of thermodynamic activities of alloys at high temperatures. Moreover, there is room for improvement of the method in many aspects, so that an increase in performance can be expected in the foreseeable future.

ACKNOWLEDGEMENT

The author acknowledges valuable discussions with Univ.-Doz.Dr. J.Tomiska. The mass spectrometric investigations have been supported by the "Fonds zur Förderung der wissenschaftlichen Forschung" of Austria (Project No. P6145c).

REFERENCES

[1] D.W.Bonnell and J.W.Hastie, in Characterization of High Temperature Vapors and Gases, ed. J.W.Hastie, NBS SP 561/1 and 2, U.S.Govt.Printing Office, Washington, DC (1979)

[2] J.W.Hastie, Pure and Appl.Chem. 56, 1583 (1984)

[3] A.Fuwa, K.Sugawara, and E.Kato, J.Japan Inst.Metals 48, 1168 (1984)

[4] J.Drowart and P.Goldfinger, Angew.Chem. 79, 589 (1967)

[5] R.T.Grimley, in The Characterization of High Temperature Vapors, ed. J.L.Margrave, John Wiley, 1967, p.195

[6] D.J.Fabian, Metall.Rev. 12, 27 (1967)

[7] F.K.Raychaudhuri and F.E.Stafford, Mat.Sci.Eng. 20, 1 (1975)

[8] F.E.Stafford, High Temperatures-High Pressures 3, 213 (1971)

[9] K.A.Gingerich, J.Cryst.Growth 9, 31 (1971)

[10] K.A.Gingerich, in Current Topics in Materials Science, Vol.6, ed. E.Kaldis, North Holland Publ.Company, 1980, p.345

[11] L.N.Sidorov, L.V.Zhuravleva, and I.D.Sorokin, Mass Spectrometry Reviews 5, 73 (1986)

244

[12] C.Chatillon, A.Pattoret, and J.Drowart, High Temperatures-High Pressures 7, 119 (1975)

[13] J.Drowart, in Advances in Mass Spectrometry 1985, ed. J.F.J.Todd, John Wiley, 1986, p.195

[14] J.Tomiska, J.Phys.E: Sci.Instrum. 17, 1165 (1984)

[15] S.S.Lin and F.E.Stafford, J.Chem.Phys. 48, 3885 (1968)

[16] J.B.Mann, J.Chem.Phys. 46, 1646 (1967)

[17] J.B.Mann, in Recent Developments in Mass Spectrometry, ed. K.Ogata and T.Hayakawa, University Park Press, Baltimore, 1970, p.814

[18] T.D.Märk, Int.J.Mass Spectrom.Ion.Phys. 45, 125 (1982)

[19] T.D.Märk, in Electron Molecule Interactions and their Applications, ed. L.G.Christophorou, Academic Press, New York, 1984, p.251

[20] T.D.Märk and G.H.Dunn, ed., Electron Impact Ionization, Springer, Vienna, 1984

[21] R.Kematick, J.Anderegg, and H.Franzen, High Temperature Science 19, 17 (1985)

[22] A.Büchler and J.L.Stauffer, in Thermodynamics, Vol.1, IAEA, Vienna, 1966, p.271

[23] A.Pattoret, S.Smoes, and J.Drowart, in Thermodynamics, Vol.1, IAEA, Vienna, 1966, p.377

[24] G.DeMaria and V.Piacente, Bull.Soc.Chim.Belg. 81, 155 (1972)

[25] R.T.Grimley and T.E.Joyce, J.Phys.Chem. 73, 3047 (1969)

[26] T.E.Joyce, T.Y.Ridley, and R.T.Grimley, Rev.Sci.Instrum. 41, 1789 (1970)

[27] R.W.Jones, F.E.Stafford, and D.H.Whitmore, Metall.Trans. 1, 403 (1970)

[28] C.Bergman, M.Lafitte, and Y.Muggianu, High Temperatures-High Pressures 6, 53 (1960)

[29] C.Chatillon, C.Senillou, M.Alibert, and A.Pattoret, Rev.Sci. Instrum. 47, 334 (1976)

[30] A.Neubert, J.Chem.Thermodynamics 11, 971 (1979)

[31] L.Martin-Garin, C.Chatillon, and M.Alibert, Journ.Less-Common-Metals 63, P9 (1979)

[32] G.R.Johnston and L.D.Palmer, High Temperatures-High Pressures 12, 261 (1980)

[33] M.E.Paulaitis and C.A.Eckert, J.Chem.Thermodynamics 15, 55 (1983)

[34] M.Hoch and R.J.Usell,Jr., Met.Trans. 2, 2627 (1971)

[35] J.V.Hackworth, M.Hoch, and H.L.Geyel, Met.Trans. $\underline{2}$, 1799 (1971)

[36] A.Neckel and S.Wagner, Ber.Bunsenges.physik.Chem. $\underline{73}$, 210 (1969)

[37] S.Wagner and G.R.St.Pierre, Metall.Trans. $\underline{3}$, 2873 (1972)

[38] G.Sodeck, P.Entner, and A.Neckel, High Temperature Science $\underline{2}$, 311 (1970)

[39] S.Wagner, G.Sodeck, and A.Neckel, High Temperature Science $\underline{3}$, 481 (1971)

[40] G.R.Belton and R.J.Fruehan, J.Phys.Chem. $\underline{71}$, 1403 (1967)

[41] G.R.Belton and R.J.Fruehan, Trans. TMS-AIME $\underline{245}$, 113 (1969)

[42] J.Tomiska, H.Nowotny, L.Erdelyi, and A.Neckel, Z.Metallkd. $\underline{68}$, 350 (1977)

[43] M.Margules, Sitz.Ber.Akad.Wiss.Wien, Math.Naturw.Kl. II, $\underline{104}$, 1243 (1895)

[44] E.A.Guggenheim, Trans.Farad.Soc. $\underline{33}$, 151 (1937)

[45] O.Redlich and A.T.Kister, Ind.Eng.Chem. $\underline{40}$, 345 (1948)

[46] J.Tomiska, CALPHAD $\underline{10}$, 91 (1986)

[47] J.Tomiska, CALPHAD $\underline{10}$, 235 (1986)

[48] J.Tomiska and A.Neckel, Z.Metallkd. $\underline{77}$, 649 (1986)

[49] N.N.Sirota and M.B.Breslav-Maslennikov, Vestsi Akad.Navuk BSSR, Ser.Fiz.Mat.Navuk $\underline{1977}$, 81

[50] N.Maruyama and S.Banya, Nippon Kinzoku Gakkaishi $\underline{42}$, 992 (1978)

[51] B.Predel and R.Mohs, Arch.Eisenhüttenwes. $\underline{41}$, 143 (1970)

[52] Y.Tozaki, Y.Iguchi, S.Banya, and T.Fuwa, Chemical Metallurgy of Iron and Steel, Proc.Int.Symp.Metal.Chem. - Appl.Ferrous Metall. 1971, The Iron and Steel Institute, 1973

[53] G.I.Batalin, N.N.Minenko and V.S.Sudavtsova, Izv.Akad.Nauk SSSR, Met.(5), $\underline{1974}$, 99

[54] Y.Iguchi, S.Nobori, K.Saito, and T.Fuwa, Tetsu to Hagane $\underline{68}$, 633 (1982)

[55] J.Tomiska, Z.Metallkd. $\underline{77}$, 97 (1986)

[56] A.Neckel and S.Wagner, Mh.Chem. $\underline{100}$, 664 (1969)

[57] G.R.Belton and R.J.Fruehan, Met.Trans. $\underline{1}$, 781 (1970)

[58] G.R.Belton and R.J.Fruehan, Met.Trans. $\underline{2}$, 291 (1971)

[59] J.Tomiska, High Temperatures-High Pressures $\underline{14}$, 417 (1982)

[60] T.Ohashi, Z.Kozuka, and J.Moriyama, Nagoya Kogyo Daigaku Gakuho $\underline{17}$, 354 (1965)

[61] R.J.Fruehan, Met.Trans. $\underline{2}$, 1213 (1971)

[62] J.Berkowitz and W.A.Chupka, Ann.N.Y.Acad.Sci. 79, 1073 (1960)

[63] G.Rieckert, P.Lamparter, and S.Steeb, Z.Naturforsch. 31a, 711 (1976)

[64] J.W.Hastie and J.L.Margrave, High Temp.Sci. 1, 481 (1969)

[65] St.M.Howard and J.Hager, Met.Trans. 9B, 51 (1978)

MODERN VAPOR PRESSURE MEASUREMENTS BASED ON THE KNUDSEN-EFFUSION

Josef TOMISKA
Institute of Physical Chemistry,
University of Vienna
Waehringerstraße 42
A-1090 Vienna (Austria)

ABSTRACT. The Knudsen method of molecular effusion is employed successfully to determine small vapor pressures (< 20 mbar) - even at temperatures higher than 1500 K. The various techniques are rewiewed briefly.
Special emphasis is placed (i) on two newer automatic apparatus for
pressure measurement by the TORKER-method (Torkometer and Computer-
operated dynamic torsion pendulum) as well as (ii) on the newer PENKER-
method (Pendulum electronically balanced Knudsen-effusion recoil).

1. INTRODUCTION

The gas phase of a substance in thermodynamic interaction (exchange of
energy and mass) with a condensed phase (liquid or solid) of the same
substance may be called "vapor phase". Over each condensed sample exists
a vapor phase which contains pressures of the vapors of each component
of the sample. The partial pressure of each component vapor changes
therefore with temperature and composition of the sample, maintaining
the vapor-condensed phase equilibrium.

The density of vapors (gases) may be characterized i) either by the
pressures under which they are standing or ii) by means of the "mean
free path" of the vapor molecules. Many interesting substances - specifically metal alloys - show at actual temperatures vapor pressures lower
than 20 Pa. Those pressures are called "small", and they correspond to
mean free paths greater than 1 cm. Such lengths of the mean free path
are already comparable with the characteristic dimensions of experimental equipments. "Small" vapor pressures implicate therefore that the
probability of a molecule impinging upon a wall is considerably higher
than the probability of collision between two molecules. With the logical consequence that these "dilute" (also: "very dilute") vapors (gases)
will not obey any more the laws of hydromechanics, and their behaviour
must be described by means of the kinetic theory [1].

The measurement of small vapor pressures require therefore experimental techniques different to those employed for determing "common"
vapor pressures. Several effectful ways are known to determine small
vapor pressures at lower temperatures (< 1000 K), but there is a tremen-

H. Brodowsky and H.-J. Schaller (eds.), Thermochemistry of Alloys, 247–260.

duous lack of convenient methods applicable in high temperature chemistry [2-4].

The experimental difficulties of vapor pressure measurements increase considerably with the temperature of the condensed samples; E.g., at temperatures higher than 1500 K materials are limited to make inert, but still compact cell liners. Also applicable sensors are rare, as well as suitable materials for convenient supporting-, handling- and protecting systems.

2.   KNUDSEN EFFUSION METHOD

One of the few methods, which are still applicable in high temperature chemistry is the molecular effusion after Knudsen [5]. Following this method, small vapor pressures are determined by means of the effusion of vaporized sample out of an isothermal vessel which is called "Knudsen cell". This is a (cylindrical) crucible with a small knife-edge shaped orifice (0.5 - 1.5 mm diameter) in the lid. Effusion through the orifice gives a molecular beam which spreads out in isotropic distribution over a sphere ("cosine law"). If inside the Knudsen cell thermodynamic equilibrium is established between the condensed sample and its vapor phase, then the pressure of the escaping molecular beam can be calculated from the equation for steady-state effusion of dilute gases [5],

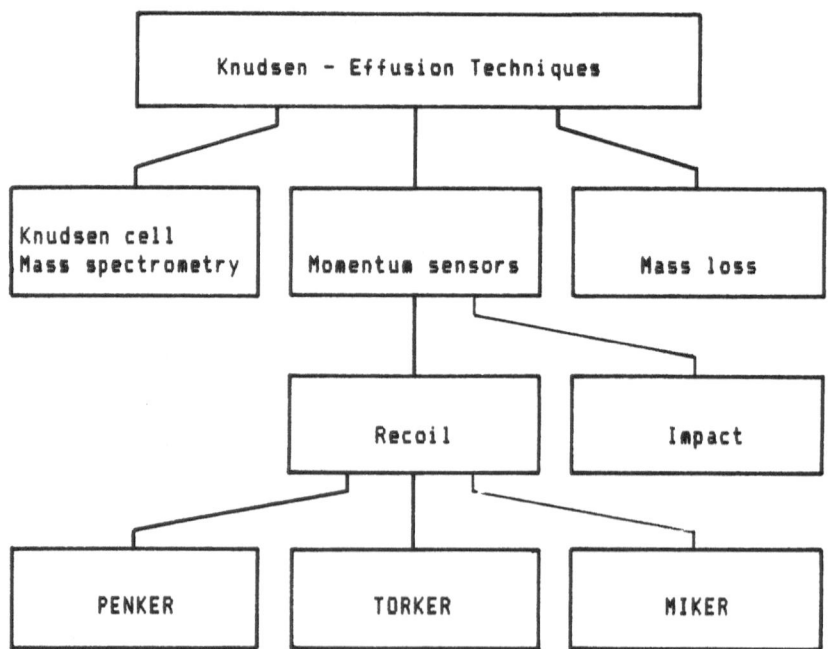

Figure 1.   Block diagram of the Knudsen-effusion techniques.

$$p_j = dm_j/dt \ A_o^{-1} \sqrt{2 \pi k_B \ T/M_j} \tag{1}$$

where p_j = equilibrium partial vapor pressure of the species j, dm_j/dt = mass rate of effusion of the species j, A_o orifice area, k_B = Boltzmann's constant, T = temperature in K, M_j = mass of a molecule of the species j in the vapor. Vapor pressure measurements based on the Knudsen effusion can be performed in three different ways (compare figure 1).

2.1. Mass-loss technique

If the mass M of the molecules in the vapor is known, the direct determination of the mass rate of effusion dm/dt allows the evaluation of the equilibrium vapor pressures of the actual sample by Eq. 1 [5]. Newer application employs collection methods, however, most sensitive analytical techniques like neutron activation, microprobe, radiochemical counting, etc. are required. Herzig and co-workers [6,7] employ Knudsen twin cells and condensation on a collecting plate. The ratio of the vapor pressure of pure species (e.g., Au) and of the partial pressure of this species over the alloy (e.g., Au-Pd, Au-Ag, Au-Ag-Pd) is then determined by analysing the decay rate of 2 radioisotopes of this species (e.g., 195Au, 198Au) in an intrinsic germanium well-type detector. This method yields directly the partial mixing functions of the investigated component (e.g., Au) of the investigated alloy.

2.2. Knudsen cell mass spectrometry

Following this technique, the Knudsen cell is employed as the "gas source" of a high-temperature mass spectrometer, and the effusing molecular beam is directed into the ionisation chamber of the connected mass spectrometer [compare 8,9]. Detecting the ionized vapor beam by means of an electron multiplier yields then the intensities of the composing species of the vaporized sample – may be the most convenient technique to determine the composition of vapor phases. Thermodynamic evaluation can then be based upon the relation between the partial vapor pressure p_j of the species j (inside the Knudsen cell) and the ion current intensities J^l_j of the isotopes l of this species j [2,8-11]:

$$p_j(x_j,T) = J^l_j(x_j,T) \ T \ / \ (C \ \sigma_j \ \gamma_i N^l_j) \tag{2}$$

where x_j = mole fraction of the component j; C = sensitivity factor ; σ_j = ionisation cross section; γ_i = multiplier gain; N^l_j = abundance of isotope l of the species j.

Modern Knudsen cell mass spectrometry is a well established method for high-temperature investigations of the thermodynamic properties of both gaseous and condensed phases and is employed specifically in alloy thermodynamics with great success. However, direct determination of the absolute values of vapor pressures are beset with difficulties associated with indeterminate changes in the sensitivity factor C (for detailed formula see [8]) when sample is changed. Employing the A.I.R.-(algebraic intensity-ratio) method makes independent of the troublesome sensitivity factor C, and yields directly the values of the thermodynamic mixing functions of binary or ternary alloy systems [11-13].

2.3. Momentum sensors techniques

Due to the fact that the pressure of a gas may be defined as the average force per unit area, caused by the momentum which is transferred by the gas to the unit surface per unit time, vapor pressures can be determined by means of suitable impact momentum or recoil momentum sensors.

Impact momentum sensors have been used rather infrequently, because in general they are not sufficiently sensitive. For vapor pressure measurements they have, when compared with recoil momentum sensors, the considerable disadvantage that the fate of the impinging molecules, i.e., whether they condense on, or revaporize from, the target, must be known [2].

In literature three different techniques have been described for vapor pressure measurements based on the recoil momentum of the molecular beam effusing the Knudsen cell (see figure 1).

3. MIKER-TECHNIQUE

The microbalance - inverted Knudsen effusion - recoil-technique (MIKER) was suggested by Margrave [14,15]. This author has pointed out that it is possible to determine both vapor pressure and the average molecular weight if an inverted Knudsen-cell, with an orifice pointing downward, is suspended from a vacuum microbalance. The effusing vapor beam is directed away from the cell suspension and the mass of the cell is determined not only with but also without effusion either by measuring the cell mass both when hot and when cold or by opening and closing the orifice while the cell is on the microbalance [2].

4. TORKER-METHOD

Since Volmer [16] and his co-workers [17] introduced the "torsion - Knudsen effusion - recoil" (TORKER) technique for vapor pressure measurements this method has been used by many groups [2,4]. In the TORKER method the measured vapor pressure is independent of the molecular weights of the effusing molecules. Detailed descriptions are given in Margrave [2] by Carter and Freeman and in Rapp [18] by Cater.

The Knudsen cells applied in the TORKER-technique are often called "TORKER cells" and show not one but 2n orifices (n=1,2,...), with pairs pointing in opposite directions (compare figure 2). The recoil of the 2n antiparallel escaping vapor beams causes the momentum of the TORKER cell, which twists the supporting torsion fibre until it is counteracted by the elastic torsion momentum of the fibre (see figure 2). The deflection angle in rad is measured by means of a light beam, and the vapor pressure p is given by the equation

$$p = \alpha \cdot M_T \cdot F_{2n}, \qquad (3)$$

where M_T is the torsion constant in N.m/rad, and the TORKER cell factor F_{2n} is defined by

$$F_{2n} = (0.25\pi \sum_{i=1}^{2n} d_i{}^2 \; l_i \; f_i)^{-1}, \qquad (4)$$

where d_i = the effusion orifice diameter in m; l_i = the moment arm of d_i in m; f_i = the correction factor after Freeman and Searcy [19].

For description of TORKER apparatus of the simple first generation see [2,4]. Although the principle of the TORKER technique is very simple, in actual application many difficulties can arise [2,20].

4.1. Torsion elements

Measuring range and sensitivity of TORKER apparatus depend basicly on the mechanical quality of the torsion fibres, and on the length of the light beam used. The most common fibres are made from quartz, phosphor bronze or tungsten [2,21], and can be used without greater difficulties for measurements of vapor pressures not lower than 0.1 Pa, a typical TORKER arrangement supposed (TORKER cell: 2 orifices, cell factor $F_2 = 124*10^6$ m^{-3} (d = $8*10^{-4}$ m, l = 0.01 m, f = 0.8); length of the light beam: 2 m). Replacement of the torsion fibres by Pt - Ni tension ribbons allows measurements of vapor pressures of 0.03 Pa and, for TORKER arrangements not heavier than 15 g, even as low as 0.003 Pa [20,22].

4.2. Traditional damping and compensation systems

Unfortunately small vibrations of the torsion system cannot be prevented. Most authors use magnetic damping to control these unavoidable oscillations, whereas Spencer and Pratt [23] and later Cunat et al. [24] successfully damped the vibrations by means of a pair of small paddles partially immersed in vacuum oil contained in a shallow angular trough. Both kinds of damping have the disadvantage that they must always be removed for calibration of the torsion element, if the method of undamped oscillations is to be used.

Compensation of the deflection angle due to effusion with mechanical or electromagnetic forces can be done with less complicated and less expensive experimental equipment than is required for precise angular measurements over large deflections, and long light beams are possible. Precise measurement of torsion pendulum angle is necessary only at the null position. By using an optical lever of 2 m, one can achieve a precision of angular measurement of 1 * 10^{-4} rad with simple procedures and inexpensive equipment. Another advantage of a compensation technique is the much broader range of vapor pressure measurements without changing torsion elements.

252

Munir and Searcy [25] report a mechanical null method (figure 2-A). These authors made use of a goniometer to compensate manually for the angle of deflection due to effusion. The disadvantages of such a mechanical drawing back are evident (see figure 2-A): (i) The goniometer must be in an easy-to-handle position, (ii) Determination of lower vapor pressures demand expensive high vacuum rotary transmission leadthroughs (with respect to the requirements to maintain a sufficient high working vacuum). (iii) Adjusting of the TORKER cell in the high temperature furnace is beset with difficulties with respect to the fixed rotary transmission leadthrough. (iv) Resolution and sensitivity of TORKER arrangements are limited.

Following the traditional electromagnetical compensation techniques [23,26,27], a magnet is rigidly attached to the suspension system, and surrounded by suitable coils (figure 2-B). The torsion momentum due to

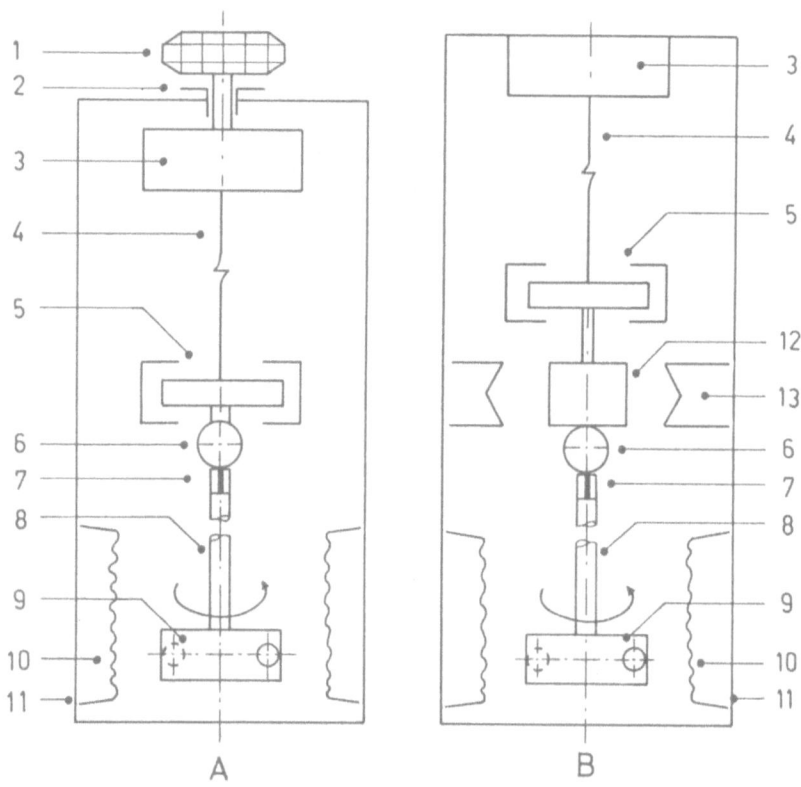

Figure 2. Traditional compensation methods for TORKER apparatus. A: Mechanical null method [25]; B: Electromagnetical compensation [23,26,27]; 1, goniometer; 2, high vacuum rotary transmission leadthrough; 3 adjusting system; 4, torsion fibre; 5, damping system; 6, mirror; 7, bayonet socket of suspension rod (8); 8, suspension rod; 9, TORKER cell; 10, high temperature furnace; 11, UHV housing; 12, magnet; 13, solenoids.

the effusion is exactly counter-balanced by current passing through the coils needed to maintain null rotation. But attachment of a magnet to the torsion fibre, as cited above, limits the sensitivity of TORKER measurements: Light magnets are too weak for compensation and heavier ones require stronger and therefore less sensitive torsion fibres [20].

4.3. Newer TORKER arrangements

4.3.1. Torkometer. This device is a simple, inexpensive null method TORKER apparatus to measure a broad range of vapor pressures without changing the torsion fibre [20,22]. The compensation system of the "Torkometer" can be operated manually or automatically and is based on an industrial core-magnet moving-coil galvanometer with two windings.

A meter movement of a core-magnet moving-coil galvanometers [20] is schematised in figure 4: In contrast to common moving-coil galvanometer the permanent magnet (8) of a core-magnet meter movement is cylindrical, connected rigidly to the support system (2), and surrounded by the moving coil. The winding support (5) of the moving coil is attached to the mirror (10) by the suspension rod (9), and to the lower end of the upper tension ribbon (3). The tension ribbons (3,11) are connected to the support system (2) via curved plate springs (1). The weight of the entire moving coil (including the winding support, both windings (7), and the connection pieces (4,6)) is less than o.5 g [20].

A meter movement such as the one described can be easily added to TORKER arrangements (compare figures 2-B and 3): Enlarge suitably the upper tension ribbon (figure 3, (3)), cut off the lower parts of the supporting system (figure 3, (2)), and replace the lower tension ribbon (figure 3, (11)) by the bayonet socket(figure 2-B, (7)) of the suspension rod (figure 2-B, (8)) of a high temperature TORKER cell. Connecting the adapted meter movement to a precise adjusting system yields then the "Torkometer" [20]. For a suitable precise adjusting system see [22].

Either winding of the moving coil can be used to compensate for the recoil momentum caused by the molecular beams escaping the Knudsen cell, returning the mirror (figure 3, (10)) to its original position. The null point is sensed by means of a 2-meter light beam and the mirror attached to the Knudsen cell suspension. Using a laser beam and 3 photodetectors, the current necessary to maintain null rotation can be automatically regulated via 2 operational amplifiers [20]. The vapor pressure can then be calculated from the current measured by an ampmeter, either manually or automatically by means of a simple microprocessor. Short circuit of the winding not used for actual compensation allows magnetic damping of unavoidable vibrations of the torsion system with the advantage of cutting out the damping at any moment desired. This enables calibration of the torsion fibre under high vacuum with improved accuracy (0.1 %).

In the null method, where the torque due to the vapor is counteracted by electromagnetic interaction, the vapor pressure is related to the current i through the moving coil necessary to maintain zero rotation, and Eq.(3) must be substituted by

$$p = i \cdot C_w \cdot M_T \cdot F_{2n}, \tag{5}$$

where C_w is the winding calibration factor ($C_w = \alpha / i$). The low resistance winding I may be employd to determine small vapor pressures
whereas the high resistance winding II allows the measurement of higher
pressures without using excessive current. By means of such a Torkometer
it is possible to determine continuously vapor pressures within the
range 0.005 Pa to the upper limit of Knudsen effusion conditions (about
20 Pa). Resolution is limited by the current measurement and is about
0.6%-0.9% of the picoammeter range.

The significant advantages of the Torkometer are: (a) there is constant sensitivity over the entire pressure range of Knudsen effusion
conditions; (b) there is negligible mass load of the torsion fibre (max.
0.5 g); (c) calibration of the torsion element can be done under high
vacuum with improved accuracy (0.1%).

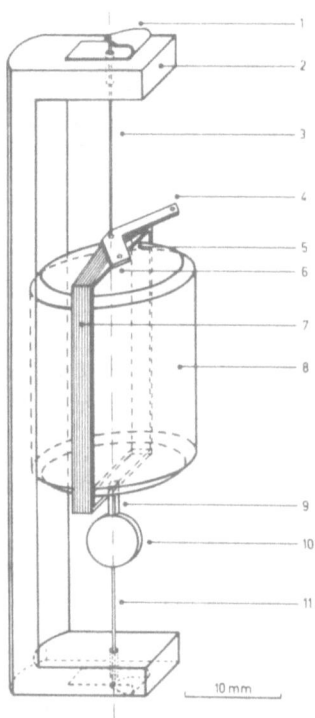

Figure 3. Meter movement of a core-magnet moving-coil galvanometer. 1,
curved plate spring; 2, supporting system; 3, upper tension ribbon; 4,
connection piece of winding I; 5, winding support; 6, connection piece
of winding II; 7, winding I(10 Ohm) and II (4000 Ohm); 8, permanent core
magnet: 9, suspension rod; 10, mirror; 11, lower tension ribbon.

4.3.2. Computer-operated dynamic torsion pendulum. Edwards and co-worker [28,29] have developed a sophisticated computer automated data aquisition system for a dynamic Knudsen torsion pendulum applied to vapor pressure measurements with a precision of +- 0.0035 rad angular deflection. The objective of the automatic data acquisition system is to observe the times at which the oscillating torsion pendulum reaches preset positions, to record those data as matched position versus time pairs, and to use those data to obtain the integral equation of motion of the torsion pendulum.

Following this technique (figure 4-A), laser beams (1) reflected from mirrors (3) on the torsion pendulum are detected by a bank of photodetectors (2) which, in turn, transmit signals representing pendulum positions through an encoder and interface to a personal computer with an internal clock. The computer uses torsion pendulum position and time data with stored information to solve the equation of motion of the torsion pendulum and to calculate vapor pressures. The data acquisition system operates undamped in the dynamic mode, the swing is alternately enhanced and retarded by the recoil momentum of the effusing vapor. The sources of the light beams are four inexpensive, low-power, He-Ne lasers on adjustable mounts which allow them to be directed individually during alignment of the system. The beam detection system consists of a hemicylindrical (radius 0.3 m) bank of 144 phototransistors arranged in four levels, with 36 on each level, to correspond with

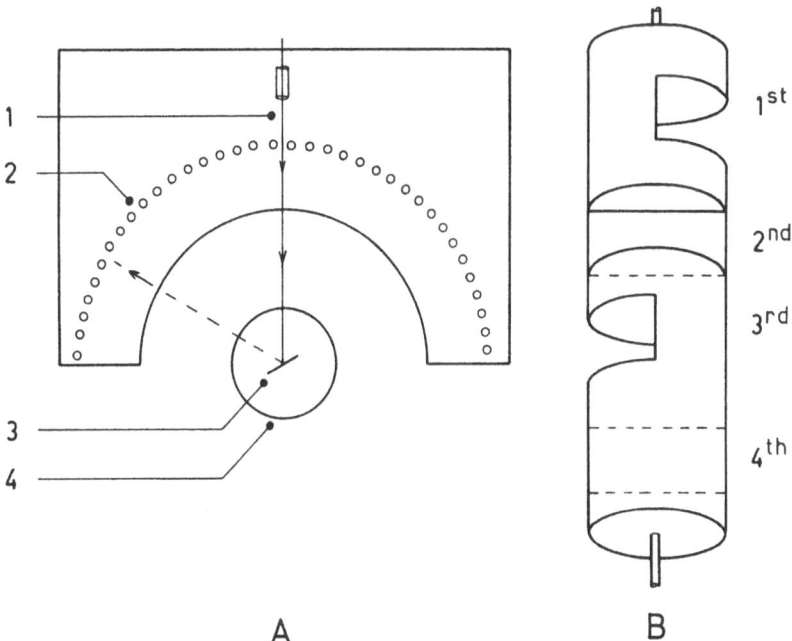

A   B

Figure 4. Computer-operated dynamic torsion pendulum [28,29]. A: Top view of one level of the photodetector assembly: 1, laser beam; 2, phototransistor; 3, mirror; 4, vacuum wall (glass); B: Mirror mount.

the 4 mirrors and the 4 laser beams. The beams are introduced radially along the symmetric bisector of the bank of detectors; thus only the acutely reflected beam falls on the detectors. The detectors are set in 36 vertical mounting posts, each containing 4 detectors, one on each level. The system does not interact physically or through any field with the torsion pendulum; thus no additional torque or damping required compensation [28]. The data acquisition system may be added to existing TORKER apparatus with only minor modification being required. E.g., taking the TORKER arrangement on figure 2-B, the magnet (12) and the solenoids (13) must be omitted, the mirror (6) substituted by the 4-mirror mount (figure 4-B), and a glass vacuum wall must be incooperated.

For vapor pressure measurements at high temperatures the equation of motion when the TORKER cell is at the experimental temperature is compared with the equation of motion when the cell is at room temperature to determine the torque due to effusing vapor and thence the vapor pressure inside the TORKER cell [28].

4.4. Discussion of TORKER technique

The important advantage of the TORKER method lies in the high sensitivity and in the possibility of quick measurements of the total vapor pressures at different temperatures. But this is balanced by difficulties, discussed in detail by Freeman in [2]. Perhaps the most important problem of the TORKER technique arises from the unsymmetrical furnace-TORKER cell configuration which causes thermomolecular flow forces especially at high temperatures [2]. Czanderna in [30], and Thomas and Poulis in [31] estimated that this effect can falsify the torque caused by the torsion momentum which is produced by the effusing vapor beams.

Reduction of uncertainties in vapor pressure measurements due to insufficient vacuum surrounding the TORKER cell, unstable furnace temperatures, and temperature gradients is no longer a question of technical knowhow, but only of available money and resources.

5. PENKER-METHOD

"PENKER" stands for Pendulum electronically balanced Knudsen effusion - recoil and represents a new recoil momentum sensor technique [4,32]. As figure 5 shows the PENKER-method is an automated null method employing a linear pendulum, suspended near the centre of mass, instead of the torsion pendulum which is used in the TORKER method.

Following the PENKER-method [32] the determination of vapor pressures is also independent of the molecular weights of the effusing molecules: the displacement of a linear pendulum, caused by the recoil of the escaping molecular beam which is automatically counteracted by an electromagnetic compensation system which is electronically controlled. Using a linear pendulum for vapor pressure measurements, the effusion orifice must be drilled into the side-wall of the Knudsen cell - mnemonically called "PENKER cell" (see sector drawing in figure 5). To double or multiply the force due to the escaping vapor beam two or more PENKER cells can be used together (compare figure 5) [32].

A PENKER apparatus is schematised in figure 5. The pendulum (3) is
supported by a sapphire knife-edge (4) and an agate edge bearing. Therefore the bearing friction loss is negligible small. The upper arm of the
pendulum consists of an alumina capillary with a thermocouple (3). The
PENKER cell (2) is connected to the upper arm of the pendulum (3) via a
bayonet socket and is in the centre of the high temperature furnace (1).

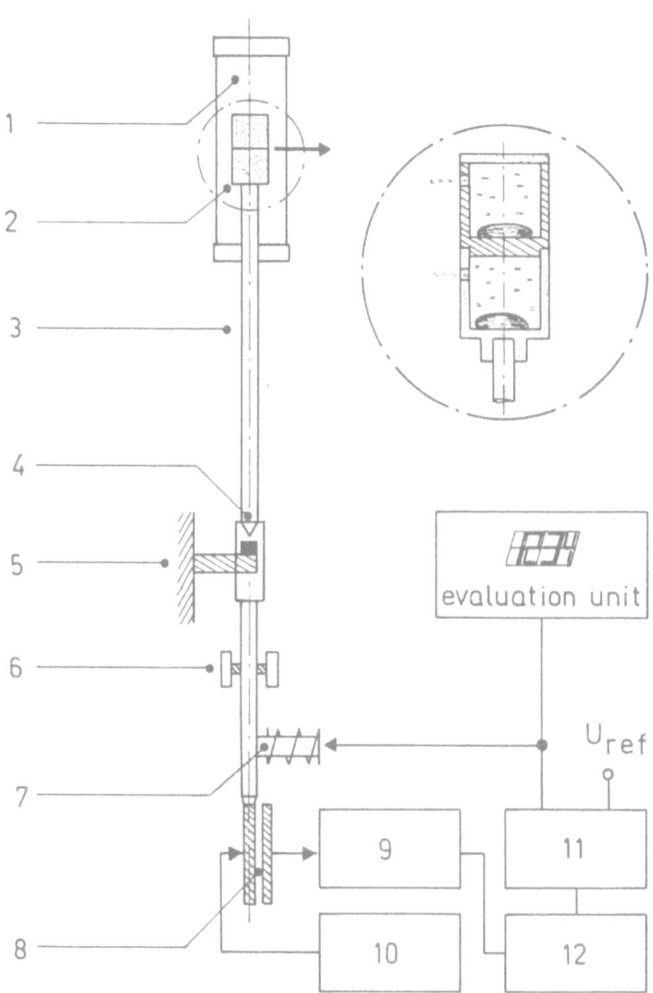

Figure 5. Schematic detail of a PENKER apparatus (Pendulum electronically balanced Knudsen-effusion recoil). 1, high tem-perature furnace;
2, PENKER cell; 3, pendulum arm; 4, edge of the pendulum (sapphire); 5,
adjusting and catching system; 6, counterweight; 7, compensation coil;
8, plate condenser; 9, bridge amplifier; 10, sine-oscillator; 11, regu-
lation amplifier; 12, rectifier.

The temperature is measured directly below the PENKER cell (2) by the thermocouple or by an optical pyrometer (not in figure 5). When sample is changed, the sensitive pendulum system must be protected. This requires a robust, but precise adjusting and catching system (5): A catching-motor lifts the pendulum from the edge bearing until it is caught within a conical guide (not in figure 5). The adjusting system (5) makes sure that the knife-edge (4) will be positioned reproducibly [32].

5.1. Control of the pendulum position

The control of the pendulum position together with precise counteracting of pendulum displacement is the most sensitive part of PENKER apparatus. Convenient devices may be obtained by simple adapting of linear susceptibility meter [32]. It is only necessary, to replace the electromagnet and sample arrangement with the high temperature furnace and the PENKER cell. Fully automated systems with high resolution (about 2×10^{-13} SI units) have been reported by several authors [33-35].

Test measurements [32] on cadmium and on lead proved the efficiency of the compensation device constructed in cooperation between Sobczak and the A.PAAR KG, Graz (Austria) [34]: The deflection of the linear pendulum is determined capacitively by means of a plate condensor (figure 5; (8)). One plate of the condensor is rigidly attached to the lower part of the pendulum (figure 5; (3)), the second plate is rigidly attached to the outer framework (not in figure 5). The actual capacity of the plate condensor (figure 5; (8)) is determined electronically and converted into the corresponding intensity of the compensation current traversing the compensation coil (figure 5; (7)). The required current intensity is proportional to the vapor pressure and is indicated by the four 1/2 digits evaluation unit with various ranges of sensitivity [32]. With the counterweight (figure 5; (6)) the sensitivity of the balance can be raised to a desired value.

PENKER apparatus may also be based on the compensation system developed by Müller and Güntherodt [35]: The spot of a laser beam, which is reflected from a mirror fixed to the frame of the pendulum gives rise to an error signal on a photopotentiometer. This signal is amplified and controls a current source. The controller increases the current through the compensation coil until the pendulum is returned to null position.

5.2. Discussion of PENKER technique

The difficulties arising from the application of the TORKER method, as described in point 4.4., can be dealt with by employing the PENKER technique. At present PENKER apparatus are less sensitive the Torkometer, nevertheless the PENKER method has clear advantages over the TORKER technique: (a) The symmetrical furnace-cell configuration together with the fact that the PENKER cell is rigidly attached to the arm of a linear pendulum, prohibit electromagnetically induced cell rotations. (b) Neither oscillations of the pendulum nor a null point shift may be expected - even at high temperatures (compare [32] and [35]). (c) The cylindrical PENKER cells enable smaller diameter of the cylindrical furnace, and the necessary electric input is clearly reduced too.

Another clear advantage over the TORKER method is (d) the better mechanical stability of the used linear pendulum. (e) PENKER cells are simpler manufactured than TORKER cells, and the same cells can be used for measuring either the vapor pressure of one sample or the difference in vapor pressure of two samples: The effusion orifices must only be turned either to the same direction or to opposite ones.

Acknowledgments

Grateful acknowledgment is made to the financial support of the "Fonds zur Förderung der wissenschaftlichen Forschung in österreich".

References

1. L.D.Landau, and E.M.Lifschitz, Lehrbuch der Theoretischen Physik Band X Physikalische Kinetik (Akademie, Berlin, 1983).

2. J.L.Margrave, editor,The Characterization of High-Temperature Vapors (Wiley, New York, 1967).

3. O.Kubaschewski, and C.B.Alcock, Metallurgical Thermochemistry 5th edn (Pergamon, Oxford, 1979).

4. K.L.Komarek, Ber.Bunsenges.Phys.Chem. 87 (1983) 709.

5. M.Knudsen, Ann.Physik 28 (1909) 75; Ann.Physik 28 (1909) 999.

6. H.E.Peltner, and Chr.Herzig, Acta metall. 29 (1981) 1107.

7. R.Höhn, and Chr.Herzig, Z.Metallkde. 77 (1986) 291.

8. J.Tomiska, J.Phys.E: Sci.Instrum. 17 (1984) 1165.

9. J.Tomiska, Z.Metallkde. 76 (1985) 532.

10. A.Neckel, and S.Wagner, Ber.Bunsenges.Phys.Chem. 73 (1969) 210.

11. J.Tomiska, H.Nowotny, L.Erdélyi, and A.Neckel, Z.Metallkde. 68 (1977) 350.

12. J.Tomiska, High Temp.-High Pressures 14 (1982) 417.

13. J.Tomiska, and A.Neckel, Z.Metallkde. 77 (1986) 649.

14. J.L.Margrave, J.Chem.Phys. 27 (1957) 1412.

15. J.L.Margrave, Physico-Chemical Measurements at High Temperatures, editors Bockris, White and Mackenzie, (Butterworth Scientific Publications, London, 1959).

16. M. Volmer, Z.Phys.Chem. Bodenstein Festband 156A (1931) 863.

17. W.Teuscher, Dissertation Technische Hochschule, (Berlin, 1923);
    S.Heller, Dissertation Technische Hochschule, (Berlin, 1930);
    K.Neumann, Dissertation Technische Hochschule, (Berlin, 1930).

18. R.A.Rapp, editor, Physicochemical Measurements in Metal Research
    Vol.IV, Part 1, of Techniques of Metal Research, editor
    R.F.Bunshah, (Interscience, New York, 1970).

19. R.D.Freeman, and A.W.Searcy, J.Chem.Phys. 22 (1954) 762.

20. J.Tomiska, Rev.Sci.Instrum. 52(5) (1981) 750.

21. R.D.Freeman, and P.D.Gwinup, Rev.Sci.Instrum. 37 (1966) 773.

22. J.Tomiska, Dissertation Universität Wien, (Wien, 1973).

23. P.J.Spencer, and J.N.Pratt, Br.J.Appl.Phys. 18 (1967) 1473.

24. Ch.Cunat, G. Chervet, J.P.Hilger, and J.Hertz, Mem.Sci.Rev.Metall.
    73 (1976) 185.

25. Z.A.Munir, and A.W.Searcy, J.Chem.Phys. 42 (1965) 4223.

26. C.L.Rosen, Rev.Sci.Instrum. 31 (1960) 837.

27. R.R.Hammer, and J.A.Pask, J.Am.Ceram.Soc. 47 (1964) 264.

28. J.G.Edwards, M.K.Heckler, and H.B.Thompson, Rev.Sci.Instrum. 50
    (1979) 374.

29. H.B.Thompson, and J.G.Edwards, J.Vac.Sci.Technol. 17 (1980) 117.

30. M.J.Katz, editor, Vacuum Microbalance Techniques, Vol 1 (Plenum,
    New York, 1961).

31. K.H. Behrndt, editor, Vacuum Microbalance Techniques, Vol 13 (Ple-
    num, New York, 1963).

32. J.Tomiska, J.Phys.E: Sci.Instrum. 14 (1981) 420.

33. E.Wachtel, and W-U.Kopp, Metall 27 (1973) 1174.

34. Anton PAAR KG Graz (Austria), Suszeptibilitätsmeßgerät SUS 10
    (Graz, 1977); private information to J.Tomiska (1979);
    R.Sobczak, and H.Bittner, Rev.Chim.min. t 6 (1969) 983.

35. M.Müller, and H-J.Güntherodt, J.Phys.E: Sci.Instrum. 14 (1981) 453.

THERMODYNAMIC ACTIVITY FROM KNUDSEN MEASUREMENTS IN   α - Ag-As AND

α - Ag-Bi ALLOYS

D. Schmid and Th. Hehenkamp

Institut für Metallphysik and SFB 126, Universität Göttingen,

Hospitalstr. 3/5, 3400 Göttingen, F.R. Germany

Abstract

In copper alloys two major contributions influence the chemical excess potential $\Delta \mu^{xs}$ in dilute alloys, repulsive electronic and attractive elastic impurity-impurity interactions. In some cases this led to a distinct minimum of $\Delta \mu^{xs}$ at rather low concentrations. It is the purpose of the present paper to check silver alloys in this respect too. In order to see the influence of misfit clearly two silver systems were chosen, one with almost no (Ag-As) and one with a large misfit (Ag-Bi). The results indicate that the silver systems behave in the same way as the respective copper systems.

H. Brodowsky and H.-J. Schaller (eds.), Thermochemistry of Alloys, 261–275.

Introduction

The Knudsen cell is frequently used to determine absolute vapor pressures of pure elements and alloys. Knowledge of these data directly yields the thermodynamic activity a_i and the chemical excess potential $\Delta \mu_i^{xs}$ of a component i, which are essential for the elaboration and the improvement of solution models. The excess potential of noble metal alloys with impurity atoms of positive excess charge has been studied extensively by Hehenkamp et al [1 - 5] . From the previous investigations particularly in copper alloys it is well known that in dilute alloys at least two major effects contribute quite generally to the chemical excess potential, repulsive electronic impurity-impurity interactions and attractive elastic interactions. The concentration dependence of both effects is opposite in direction and different in order leading to a minimum in $\Delta \mu_i^{xs}$ at low concentrations. The purpose of the present investigation is to get more information about the misfit influence in silver alloys. Ag-As and Ag-Bi alloys are particularly suitable since they have the same electronic excess charge, whereas the misfit in Ag-As is negligible in contrast to Ag-Bi, where it is large.

Therefore, it is necessary to measure activities at dilute impurity concentrations to separate misfit from electronic interactions.

Experimental methods

For alloy production and activity measurements knowledge of the solidus curve is of vital importance. Since the maximum solubility of Bi in solid Ag is only a few at%, old solubility data from Hansen [7] have been redetermined. The method employed involves quenching of the alloys from the two phase liquid-solid-region and subsequent electron microprobe analysis of the α- grains [8] . The results are shown and compared with those reported by Hansen in Fig. 1 . The accuracy of our data is about ± 5 K in temperature.Data given by Hansen have to be substantially corrected by up to 50 K.

The vapor pressures have been measured using a Knudsen apparatus which is described in principle in previous papers [2,5] . A Knudsen cell made of graphite is directly heated by an electric current passing through it.

The vapor is condensed on Cu-substrates which are fixed in front of the effusion orifice. The possibility to accumulate material on the substrate for a long period of time and analysis of the vapor elements afterwards are the great advantages of this technique, since it permits long "noise free" integration times. In order to optimize the apparatus for the present measurements some modifications were necessary. The condensation of the vapor can be improved in general by cooling the substrates. This has been realized by water cooling so far. However, the condensation factor of As on Cu substrates was only about 0.2 %. In order to improve the condensation factor the water-cooling system was replaced by a LN_2-cooling system. Additionally the oil diffusion pump had to be replaced by a combination of a turbo molecular pump and a titanium sublimation pump to achieve an almost oil free vacuum (see Fig. 2). Otherwise oil would condense on the substrates and prevent proper sticking of the evaporated layers. In this way a relative improvement in condensation about two orders of magnitude could be obtained for As. In case of Bi the relative improvement was smaller but the absolute value increased from 20 % to more than 80 % at least. For activity measurements in very dilute alloys it is necessary to minimize the loss of impurity component between evaporation intervals. In particular this is important for alloys with high vapor pressure, such as Ag-As and Ag-Bi. Therefore a mechanism for sealing the Knudsen cell during the warm up period has to be installed. In order to obtain equilibrium conditions between gas and solid inside the cell the surface of the sample material has to be a factor of about a hundred or more larger than the area of the orifice of the Knudsen cell. This was realized by using as many fine alloy chips as possible without disturbing the vapor flow. In this way the range of measurements could be extended to silver alloys with Bi concentrations as low as 0.1 at%.

The vapor pressure P_B of component B can be easily determined by Knudsen experiments using the Hertz-Knudsen-Langmuir-equation [9]

$$\frac{m_B}{t} = P_B \cdot A \cdot \alpha \cdot \beta \cdot k \ \sqrt{\frac{M_B}{2 \pi \ RT}} \tag{1}$$

m_B is the total mass of component B deposited on the substrate during evaporation time t, A the area of the effusion orifice, α and β the coefficients of evaporation and condensation respectively, k the Clausing factor [10] , which takes the impedance for the gas flow through the orifice into account and is determined by the ratio of the length of the orifice channel and the orifice diameter. M_B is the molar mass of B, R the gas constant and T the absolute temperature.

Many of the alloying elements like As and Bi form different molecular associates in the gas phase. In order to derive correct data for the actual pressure from Knudsen measurements it is necessary to know the distributions of the different associates in the vapor. These, however, cannot be measured directly but may be calculated from their equilibrium constants K (T) following the law of mass action

$$P_{B,j} \Big/ (P_{B,1})^j = K_j (T) \quad (j = 2,3,4 \text{ in case of As}) \tag{2}$$

$P_{B,1}$ is the partial pressure of the monoatomic species and $P_{B,j}$ the partial pressure of the j-th associate of component B, respectively. According to Hehenkamp and Lüdecke [2] equations (1) and (2) lead to

$$\frac{m_B}{t} = A \; \alpha \; \beta \; k \; \sqrt{\frac{M_B}{2 \pi RT}} \; [P_{B,1} + \sqrt{2} \; K_2 \; (P_{B,1})^2 +$$

$$\sqrt{3} \; K_3 \; (P_{B,1})^3 + \sqrt{4} \; K_4 \; (P_{B,1})^4] \tag{3}$$

The solution of equation (3) gives $P_{B,1}$ (T), if the equilibrium constants are either measured as function of temperature or known from the literature. In case of As and Bi the data for the equilibrium constants have been taken from Barin and Knacke [11] . The total pressure P_B will be obtained by using the law of mass action and summing up the partial pressures. Moreover the equilibrium between these different molecular aggregates has to be taken properly into account calculating the thermodynamic activity a_B. For example the arsenic activity is given by [2,12]

$$a_{As} = \frac{P_{As,1}}{P^o_{As,1}} = \sqrt[2]{\frac{P_{As,2}}{P^o_{As,2}}} = \sqrt[3]{\frac{P_{As,3}}{P^o_{As,3}}} = \sqrt[4]{\frac{P_{As,4}}{P^o_{As,4}}} \tag{4}$$

Here $P_{As,i}$ and $P^o_{As,i}$ are the vapor pressures of the i-th associate above the alloy and above the pure solute, respectively, at the same temperature.

The surface layer deposited on the substrates has been analyzed with an electron microprobe. This is possible if the thickness of the layer is less than the penetration of the electron beam. A typical distribution profile of a bismuth layer condensed on copper substrates obtained with an ARL EMX-SM electron microprobe is shown in Fig. 3. The axis of ordinate is calibrated versus maximum counting rates. Data for an ideal and a real orifice are calculated according to Clausing's cosine law [13]

$$\frac{m}{t} = \frac{m}{t} \ (\ x = o \ ) \ \cdot \ S \ \cdot \ \cos^4 \theta \qquad (5)$$

where θ is the angle between the center of the orifice and the point of impact on the target, S a factor which takes the nonideality of the orifice into account.

The good agreement between theoretical and experimental data makes evident that the vapor flow was solely determined by the flow out of the effusion orifice without a possible disturbance by surface diffusion. This permits a quick check, whether the operating conditions of the Knudsen cell are well maintained or not. Since the profile in Fig. 3 is radially symmetric, the total mass m_B deposited on the substrate can be easily obtained by integration. In order to get absolute values for m_B, however, the counting rates have to be calibrated versus mass per unit area. Four calibration procedures will be presented here; three experimental, using pure elements as standard states, and a theoretical method. One possibility is to form homogeneous thin layers of known mass per area, for example by vapor plating [2]. The thickness of this layer may be determined by resonance vibration. Another method uses the Knudsen apparatus itself [12] . Half of a substrate is vapor plated whereas the other part is covered by a thin tungsten foil. In this way a step is produced which enables measurements with an interference microscope based on Tolansky interference. The great advantage of this procedure is that several calibration points can be got by one measurement. The direct determination of the mass of the film by weighing is possible too, the substrates have to be weighed before and after vapor plating. In this case several standards are necessary to eliminate the drifts depending on the balance and the humidity of the air. This method was applied for Bi using a mechanical microbalance. The result is given in Fig. 4. k_{max} is the ratio of the maximum counting rate on the substrate and the counting rate on pure massive Bi. The relation k_{max} versus mass of condensed Bi shows the typical slight S-form depending on the film-substrate combination and the electron microprobe parameters acceleration voltage and wavelength. Moreover the mass per unit area can be estimated by theoretical considerations. A simple method for calculating by hand is proposed by Bishop and Poole[14]. Using a model describing the electron penetration and energy loss in the specimen, they have obtained a set of curves by Monte Carlo calculations. These curves are functions of the mean atomic number of the target and also of the beam energy and the critical ionization potential of the element under investigation. They represent the fraction of characteristic ionization produced within a surface layer, which is expressed as a fraction of the Bethe range. From these curves the thickness of surface layers ranging from less than $1 \mu g \quad cm^{-2}$ up to several mg cm^{-2} may be rapidly estimated from microprobe measurements. Three values determined in this way are shown in Fig. 4 too. Deviations of nearly 30 % are found. This depends mainly on the fact, that the depth distribution function in the film is different from that in the pure standard because of the influence of the substrate. This difference is small if the mean atomic numbers of the surface film and the substrate are similar but becomes significant if there is a large discrepancy in atomic numbers as in the case of Bi film ($Z = 83$) and Cu substrate ($Z = 29$). Accordingly the deviations for the combination As film ($Z = 33$) and Cu substrate are much smaller [12].

It can be concluded that the method of Bishop and Poole gives quantitative information about the thickness of surface films quickly without experimental expenditure. However, in case of adverse film-substrate combinations suitable corrections are necessary.

Results and Discussion

The solution of the modified Knudsen equation (3) gives the vapor pressure of each molecular associate of the solute component. The degree of association α_j can be determined therefrom, which is defined as the ratio of vapor pressure of the j-th associate to the total pressure.

$$\alpha_j = P_{B,j} \bigg/ \sum_{j=1}^{n} P_{B,j} \qquad \left(\begin{matrix} n = 2 \text{ for Bi} \\ n = 4 \text{ for As} \end{matrix} \right) \qquad (6)$$

In case of Ag-As and Ag-Bi alloys the trend of the calculated values for α_j is in excellent agreement with those reported for Ag-Sb alloys [2]; the degree of association strongly increases upon alloying and with decreasing temperature.

The thermodynamic activity of As and Bi in solid Ag alloys is shown in Fig. 5 .

Obviously the activity of Bi in Ag is somewhat higher than the activity of As in Ag. In both cases the values corresponding to Raoult's law are drastically exceeded, however. The deviations from ideal behavior can be described more clearly by the chemical excess potential $\Delta \mu_B^{xs}$ defined by

$$\Delta \mu_B^{xs} = RT \ln \left(\frac{a_B}{x_B} \right) \qquad (7)$$

R being the gas constant , a_B the thermodynamic activity and x_B the molar fraction of solute component B. The excess potential was calculated by fitting polynomials of different orders to the data. For Ag-As alloys a second order fit is satisfactory whereas for Ag-Bi alloys only a polynomial of third order can meet the boundary condition $a_B(x_B = 0) = 0$ sufficiently. The results are given in Fig. 6. $\Delta \mu_B^{xs}$ of Ag-As increases monotonically with As concentrations whereas a flat decrease gives an indication for a minimum in Ag-Bi. For further analysis it is useful to introduce the relative chemical excess potential $\Delta \mu_{B,rel}^{xs}$ and to separate its different contributions [1]

$$\Delta \bar{G}_{B,rel}^{xs} (x_B) = \Delta \bar{G}_B^{xs} (x_B) - \Delta \bar{G}_B^{xs} (x_B = 0) = \Delta \bar{G}_{B,rel}^{xs,str} + \Delta \bar{G}_{B,rel}^{xs,phon}$$
$$+ \Delta \bar{G}_{B,rel}^{xs,el} \qquad (8)$$

$$\Delta \bar{G}_{B,rel}^{xs,str} \quad , \quad \Delta \bar{G}_{B,rel}^{xs,phon} \quad , \text{ and } \Delta \bar{G}_{B,rel}^{xs,el}$$

are the relative contributions due to elastic strain, phonons and electrons respectively.

For a rough calculation of the phonon contribution a simple Debye model can be employed yielding that the vibrational entropy is correlated with the change of the Debye temperature θ_D upon alloying. For some noble metal systems these changes have been published by Massalski and Mizutani [15] (see Fig. 7). In case of several Ag alloys negliglible changes of θ_D were found within the concentration range of the present measurements. Specifically in Ag-As and Ag-Bi the Debye temperatures have not been measured yet but it seems to be a good approximation to assume a similar behavior and to neglect the phonon contribution.

The elastic energy due to the misfit of the impurity atoms can be estimated employing the model of a dilatation center in an isotropic elastic medium. According to Eshelby and Friedel [16,17] the enthalpic part of the strain energy is given by

$$\Delta \bar{H}_{B,rel}^{str} = \frac{2 G_A \Omega_A \delta^2}{9 \frac{(1-\nu)}{(1+\nu)}} \quad [\,(1 - x_B^2)^2 - 1\,] \tag{9}$$

G_A being the shear modulus of the matrix A, Ω_A its atomic volume, ν its poisson ratio.

$$\delta = \frac{1}{\Omega} \frac{d\Omega}{dx_B} \tag{9a}$$

is the atomic size or misfit factor. Calculated values for the relative partial misfit energies of silver alloys are summarized in Fig. 8.

The electronic contribution of the relative excess potential can be estimated now. In accord with earlier investigations, summarized in Fig. 9 [2, 18] the slopes of $\Delta \bar{G}_{B,rel}^{xs,el}$ are flat at lowest impurity concentrations becoming steeper as concentration rises. This agreement corroborates the formalism presented here. Moreover it gives support to the conclusion that only big misfits will lead to an observable minimum in the excess potential.

A similar result was already found in case of copper alloys with electropositive solutes [3, 19, 20] . In Cu-Ge, an alloy where misfit is small, a monotonic increase in the excess potential was observed whereas in Cu-Sb, a system with large misfit, a distinct minimum was found. Furthermore it has been discussed in an earlier paper [21] that this systematic behavior might be valid for Cu-In and Cu-Sn alloys as well [22, 23] . Fig. 10 shows the calculated electronic contribution to the excess potential for all these copper alloys. The qualitative and quantitative correspondence to the silver alloys becomes obvious now.

Finally it can be concluded that the approach for the explanation of the changes in the excess potential upon alloying presented here is justified by the results, though it neglects possible changes of the excess entropies.

References

[1] Th. Hehenkamp and V. Schlett, Acta metall. 28, 1721 (1980)

[2] Th. Hehenkamp and D. Lüdecke, Acta metall. 29, 939 (1981)

[3] Th. Hehenkamp, Mikrochimica Acta (Wien), Suppl. 9, 15 (1981)

[4] D. Lüdecke, C. Lüdecke and Th. Hehenkamp, Acta metall. 31, 95 (1983)

[5] Th. Hehenkamp, Ber. Bunsenges. Phys. Chem. 87, 806 (1983)

[6] W. B. Pearson, Handbook of Lattice Spacings and Structures of Metals and Alloys, Pergamon Press, London (1958)

[7] M. Hansen and K. Anderko, Constitution of Binary Alloys, McGraw Hill, New York, Toronto (1958)

[8] Th. Muschik and Th. Hehenkamp, Z. Metallkunde 78, 358 (1987)

[9] A. N. Nesmeyanov, Vapor Pressures of the Chemical Elements, Elsevier, Amsterdam (1963)

[10] R. D. Freeman and J. G. Edwards, in: Condensation and Evaporation of Solids, E. Rutner, P. Goldfinger, J. P. Hirth (Editors), New York (1964)

[11] I. Barin, O. Knacke and O. Kubaschewski, Thermochemical Properties of Inorganic Substances, Springer, New York (1977)

[12] V. Behrens, Thesis, Göttingen (1985)

[13] P. Clausing, Z. Physik 66, 471 (1930)

[14] H. E. Bishop and D. M. Poole, J. Phys. D. Appl. Phys. 6, 1142 (1973)

[15] T. B. Massalski and U. Mizutani, Prog. Mater. Sci. 22, 151 (1978)

[16] J. D. Eshelby, Solid St. Phys. 3, 79 (1956)

[17] J. Friedel, Phil. Mag. Suppl. 45, 456 (1954); Adv. Phys. 3, 446 (1954)

[18] C. B. Alcock, K.T. Jacob and T. Palamutcu, Acta metall. 21, 1003 (1973)

[19] Th. Hehenkamp, W. Schmidt and V. Schlett, Acta metall. 28, 1731 (1980)

[20] V. Behrens, Diploma Thesis, Göttingen (1980) (see also [3])

[21] D. Schmid, V. Behrens and Th. Hehenkamp, Acta metall. to be published

[22] D. Bhattacharya and D. Bruce Masson, Metall. Trans. 5, 1357 (1974)

[23] F. Sommer, W. Balbach and B. Predel, Thermochim. Acta 33, 119 (1979)

[24] K. T. Jacob, C. B. Alcock and J. C. Chan, Acta metall. 22, 545 (1974)

Acknowledgments

The financial support of this work by Deutsche Forschungsgemeinschaft is gratefully acknowledged. Computer calculations have been performed at Gesellschaft für wissenschaftliche Datenverarbeitung Göttingen.

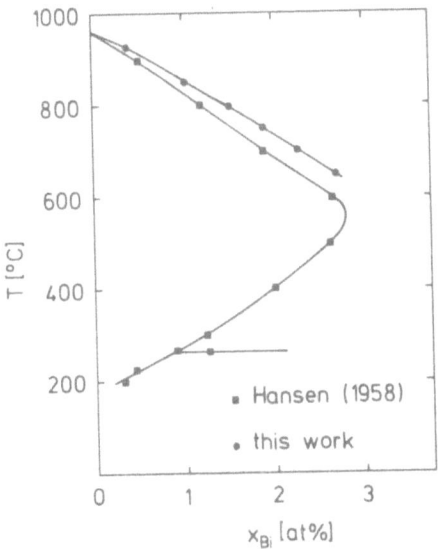

Fig. 1.   Solubility of Bi in solid Ag according to this work
and (. 7)

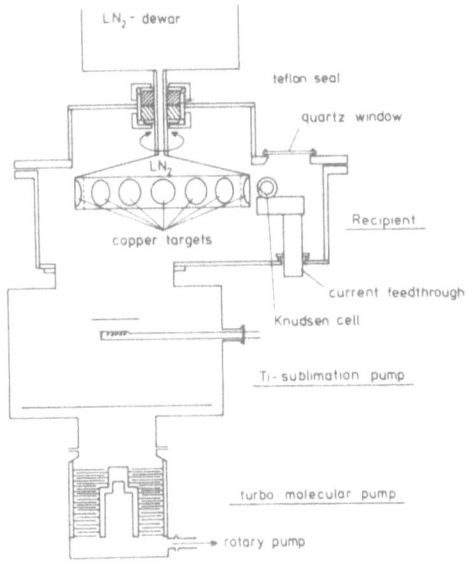

Fig. 2   The modified effusion apparatus

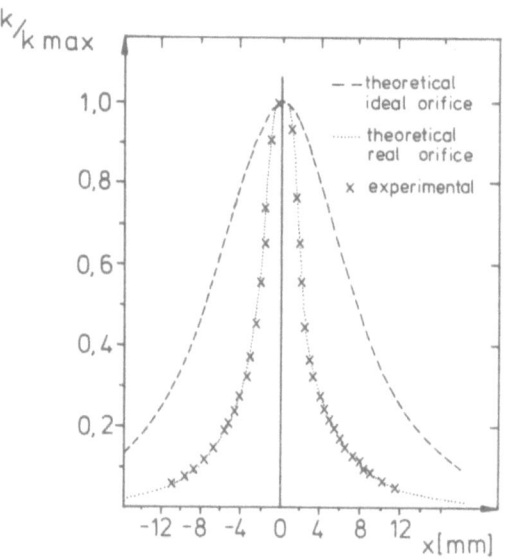

Fig. 3   Distribution of bismuth on a copper-substrate in compa-
          rison with calculated data according to Clausing-equation (5)

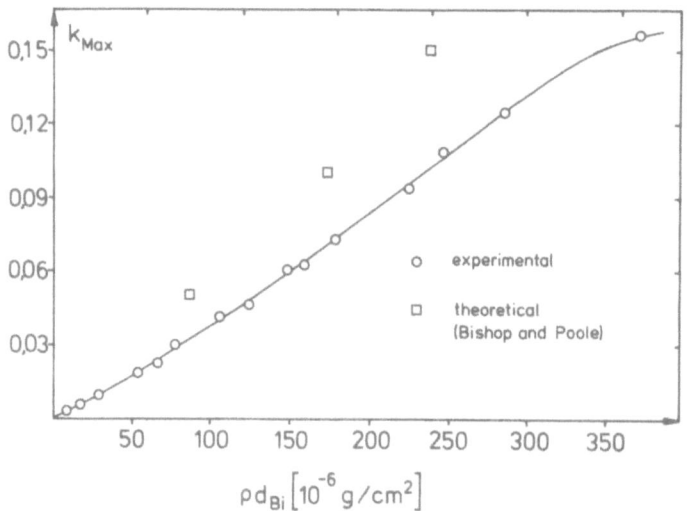

Fig. 4   Calibration of the electron microprobe: Experimental and
          theoretical (14) values for a Bi-film on a Cu-substrate

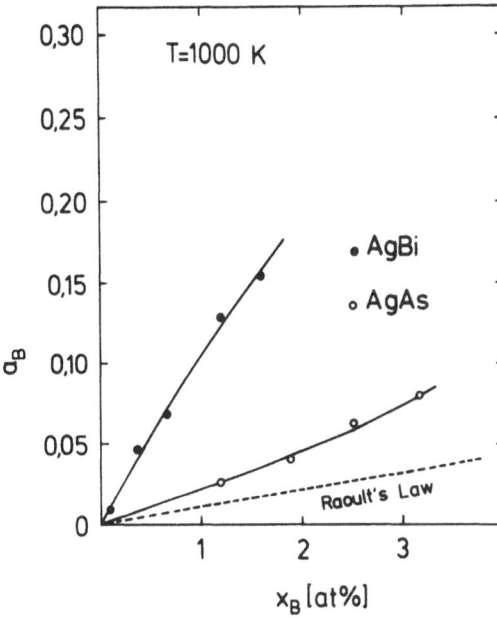

Fig. 5   Thermodynamic activities of As and Bi in α-Ag-As and
          α - Ag-Bi alloys

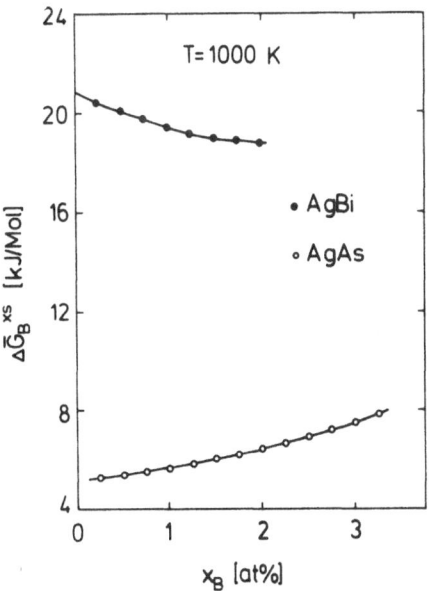

Fig. 6   Chemical excess potentials of As and Bi in α -Ag-As
          and   α-Ag-Bi alloys

274

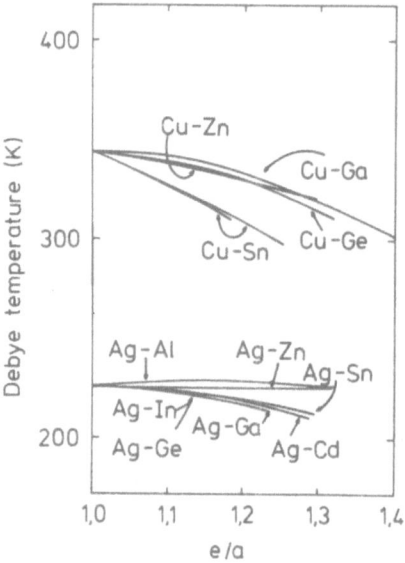

Fig. 7   Changes of the Debye temperature upon alloying in various
          noble metal systems   (15)

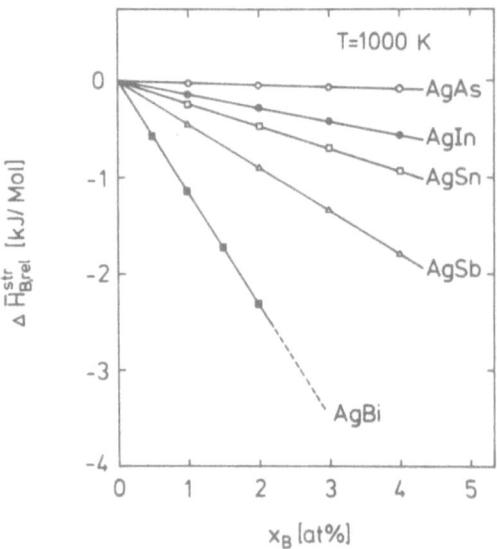

Fig. 8   Relative partial misfit energies for several silver alloys

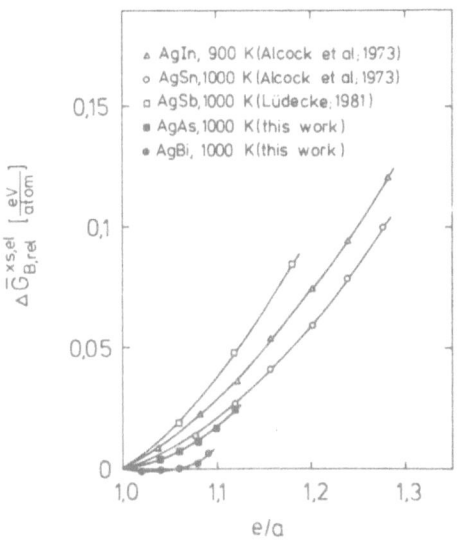

Fig. 9   Electronic contributions to the relative chemical excess potentials of Ag alloys vs electron to atom ratio $\frac{e}{a}$ (2, 18)

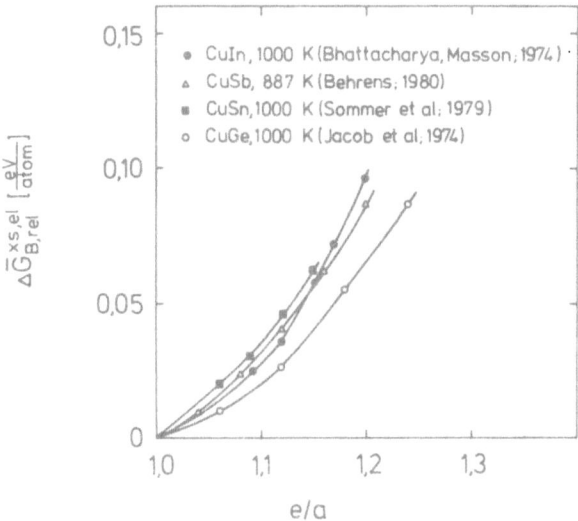

Fig. 10   Electronic contributions to the relative chemical excess potentials of Cu alloys vs electron to atom ratio $\frac{e}{a}$ (20, 22, 23,
24)

TORSION EFFUSION MEASUREMENTS ON Fe-Mn, Co-Mn and Fe-Co-Mn ALLOYS

F. H. Hayes & G. McHugh*
Metallurgy and Materials Science Dept.,
University of Manchester/UMIST,
Grosvenor Street, Manchester M1 7HS,
United Kingdom.

ABSTRACT. The aim of this work is to obtain the thermodynamic
properties of the solid alloy phases of binary and ternary combinations
of iron, cobalt and manganese. This has been achieved by measuring
equilibrium manganese vapour pressures using the Torsion Effusion
Technique applied to Fe-Mn, Co-Mn and Fe-Co-Mn alloys at temperatures
in the range 1250K to 1450K. The design and construction of the
torsion effusion equipment, suitable for studies at temperatures up to
2000K, are described. Results obtained for the Fe-Mn and Co-Mn systems
are presented and compared in the form of activity-composition diagrams
with results of previous workers. Results for Fe-Co-Mn alloys with
Fe/Co ratio of 1.05 are given in the form of manganese vapour pressure-
temperature plots and as plots of integral Gibbs energies, integral
enthalpies and integral entropies of mixing versus manganese content at
1350K. It is found that the Bonnier equation applied in the direction
of manganese gives values which agree closely with the present Fe-Co-Mn
results.

1. INTRODUCTION

Up to the present time very few systematic thermodynamic investigations
have been conducted on ternary or multicomponent alloy systems,
especially among the more complex high melting-high affinity transition
metal systems (1). A group of transition metals of the first period,
of considerable practical and theoretical interest, namely manganese,
iron, cobalt and nickel has therefore been selected for study. For
manganese-containing alloys the dominance of manganese in the vapour
phase in equilibrium with condensed phases of these alloys makes them
amenable to thermodynamic study by total vapour pressure measurements.
A convenient experimental technique which enables such measurements to
be made is provided by the Torsion Effusion Method. This method was
originally introduced by Volmer (2) and has subsequently been applied
by Aldred and Pratt to Ag-Pb (3) and to Ag-Bi (4), by Aldred and

* present address: Harwell Laboratory, Oxon OX11 ORA, U.K.

H. Brodowsky and H.-J. Schaller (eds.), Thermochemistry of Alloys, 277–292.

Myles to Cr-V (5) and Fe-V (6), by Spencer and Pratt to alloys of Mn
with Cu (7), with Sn (8) and with Au (9), and by Roy and Hultgren to
Fe-Mn alloys (10). A particular objective of the present work was to
obtain, in addition to binary thermodynamic data, integral Gibbs
energies of mixing of ternary alloy phases so that these measured
results could be compared with equivalent values estimated from edge
binary data by means of the various proposed mixing equations (1).
In this paper we present results for Fe-Mn, Co-Mn and Fe-Co-Mn alloys.

2. EXPERIMENTAL

2.1. Torsion Effusion Apparatus

The design of the apparatus, based largely on that of Spencer and Pratt
(11) is shown in Figure 1. The nucleus of the apparatus is housed in
a brass vacuum chamber (A), which is constructed in three sections.
The lower section of the chamber contains the furnace, which is of the
resistance-heated radiation type. The heating element (B) is a hollow
tantalum cylinder, 32 mm in diameter, 64 mm in length and 0.3 mm wall
thickness. It is supported by the power supply leads (C), which are
made from seven thicknesses of 12 mm wide tantalum strip of the same
thickness as the heating element. These strips are spot welded together
and to the heating element. The points of attachement to the heating
element are at opposite ends of a diameter at the base of the element.
Two vertical slits are made in the heating element 38 mm in length from
the base, and at opposite ends of a diameter which is perpendicular to
the diameter joining the points of attachment of the leads. These
slits improve the temperature distribution in the upper region of the
furnace. Electrical power is supplied to the element by two Edwards
9A water-cooled leadthroughs (D), which enter the vacuum system through
the baseplate (E). The tantalum leads to the element are attached to
the conductors of the leadthroughs by adjustable copper clamps (F).
     The heating element is surrounded by radiation shields (G). The
side shields consist of eleven concentric cylinders of increasing
height from the inner shield. These shields stand on a molybdenum
platform (H), which is supported by a stainless steel tripod (I). Also
standing on this platform and within the side shields, are eleven lower
radiation shields. These consist of discs of constant diameter, with
cut-out portions to allow the furnace leads to pass through. The upper
shields are discs of increasing diameter, supported by a tantalum tube
(J) of 6 mm inside diameter, and 3 mm wall thickness. This tube is in
turn clamped into a stainless steel support (K), which protrudes from
the middle section of the vacuum system. The inner shield in every
case is made from 0.3 mm thick tantalum, and the others from 0.2 mm
molybdenum.
     Temperature measurement in the hot zone of the furnace is
accomplished by means of a platinum/platinum-13% rhodium thermocouple
(L), which enters the vacuum system through the baseplate, and passes
up through the lower radiation shields, enclosed in an alumina sheath.
The junction is situated in the furnace approximately 3 mm below the

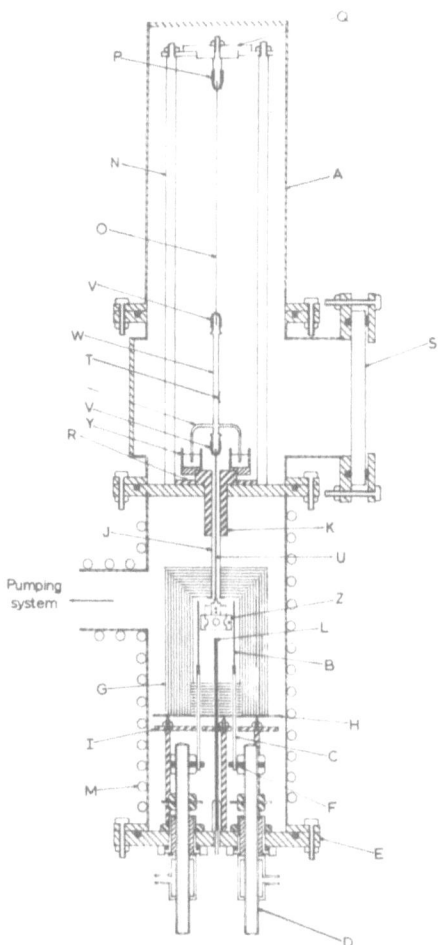

Figure 1. Torsion Effusion Apparatus. A-brass vacuum chamber; B-tantulum heating element; C-tantalum leads; D-water-cooled copper lead-throughs; E-brass base plate; F-adjustable copper clamps; G-tantalum radiation shields; H-molybdenum platform; I-stainless steel tripod; J-tantalum tube; K-stainless steel support; L-Pt/Pt-13 % Rh thermocouple; M-copper water cooling tube; N-stainless steel tripod; O-tungsten torsion fibre; P-steel pin chuck; Q-stainless steel plate; R-stainless steel locating ring; S-plate glass window; T-galvanometer mirror; U-tungsten rod; V-brass pin chucks; W-brass rod; X-copper paddles; Y-annular copper oil-bath; Z-tantalum effusion cell.

the position of the effusion cell, and is thus heated by radiation in the same way as the cell. The e.m.f. of the thermocouple is measured by means of a precision Cambridge Vernier Potentiometer, employing a Weston standard cell.

Despite the comprehensive radiation shielding, some heat is lost from the furnace by conduction. Consequently, in order to protect the "O" ring joints of the vacuum system from overheating, this lower section of the chamber is water cooled (M).

The middle section of the chamber has a solid base, upon which is mounted a stainless steel tripod stand (N), which is the support for the suspension system. This tripod extends into the top section of the vacuum chamber, in order to give adequate length to the torsion fibre (O). The fibre is held in a pin chuck (P), fixed at the centre of a plate (Q), which in turn can be rotated about the axis of the fibre in a bearing at the top of the stand. In order that the pin chuck can be centred at the axis of the vacuum chamber, the tripod is held in position on the base by a locating ring (R). This section of the vacuum chamber also houses the window (S), through which a light beam passes in and out of the system via a galvanometer mirror (T) attached to the suspension system, enabling the rotation of the suspension system to be measured by means of a transparent scale. The window is a 13 mm thick plate glass disc, 150 mm in diameter.

The overall length of the vacuum chamber is 0.7 metre, and the diameter 130 mm. It is mounted on a steel plate which is bolted to the upper flange of the middle section. This plate is supported at three points by levelling screws, attached to a rigid framework. The whole chamber may thus be moved slightly in order to align the suspension system. This movement is facilitated by a flexible coupling in the vacuum pipework. A pulley and counterweight system is incorporated, in order that the lower section of the vacuum chamber may be raised and lowered with a minimum amount of disturbance to the furnace or suspension system.

The chamber is evacuated through a sidearm on the lower section by an Edwards EO4 water cooled oil diffusion pump, fitted with a water cooled baffle, and a liquid nitrogen cooled vapour trap. This pump is normally backed by a rotary oil pump, but a liquid nitrogen cooled molecular sieve pump, Edwards type EZ500 is provided so that the whole pumping system may be operated free from mechanical vibrations during a run. The background pressure in the system is indicated by a hot filament ionization. A Pirani gauge is also incorporated for the measurement of pressures above the range of the ionization gauge. The background pressure in the chamber during a run is generally less than 2×10^{-5} torr at all temperatures below 1500K.

2.2. Suspension System

The lower end of the torsion fibre is clamped into a brass pin chuck (V) machined as an integral part of the top of the brass rod (W). This rod also supports the galvanometer mirror (T) via a copper clip. The lower end of this central brass rod (W) has a second pin chuck (V) to

which is attached an 0.5 mm tungsten rod (U). A tantalum spade welded
to lower end of this tungsten rod holds the effusion cell (Z) in
position by means of two pins. The tips of two small copper paddles
(X) attached to the brass rod are just immersed in an annular oil bath
(Y) positioned on the stainless steel support on the base of the middle
section of the vacuum chamber.

2.3. Torsion Fibres

For the present work the fibres generally used were 0.05 mm diameter
tungsten wire, of length approximately 250 mm, having a torsion
of approximately 3×10^{-7} Nm rad^{-1}.
    It was found that tungsten torsion fibres gave better performance
if their surface was thoroughly cleaned of the residue from the drawing
process before use. This is accomplished electrolytically by making
the fibre one electrode in an electrolyte of saturated aqueous
potassium hydroxide solution, and applying a small a.c. voltage across
the cell (approximately 10V, 50Hz). After cleaning, the ends of the
fibres are electroplated with copper to provide a more substantial grip
for the pin chucks.
    The torsion fibres are calibrated by a dynamic method. With the
damping oil drained, the time period of oscillation of the suspension
system (t_s) is determined. a small mass of known moment of inertia (I)
is attached to the suspension, and the time period of oscillation redetermined (t_{m+s}). The torsion constant (τ) is then given by the
expression,

$$\tau = (4\pi^2 I)/(t_{m+s}^2 - t_s^2)$$

    Determined by this method, the torsion constant of tungsten fibres
is found not to change substantially with time, over very long periods-
several months continual use.
    An optical lever was used to measure the deflections of the
system. Since a flat scale situated one metre from the mirror was used,
small angle assumptions are not valid. Consequently the angular
deflection was calculated from the following expression

$$\tan 2\theta = dL/(L^2 - y(d-y))$$

where L = shortest distance between mirror and plane of the scale,
y = distance from rest point of light spot on scale to point at which
L is measured, d = linear deflection of light spot.

2.4. Effusion Cells

For the intial testing of the apparatus with pure silver, a well-
characterised material, effusion cells made from graphite were used.
For work with manganese and manganese alloys it was originally intended
to use cells made from boron nitride since graphite is unsuitable.
However, considerable difficulty was experienced in obtaining boron

nitride of suitable quality which led instead to the use of tantalum
as a cell material for the present work. Cells were fabricated from
seam-free tantalum tube, 9.25 mm inside diameter, 22.0 mm in length and
0.254 mm wall thickness. Lines were scribed along the length of the
tube at opposite ends of a diameter, and a centre line was marked around
the tube on a lathe. Orifices were very accurately drilled in
symmetrical positions about the centre line by spark erosion using a
Metaserv spark machiner which produces a good circular hole. End caps
were drawn from 0.12 mm thick tantalum sheet to be a tight push-on fit.
The cell was attached to the suspension system using a harness made
from 0.12 mm tantalum sheet, with a socket to accept the tantalum spade
and holes to accommodate the locking pegs. Orifice areas, orifice
radii (0.2-0.35 mm) and channel lengths (0.2-0.3 mm) were determined by
measurements on photographs taken with a projection microscope;
distances of orifices from the suspension axes of the different cells
used (7.5-9.5 mm) were measured using a travelling microscope.

2.5. Temperature Measurement

Temperature does not enter into the calucation of the vapour pressure
in the torsion effusion method unlike the direct effusion method.
However to have any value specific temperatures have to be ascribed
to individual measured vapour pressure points. A fixed Pt/Pt-13% Rh
thermocouple, positioned approximately 3 mm below the cell in the
furnace hot zone, was used to measure the effusion cell temperature.
To test the apparatus, the vapour pressures of pure silver and pure
manganese were measured over the temperature range 1100 - 1250K for
silver and 1250 - 1450K for manganese. Effusion cells for the
measurements with silver were made from high density reactor grade
graphite (UKAEA) and for those with manganese from tantalum (Metalwork-
Plansee). For the measurements with silver, the thermocouple (L in
Figure 1) was calibrated in a separate experiment against a second
thermocouple attached to the effusion cell. For the measurements with
manganese a different approach was used. The gamma- to delta-manganese
allotropic transformation temperature was located using a freely
suspended cell. The difference between this temperature, 1410K (12)
and the temperature indicated on the measurement thermocouple amounted
to 10 degrees, this was assumed to be constant over the experimental
range. Excellent agreement was found between the present results and
the values given in the silver and manganese assessments given by
Hultgren et al (12).

2.6. Calculation of vapour pressures

The following expression was used to calculate vapour pressures, P,
from the observations

$$P = 2\tau\theta/(A_1 d_1 f_1 + A_2 d_2 f_2)$$

where A_1, A_2 are orifice areas, d_1, d_2 are distances of orifices from
axis, f_1, f_2 are recoil force corrections, τ is the torsion constant

of the suspension fibre and θ is the angular deflection.

Recoil force correction factors f_1 and f_2 were calculated using the following expression given by Freeman and Searcy (13)

$$1/f = 0.9982 + 0.3490(l/r) + 0.0147(l/r)^2$$

where l and r are the length and radius of the orifice respectively.

2.7. Surface Depletion

In order to estimate the extent to which surface depletion is likely to be a problem in the present work a separate series of isothermal experiments were carried out using finely divided Fe-Mn samples containing 8.6 at % Mn. In these experiments extended times at temperatures in excess of 1400K were used together with effusion cells with orifices up to four times the normal area in order to deliberately cause measurable surface depletion to occur. Mathematical modelling of the depletion process (14) based upon the analysis of Schroeder and Elliot (15) applied to the present results indicate a depletion of between 2-6% for those experiments in which normal orifice sizes were used with iron-manganese alloys for 30 minutes at temperature. In a normal run however, samples are only at the high temperatures for a small part of the total run time. These considerations coupled with the observation that there is no consistent difference between activities measured with cells of different orifice sizes show that little depletion occurs during normal runs.

2.8. Materials

The following materials were used: Ag grade 1 with less than 10ppm metallic impurities, Johnson-Matthey Chemicals Ltd; Mn broken cathode chips, Alpha (USA) Ltd. 99.97% principal impurities S (200ppm), Fe (10ppm) and Si (10ppm); Fe powder, Hoganas Ltd, impurities Mn (500ppm) Si (200ppm) and C, S and P (100ppm); Co sponge, grade 1, Johnson-Matthey Ltd., impurities Si (5ppm) and Fe (2ppm). Alloys were made either by arc-melting under argon or by powder metallurgy. Homogeneity of alloys was confirmed by optical metallography and X-ray powder diffraction. A sample of each alloy was chemically analysed using atomic absorption spectrophotometry.

3. RESULTS

The results are presented in the form of diagrams of log P_{Mn} versus 1/T plots for iron-manganese alloys in Figure 2, for manganese-cobalt alloys in Figure 3 and for iron-cobalt-manganese alloys (Fe/Co = 1.05) in Figure 4.

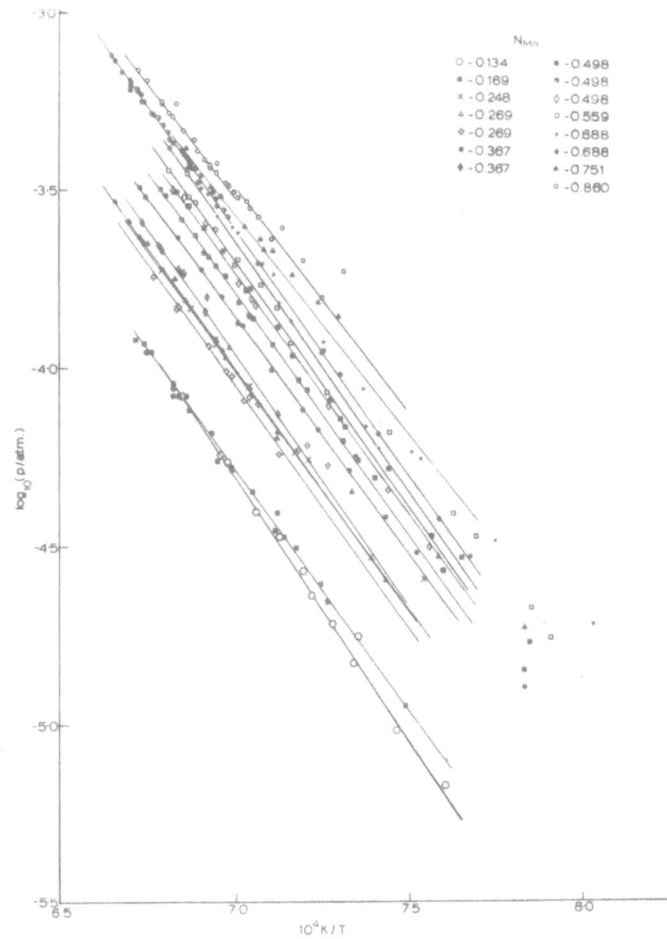

Figure 2. Manganese Vapour Pressure results for the Iron-Manganese system. Ten different alloys were studied with compositions lying in the range from 13.4 to 86.0 atomic percent manganese. For each alloy composition the vapour pressure data have been fitted by a least squares technique to an equation of the following form:

$$\log (p_{Mn}/atm) = 10^4 A/T + B$$

where A and B are constants.

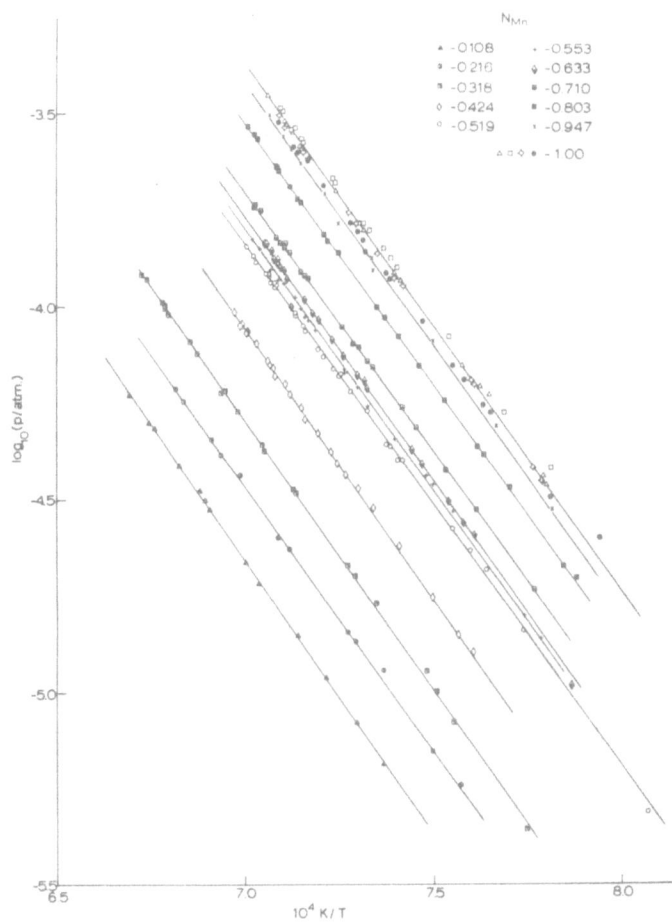

Figure 3.   Manganese Vapour Pressure results for the Cobalt-Manganese system.   Ten different alloys were studied with compositions lying in the range 10.8 to 94.7 atomic percent manganese.

286

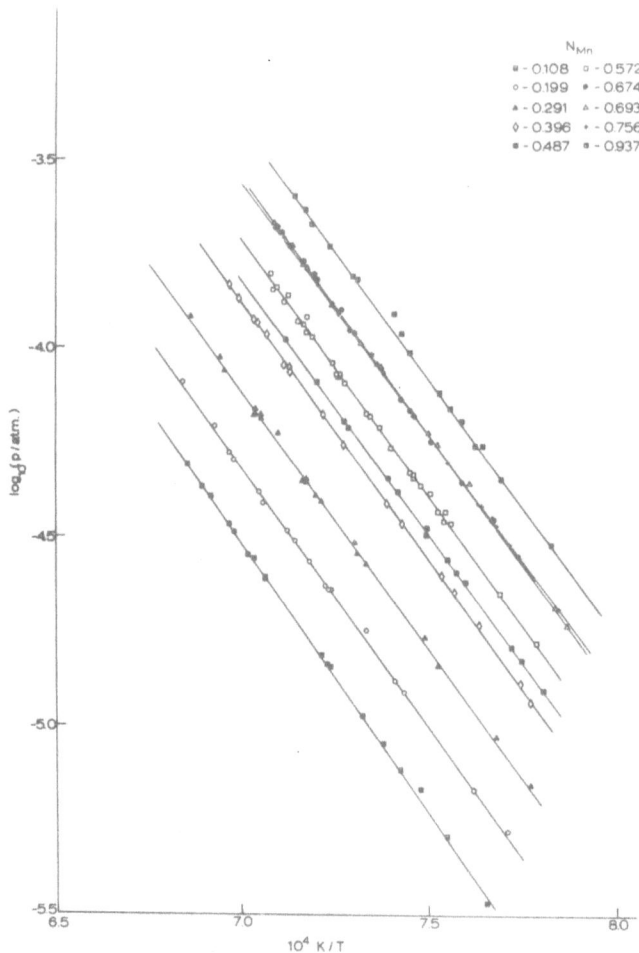

Figure 4.   Manganese Vapour Pressure results for Iron-Cobalt-Manganese
alloys with constant iron to cobalt atomic fraction ratio of 1.05 to 1.
Manganese contents range from 10.8 to 93.7 atomic percent.

4. DISCUSSION

4.1. Reliability of Experimental Data

Experimental errors fall into two major categories, random errors, and systematic errors. Random errors manifest themselves as a scatter in experimental data. A quantitative estimate of the precision of measurements subject to such errors may be obtained from a statistical treatment of the data. Systematic errors, however are more difficult to identify, and directly affect the accuracy, that is the closeness to true value, of measurements. An attempt to quantify such errors is at best only an estimate. Such errors do not generally affect precision, highly precise measurements may be inaccurate. An assessment of the individual errors in the quantities measured in the expression for the vapour pressure show that the maximum total error for any given run is approximately $\pm 1.0\%$.

Any error in the measurement of θ will generally be random for a given run, and can be accounted for in the standard deviation of the log p vs 1/T plot. For the present results, this represents a precision which is typically $\pm$ 1.5% (95% confidence limits), in the measured pressure. In ascribing vapour pressures to particular temperatures the uncertainty in the actual temperature of the effusion cell is involved. Since there may be a difference of up to 10K between the measured temperature and the true temperature the total uncertainty and excluding any error due to a change in surface composition, leads to a total uncertainty associated with measurements of absolute vapour pressures, ascribed to quoted temperatures, of 18.5%.

For the calculation of manganese activities a procedure was adopted to eliminate the large systematic error associated with absolute vapour pressure measurements.

For a given group of experiments involving both alloys and pure manganese, in which the only change to the apparatus between runs involved changing the effusion cell, it was found that the same line described the log p^o_{Mn} vs 1/T results within normal experimental precision even though experiments on alloys interspersed those on pure manganese. This result demonstrates that provided only the cell is changed the same true temperature - measured temperature calibration applies for a group of measurements. Any other change such as support rod, replacement of radiation shield, furnace element etc. requires a re-calibration because of changes in the temperature profile. Again the problem is reduced to one of ascribing a temperature to an accurately known activity. The value of the vapour pressure of pure manganese at the required temperature was obtained from Hultgren's assessment (12). The value of reciprocal temperature required to reproduce this pressure was obtained from the appropriate vapour pressure equation for pure manganese. This reciprocal temperature was then substituted into the vapour pressure equation for the alloy to obtain the value of log p for for the vapour pressure of manganese over that alloy. The activity of manganese is then obtained from the expression,

$$\log a_{Mn} = \log p_{Mn} - \log p^o_{Mn}$$

where p^o_{Mn} is the pressure obtained from Hultgren's assessment (12). This method of calculating the manganese activity eliminates systematic errors, and, as the vapour pressure of pure manganese is now relatively well characterised, it corrects the activity to that of the desired temperature.

Combining the standard deviations obtained from the alloy and pure

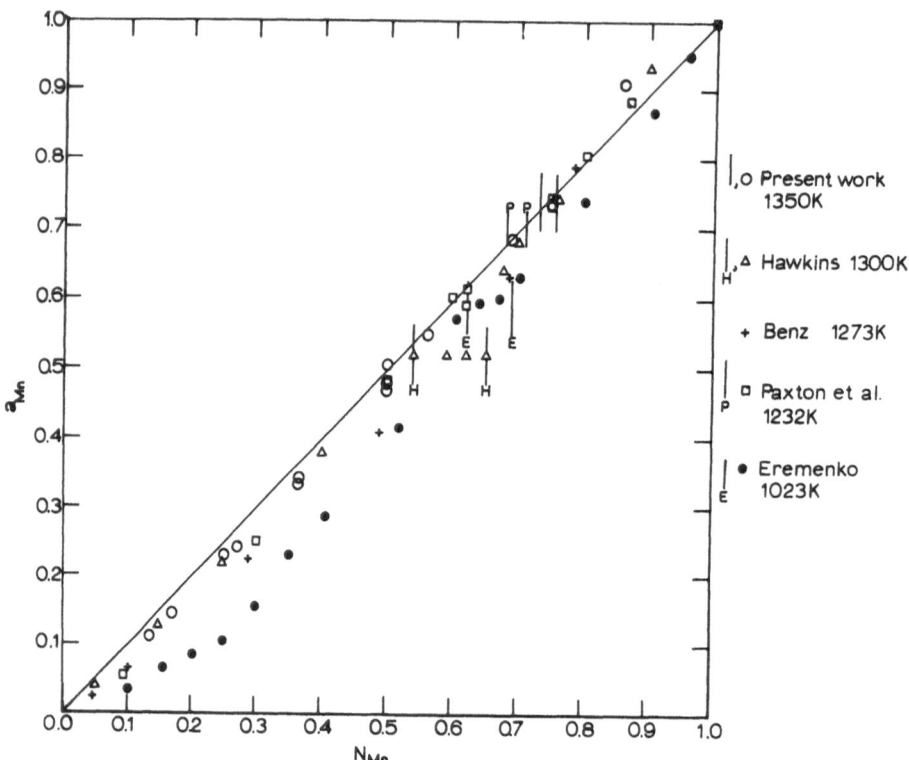

Figure 5. Comparison of manganese activities at 1350K obtained in the present study for Fe-Mn alloys with results of previous authors. Most investigations indicate a slight negative departure from Raoult's Law in the fcc solid solution range. Maximum scatter occurs near the two-phase region, boundaries of which used by different authors are indicated with vertical lines.

manganese vapour pressure measurements, the following average probable errors are obtained for Mn activity in Fe-Mn 6.3%, in Co-Mn 4.0% and in Fe-Co-Mn 4.0%.

4.2. Comparison with previous work

Several previous studies of the high temperature thermodynamic properties of the iron-manganese binary system in the solid state have been made, by a variety of methods. Almost invariably the partial molar free energy of manganese or the manganese activity have been the subject of the experimental study. One exception to this is the work of Lyubimov et al (16). Their Langmuir free evaporation technique was

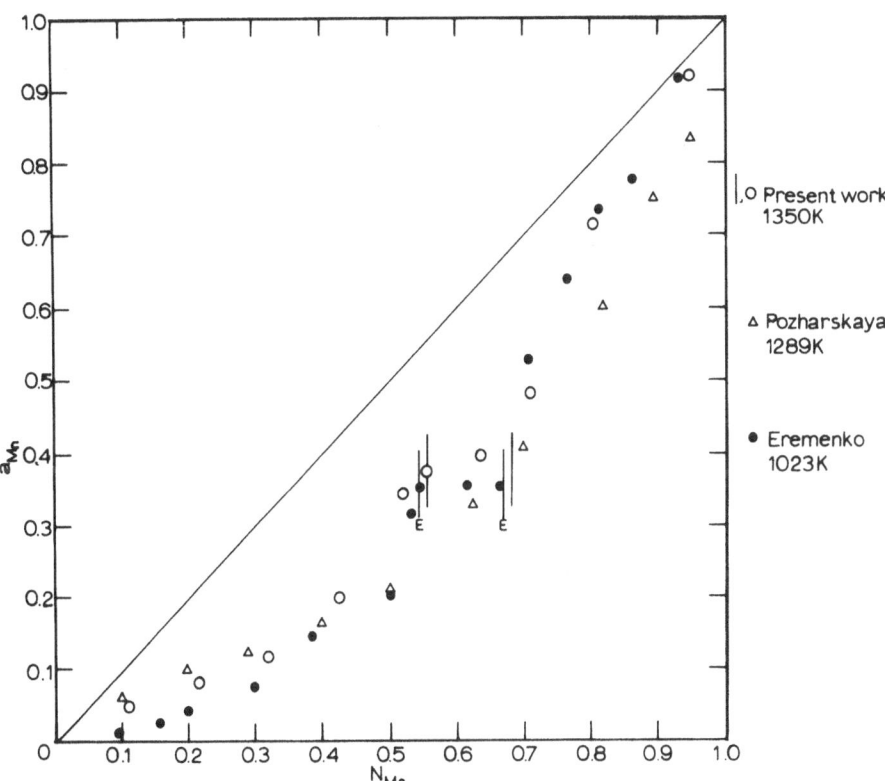

Figure 6. Comparison of manganese activities at 1350K from the present work on Co-Mn alloys with results of previous workers.

designed to yield the partial vapour pressures of both components. There have been four other investigations of the vapour pressure of the system, three by the Knudsen effusion technique (17-19) and one by torsion effusion (10). Eremenko et al (20) used electrochemical cells employing molten salt electrolyte whereas Hawkins (21) performed e.m.f. measurements with an oxygen potential cell with a solid oxide electrolyte and Benz (22) with solid CaF_2.

As seen in Figure 5, most investigations indicate a slight

negative departure of the activity of manganese from Raoult's Law in
the f.c.c. solid solution range. The present work shows good agreement
with Butler et al (17), Smith et al (18) and Hawkins (21) as to the
magnitude of this departure. The activity coefficient of 0.3 obtained
for extremely dilute solutions (N_{Mn} = 3 x 10^{-6}) by Smith and
Shuttleworth (19) is consistent with this. There is also agreement
between these studies that the manganese activity in β-Mn solid
solution at high manganese concentrations exhibits a slight positive
departure from ideality. The present results are not in agreement with
those of Roy and Hultgren (10).

Present results for Co-Mn alloys are compared with previous
thermodynamic studies in Figure 6. Pozharskaya and Evseev (23) measured
the manganese vapour pressure by a Knudsen effusion technique, in which
the effusion rate was determined by a photometric method. Eremenko
et al (24) performed e.m.f. measurements on a concentration cell with
a molten salt electrolyte. It is difficult to reach firm conclusions
about their consistency as the studies cover different temperature
ranges, and since reliable entropy data are not available. However the
agreement between the manganese activities appears to be reasonable
except at the extremes of manganese concentrations. Also the work of
Pozharskaya does not show the two phase region, which is now considered
to exist (25,26) at these temperatures with extensive f.c.c. cobalt and
Mn solid solution ranges on either side. There is some doubt about the
exact width and position of this two phase region (25,26). However,
one striking feature of the present work is the agreement between the
position of the phase boundaries, deduced from the present activity
measurements, and those found by Eremenko. This adds weight to the
belief that the two phase region is wider than suggested by
Hellawell (25).

4.3. Fe-Co-Mn Alloys

The equilibrium manganese vapour pressures shown in Figure 4 for
Fe-Co-Mn alloys have been used to obtain the integral Gibbs energies,
the integral enthalpies and the integral entropies of mixing given in
Figure 7. These were obtained using the Darken ternary integration
procedure (27). This integration can be carried out in two directions
for a given Fe:Co ratio - either from the pure manganese apex or from
the intersection point on the iron-cobalt boundary. The results of both
integration routes are given in Figure 7. Gibbs energy data for the
iron-cobalt system was taken from Kaufman and Nesor (28). When data
given in Figure 7 are compared with those obtained by applying various
mixing equations to edge binary data it is found that the Bonnier
equation at constant manganese gives the best agreement. The results
are in keeping with the conclusions of Spencer, Hayes and
Kubaschewski (1).

ACKNOWLEDGEMENT

The authors would like to thank Prof.K.M.Entwistle and Prof.Edwin
Smith for the provision of laboratory facilities.

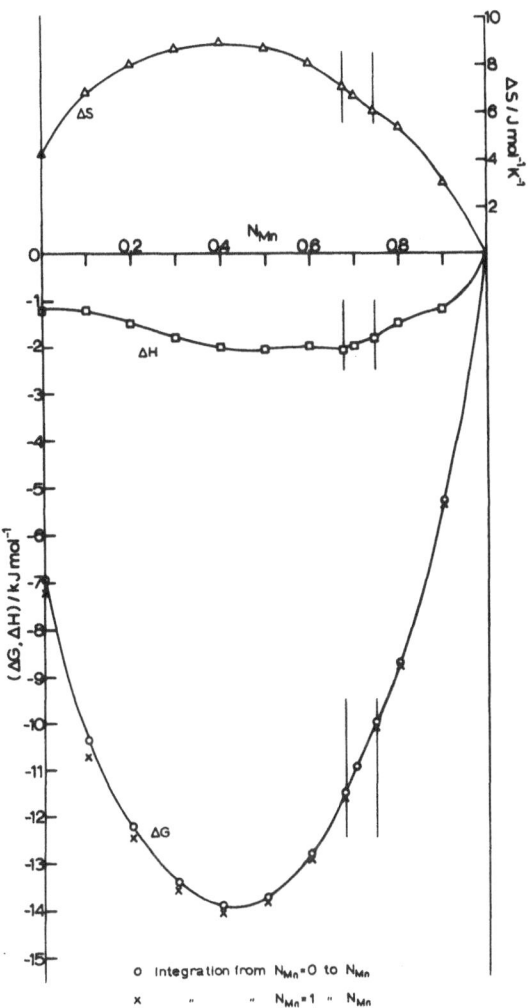

Figure 7.   Integral thermodynamic mixing properties for f.c.c. -
Fe-Co-Mn (Fe/Co = 1.05) at 1350K.

5.   REFERENCES

1. P.J.Spencer, F.H.Hayes, and O.Kubaschewski, Rev.Chim.Minerale,
   9, 13, (1972)
2. M̄.Volmer, Z.Phys.Chem., Bodenstein Festband, 863, (1931)
3. A.T.Aldred and J.N.Pratt, Trans.Farad.Soc., 57, 611, (1961).
4. A.T.Aldred and J.N.Pratt, Trans.Farad.Soc., 59, 673, (1963).
5. A.T.Aldred and K.M.Myles, Trans A.I.M.E., 230, 736, (1964).

292

6. A.T.Aldred and K.M.Myles, J.Phys,Chem., 68, 64, (1964).
7. P.J.Spencer and J.N.Pratt, Trans.Farad.Soc., 64, 1470, (1968).
8. P.J.Spencer and J.N.Pratt, Trans.A.I.M.E., 242, 1709, (1968).
9. P.J.Spencer and J.N.Pratt, Rev.Int.Hautes Temp.et.Refract., 5, 155, (1968).
10. P.Roy and R.Hultgren, Trans.A.I.M.E., 233, 1811, (1965).
11. P.J.Spencer and J.N.Pratt, Brit.J.Appl.Phys., 18, 1473, (1967).
12. R.Hultgren, P.D.Desai, D.T.Hawkins, M.Gleiser, K.K.Kelley and D.D.Wagman "Selected Values of the Thermodynamic Properties of the Elements," ASM, Metals Park, Ohio (1973).
13. R.D.Freeman and A.W.Searcy, J.Chem.Phys., 22, 762, (1954).
14. G.McHugh and F.H.Hayes (to be published).
15. D.L.Schroeder and J.F.Elliott, Trans.A.I.M.E., 236, 1091, (1966).
16. A.P.Lyubimov, A.A.Granovskaya, L.E.Berenshtein, Zh.Fhiz.Khim., 32, 1591, (1958).
17. J.F.Butler, C.L.McCabe, H.W.Paxton, Trans.A.I.M.E., 221, 479, (1961).
18. J.H.Smith, H.W.Paxton, C.L.McCabe, Trans.A.I.M.E., 221, 895, (1961).
19. R.Smith, R.Shuttleworth, Acta Met., 13, 623, (1965).
20. V.N.Eremenko, G.M.Lukashenko, V.R.Sidorko, Izv.Akad.Nauk.SSSR Metal, 2, 170, (1969).
21. R.J.Hawkins, "Chemical Metallurgy of Iron & Steel", Sheffield University 1971. (Iron and Steel Institute London).
22. R.Benz, Met.Trans., 5, 2217, (1974).
23. G.V.Pozharskaya, A.M.Evseev, Vestnik Moskov Univ., 15, 15, (1960).
24. V.N.Eremenko, G.M.Lukashenko, V.R.Sidorko, Izv.Akad.Nauk.SSSR Metal, 3, 192, (1967).
25. A. Hellawell and W.Hume-Rothery, Phil.Trans.Roy.Soc., A249, 417, (1957).
26. Yu.B.Kuz'ma, Zh.Neorg.Khim., 7, 225, (1962).
27. L.S.Darken, J.Am.Chem.Soc., 72, 2909, (1950).
28. L.Kaufman and H.Nesor, Z.Metallkde., 64, 249, (1973).

THE ISOPIESTIC METHOD AND ITS APPLICATION TO A THERMODYNAMIC STUDY OF THE Au-Zn SYSTEM

Herbert Ipser, Regina Krachler, and Kurt L. Komarek
Institute of Inorganic Chemistry, University of Vienna,
Währingerstraße 42
A-1090 Wien
Austria

ABSTRACT. Zinc vapor pressures of liquid and solid gold-zinc alloys were determined oy an isopiestic method. The principles of this method are outlined and possible errors are analyzed. The application of the nonisothermal isopiestic method at very low vapor pressures and the possibly necessary corrections are discussed. Activities and partial molar enthalpies of zinc were derived from the vapor pressures, and integral thermodynamic data for liquid gold-zinc alloys were obtained by Gibbs-Duhem integration. The thermodynamic behavior of the ordered B2-phase (beta'-AuZn) is analyzed in terms of a theoretical model.

1. INTRODUCTION

The ordered CsCl-(B2-) phase in the binary Au-Zn system, ß'-AuZn, which exists around the equiatomic composition with a wide range of homogeneity (1) has traditionally been considered to be of the anti-structure type. Its thermodynamic properties have been evaluated repeatedly in terms of statistical-theoretical models (2-4) based on the experimental results by Pemsler and Rapperport (5). Recently Neumann (6) investigated the correlation between enthalpy of formation and defect mechanism (anti-structure or triple defects) for a number of B2-phases; he pointed out that some of them should be expected to exhibit an intermediate behavior, i.e. both types of defects should occur more or less simultaneously, and ß'-AuZn was one of the prominent candidates. This is in agreement with the results of diffusion measurements in this phase (7) from which the authors concluded that the vacancy concentration increases considerably on the zinc-rich side placing ß'-AuZn actually close to the group of the typical triple-defect B2-phases like ß'-CoAl and ß'-NiAl, e.g. (4).

In order to determine the importance of triple defects in ß'-AuZn, it was decided to apply a new statistical mcdel, developed by the present authors, which can account for the simultaneous presence of both substitutional and triple defects in B2-phases (8). However,

H. Brodowsky and H.-J. Schaller (eds.), Thermochemistry of Alloys, 293–306.
© *1989 by Kluwer Academic Publishers.*

since the data of Pemsler and Rapperport (5) do not agree with the results of previous investigators (9-12) a new thermodynamic investigation of the system was started using an isopiestic vapor pressure method. The extension of these measurements into the liquid phase allowed an additional comparison of our results with those of Gerling and Predel (13), e.g., who had studied liquid Au-Zn alloys by an emf-method.

2. THE ISOPIESTIC METHOD

2.1. Historical Development

Originally the isopiestic method had been used isothermally for the determination of vapor pressures of aqueous solutions: the solution to be investigated was equilibrated in a closed system with a standard solution, whose water vapor pressure was known as a function of concentration and temperature (see Refs. (14-16), e.g.). However, to extend the pressure range to lower values, it was tried very soon to substitute the standard solution by pure water, the vapor pressure of which could be varied very conveniently by changing its temperature (17). Thus the nonisothermal isopiestic method (sometimes called pseudoisopiestic method (18)) was born.

The method was applied to metallic systems (Cu-Zn) for the first time in 1938 by Seith and Krauss (19) and by Hargreaves (20) who used a variation commonly known as dew point method. Some twenty years later several authors (21-23) took up the isopiestic method again to use it for the study of binary metallic systems, mostly in a nonisothermal way. Herasymenko (21) was the first one to place a certain number of samples in a temperature gradient and to equilibrate them with the vapor from a reservoir of the volatile component.

2.2. Principle of the Method

The principle of the (pseudo-)isopiestic method is that a sample (which may be the pure nonvolatile component A or a master alloy $A_{1-x}B_x$) at the sample temperature T_S is equilibrated in a closed system with vapor from a reservoir of the pure volatile comoponent B at the reservoir temperature T_R. Of course, this reservoir has to be at the temperature minimum. The vapor pressures of the pure components A and B should differ by at least three orders of magnitude, and usually the vapor pressure of the nonvolatile component A is low enough to be neglected altogether.

The equilibrium condition usually applied requires that after sufficient time pure B from the reservoir will have reacted with the sample in such a way that an alloy is formed with exactly that composition for which the partial pressure of B over the sample at T_S is equal to the vapor pressure of B over pure B at T_R:

$$p_B(T_S) = p_B^{\bullet}(T_R);\qquad\qquad [1]$$

thus one has to assume a constant pressure of B in the system
independent of temperature (see below). From this assumption the
thermodynamic activity of B in the sample can be obtained in a simple
way:

$$a_B(T_S) = \frac{p_B(T_S)}{p_B^{\circ}(T_S)} = \frac{p_B^{\circ}(T_R)}{p_B^{\circ}(T_S)}.$$ [2]

All one needs to know is the temperature dependence of the vapor
pressure of pure B. Additionally, one has to determine the composition
of the sample (which has changed during the experiment), either simply
from its weight change, or by chemical analysis.

The pressure range over which the isopiestic method has been used
is nearly ten orders of magnitude, from about 10^{-10} bar (Al-systems,
see for example (24)) to 10^{-1} bar (Se-systems, see for example (25)).

The method, as it was applied in the present study, differs from
the principle described above only in the number of samples.
Approximately 15 to 20 samples contained in graphite crucibles and
placed in a temperature gradient (as described in (26)) were
equilibrated with zinc vapor from a reservoir. Thus a corresponding
number of data points (activity at the individual sample temperature)
was obtained from one single experiment.

2.3. Error Sources

The accuracy of isopiestic measurements is determined by a number of
different factors. Some of the possible errors can be minimized with
reasonable effort, others, however, are very hard to control, and in
some instances it may be even difficult to estimate their possible
influence on the results. The most important error sources, divided
into statistical and systematic errors, are listed below.

2.3.1. <u>Statistical errors.</u> The main statistical errors are caused by
inaccuracies in the <u>temperature measurement</u> (since the vapor pressure
of the pure volatile component B is usually expressed as some function
of T) and in the determination of the <u>sample compositions</u>. Generally
we think that the temperatures we report are accurate within ± 2 K
where this possible error contains the error of the measuring device
(thermocouple) itself, but also the difference between the temperature
at the position where it is measured and the actual sample
temperature. It is estimated that an error of ± 2 K in temperature
causes an error of about ± 0.05 in the value of $\ln a_B$ (at T_S), and
since $\Delta \bar{H}_B$ is determined from the temperature dependence of $\ln a_B$ one
can similarly estimate an error of less than ± 5 kJ g-atom^{-1} for the
partial enthalpy values. This results in a maximum error of about
± 0.15 in $\ln a_B$ after conversion to one common temperature using these
enthalpy values.

The sample compositions can usually be obtained from the weight
change during the experiment much more accurately than by analytical
methods. For the Au-Zn system with sample masses of about 0.2 g it can

be shown that a weighing error of ± 0.1 mg leads to an error of about ± 0.1 at% in the compositions.

2.3.2. Systematic errors. Errors in the results of isopiestic measurements may also be caused by incomplete equilibration. Such errors, however, can readily be eliminated by repeating experiments under identical conditions, but with increasing reaction periods until the sample compositions remain constant. Similarly, unwanted side reactions must be excluded as far as possible. Transport reactions, e.g., which may be caused by minute amounts of impurities, may simulate incorrect compositions, particularly if these are computed from the weight change of the samples during the experiment. Another possible error in the activity values may arise from any uncertainty about the molecular composition of the volatile component in the gas phase. This is certainly not a problem in the Au-Zn system, where it can be assumed that zinc vapor consists solely of zinc atoms. However, the situation is very much different, if the gas phase contains different molecules as in the selenium systems, e.g., where all possible species between Se_2 and Se_8 are present (25). Then one has to know the different equilibrium constants as a function of temperature in order to derive correct activity values. An effect whose influence is extremely difficult to estimate is thermodiffusion. Errors of the order of about 1.5 % in ln a_B are thought to be possible under unfavorable conditions. Any errors in the literature data needed for the evaluation of isopiestic experiments will also lead to corresponding errors in the final results, and their magnitude is completely unknown.

    Finally it is sometimes argued that the assumption of constant pressure in the equilibration tube independent of temperature, as used in our equilibrium condition in eq. [1], may cause an error in the values of the thermodynamic activities. This final point will be discussed below.

2.4. Equilibration at Very Low Pressures

When employing a nonisothermal isopiestic (pseudoisopiestic) method, one has to distinguish basically between two limiting cases, as pointed out by Lüdecke et al. (18): the very low pressure range, for which the so-called Knudsen number $K = \lambda/d$ (λ = mean free path, d = dimension of the reaction vessel) is large, i.e. $K \gg 1$, and we have Knudsen flow in the gas phase, and the range of relatively high pressures, where $K \ll 1$. In the latter case λ is very small compared to the dimension of the reaction tube; this means that the gas atoms or molecules collide frequently enough with each other so that a continuous exchange of momentum is warranted. Then one can reasonably assume that the pressure is the same over the whole apparatus, and equations [1] and [2] are valid without any restriction.

    If, on the other hand, the pressure is very low (resulting in large values for the mean free path), then one has to look for a different equilibrium condition; usually one takes the flux J and assumes that a stationary state is reached when there is no net flux in the tube, i.e. when the number of particles moving in the one direction is equal to the number moving in the opposite one:

$$\overrightarrow{J} = \overleftarrow{J}. \qquad [3]$$

Then one can show by using the equation for Knudsen effusion

$$\overrightarrow{J} = (1/4)A\nu\overline{c}, \qquad [4]$$

(where A is the area of the cross section, ν is the density of the gas particles and $\overline{c}$ their mean velocity) and by inserting $\overline{c} = \sqrt{8RT/\pi M}$ from the kinetic gas theory and by obtaining an expression for ν from the ideal gas law, that

$$p_B(T_S) = \sqrt{T_S/T_R}\ p_B^o(T_R), \qquad [5]$$

i.e. the pressure over the sample is higher by the factor $\sqrt{T_S/T_R}$ (which amounts to 1.27 in the extreme case, that $T_S = 1300$ K and $T_R = 800$ K, e.g.).

For the Au-Zn system one can estimate, that the mean free path under our experimental conditions should be of the order of about 10^{-5} m which is certainly much smaller than the dimensions of the reaction tube. But even for other cases, where the vapor pressure of the volatile component is considerably lower (like in the manganese systems (27), e.g.) it is thought that the presence of other gases like Ar, N_2, O_2... should increase the total pressure so much that the mean free path approaches the dimension of the reaction tube (or becomes even smaller). This means that a possible error caused by applying the equilibrium condition [1] will probably be much smaller than the factor $\sqrt{T_S/T_R}$. These other gases will consist of the gas atmosphere which remains in the apparatus after evacuating and sealing, but also of an unknown amount diffusing into the apparatus through the walls, which appears to be inevitable if quartz is used, and if the experiment lasts for several weeks.

2.5. Total Error

Summarizing all possible errors one can observe two opposing effects. Whereas for low concentrations of the volatile component B the systematic errors play an important role because of the high sample temperatures, it is obvious that the relative statistical errors increase with growing content of B since the absolute values of ln a_B become smaller. It is expected that the uncertainty of the values of ln a_B after conversion to one common temperature will be less than 10 %, under favorable conditions probably even less than 5 %.

3. EXPERIMENTAL RESULTS AND DISCUSSION

The equilibrium curves (i.e. sample temperature vs sample composition) of eleven isopiestic experiments are shown in Fig. 1. These curves are

actually isobars, curves of equal zinc vapor pressure which is fixed
by the corresponding reservoir temperature. Very frequently one
obtains valuable information on the phase diagram from these curves,
as for example on the shape of solidus and liquidus line in the Au-Zn
system between about 30 and 50 at% Zn (cf. Fig. 1). These results are
in excellent agreement with independent DTA-measurements (31).

Using eq. [2] one can calculate for each sample the thermodynamic
activity of zinc at the corresponding temperature based on the vapor
pressure data given by Hultgren et al. (28). Plotting $\ln a_{Zn}$ vs. $1/T_S$
yields a straight line for each isopiestic run, and by interpolation
at fixed compositions the partial molar enthalpy of zinc can be
derived (usually by linear regression) according to

$$\frac{\partial \ln a_{Zn}}{\partial (1/T_S)} = \frac{\Delta \overline{H}_{Zn}}{R}.$$

[6]

3.1. The Phase ß'-AuZn

With $\Delta \overline{H}_{Zn}$-values obtained for the ß'-AuZn phase by the procedure
described above one can convert the zinc activities to a common
temperature of 700 K. They are shown in Fig. 2. It should be pointed
out that this temperature was selected (although it is outside our
experimental temperature range) in order to compare our activity
values with those by Pemsler and Rapperport (5) which were reported
for 700 K. One can see that the agreement is perfect on the gold-rich
side while some deviations occur on the zinc-rich side. Nevertheless
it is thought that the present results confirm the data by Pemsler and
Rapperport (5).

The full line in the figure represents the activity curve
obtained from our theoretical model (8). A slightly "asymmetrical"
curve was used, which is obtained when the parameters a and b are
different indicating the occurrence of triple defects to a small
extent. (The parameter a gives the probability for an Au-atom on the
Au-sublattice to change place with a vacancy on the Zn-sublattice, and
b is the probability for a Zn-atom on the Zn-sublattice to change
place with a vacancy on the Au-sublattice; the parameter c gives the
total vacancy concentration at the stoichiometric composition.) If
these probabilities a and b are equal, one has the case of an anti-
structure defect mechanism; if they are different, then the vacancy
concentrations on the two sublattices will be different too, which
means that there is a certain contribution of triple defects which
may become, with increasing difference between a and b, the dominating
mechanism.

In the Au-Zn system the agreement between theoretical curve and
experimental results is very reasonable too, if a substitutional model
is used, i.e. equal a and b, resulting in a "symmetrical" curve.
However, it is thought that the results of the diffusion measurements
in Ref. (7) do really indicate some amount of triple defects, and
that therefore a corresponding curve which considers their presence

would be more meaningful (although a real decision on the basis of the different model curves seems to be difficult).

On the gold-rich side one can observe in Fig. 2 an increasing deviation of the theoretical activity curve from the experimental data. A similar effect was observed with the original theoretical models by Chang (29) and Libowitz (2). In order to obtain the correct slope, Libowitz had to assume additional interactions between anti-structure Au-atoms (2), and Chang's model had to be extended by including second-nearest neighbor interactions into the enthalpy term (3). It is not known, why ß'-AuZn behaves in such a special way, since it is - to our knowledge - the only case where these "corrections" are really necessary.

It is interesting to compare the variation of the vacancy concentration, as predicted from our model, with experimental values which are again taken from the investigation by Pemsler and Rapperport (5) (see Fig. 3). Although vacancy concentrations obtained from density measurements on quenched samples will always exhibit a high uncertainty, it is hoped that the good agreement in Fig. 3 is not purely accidental but will rather support the validity of the theoretical model.

3.2. The Liquid Phase

Zinc activities and partial molar enthalpies for liquid alloys were obtained in an analogous way as described above. From these the composition dependence of $\ln a_{Zn}$ was calculated at a temperature of 1173 K. The result is shown in Fig. 4; also shown are the data by Gerling and Predel (13). The agreement is very good despite a small systematic but unexplained deviation over the whole composition range.

Fig. 5 shows the integral Gibbs energy of formation of liquid Au-Zn alloys obtained by Gibbs-Duhem integration. Again the agreement with the results of Gerling and Predel (13) is good, the small difference may be partially caused by different extrapolation procedures in the composition range between 20 and 0 at% Zn which will certainly influence the shape of the integral curve. However, both curves exhibit a minimum which is slightly shifted to the gold-rich side. If one assumes that the minimum in ΔG for the ordered B2-phase is exactly at the equiatomic composition (as one would expect), then the congruent melting point of ß'-AuZn should be shifted to the zinc-rich side. This is in contradiction to the results of our phase diagram studies (30, 31) which show that the opposite is true: the melting point maximum lies at about 48.5 at% Zn. This discrepancy needs further investigation, but it is well possible, that it is only introduced by the Gibbs-Duhem integration.

4. ACKNOWLEDGMENTS

Partial financial support of this work by the "Hochschuljubiläums-stiftung der Stadt Wien" is gratefully acknowledged.

300

5. REFERENCES

(1) W.G. Moffatt, The Handbook of Binary Phase Diagrams, General
    Electric Company, Schenectady, N.Y.
(2) G.G. Libowitz, Met. Trans. 2 (1971), 85.
(3) H. Ipser, J.P. Neumann, and Y.A. Chang, Monatsh. Chem. 107
    (1976).
(4) Y.A. Chang and J.P. Neumann, Progr. Solid State Chem. Vol.14,
    W.L. Worrell and G.R. Rosenblatt (eds.), Pergamon Press, Oxford
    (1982), 221.
(5) J.P. Pemsler and E.J. Rapperport, Met. Trans. 2 (1971), 79.
(6) J.P. Neumann, Acta Met. 28 (1980), 1165.
(7) D. Gupta and D.S. Lieberman, Phys. Rev. B4 (1971), 1070.
(8) R. Krachler, H. Ipser, and K.L. Komarek, J. Phys. Chem. Solids,
    submitted for publication; also R. Krachler, Ph.D.-Thesis,
    University of Vienna (1984).
(9) J. Terpilowsky, Zeszyty Nauk. Politech. Wroclaw., Chem. IV, 17
    (1957), 13.
(10) R. Alderdice, R.A. Connell and D.B. Downie, Acta Met. 21 (1973),
     485.
(11) D.B. Masson, Met. Trans. 2(1971), 919.
(12) A. Yazawa and A. Gubcová, Trans. Jap. Inst. Met. 11 (1970), 419.
(13) U. Gerling and B. Predel, Z. Metallk. 71 (1980), 79.
(14) D.A. Sinclair, J. Phys. Chem. 37 (1933), 495.
(15) R.A. Robinson and D.A. Sinclair, J. Amer. Chem. Soc. 56 (1934),
     1830.
(16) A.R. Gordon, J. Amer. Chem. Soc. 65 (1943), 221.
(17) R.H. Stokes, J. Amer. Chem. Soc. 69 (1947), 1291.
(18) D. Lüdecke, C. Lüdecke, and T. Hehenkamp, Acta Met. 31 (1983),
     95.
(19) W. Seith and W. Krauss, Z. Elektrochem. 44 (1938), 98.
(20) R. Hargreaves, J. Inst. Met. 64 (1939), 115.
(21) P. Herasymenko, Acta Met. 4 (1956), 1.
(22) H.W. Rayson and W.A. Alexander, Can. J. Chem. 35 (1957), 1571.
(23) F.D. Rosenthal, G.J. Mills, and F.J. Dunkerley, Trans. Met. Soc.
     AIME 212 (1958), 153.
(24) A. Steiner and K.L. Komarek, Trans. Met. Soc. AIME 230 (1964),
     786.
(25) H. Jelinek and K.L. Komarek, Monatsh. Chem. 105 (1974), 689.
(26) M. Ettenberg, K.L. Komarek, and E. Miller, J. Solid State Chem. 1
     (1970), 583.
(27) R. Krachler, H. Ipser, and K.L. Komarek, Z. Metallk. 75 (1984),
     724.
(28) R. Hultgren, P.D. Desai, D.T. Hawkins, M. Gleiser, K.K. Kelley,
     and D.D. Wagman, Selected Values of the Thermodynamic Values of
     the Elements, American Society for Metals, Metals Park, Ohio
     (1973).
(29) Y.A. Chang, Treatise on Materials Science and Technology, Vol. 4,
     H. Herman, (ed.), Academic Press, New York (1974), 173.

(30) H. Ipser, A. Mikula, and P. Terzieff, <u>Monatsh. Chem.</u> **114** (1983),
     1177.
(31) H. Ipser, R. Krachler, and K.L. Komarek, publication in
     preparation.

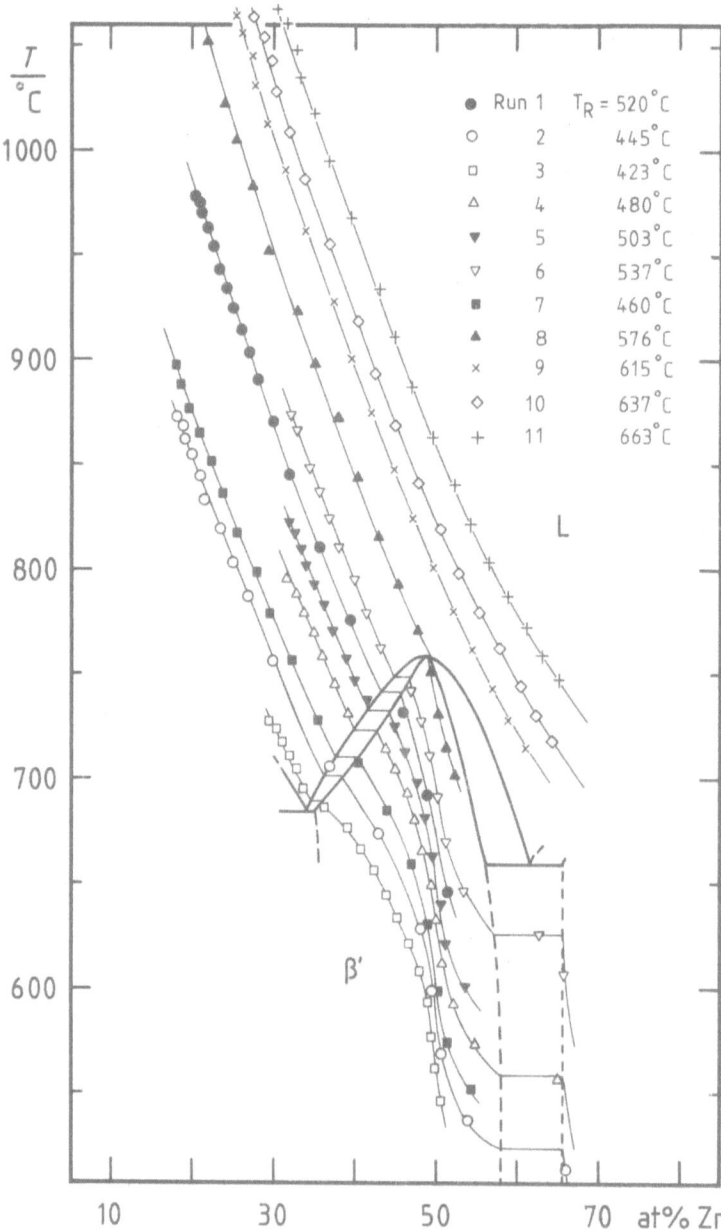

Fig. 1. Sample composition versus sample temperature and partial gold-
zinc phase diagram.

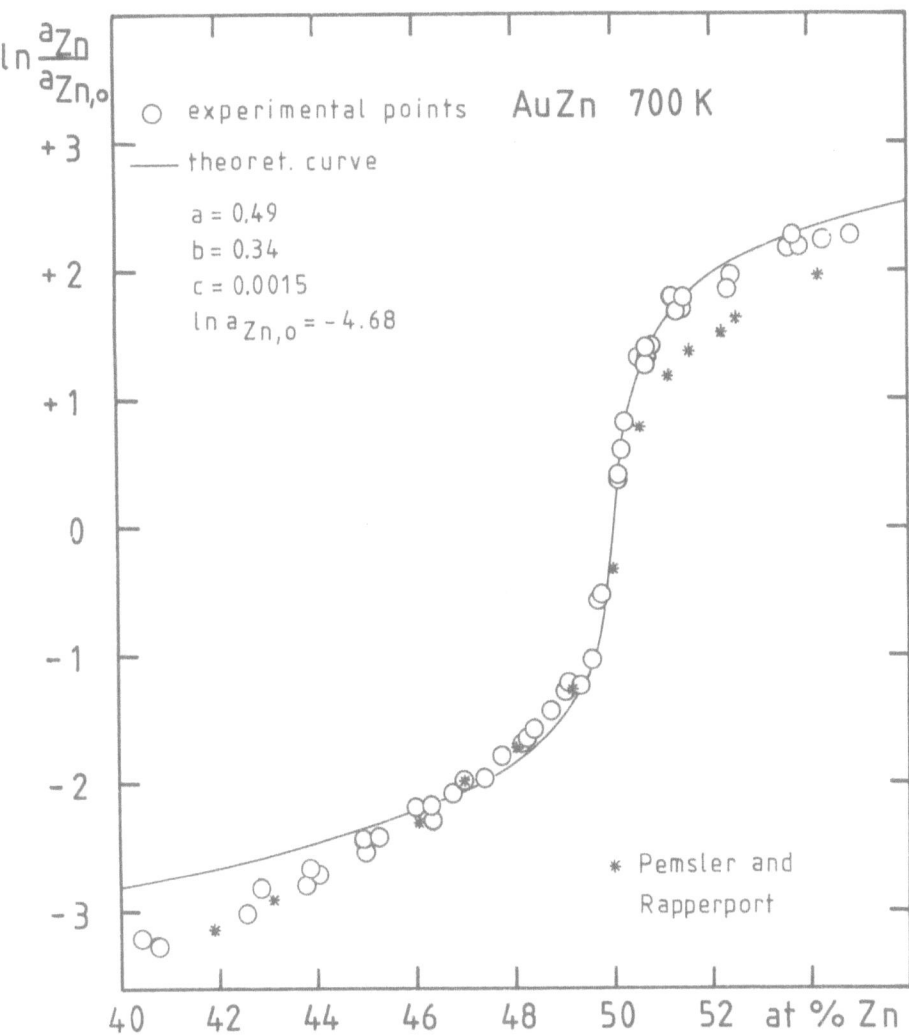

Fig. 2. Activities of zinc at 700 K in the ß'-AuZn phase at 700 K.
Standard state: Zn(1). The theoretical curve was calculated
with the parameter values given; $a_{Zn,0}$ is the activity at the
stoichiometric composition.

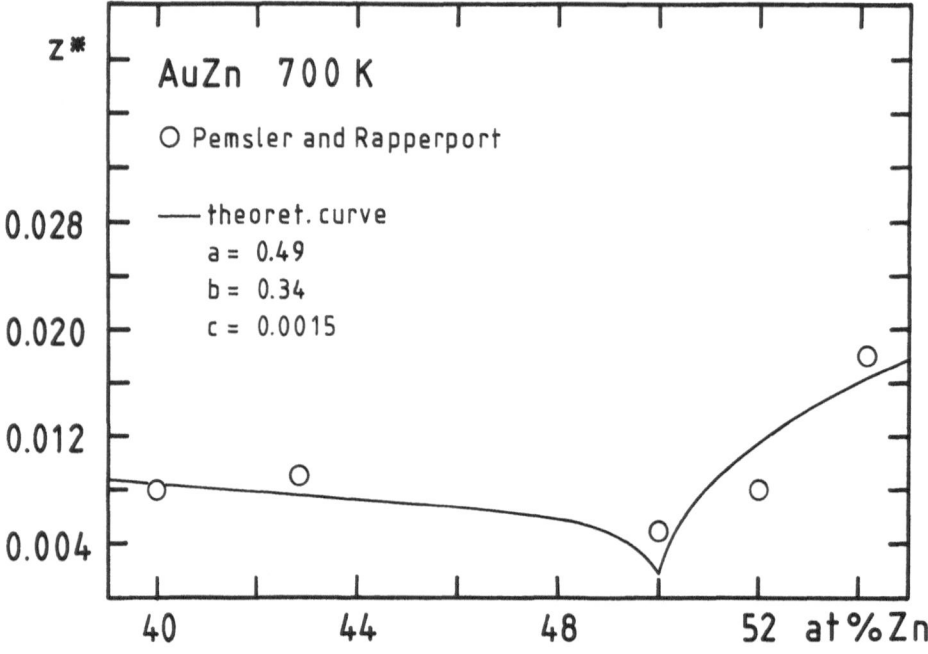

Fig. 3. Experimental vacancy concentrations at 700 K from Pemsler and Rapperport (5) compared to the theoretically predicted curve.

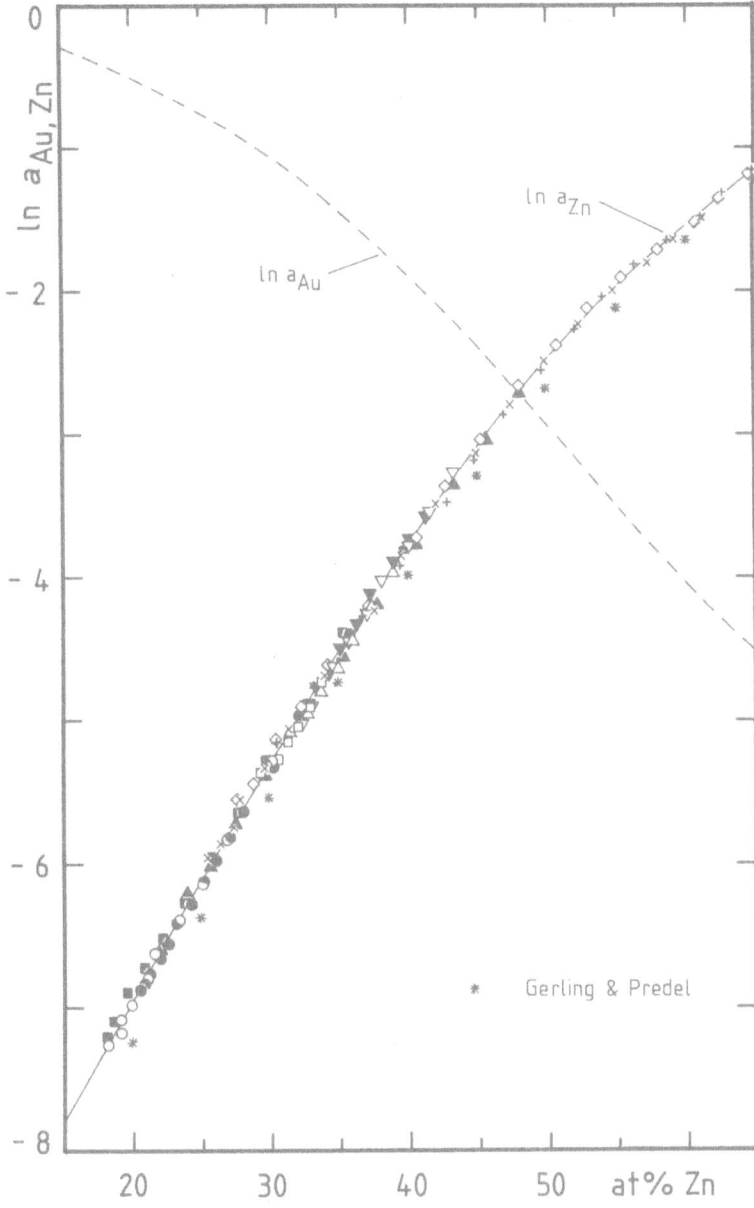

Fig. 4. Activities of zinc and gold at 1173 K for liquid alloys.
Standard states: Zn(1), Au(1).

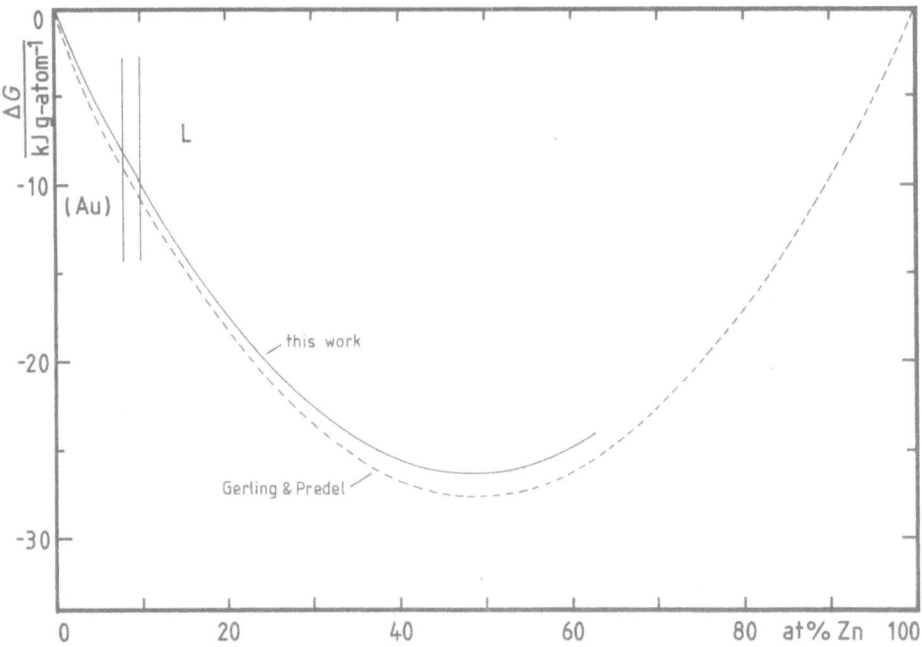

Fig. 5. Integral Gibbs energies for liquid gold-zinc alloys at
1173 K. Standard states: Zn(l), Au(l).

SOME METHODS OF DETERMINING THE THERMODYNAMIC PROPERTIES OF ALLOYS

J. Hertz
Laboratory of Metallurgical Thermodynamic
University of Nancy I - U.A. C.N.R.S. 1108
B.P. 239 - 54506 Vandœuvre-lès-Nancy Cédex
France

ABSTRACT. On the basis of the published data of my coworkers I intend to present four strategies to measure the thermodynamic characteristics of alloys. Applications are given by **M. NOTIN** for E.M.F. and liquid calorimetry on calcium base alloys ; **J.C. GACHON** for direct reaction calorimetry at 1800 K on transition-transition compounds ; **J. CHARLES** for the phase diagram optimisation : **J.P. HILGER** for irreversible processes in steels ; **Ch. CUNAT** and **F.A. KUHNAST** for irreversible processes in amorphous alloys.

MISTER CHAIRMAN, LADIES AND GENTLEMEN, I WOULD LIKE TO THANK THE ORGANISATION COMMITTEE
for its perfect organisation of this Summer-Institute, for the nice trip to Schleswig-village with the very interesting visit of the Archeological Museum and also to thank this committee for giving me the opportunity to present, to this distinguished audience, some of the methods we are using in my laboratory, in Nancy, to measure new thermodynamic data in the field of metallurgy.

BUT FIRST LET ME RECALL BRIEFLY SOME ELEMENTS OF FORMALISM IN THERMODYNAMICS
and the most important formulae used by experimenters. My talk is often devoted to binary alloys but the main formulae can be extended simply to multicomponent systems.
Starting from N_1 moles of pure component 1* and N_2 moles of pure component 2* we are going to synthetise an alloy. Any Y extensive quantity of this system is, generally speaking, highly modified during the alloying process. This increment is called the **integral extensive quantity of formation or mixing** of the alloy :
$$Y^m = Y - N_1 y_1^* - N_2 y_2^*$$
where y_1^* and y_2^* are the molar quantities of the pure components. Generally for the metallurgists a mole corresponds to one atom-gram of matter. The most important extensive quantity in the thermochemistry-science is the Gibbs function, which does minimise, in the equilibrium conditions, for a system containing a constant number of moles of

H. Brodowsky and H.-J. Schaller (eds.), Thermochemistry of Alloys, 307–328.

components, at constant temperature and constant pressure

$$G^m = G - N_1\,\mu_1^* - N_2\,\mu_2^*$$

μ_1^* and μ_2^* are called the chemical potential of the pure components.
The **molar** integral quantities of mixing are the integral quantities divided by the total number of moles in the system :

$$y^m = \frac{Y^m}{N_1+N_2} = y - x_1 y_1^* - x_2 y_2^*$$

where $x_i = \dfrac{N_i}{\Sigma N_i}$ is called the mole fraction of component i.

Now I would like to underline two important relations attributed simultaneously to **Gibbs** and **Helmholtz**

$$G^m = H^m - T\,S^m$$

is the first one and indicates the two different contributions in the Gibbs function : the **enthalpy** H^m which tends to minimise and the **entropy** S^m which tends, on the contrary, to maximise. Taking into account that, at constant pressure, $dH^m = TdS^m$, the equilibrium condition corresponds to the balance between these two trends. H^m represents the calorific energy exchanged with the outside during alloying and can be measured experimentally by calorimetric technics, whereas S^m corresponds to the sum of all the processes of disordering during the alloying and, generally speaking, S^m can't be measured directly. For this reason, the second Gibbs-Helmholtz'relation is often used by experimenters to measure the entropies, differentiating the Gibbs function :

$$\left(\frac{\partial G^m}{\partial T}\right)_P = - S^m$$

And now I have to define the **partial molar quantities of mixing** which are the first partial derivatives in the integral quantities by the number of moles of one unique component

$$y_1^m = \left(\frac{\partial Y^m}{\partial N_i}\right)_{P,T,N_j}$$

Note that, to measure a partial quantity of a system, we have to introduce a small quantity, dN_i, of this unique pure component into the system and to determine the variation dY^m during the mixing, at constant pressure and temperature. This is often easy to realise in practical conditions with liquid systems.

The partial molar quantities of the Gibbs function are called the **chemical potential of mixing**

$$\mu_1^m = g_1^m = \left(\frac{\partial G^m}{\partial N_i}\right)_{P,T,N_j} = \mu_i - \mu_1^*$$

with the Gibbs-Helmholtz'decomposition we analyse :

$$\mu_1^m = h_1^m - T\,s_1^m$$

h_1^m and s_1^m are respectively the **partial enthalpy of mixing** and the **partial entropy of mixing**.

The **activity scale** a_i of one component is an exponential picture of the chemical potential of mixing

$$\mu_1^m = RT\ln a_i = \mu_i - \mu_1^*$$

in which each reference state of the pure component μ_1^*, liquid state or individual crystal type, defines one particular scale of activity. Considering that all the integral quantities are linear homogeneous functions of the number of moles of each component, there exists an **Euler's relation** between the partial and the integral quantities

$$Y^m = N_1 y_1^m + N_2 y_2^m = \Sigma N_i y_i^m$$

$$y^m = x_1 y_1^m + x_2 y_2^m = \Sigma x_i y_i^m$$

For experimenters, the **Gibbs-Duhem's** isobar-isothermal relations which we are concerned with below, are very precious because they allow the determination of the second partial quantity starting from knowledge of the first one (this property can also be extended to multicomponent systems) :

$$x_1 \, dy_1^m + x_2 \, dy_2^m = 0 \quad \text{at constant P and T}$$

By integration we obtain

$$y_2^m = - \int_{x_2 = 1}^{x_2} \frac{x_1}{x_2} \, dy_1^m$$

The experimental difficulty often encountered is to measure y_1^m up to the **pure component 2*** ($x_2 = 1$) in view of obtaining the zero constant in Gibbs-Duhem's integration.

Starting from the two Gibbs-Helmholtz's relations it is easy to establish **Van't Hoff's** derivative

$$h^m = [\frac{\partial(g^m/T)}{\partial(1/T)}]_{P=Po}$$

which allows experimenters to determine the enthalpic contribution in the Gibbs function.

On the other hand, the variation of h^m with temperature is linked to the excess heat-capacity at constant pressure of the alloy by **Kirchhoff's** law :

$$(\frac{\partial h^m}{\partial T})_P = \Delta C_P^E = C_p(\text{alloy}) - x_1 C_{P1}^* - x_2 C_{P2}^*$$

For many systems we don't have any reliable information on ΔC_P^E, in particular for its variations versus temperature. For this reason, experimenters often use **Kopp-Neumann's** rule : in this rough approximation $\Delta C_P^{E*} = 0$ and consequently Gibb's function varies linearly with the temperature : $h^m(T) = $ constant and $s^m(T) = $ constant.

The last very important source of thermodynamic experimental data is linked to **Nernst's** postulate $S \to 0$ when $T \to 0$ and consequently :

$$s^m = \int_o^T \frac{\Delta C_P^E}{T} \, dT$$

the difficulty is to measure the specific-heat up to high temperatures, starting from very low ones (in general at 4 K in liquid helium). Only a few laboratories are able to realise such a performance.

To conclude this introduction, I would like to summarise simply the objective of experimental works in the field of the thermochemistry. **In condensed systems we always have to determine Gibbs function and often to analyse its two contributions separately : the enthalpy of mixing h^m and the entropy of mixing s^m.**

AND NOW LET US HAVE A LOOK AT THE PRACTICAL PROBLEM OF MEASURING

effectively such quantities. From my point of view there is no universal technic nor any universal apparatus to give us, in any case, Gibb's function with good accuracy. I prefer to speak about **"stategies of hunting"**. And I would like to describe four **principal types of experimental strategies** valide with a lot of experimental devices. In **the first strategy** the experimental primary information is given by the measure of **only one chemical potential** of mixing, in the whole range of concentrations and, at least, for two different temperatures, separated about one or two hundred Kelvin. We get this chemical potential of mixing by a chemical equilibrium technique involving the chemical analysis of a mixture, or measuring a partial pressure or an E.M.F. in an electrolytic cell... The following way is described in table I :

Table I . Strategy number I

Primary experiment $\mu_1^m(T_1$, measured $\forall x$)

$\mu_1^m(T_2$, measured $\forall x$)

Gibbs Duhem's integration $\mu_2^m(T_1, \forall x)$

$\mu_2^m(T_2, \forall x)$

Gibbs Helmholtz derivative $\forall x$

$$s_1^m(x) = \frac{\mu_1^m(T_1,x) - \mu_1^m(T_2,x)}{T_1 - T_2} \quad \text{and item for } s_2^m(x)$$

Van't Hoff's derivative $\forall x$

$$h_1^m(x) = \frac{\mu_1^m(T_1,x)/T_1 - \mu_1^m(T_2,x)/T_2}{1/T_1 - 1/T_2} \quad \text{and item for } h_2^m(x)$$

Euler's relation for integral quantities $\forall x$

$$h^m(x) = x_1 h_1^m(x) + x_2 h_2^m(x)$$

and

$$s^m(x) = x_1 s_1^m(x) + x_2 s_2^m(x)$$

Kopp-Neumann's rule associated with Gibbs-Helmholtz'relation give

$$g^m(T,x) = h^m(x) - Ts^m(x)$$

The most important difficulty encountered with this strategy is linked to the rough accuracy we get when measuring the slopes to obtain s_1^m or h_1^m.

If one of these two quantities is highly-predominant in the corresponding chemical potential of mixing the other will be determined with a bad uncertainly. If the two enthalpic and entropic contributions are of the same order of magnitude the uncertainty on the slopes affects simultaneously the determination of these two contributions. This is the reason why we often use a double experimental primary information, the first one for the chemical potential, the second for the enthalpy, using a calorimetric technique and this is my "second strategy" that I have described in table II.

Table II . Strategy number II

Primary experiments $\quad \mu_1^m(T_1, \forall x)$

and $\qquad h_1^m(T_2, \forall x) = h_1^m(x)$

Kopp-Neumann's approximation

$h_1^m(x) = h_1^m(T_2, x)$ and consequently

Gibbs-Helmholtz' derivative

$$-s_1^m(x) = \frac{\mu_1^m(T_1, x) - h_1^m(T_2, x)}{T_1}$$

Gibbs Duhem's integration

$h_2^m(x) \quad$ and $s_2^m(x)$

Euler's relations

$h^m(x) = x_1 h_1^m(x) + x_2 h_2^m(x)$

$s^m(x) = x_1 s_1^m(x) + x_2 s_2^m(x)$

Gibbs-Helmholtz formula

$g^m(T, x) = h^m(x) - Ts^m(x)$

In many cases chemical potential and calorimetric measurements are not realised at the same temperature, depending on each particular system and on the technique used. But, in this strategy, Kopp-Neumann's rule is also invoked.

The **third strategy** concerns only the **stoichiometric compounds** or a solution studied at a given particular concentration. In this case the integral enthalpy of formation can be measured directly in a calorimeter. Two general technics are described in table III.

Unfortunately, in most cases ΔC_p^E is not really measured, or is measured starting only at room temperature. In these two cases s^m has to be estimated (see below CALPHAD's strategy).

The **fourth strategy** is based on the calculation of the phase diagram, it consists in using the experimental phase diagram as a source of thermodynamic data. Many different automatic programs are used and often analysed in the CALPHAD group (see CALPHAD journal). In a general way, each solution is described by its Gibbs function of mixing.

$$g^m = x\mu_A^* + (1-x)\ \mu_B^* + RT\ [x \ln x + (1-x)\ \ln(1-x)] + g^E$$

In this expression, the two first terms are related to the reference states of the pure components, the third term expresses the entropy of

Table III. Strategy number III : stoichiometric compounds

a) Measurement of the enthalpy of formation at T_0

Direct reaction calorimetry	Liquid bath dissolving calorimetry

mixture
$nA* + (1-n)B*$ at T_0 $nA*$ or $(1-n)B*$ or $A_n B_{1-n}$ at T_0

Reaction in the crucible of the   dissolving in the bath
calorimeter at temperature T_0 h_1 h_2 h_3

$A_n B_{1-n} + h^m(T_0)$ $h^m(T_0) = h_1 + h_2 - h_3$
T_0 temperature of the crucible T_0 temperature of the initial state

b) C_p automatic measurements $T \in (0, T)$

Calculations

$$s^m(T) = \int_0^T \frac{\Delta C_p^E}{T} \, dT \qquad\qquad h^m(T) = h^m(T_0) + \int_0^T C_p^E \, dT$$

Gibbs-Helmholtz'relation

$$g^m(T) = h^m(T) - T\, s^m(T)$$

configuration of an ideal substitutional solution and the last "excess" g^E term can be analysed mathematicaly by a polynomial development

$$g^E = x(1-x) \sum_1^n P_i(T,x)$$

In our group, Charles (1) uses orthogonal Legendre's basis

$$g^E = x(1-x) \sum_1^n (a_i + b_i T) L_i(x)$$

where $L_i(x)$ are Legendre's polynomials.
With Kopp-Neumann's rule :

$$h^m = x(1-x) \sum_1^n a_i L_i(x)$$

and $\quad -s^E = x(1-x) \sum_1^n b_i L_i(x)$

In Charles'program (1) the adjustable parameters are the a_i and b_i coefficients of the liquid solution and also the enthalpies and entropies of formation of each compound. In view of determining these parameters a set of linear equations is established, translating :

1) some equilibrium conditions between two phases as can be obtained from the experimental phase diagram.

2) all the experimental data concerning the enthalpies, entropies or Gibbs function available in the literature but considered only as proposal values.

3) the entropies and enthalpies of melting of the pure components as definitive values.

This set of equations is about twice over-determined in comparison with the number of unknown parameters. Finally, a least square deviation method, using weighing factors for each information, is used to optimise the system. At the end of the calculation all the unknown data are estimated and all the known data are "criticised" in consistency with the other ones. The calculated diagram is also drawn. This fourth strategy is presently one of the most efficient processes and avoids too many experiments.

BUT LET US GO NOW TO THE ILLUSTRATION OF THESE FOUR STRATEGIES ON DIFFERENT EXAMPLES TAKEN IN MY LABORATORY.

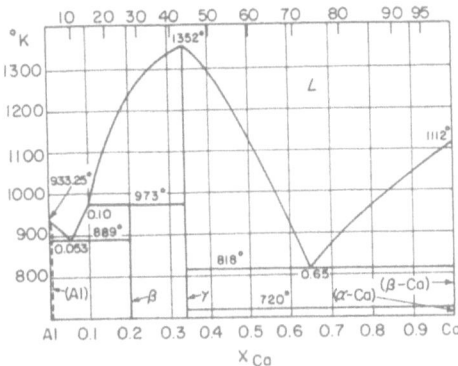

Fig.1:(Al,Ca) MATSUYAMA'S (2) diagram as drawn in HANSEN'S compilation book (3).

I would like to choose the **Aluminium-Calcium** system, for which three different strategies have been tested in our group.
Figure 1 gives the Al-Ca phase diagram according to a rather old publication by Matsuyama (2) as taken in the Hansen compilation book (3). This system appears to be very interesting for its thermodynamic properties. In fact, in the Ca-poor part of the diagram the system exhibits attractive properties between the unlike atoms and stabilises the solid compounds $\beta = Al_4Ca$ and chiefly $\gamma = Al_2Ca$.
On the contrary, in the Ca-rich part the system shows a deep eutectic plateau that means that the liquid phase is stabilised and perhaps the system exhibits repulsive effects on unlike atoms.

The first technic we used to study this system is the measurement of the μ_{Ca}^m

Chemical potential by the E.M.F. technic.

The original device has been well-described by Lefebvre, Notin, Hilger and Hertz (4). It contains two different parts (see figure 2).

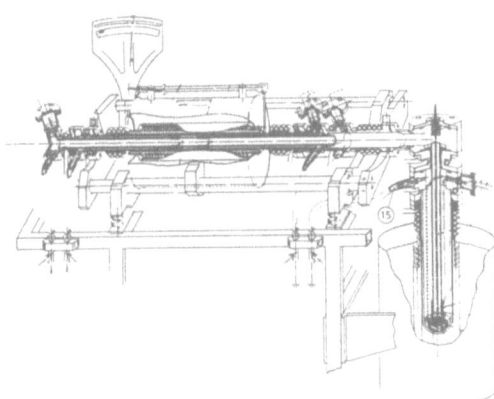

Fig.2: The pin-point E.M.F. device: left, the oxygen pump with the zirconia active tube, right the measuring cell showing here one single metallic point-electrode touching a solid electrolyte (4).

314

An oxygen pump in horizontal furnace is essentially a zirconia tube : an electrical current flows inside its wall and provides inside the tube, a high oxygen purity to the argon gas of about $P_{O_2} = 10^{-20}$ atm. An electronic regulator controls the purification process. This highly purified argon gas flows to the measuring cell contained in an alumina-closed tube inside a vertical furnace. In this cell, (see figure 3 for more details), three different metallic points called "pin-point electrodes" are in contact with a solid electrolyte-pellet. On the other side of this electrolyte, a mixture of metal and metal oxyde provides a stable reference E.M.F. potential. For the present Al-Ca study, the electrolyte is a mixture of Calcium fluoride and lime. The depolarisation technics used has been previously described by Notin and Hertz (5). By cathodic electrolysis, a small amount of calcium is introduced at the tip of an electrode. During the depolarisation process (after switching off the current), we can observe different waves on the variation of the cell-potential versus time. Each wave corresponds to a Calcium constant activity at the point of the electrode. In general, this corresponds to one biphase domain in the diagram. The zero of the E.M.F. scale is obtained with an iron point-electrode which always exhibits activity equal to one for the pure calcium just after its polarisation.

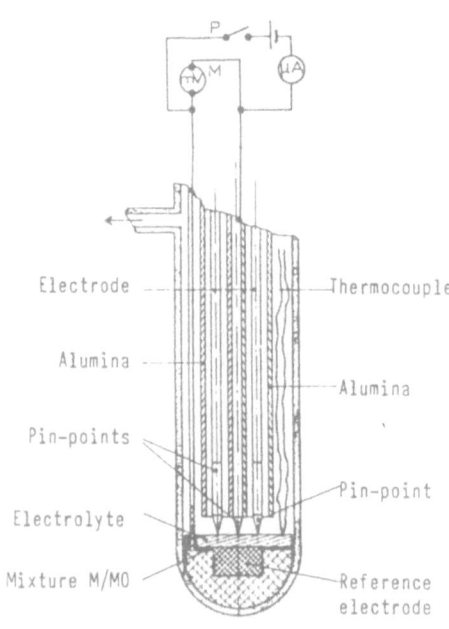

Fig.3: Three pin-point electrodes in touch with a solid electrolyte itself carried by a stable

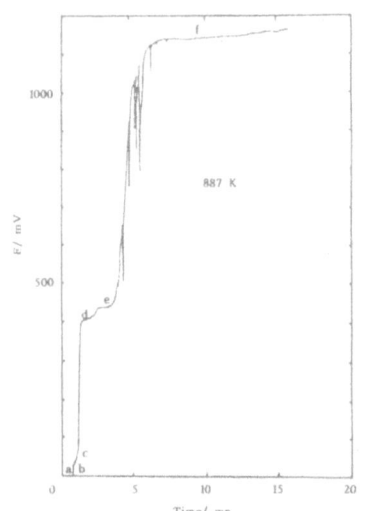

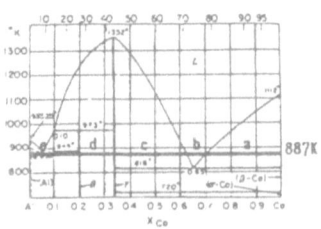

Fig.4: The depolarization curve obtained with an aluminium pin-point electrode after a calcium deposit by electrolysis. The potential zero is taken for pure solid calcium.

Figure 4 shows the depolarisation curve obtained with an aluminium-electrode : the "d" and "e" waves are used to determine the chemical potential of calcium in the (γ+β) mixture and in the (β+Al) mixture respectively. As Notin has demonstrated in his thesis (6) an electrolyte made of a mixture (CaF$_2$ + CaO) is able to characterise equally as well the fluor activity, the oxygen activity or the calcium activity of the microsystem at the point of the electrode.

$$- E = \frac{\mu_{Ca}^m}{2F} = \frac{RT \ln a_{Ca}}{2F}$$

The slopes of the "d" and "e" wave-potentials versus temperature are given in figure 5 : the "d" wave is a staight line in good consistency with Kopp-Neumann's approximation whereas the "e" wave exhibits a strong curvature, in the vicinity of the eutectic plateau, linked to the variation of calcium concentration in the primary aluminium solution. The slopes of these curves can be interpreted as the partial entropies of mixing

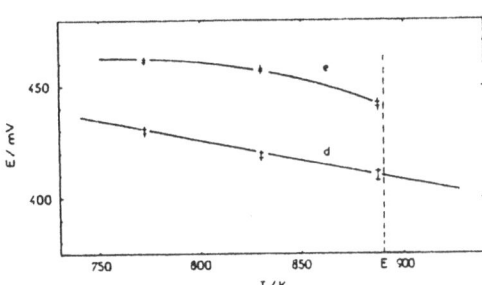

Fig.5:Variation of the "d" and "e" wave-potentials versus temperature.

$$s_{Ca}^m = + 2F \frac{dE}{dT}$$

The results obtained by Notin et al can be analysed briefly (for more details see publication 7).

At 800 K   E(d) = 460 ± 2 mV μ_{Ca}^m(d) = - 82000 ± 400 J/mol

E(e) = 426 ± 2 mV μ_{Ca}^m(e) = - 88800 ± 400 J/mol

Gibbs Duhem's integration is here very simple starting from the e = (Al+β) equilibrium with μ_{Al}^m = 0

$$g^m (\beta,800 \text{ K}) = 0.2 \ \mu_{Ca}^m(e) = - 17760 \pm 80 \text{ J/mole}$$

Taking into account the equilibrium of this β phase with γ we can also write Euler's equation

$$g^m (\beta,800 \text{ K}) = 0.2 \ \mu_{Ca}^m(d) + 0.8 \ \mu_{Al}^m(d)$$

from this relation, we found μ_{Al}^m(d) = - 1690 J/mole
and

$$g^m (\gamma,800 \text{ K}) = 0,33 \ \mu_{Ca}^m(d) + 0,67 \ \mu_{Al}^m(d) = - 28470 \pm 300 \text{ J/mole}$$

Exactly the same procedure of calculation is used for the entropies starting from the $\frac{dE}{dT}$ slopes :

with $\frac{dE}{dT}$(d) = - 0.175 ± 0.030 mV/K

and (for T < 850 K) $\frac{dE}{dT}$(e) = - 0.062 ± 0.020 mV/K

we obtain $s^m(\beta)$ = - 2.2 ± 0.8 J/mole K

and $s^m(\gamma)$ = - 7.3 ± 1.5 J/mole K

The uncertainty on the slopes is not very high and corresponds to the representative deviation in such experiments. If the absolute error on these two entropies is not too high, the relative precision is very bad in consideration of their very small values.

Strategy number I can be achieved by subtracting the entropic terms (of the two compounds) from the Gibbs functions :

$$h^m(\beta) = g^m(\beta, 800 \text{ K}) + 800 \; s^m(\beta) = -19500 \pm 750 \text{ J/mole}$$

and

$$h^m(\gamma) = g^m(\gamma, 800 \text{ K}) + 800 \; s^m(\gamma) = -34300 \pm 1500 \text{ J/mole}$$

The second strategy can be introduced to better the accuracy

on the enthalpic and entropic terms respectively. Above 973 K (see diagram 1) introduction of pure calcium in an aluminium bath, will precipitate the γ phase. Under 973 K, the β phase. The direct precipitation reaction will be realised in a Calvet type calorimeter. In such a device containing about 400 junctions in the thermopile, almost all the heat flow between the crucible and the isothermal block is conducted by the thermopile and translated in the calorimetric signal. For this reason, the duration of the integration of the thermal effect, associated to the reaction process, can be very long (even a few hours), without any trouble for the standard deviation of the measure.

Figure 6.1 gives the variations of h_{Ca}^m versus Ca concentration. Each rectangle represents one measurement, when introducing a small amount of pure calcium in the aluminium-rich bath :

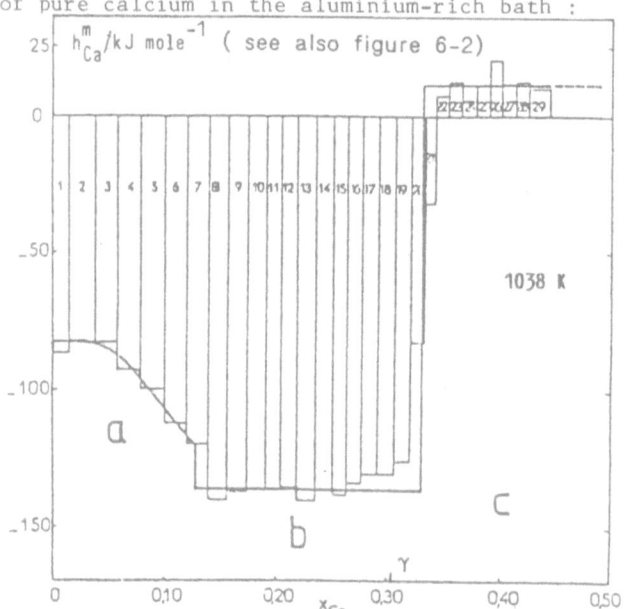

$$h_{Ca}^m = \frac{\Delta H}{\Delta N_{Ca}}$$

where ΔH is the total heat effect during the reaction (when subtracting the enthalpy needed to heat the calcium during the drop) and ΔN_{Ca} the number of moles of Ca in this sample. Experiments were conducted at 1038 K and, for this reason, after saturation of the homogeneous liquid bath (part **a** in figure 6.1) we observe the precipitation of the γ phase (part **b**). Note the jump of h_{Ca}^m at each boundary of a biphase domain. In particular when crossing the stoichiometric composition of the γ phase ($x_{Ca} = 0.33$), h_{Ca}^m becomes positive.

Fig.6-1: Experimental determination at 1038K of the partial molar enthalpy of mixing for calcium in the (Al,Ca) system: a: homogeneous liquid- b: (liquid +γ) c:(γ+liquid)

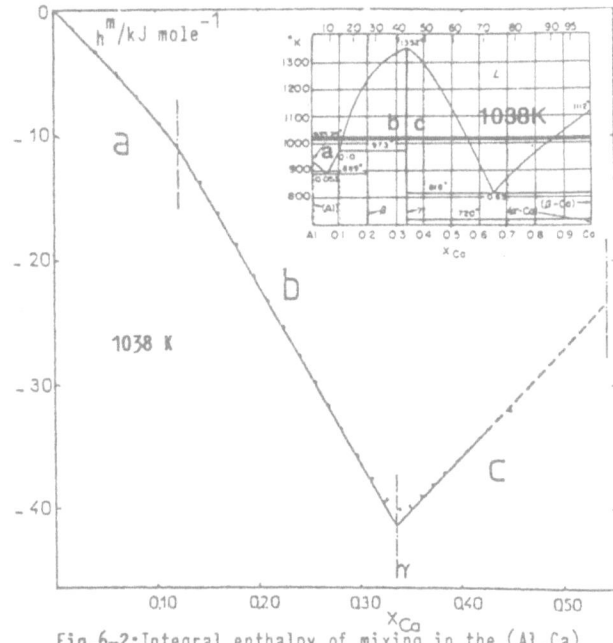

Integration of h_{Ca}^m gives the integral enthalpy of mixing (figure 6.2)

$$h^m = \frac{\Sigma \bar{H}_i}{N_o(Al)+\Sigma N_i(Ca)}$$

Thus we obtain the enthalpy of formation of the compound. The same type of experiment was conducted by Notin et al also at 953 K obtaining the precipitation of the β compound (8) :

h^m (β , 953 K) = -18700 ± 300J/mole

h^m (γ , 1038 K) = -33400 ± 600J/mole

Fig.6-2:Integral enthalpy of mixing in the (Al,Ca) system at 1038K.

Discussion :

In table IV, we recall the main results obtained by the first strategy (only by E.M.F. measurement) and we compare the enthalpic results of this method with the direct precipitation calorimetry.

Table IV	
Strategy n° I **E.M.F. only** J/mole	**Direct precipitation** **calorimetry** J/mole
g^m (800 K, β) = - 17760 ± 80	
g^m (800 K, γ) = - 28470 ± 300	-
h^m (β) = - 19500 ± 756 ⎫	h^m (β) = - 18700 ± 300
h^m (γ) = - 34300 ± 1500 ⎭	h^m (γ) = - 33400 ± 600
	Strategy II **crossing the results**
	$s^m = \dfrac{h^m - g^m \ (800 \ K)}{800}$
s^m (β) = - 2.2 ± 0.8 J/mole K	s^m (β) = -1.2 ± 0.5
s^m (γ) = - 7.3 ± 1.5 J/mole K	s^m (γ) = - 6.2± 1 J/mole K

The consistency of the enthalpic results we got by these two methods is rather good : this illustrate the fact that g^m (800 K) contains about 80 % of enthalpic contribution. For this reason, the measurement of g^m is a perfect information for h^m. Nevertheless the accuracy increases by the calorimetric technic : the uncertainty of h^m is about 5 % for E.M.F. and 2 % for calorimetry. On the contrary, in this particular case, at 800 K, g^m is really poor information for the entropies, and for this reason, strategy II (crossing E.M.F. and calorimetry) gives a more valid information for the entropies.

Presently, let us look at a CALPHAD strategy

applied to the optimisation of the same system. Taking into account all the available experimental data, such as the enthalpy of mixing of the liquid phase measured by Sommer, Lee and Predel (9), the Gibbs'function for the liquid by Schurmann, Funders and Litterscheidt (10), the entropies of melting of β and γ compounds by Kocherov and Geld (11), our one E.M.F. measurements for β and γ compounds (7) and simultaneously the phase diagram of Matsuyama (2), Charles (12) has developed an optimisation process. Table V.b gives the results he got. The consistency between all the results appears not too bad after this optimisation. Nevertheless, the phase diagram restitution presents some important deformation in the temperature scale. Another calculation (Table V.c) conducted **without any experimental data for the liquid phase** gives a more valid answer for the diagram and also restitutes exactly the enthalpies obtained by calorimetry for β and γ compounds. Evaluation of the Gibbs'function of the liquid phase is correct, but its decomposition between enthalpic and entropic contributions attributes too large an influence to the entropy. This example illustrates the good information it is possible to obtain by the CALPHAD method and the **experimental economy** it can provide.

So, what about the third strategy ?

I will take my examples from a series of recent measurements developed by **Gachon** (13) in the field of **intermetallic transition-transition compounds**. These phases are largely focused by physicists for the reason that the bonding energy is essentialy linked to the d-band interactions.
The technic used is the direct reaction calorimetry, dropping compacts of mixed powdered components directly into the hot crucible of the calorimeter. In view of obtaining complete reaction it is often necessary to choose the temperature of the calorimeter just below the solidus line and for this reason, we use a high temperature device valid up to 1800 K.
There even exists on the market one high-temperature calorimeter developed by Mercier in the SETARAM company (see figure 7). We have modified this device in view of conducting through the thermopile a more important part of the heat-flow. We use a sensor made up of two-rows of thermocouple junctions disposed around the reaction crucible. Nevertheless this device remains a "partial" heat-fluxmeter

TABLE V: CALPHAD'S method to optimise the (Al,Ca) system.

a)- EXPERIMENTAL DATA

COMPOUNDS	LIQUID x_{Ca}	G (kJ) (10)	H (kJ) (9)	S (J/K)	DIAGRAM x_{Ca}	T (K)
	0,0	0	0	0	Al/Al$_4$Ca	
Al$_4$Ca	0,1	-9,7	-9,9	-0,19	0,053	
h_1^f=-19,4kJ	0,2	-16,4	-17,7	-1,14	(2)	889%
(7)	0,3	-21,2	-23,1	-1,62	Al$_4$Ca/Al$_2$Ca	
s_1^f= -2J/K	0,4	-22,9	-23,8	-0,76	0,10	
s_1^m=+12J/K (?)	0,5	-22,1	-21,9	0,19	(2)	973%
Al$_2$Ca	0,6	-19,8	-18,4	1,14	Al$_2$Ca/Ca	
h_2^f=-34,4kJ	0,7	-16,4	-14,2	1,91	0,65	
(7)	0,8	-12,3	-9,8	2,19	(?)	818%
s_2^f= -7,3J/K	0,9	-6,9	-4,8	1,91		
s_2^m=+16 J/K (11)	1,0	0	0	0		

b)- OPTIMISATION WITH ALL THE AVAILABLE EXPERIMENTAL DATA

COMPOUNDS	LIQUID x_{Ca}	G (kJ)	H (kJ)	S (J/K)	DIAGRAM x_{Ca}	T (K)
	0,0	0	0	0	Al/Al$_4$Ca	
Al$_4$Ca	0,1	-9,5	-9,4	0,08	0,055	860
h_1^f=-20,3kJ	0,2	-16,1	-16,5	-0,31		
s_1^f=-2,4J/K	0,3	-20,4	-21,0	-0,49	Al$_4$Ca/Al$_2$Ca	
s_1^m=12,8J/K	0,4	-22,3	-22,7	-0,37	0,11	1006
	0,5	-21,8	-21,8	0,02		
Al$_2$Ca	0,6	-19,4	-18,7	0,59	Al$_2$Ca/Ca	
h_2^f=-35,1kJ	0,7	-15,4	-14,0	1,20	0,65	814
s_2^f= -7,4J/K	0,8	-10,4	-6,5	1,52		
s_2^m=17,1J/K	0,9	-5,2	-3,4	1,53	Fusion Al$_2$Ca	
	1,0	0	0	0		1357

c)-OPTIMISATION WITHOUT THERMODYNAMIC MEASUREMENT FOR LIQUID

COMPOUNDS	LIQUID x_{Ca}	G (kJ)	H (kJ)	S (J/K)	DIAGRAM x_{Ca}	T (K)
	0,0	0	0	0	Al/Al$_4$Ca	
Al$_4$Ca	0,1	-9,2	-8,3	2,02	0,053	835
h_1^f=-18,7kJ	0,2	-15,1	-14,2	0,79		
s_1^f=-2,4J/K	0,3	-18,5	-19,7	-0,97	Al$_4$Ca/Al$_2$Ca	
s_1^m=13,1J/K	0,4	-20,1	-22,2	-1,81	0,103	983
	0,5	-20,0	-21,2	-1,04		
Al$_2$Ca	0,6	-18,3	-15,8	1,31	Al$_2$Ca/Ca	
h_2^f=-33,4kJ	0,7	-15,5	-10,3	4,49	0,65	817
s_2^f= -7,3J/K	0,8	-11,5	-3,4	7,33		
s_2^m=16,7J/K	0,9	-8,6	1,1	6,70	Fusion Al$_2$Ca	
	1,0	0	0	0		1348

320

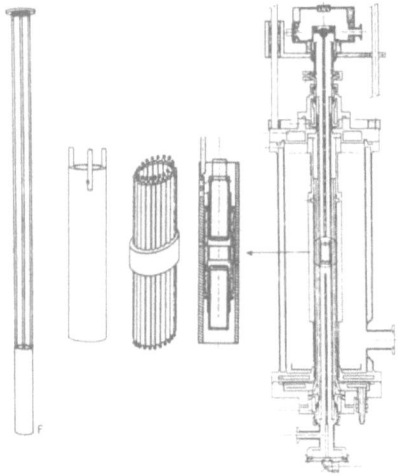

Fig.7

SETARAM'S 1800K calorimeter

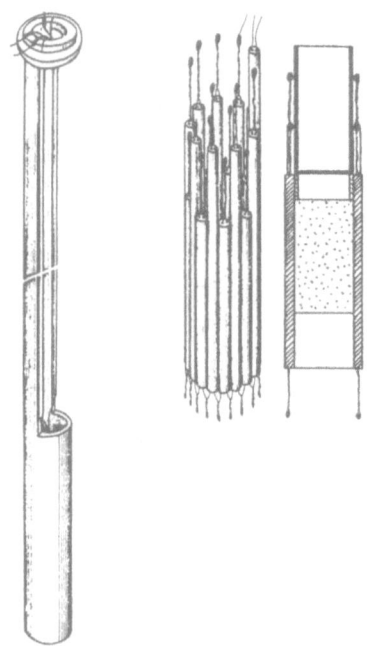

Our modification of the block-support, of the thermopile and of the crucible's disposal. Note the two rows of thermocouple junctions arround the work crucible.

in consideration of the radiation heat-transfers at high temperature. For this reason, only quick reactions (a few minutes) associated with generally high thermal powers can be observed. The base line is not very stable in time in comparison with a Calvet calorimeter. The standard deviation, in good operating conditions, is about or less than 6 %. But, accuracy on the final result can be better than the standard deviation if we work in **compensation** conditions. In fact, for **exothermal** reactions, the heat content of the sample, when introducing it at room temperature, can compensate for approximately the heat of mixing. If this condition is satisfied, the calorimetric measure is just **a small correcting term** in the final result and the accuracy is linked to the knowledge of the heat content of pure components (generally well-known) and to the standard deviation of the **weighing.** Figure 8 gives the organigram of the experimental process, which provides the synthesis of the alloy directly in the calorimeter and the verification of the crystal structure after cooling. Table VI gives the results obtained by Gachon et al for 27 compounds, generally synthesised by alloying Ti, Zr or Hf to the triade elements. More details can be found in the original papers of Gachon et al (14-18) but I would like to discuss one particular case : figure 9 gives the results we got by

TABLE VI

REVIEW OF EXPERIMENTAL DATA OBTAINED
BY HIGH TEMPERATURE CALORIMETRY

We present in the following all the results we got on binary alloys by direct reaction calorimetry at high temperatures using sintered pellets:

Compound	Enthalpy of formation $J.mol^{-1}$	Standard deviation $J.mol^{-1}$	Temperature K
$Fe_{0.50}Ti_{0.50}$	-31000	1300	1440
$Fe_{0.67}Ti_{0.33}$	-27600	1000	1514
$Co_{0.50}Ti_{0.50}$	-44300	500	1490
$Co_{0.67}Ti_{0.33}$	-34100	600	1432
$Ni_{0.33}Ti_{0.67}$	-29300	500	1202
$Ni_{0.50}Ti_{0.50}$	-34000	2000	1475
$Ni_{0.75}Ti_{0.25}$	-42900	1000	1513
$Fe_{0.67}Zr_{0.33}$	-29700	1700	1760
$Co_{0.33}Zr_{0.67}$	-33000	2000	1290
$Co_{0.50}Zr_{0.50}$	-42200	1000	1512
$Co_{0.67}Zr_{0.33}$	-41000	1600	1708
$Co_{0.80}Zr_{0.20}$	-29800	1500	1596
$Ni_{0.33}Zr_{0.67}$	-36800	1000	1230
$Ni_{0.50}Zr_{0.50}$	-51500	2000	1405
$Ni_{0.78}Zr_{0.22}$	-39500	500	1670
$Ni_{0.83}Zr_{0.17}$	-32400	3000	1479
$Pd_{0.50}Sc_{0.50}$	-106000	7000	1681
$Pd_{0.50}Ti_{0.50}$	-53000	1100	1578
$Pd_{0.50}Zr_{0.50}$	-62000	5000	1667
$Pd_{0.50}Hf_{0.50}$	-79000	7000	1685
$Pt_{0.53}Ti_{0.47}$	-75000	7000	1613
$Pt_{0.50}Zr_{0.50}$	-90000	10000	1629
$Pt_{0.50}Hf_{0.50}$	-113000	6000	1336
$Pt_{0.23}V_{0.77}$	-33600	1000	1694
$Pt_{0.50}V_{0.50}$	-39300	300	1659
$Pt_{0.74}V_{0.26}$	-30000	4000	1259
$Fe_{0.82}V_{0.18}$	-4200	1200	1525

322

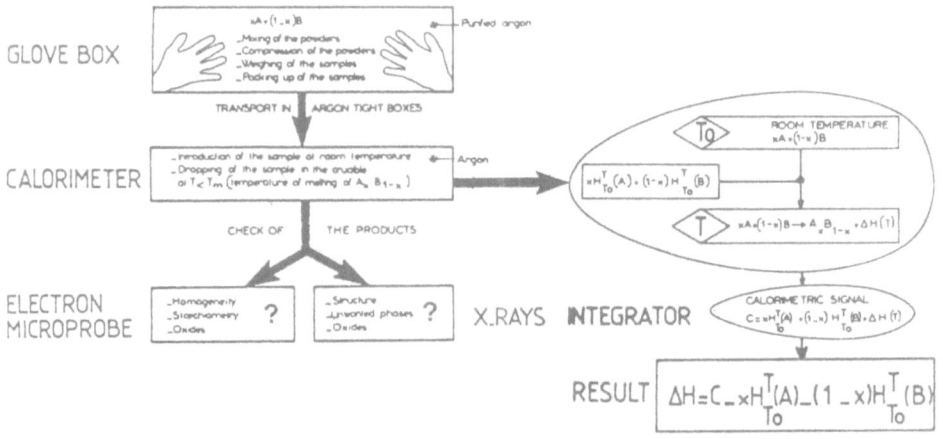

Fig.8: Diagram of the experimental process

measuring four enthalpies of formation of compounds belonging to the (Ni Zr) system. The same figure shows the temperatures used to obtain the quick reactions. We can observe, very close with our results, the values obtained by Henaff, Colinet, Pasturel and Buschow using an indirect dissolution calorimetric technic (19). Using Miedema's model (20) it was possible to estimate the same enthalpies. In this particular case, (saturation of the d-band for Ni),there exists a rather large deviation between Miedema's model and experiments. Physicists can also corroborate such experimental results when choosing valuable adjustable parameters for the **modelisation of the d-band.** In this way, figure 9 shows the results obtained by Bennet and Watson (21) with a rectangular d band approximation and by Colinet, Pasturel and Hicter (22) with a parabolic d band representation.

In another direction CALPHAD 's strategy can be applied to estimate all the missing information in this system, using only the four measured enthalpies of mixing, the two entropies of melting of the pure Nickel and the pure Zirconium and also some equilibrium conditions read on the experimental phase diagram.

Figure 10 gives such an optimisation of this system, realised by Charles, which provides good estimations of the enthalpies of formation of three other compounds together with the entropies of formation and melting of all the compounds and also Gibbs'function of the liquid phase.

Exactly the same technique (strategy III and CALPHAD's approach) can be applied for carbides. Figure 11 is related to one of our recent works (23) concerning the (C,Cr) system. For the synthesis of carbides the operating temperature does often attain **1800 K** inside the calorimeter and perhaps this temperature is **still insufficient.**

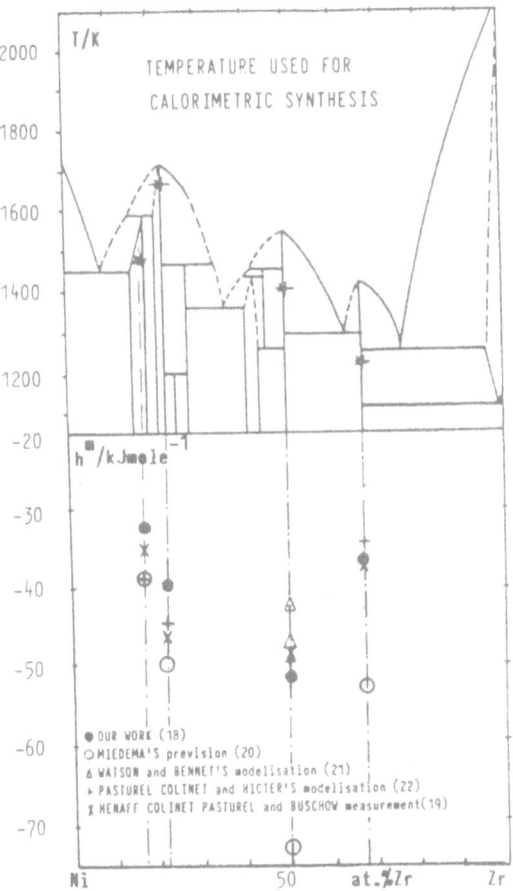

Fig.9: Direct calorimetric measurement of the enthalpies of formation of four (Ni,Zr) compounds. Comparison with various models and with the indirect experimental results of HENAFF, COLINET, PASTUREL and BUSCHOW (19).

Future studies in many laboratories will concern ternary systems.

In many cases estimations of the thermodynamic properties of ternary solutions can be deduced from the properties of the three adjacent binary systems. Nevertheless some laboratories are already engaged in measurements of thermodynamic properties of ternary systems. As an example figure 12 shows the projection of the (Al,Cu,Mg) system in the rich aluminium part where the "S" ternary compound appears. This Al_2CuMg compound forms peritectically and can't be synthetised directly starting from the pure components in the stoichiometric proportions. To measure the enthalpy of formation of this compound, we dissolved in an aluminium bath (inside the crucible of a Calvet calorimeter) an alloy A (figure 12) containing about the most important fraction of this compound we can obtain directly by solidification of a liquid ternary alloy (24). I think this example can be considered as representative of the validity of the **indirect** calorimetric method, but if your interest turns to this method, I would like to outline the beautiful lot of results recently obtained in this way by **O.J. Kleppa** in the James Franck Institute of Chicago.

A description of the coherent experimental procedures dealing with the thermodynamics of alloys can't be achieved when leaving out
THE C_p
MEASUREMENTS AND THE DIFFERENTIAL SCANNING CALORIMETRY (D.S.C.).
This
technic can be defined by the heating of a sample at linear (versus time) increasing or decreasing temperature. Automatic C_p measurements can be realised in a twice symetric calorimeter when subtracting the

Fig.10:CALPHAD'S optimisation of the Ni Zr system

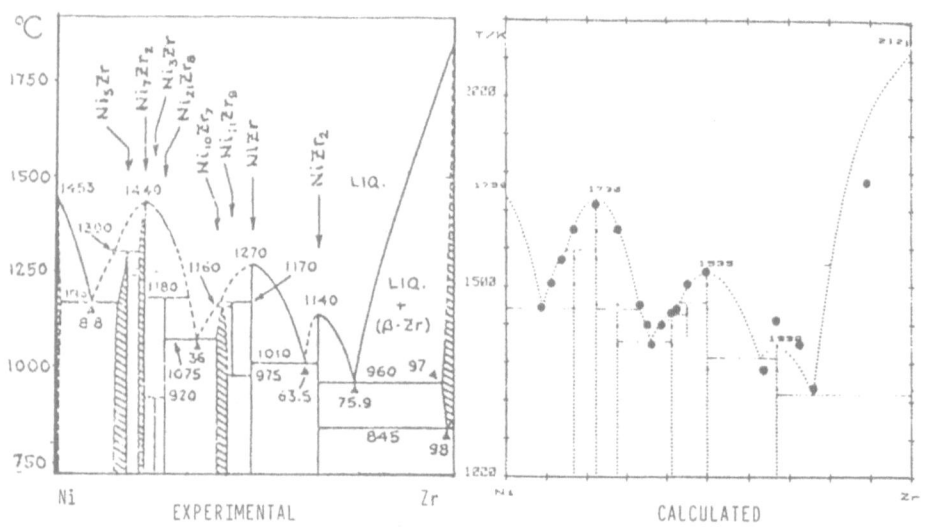

EXPERIMENTAL

CALCULATED

The points indicate the various
equilibrium conditions taken into
account for the calculation.

COMPOUND	EXPERIMENTAL		OPTIMISATION			
	h^m	$T_{melt.}$	h^m	s^m	s_{melt}	T_{melt}
Ni_5Zr	−32400		−31828	−0.34	10.46	1603
Ni_7Zr_2	−39500	1713	−40201	−0.17	10.87	1730
$Ni_{21}Zr_8$			−41571	1.01	10.38	1520
$Ni_{10}Zr_7$			−44038	3.56	9.37	1425
$Ni_{11}Zr_9$			−44063	4.09	9.10	1467
$Ni\ Zr$	−51500	1543	−51634	−0.75	14.12	1535
$Ni\ Zr_2$	−36800	1413	−36433	1.68	10.90	1339

Reference states Ni* and Zr* liquid
Liquid phase:

$$g_l^e = x(1-x)\left[-207780 +20,03T+(2x-1)(39563-3,58T)+(6x^2-6x+1)(-22653+21,58T)\right]$$

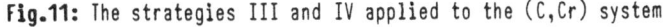

Fig.11: The strategies III and IV applied to the (C,Cr) system

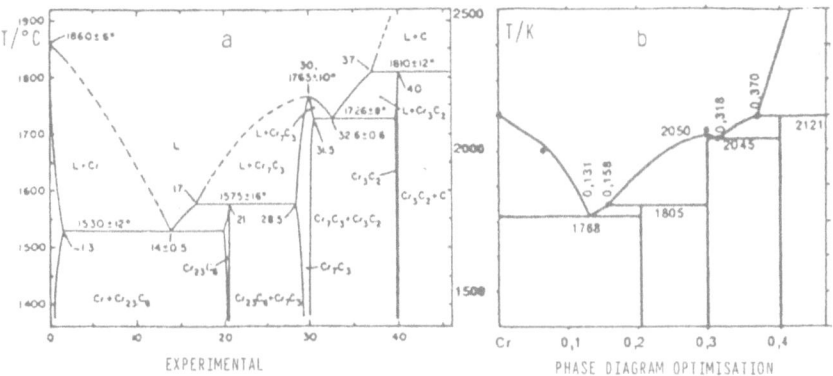

EXPERIMENTAL   PHASE DIAGRAM OPTIMISATION

Solid State : Reference states : Cr : bcc, C : Graphite.

h^m : enthalpy of formation J/mole, s^m : entropy of formation J/mole.K, $s_{melt.}$: entropy of melting J/mole.K

COMPOUND	EXPERIMENTAL		PHASE DIAGRAM OPTIMISATION			
			INPUT	OUTPUT		
	h^m	standard deviation	h^m	h^m	s^m	$s_{melt.}$
$Cr_{0.793} C_{0.207}$ $(Cr_{23} C_6)$	-9416 (1753K)	* 1000	-9416	-9248	3.79	18.1
*$Cr_{0.7} C_{0.3}$ $(Cr_7 C_3)$	-10693 (1753K)	± 1800	-10693	-10791	6.48	21.0
$Cr_{0.6} C_{0.4}$ (Cr_3C_2)	-9956 (1753K)	± 600	-9956	-10097	6.43	25.4

* Computed congruent melting point : 2050 K

Liquid State : Reference states : Cr and C are liquid. Units are : $J.mol^{-1}$ and $J.mol^{-1}.K^{-1}$

x = x graphite. Results from computation

$$Gex = x (1 - x) [(a_0 + b_0T) + (a_1 + b_1T) (2x - 1) + (a_2 + b_2T) (6x^2 - 6x + 1)]$$

$a_0 = 139\ 900$ $a_1 = 218\ 192$ $a_2 = 52\ 755$

$b_0 = -62.50$ $b_1 = -8.72$ $b_2 = +27.93$

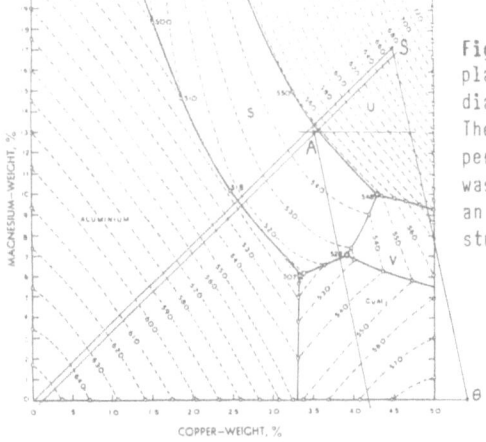

COPPER—WEIGHT, %

Fig.12: Projection on the concentration plan of the (Al,Cu,Mg) ternary phase diagram, in the rich-aluminium part. The ternary S compound (S=Al₂Cu Mg) forms peritectically. Its enthalpy of formation was measured by calorimetric dissolution in an aluminium bath. Alloy A used for this study contains more than 75% of S phase.

326

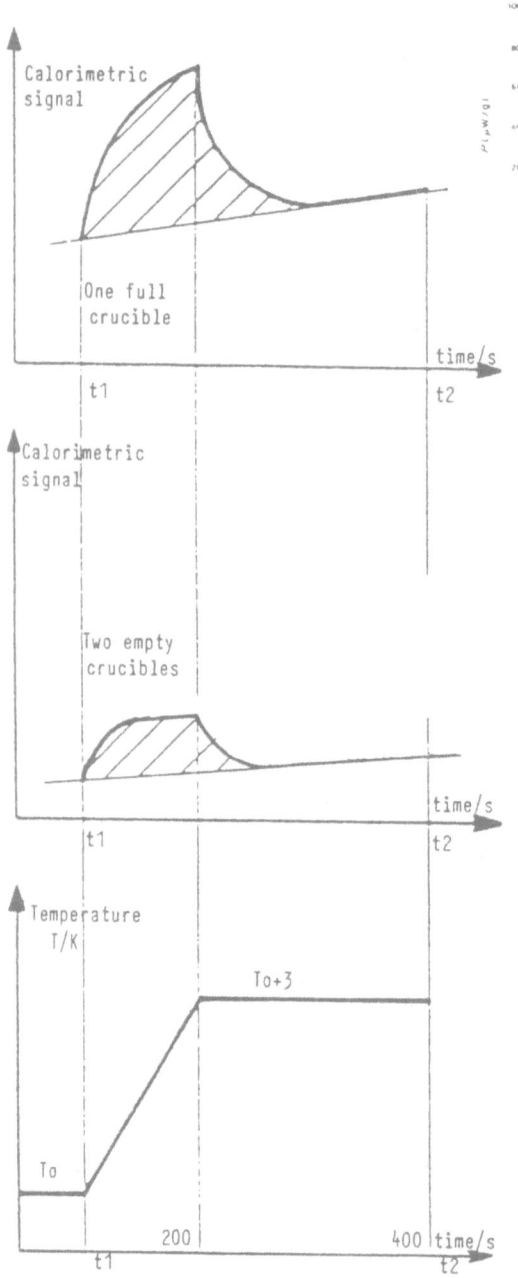

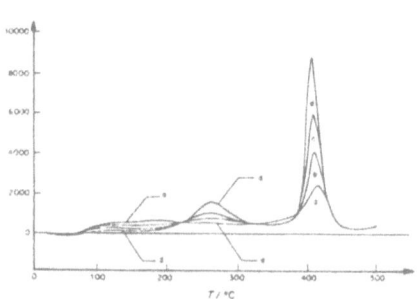

Fig.14: Irreversible D.S.C. thermograms obtained with the same steel after different heat treatments, in all cases austenisation at 850 °C for 10 minutes: **a**- water quenched.**b**- 2 minutes at 366°C before quenching.**c**- 2 minutes at 400°C before quenching.**d**- 2 minutes at 400°C before quenching. Stabilisation of the austenite by the isothermal treatment is well characterised by the D.S.C. technic.

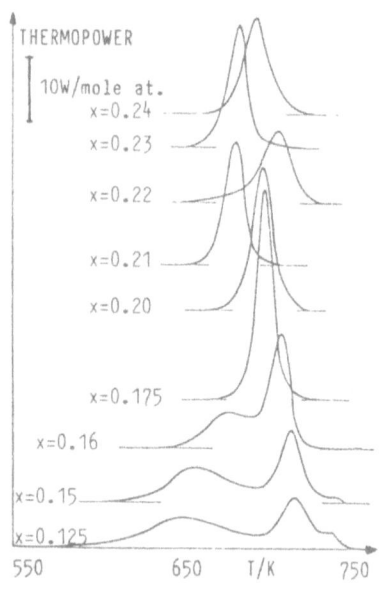

Fig.13: A step by step automatic measurement of Cp using a D.S.C. calorimeter. The comparison between the two areas (empty or full crucible) gives the information to calculate the Cp.

Fig.15: Irreversible D.S.C. thermograms obtained with a series of amorphous $Fe_{1-x}B_x$ alloys. Note the percolation-threshold when the double-peak of crystallisation becomes a single one.

two signals delivered by the calorimeter after two runs : the first one
with two empty crucibles, the second one with the substance contained
in one of the two crucibles. A more accurate procedure consists of
using a step by step D.S.C. programer providing successive jumps of
temperature of about 2 or 3 K between two isothermal plateaux. Figure
13 shows schematically the difference in the area of the two peaks
corresponding to **the same heating interval** when comparing an empty with
a full crucible. Association of a microcomputer with a D.S.C. device
allows obtaining the C_p versus T curves automatically. This technique has
been well-described by Cunat and Charles (25).
But, a D.S.C. technique can also be used to measure directly the
enthalpies associated with any phase transition in alloys. I would like
to illustrate this point with only two figures dealing with
irreversible processes consisting of the rupture of unstable states
during the heating of samples in a calorimeter. Figure 14 belongs to
the works of the group of Hilger devoted to steel heat treatment. We
can observe in this figure the evolution of the heat flow during the
annealing process. Four samples of the same alloy are here compared.
Each sample was previously austenised and water-quenched. But three of
these were partially transformed by an isothermal heat treatment, with
a duration of two minutes, at three different temperatures. This
"T.T.T." treatment increases the carbon content in austenite and for
this reason increases also the quantity of untransformed austenite
remaining after the quenching. Note the very strong difference between
the four thermograms.
Thus, the D.S.C. technic appears to be the most sensitive and valid one
to detect and correlate such a phenomenon (26-27).
Another illustration of the efficiency of the D.S.C. technics is the
measurement of **enthalpy of crystallisation of amorphous alloys.** These
works are performed in our laboratory in the group managed by
Mrs Kuhnast and Mr. Cunat. Figure 15 shows the evolution of the D.S.C.
peak of crystallisation of a series of $Fe_{1-x}B_x$ amorphous alloys
obtained by melt-spinning. Note the evolution of the shape of the peaks
with the boron content, observe the "percolation-threshold" when the
double-peak of crystallisation becomes a unique one. For more details
on these works you can see the publication of this group for (Fe,B)
(28,29,31), (Co,B)(30,31), (Ni,B)(32), (CoP)(32,33) alloys.

AS A VERY QUICK CONCLUSION,

I hope that I have convinced you that each
case of acquisition of a new thermodynamic datum is justifiable of an
original treatment always needing a **new imaginative turn of mind** rather
than a systematic technique. Only general strategies can be retained to
outline the different ways to follow.
I thank you for your kind attention.

BIBLIOGRAPHY

1. J. CHARLES, J.C. GACHON, J. HERTZ - CALPHAD **9**, n°1, (1985) 35-42
2. K. MATSUYAMA - Sc. Rep. Tohoky University, **17** (1928) 723
3. M. HANSEN and K. ANDERKO
   Constitution of Binary Alloys, MC Graw-Hill, New-York, 1958

328

4. G. LEFEBVRE, M. NOTIN, J.P. HILGER and J. HERTZ
   J. Solid. State Chem. **28**, (1979), 97-108
5. M. NOTIN and J. HERTZ - CALPHAD **6**, n° 1 (1982), 49-56
6. M. NOTIN - Thèse de Doctorat d'Etat, Université de Nancy, 30.09.83
7. M. NOTIN, J.C. GACHON and J. HERTZ
   J. of Less-Com. Metals **85** (1982), 205-212
8. M. NOTIN, J.C. GACHON and J. HERTZ - J.Chem. Therm. **14** (1982) 425-434
9. F. SOMMER, J.J. LEE and B. PREDEL - Z. Metalkde **74** (1983), 100
10. E. SCHURMANN, P. FUNDERS and H. LITTERSCHEIDT
    Arch. Eisenhüttenw. **46** (1975) 473
11. P.V. KOCHEROV and P.V. GELD - Russ. J. Inorg. Chem. **5** (1960), 861
12. J. CHARLES, M. NOTIN and J. HERTZ - Proceedings 11e Journées d'Etude
    sur les Equilibres entre phases - Marseille 22 mars 1985, 116-123
13. J.C. GACHON - Thèse de Doctorat d'Etat, Université Nancy, 26.06.86
14. J.C. GACHON, M. NOTIN and J. HERTZ
    Thermochimica Acta **48** (1981), 155-164
15. J.C. GACHON, M. DIRAND, J. HERTZ
    J. of Less-Com. Metals **85** (1982) 1-9
16. J.C. GACHON, M. DIRAND, J. HERTZ
    J. of Less-Com. Metals **92** (1983) 307-315
17. J.C. GACHON, J. HERTZ - CALPHAD **7** n° 1 (1983) 1-12
18. J.C. GACHON, J. CHARLES, J. HERTZ - CALPHAD **9** n° 1 (1985) 29-34
19. M.P. HENAFF, C. COLINET, A. PASTUREL and K.H.J. BUSCHOW
    J. Appl. Phys. **56** n° 2 (1984) 307-310
20. A.K. NIESSEN, F.R. DE BOER, R. BOOM, P.F. DE CHATEL, W.C.M. MATTENS
    A.R. MIEDEMA - CALPHAD (1983) 51
21. R.E. WATSON, L.H. BENNETT - CALPHAD **5** n° 1 (1981) 25-40 and
    CALPHAD **8** n° 4 (1984) 307-321
22. C. COLINET, P. HICTER, A. PASTUREL - CALPHAD **9** n° 1 (1985) 71-99
23. R. BERKANE, J.C. GACHON, J. CHARLES, J. HERTZ
    CALPHAD **12** n° 1 (1987) 373-380
24. M. NOTIN, M. DIRAND, D. BOUAZIZ, J. HERTZ
    C.R. Acad. Sc. t.302, série II, n° 2 (1986) 63-66
25. C. CUNAT, J. CHARLES - Mem. Et. Sc. Rev. Mét. **79** n° 4 (1982) 177-187
26. J.P. HILGER, B. GODARD, C. CUNAT, J. HERTZ
    Acta Metallurgica **31** n° 12 (1983) 2095-2101
27. A. AKEF, J.P. HILGER - Acta Metallurgica **34** n° 7 (1986) 1443-1453
28. C. CUNAT, J. CHARLES, J. HERTZ, J.M. DUBOIS, G. LE CAER
    J. de Physique Colloque C9, **43** N° 12 (1982) 191-197
29. C. CUNAT, M. NOTIN, J. HERTZ, J.M. DUBOIS, G. LE CAER
    J. of Non Cryst. Solids **55** n° 1 (1983) 45-60
30. M. LAHRICHI, F. MACHIZAUD, J. FLECHON, M. BOUROUKBA, C. CUNAT and
    J. HERTZ - Mem. Et. Sc. Rev. Mét. **83** n° 7/8 (1986) 367-378
31. C. CUNAT, M. BOUROUKBA, J. HERTZ
    Mem. Et. Sc. Rev. Mét. **83** n° 12 (1986) 619-627
32. F.A. KUHNAST, F. MACHIZAUD, J. FLECHON, C. CUNAT, J. HERTZ
    Proceeding MRS Meeting Strasbourg 5-8 june 1984 - Les Editions
    de Physique (1984) 199-204
33. J. FLECHON, A. OBAIDA, F. MACHIZAUD, F.A. KUHNAST, C. CUNAT
    J. HERTZ - J. of Non Cryst. Solids **9** (1987) 293-305

SOLID STATE GALVANIC CELLS FOR THERMODYNAMIC INVESTIGATIONS

H.-J. Schaller
Institut für Physikalische Chemie
der Universität Kiel
Olshausenstr. 40
D-2300 Kiel, FR-Germany

ABSTRACT. The use of solid oxygen and fluorine concentration cells for the study of thermodynamic alloy properties
is briefly reviewed. Experimental aspects and limiting factors of the electrochemical technique are discussed. Measurements of the thermodynamic properties of solid binary alloys have been in progress in our laboratory for several
years, and the results of electromotive force studies on
the following systems are reported here as selected examples: Pd-Sn, Pd-In, Pt-Sn, Pt-In, Pd-Al, Pt-Al, Pd-Y, Pd-
Gd, Pd-Th, and Pd-Ce. A coulometric titration technique
using CeF_3 as a fluorine conductor is described which was
used to determine the activity of Ce in Pd alloys. The reliability of the results is checked by comparison with data
obtained by other methods. The thermodynamic behavior of
these systems is characterized by extremely negative deviations from ideality and the factors responsible are discussed.

1. INTRODUCTION

In the last decade significant progress has been made in
the ability to predict thermodynamic data of alloys using
both first principle and model calculations. Examples of
these efforts are presented elsewhere in these proceedings
(see articles of Miedema and Watson). The numerical methods
applied, however, are still limited in reliability and
accuracy and have not reached the stage where basic properties can be obtained by a priori calculations. Therefore,
fresh experimental work will continue to be a necessity in
the field of alloy thermodynamics in order to improve
existing and provide new thermodynamic data of alloys.
      For direct determination of component activities which
govern the behavior of alloys in different environments
three mayor methods are available: the various vapour pres-

H. Brodowsky and H.-J. Schaller (eds.), Thermochemistry of Alloys, 329–358.

sure methods, the chemical methods, and electrochemical
techniques which are the subject of this article. In principle, the measurement of the electromotive force (emf) of
a suitable galvanic cell is one of the most accurate methods of obtaining thermodynamic properties of alloys at
elevated temperatures. However, the precision as expected
theoretically can be achieved only by careful experiments
and by ensuring that the cell reactions are well defined
and truly reversible.

The advantages of emf measurements using solid electrolytes have been outlined in two important papers published by Kiukkola and Wagner [1,2] in 1957. One of the
most important features of these papers was the introduction of doped zirconia as a solid electrolyte for thermodynamic investigations. Initiated by the work of Kiukkola and
Wagner, oxygen conducting electrolytes have been widely
used to determine the free energy of formation of oxides
and spinels and the chemical potential of oxygen dissolved
in metals or alloys.

Rapp and Maak [3] have introduced electrodes of the
second kind in solid state galvanic cells for the determination of the activity of a component in an alloy. Since
then, a number of alloys and intermetallics have been
investigated with the help of various solid electrolytes
and many valuable activity data have been obtained.

Comprehensive reviews dealing with the properties and
applications of solid electrolytes have been published
elsewhere [4-8]. This article does not aim to provide an
additional survey but to concentrate on some important aspects of the emf technique for thermodynamic investigations
using oxygen and fluorine conducting electrolytes and to
present selected examples of alloy systems studied by this
method.

2. EXPERIMENTAL ASPECTS

2.1. Galvanic cells with oxide electrolytes

The simplest applications involve the free energy of formation measurements of metal oxides on cells of the following
type:

$$A,AO \mid ZrO_2 (+CaO) \mid B,BO. \qquad (I)$$
$$\mu'_{O_2} \qquad\qquad\qquad \mu''_{O_2}$$

The quantities μ'_{O_2} and μ''_{O_2} represent the chemical potentials of oxygen on both sides of the electrolyte established by the metal, metal oxide equilibria. Provided local
thermodynamic equilibrium is maintained at the interfaces
of the cell components, the emf E of the cell is given by

the equation:

$$E = \frac{1}{4F} \cdot \int_{\mu'_{O_2}}^{\mu''_{O_2}} t_{O^{2-}} \cdot d\mu_{O_2} .$$

(1)

F denotes Faraday's constant and $t_{O^{2-}}$ the transference number of the oxygen ions which is a function of the chemical potential of oxygen [9]. In principle, a galvanic cell may be operated outside the range of predominantly ionic conduction ($t_{O^{2-}} < 0,99$) if the transference number is precisely known.

In cells with solid coexistence electrodes, however, the flow of electrons and holes, respectively, disturbs the equilibrium and gives rise to erratic emf data. With any short-circuiting current through the electrolyte, i.e. with electrons migrating from the cathode to the anode, the conditions of local charge and charge flux neutrality throughout the electrolyte require that equivalent amounts of oxygen ions are transported in the opposite direction. If the diffusional processes at the electrode-electrolyte interface cannot supply or remove oxygen rapidly enough, there will be oxide formation at the cathode and metal formation at the anode which gives rise to unstable emf's. Even when there is no layer formation, the oxygen transfer due to an electronic current can cause steady-state oxygen gradients in the electrodes, and a stable emf is obtained which includes an overvoltage contribution.

Because of these problems, electrolytes in solid state galvanic cells are unsuitable to use when the electronic conductivity exceeds a limit of about 1% of the ionic contribution. Therefore, it is important to know the temperature and oxygen partial pressure region over which the electronic conductivity is below this level for each electrolyte applied.

For calcia- and yttria- stabilized zirconia the useful oxygen partial pressure range was determined to extend from 10^{-18} bar to 10^{5} bar at 1000° C [10]. At oxygen pressures above 10^{5} bar and below 10^{-18} bar p-type and n-type electronic conductivity, respectively, become greater than 1%.

Using yttria- or lanthania- doped thoria as an electrolyte, the application range can be extended down to an oxygen pressure of 10^{-25} bar at 1000° C on the reduction side [11]. There is some evidence that the electrolytic domain possibly extends to even lower oxygen pressures; values down to 10^{-34} bar at 1000° C have been reported in the literature [12]. On the high p_{O_2} side, thoria based electrolytes are useful only up to oxygen pressures of about 10^{-6} bar at 1000° C. The electrolytic properties of thoria in particular, are strongly influenced by the impurity content of the electrolyte and the sample preparation. Thus it is not surprising that various investigators differ on the exact value of the limit. It is important, therefore, always to consider the possibility of mixed conduction whenever a thoria-based electrolyte is exposed to oxygen pressures close to the limits of pure ionic conduction.

The oxygen pressures for selected metal, metal oxide equilibria at 1000°C as calculated from literature data [13] are listed in Table I:

TABLE I log p_{O_2} (p_{O_2} in bar) of selected metal, metal oxide equilibria at 1000°C

Cu,Cu₂O	: − 6.29		Cr,Cr₂O₃	: −21.8
Pb,PbO	: − 7.82		Mn,MnO	: −23.9
Ni,NiO	: −10.36		Ta,Ta₂O₅	: −24.4
Co,CoO	: −11.88		Nb,NbO	: −25.1
Sn,SnO₂	: −13.00		Al,Al₂O₃	: −34.7
In,In₂O₃	: −13.85		Zr,ZrO₂	: −35.3
Fe,"FeO"	:− 14.84		Mg,MgO	: −39.3
Mo,MoO₂	: −14.86		Th,ThO₂	: −40.2
NbO₂,Nb₂O₅	: −17.58		Ca,CaO	: −41.5

Considering the boundaries of pure ionic conductivity mentioned above, limitations in the use of the oxide electrolytes on the reduction side become evident. Metal, metal oxide equilibria involving reactive metals such as Al, Zr, or Th cannot be investigated by means of oxide electrolytes.

On the other hand, this method may be quite applicable to the determination of the activity of these metals in alloys. For example, a thoria based electrolyte may be used to measure the activity of Th in palladium-rich Pd-Th alloys. Anticipating in part results of chapter 3, the activity of Th in a Pd-Th alloy with 10 at.% Th assumes an extremely small value ($10^{-16.6}$ at 1000°C). Considering the Th,ThO₂ equilibrium at this temperature, the oxygen pressure increases by more than 16 orders of magnitude upon replacing pure Th by the alloy and arrives at a value which is well within the electrolytic domain of a thoria electrolyte. Because of extremely negative deviations from ideality, emf measurements with oxide electrolytes may also be applied to determine the activity of Al and Zr in Pd- and Pt-rich alloys up to about 25 at.% Al and Zr, respectively.

Using thoria based electrolytes for activity measurements, attention should also be paid to the application limit on the high p_{O_2} side. For example, galvanic cells with a thoria electrolyte have been applied for the determina-

tion of the activity of Sn in Pd [14]. Indeed, a thoria
electrolyte is safe to use under the conditions of the
Sn,SnO$_2$ equilibrium. Replacing pure Sn by Pd-Sn alloys with
low Sn contents (x_{Sn} < 0.2), however, the application limit of the thoria electrolyte is exceeded by far. Thermodynamic data of Pd-Sn alloys determined with the help of a
zirconia and a thoria electrolyte, respectively, are presented in chapter 3 for comparison.

2.2. Galvanic cells with fluorine conducting electrolytes

In the recent past, fluoride electrolytes have been employed for thermodynamic investigations on alloys more and
more frequently. They are especially useful under very reducing conditions where solid oxide electrolytes become
electronic conductors. Fluoride electrolytes therefore
offer the possibility to determine the activity of strongly
electropositive metals.

The most commonly used fluoride electrolyte is CaF$_2$.
The exclusive fluorine ion conductivity in pure and doped
CaF$_2$ was demonstrated by Ure with electrical conductivity
and diffusion measurements [15]. The electrolytic behavior
of CaF$_2$ has been studied as a function of the Ca activity
by Wagner [16], Hinze and Patterson [17], and Egan [18].

Using color center migration data of Mollwo [19],
Wagner arrived at the estimate, that the electronic conductivity of CaF$_2$ is insignificant (t_e- < 10^{-2}) up to a
calcium activity of $1 \cdot 10^{-5}$ and $6 \cdot 10^{-6}$ at 600 and 840°C, respectively. The corresponding fluorine pressures calculated
from the Ca,CaF$_2$ equilibrium [13] are: $5.8 \cdot 10^{-65}$ bar
(600°C) and $2.6 \cdot 10^{-49}$ bar (840°C).

Hinze and Patterson carried out conductivity measurements on symmetrical cells with a number of metal, metal
fluoride electrodes. They concluded from these experiments
that CaF$_2$ exhibits negligible electronic conductivity even
when equilibrated with pure Ca. The increase of conductivity observed in the presence of Ca was attributed to impurity effects. However, this might also indicate the onset of
electronic conduction.

Egan studied the electronic conductivity of CaF$_2$ as a
function of the Ca activity using a polarization technique
developed by Wagner [20]. According to his results, the
electronic contribution to the conductivity can be neglected below a calcium activity of 10^{-3} in the temperature
region of 800°C which corresponds to a fluorine pressure of
$2 \cdot 10^{-48}$ bar at 800°C. This result, which is very close to
Wagner's estimate, is supposed to be the most reliable
information available at present.

The first application of the CaF$_2$ electrolyte is that
of Benz and Wagner [21] who measured the free energy of

formation of calcium silicates. Egan [22] was the first to measure the free energy of a carbide by using the following type of cell:

$$Th, ThF_4 \,|\, CaF_2 \,|\, ThF_4, ThC_2, C. \tag{II}$$

The electromotive force of the cell is directly related to the activity a_{Th} of Thorium in ThC_2 in equilibrium with carbon and to the standard free energy of formation, $\Delta G^\circ_{ThC_2}$ respectively:

$$E = -\frac{RT}{4F} \cdot \ln a_{Th} = -\frac{1}{4F} \cdot \Delta G^\circ_{ThC_2}. \tag{2}$$

Since then, CaF_2 has been used extensively as a solid electrolyte in galvanic cells designed to determine the free energy of formation of various compounds [7,23]. In contrast to intermetallics, only a few solid solution alloys have been studied with the help of fluoride electrolytes. Examples of activity measurements on these alloys are presented in chapter 3.

The most serious limitation of CaF_2 as an electrolyte is its high chemical activity at elevated temperatures. CaF_2 tends to react with quite a number of other metal fluorides in particular, forming intermediate compounds. Thus the fluoride activity is not time-independently fixed at the interfaces. Furthermore, the electrolytic domain of the electrolyte may be narrowed by the unwanted side reaction.

To avoid these problems, thermodynamic studies on Mg, Ba, and Sr compounds were carried out using the fluorides of these metals as solid electrolytes. Further applications depend upon the availability of suitable fluorine conducting electrolytes and chemically compatible electrodes in the temperature range of interest. Besides the above mentioned difluorides, tysonite-related compounds hold considerable promise for use as solid electrolytes. Tysonite is the naturally-occuring mixed fluoride of rare earth metals $(La,Ce...)F_3$. It has the same structure as the trifluorides of the lighter lanthanides and of a number of the actinides.

The Arrhenius plot in Fig.1 shows the conductivities of CeF_3 and CaF_2 for comparison. The conductivity of CeF_3 has previously been measured by Nagel and O' Keeffe [24] and by Takahashi et al. [25] up to about 700K. Our own results in the temperature range 650-1050K [26] continue the data of these authors remarkably well. The scatter of the data at high temperatures probably results from a reaction between the electrolyte and the electrodes. A comparison shows that CeF_3 is a much better conductor than CaF_2. This makes it potentially useful as an electrolyte at substantially lower temperatures, provided pure ionic conductivity

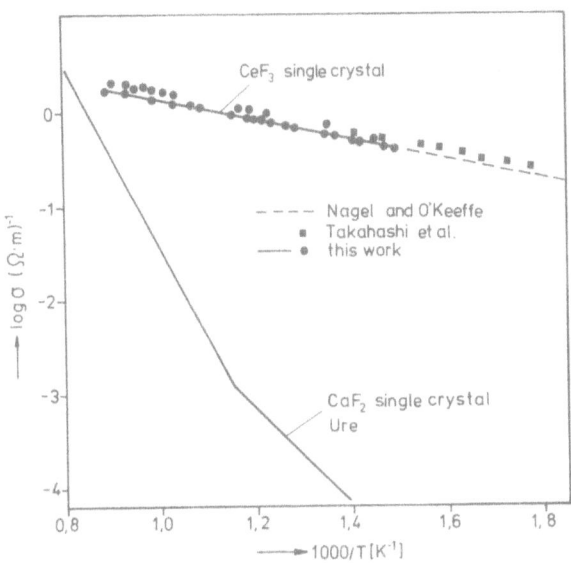

Fig.1 Conductivity of CaF₂ and CeF₃ as a function of temperature from refs. [24,25,26]

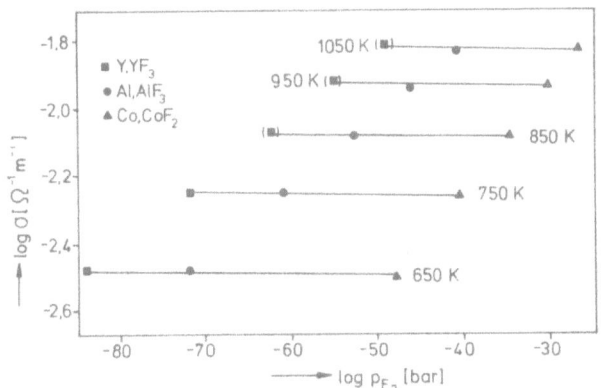

Fig.2 Total conductivity of CeF₃ as a function of fluorine
pressure [26]

prevails under the conditions of the experiment.

In the previous studies on CeF$_3$, the extent of the ionic conductivity was not particularly tested and no efforts were made to establish a well-defined cerium (or fluorine) activity during the experiments. In this investigation [26], the total conductivity of CeF$_3$ single crystals was determined from ac measurements made on symmetrical cells of the type

$$M,MF_x \mid CeF_3 \mid M,MF_x \qquad\qquad (III)$$

with the coexistence electrodes: Co,CoF$_2$, Al,AlF$_3$, and Y,YF$_3$.

The results are plotted as log σ vs. log p$_{F_2}$ isotherms in Fig.2. The isotherms do not indicate any rise such as would be expected from the onset of electronic conductivity a low fluorine pressures. The 650 and 750K isotherms remain flat all the way down to the Y,YF$_3$ coexistence. Above this temperature, reactions were observed to occur at the surface of the electrolyte in the presence of Y,YF$_3$; the corresponding data points are shown in brackets.

To supplement the foregoing evidence for the predominantly ionic conductivity of CeF$_3$ under reducing conditions, the standard free energy of formation of CeF$_3$ was determined on a cell of the following type:

$$Re \mid Ce \mid CeF_3 \mid Al,AlF_3 \mid C. \qquad\qquad (IV)$$

The Al,AlF$_3$ reference electrode was encapsuled in carbon cups. The cerium electrode was produced in situ by cathodic deposition. Performed in a protective atmosphere of highly purified argon, this procedure yields cerium electrodes free of oxygen contamination. This improved the accuracy of the measurements considerably since the direct handling of cerium metal was eliminated. As demonstrated by Kleykamp [23], oxygen contaminations of the electrodes result in erroneous readings and have to be avoided in any case. Rhenium and tungsten foils proved to be a suitable contact material for the cerium electrode.

Using literature data [13] for the Al,AlF$_3$ equilibrium, the standard free energy of formation $\Delta G^\circ_{CeF_3}$ of CeF$_3$ was calculated from the measured emf. A least squares fit to the data conformed to a straight line in the temperature range of measurements (970-1130K):

$$\Delta G^\circ_{CeF_3} = -\ (1697\pm3) + (0,235\pm0,003)\cdot T \quad [kJ/mol]. \qquad (3)$$

This finding is in excellent agreement with recommended data [27]

$$\Delta G^\circ_{CeF_3} = -\ 1695 + 0,239\cdot T \quad [kJ/mol] \qquad\qquad (4)$$

and with the results of a previous emf study between 850 and 970K with CaF₂ as the electrolyte [28]

$$\Delta G^\circ_{CeF_3} = -1718 + 0,237 \cdot T \quad [kJ/mol]. \tag{5}$$

Thus within the limits of uncertainty of the thermodynamic data, it appears that the usefulness of CeF₃ as a solid electrolyte extends down to the activity of pure Ce. Coulometric titrations using a CeF₃ solid electrolyte to study the cerium activity in palladium alloys also indicated pure ionic conduction in CeF₃ (see chapter 3).

3. EXAMPLES FROM ACTIVITY MEASUREMENTS

3.1. Pd-Sn, Pd-In

The thermodynamic properties Pd-Sn [29] and Pd-In [30] alloys were determined on cells of the following type:

$$Pt|Ni,NiO|ZrO_2 (+CaO)|Pd-Sn,SnO_2|Pd-Sn, \tag{V}$$

$$Pt|Ni,NiO|ZrO_2 (+CaO)|Pd-In,In_2O_3|Pd-In. \tag{VI}$$

The oxygen partial pressures of the electrodes fulfill the conditions of pure ionic conductivity in the zirconia electrolyte. Doped zirconia was used as a solid electrolyte. Ni,NiO rather than Sn,SnO₂ or In,In₂O₃ was used as reference electrode in order to avoid experimental problems associated with liquid tin and indium, respectively. The alloy electrodes were contacted by Pd-Sn or Pd-In sheets with compositions like the alloys under investigation. Thus concentration changes could be avoided which were observed to occur when Pt was used as a direct contact.

The arrangement of the galvanic cell is schematically shown in Fig.3. The cell was kept in a protective gas atmo-

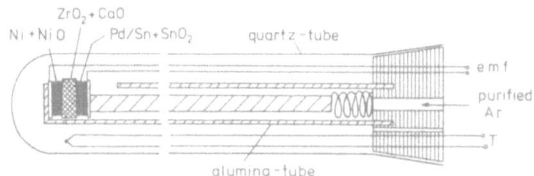

Fig.3 Cell assembly (schematic)

sphere which was maintained during the experiment by passing purified argon through the inner alumina tube over the cell. The gas stream was led back out of the assembly between the alumina and the outer silica tube, carrying away impurities that might have permeated the outer tube.

The purification of the bottled argon is an extremely important operation and requires the removal of water vapour, oxygen, and other oxidizing or reducing impurities. Bottled argon was passed through a column with $Mg(ClO_4)_2$ at room temperature and over Zr turnings at $1000°C$. In all emf studies presented in this paper, this purification method turned out to be very effective. After an initial period of equilibration (up to 8h), an emf was obtained constant for days and independent of the flow rate of argon. The emf had the same value, irrespective of whether the temperature had been approached from above or below; furthermore, the emf returned to the same value after a current was passed through the cell in either direction. These observations indicate the reversibility of the used cells.

Some authors recommend not to keep the cell in a common gas atmosphere but to separate the electrodes into two compartments, i.e. by using an electrolyte tube. A number of investigations on fluorine or oxygen concentration cells have shown that a common gaseous atmosphere has no influence on the measurements if the dissociation pressures of the oxides or fluorides involved are sufficiently low. However, an influence was observed to occur in the investigation of palladium-rich Pd-Sn alloys (x_{Sn} < 0,05) at temperatures above $900°C$. The equilibrium pressure of these electrodes assumes values of about mbar. Under these conditions, the dissociation of SnO_2 results in a noticeable increase of the tin content of the alloy during the experiment, indicated by a slightly increasing emf (about 1mV in 3h at $1000°C$). This drift corresponded to an increase of the Sn content of about 0,15 at.% in 3h. The error source could be eliminated to a great extent by operating the cells for short times only at temperatures above $900°C$ and by additional concentration controls after the experiment.

The equilibrium oxygen pressure p''_{O_2} of the alloy electrodes can be calculated from the known pressure p'_{O_2} of the Ni, NiO reference and the measured emf E by means of the Nernst equation:

$$E = \frac{RT}{4F} \cdot \ln \frac{p'_{O_2}}{p''_{O_2}} \ . \tag{6}$$

Finally, since the standard free energies of the oxides, $\Delta G°_{SnO_2}$ and $\Delta G°_{In_2O_3}$, are also known, the activities of In and Sn, a_{Sn} and a_{In}, can be calculated from p''_{O_2} by the law of mass action:

$$\ln a_{Sn} = \frac{\Delta G^{\circ}_{SnO_2}}{RT} - \ln p''_{O_2} ,$$

$$\ln a_{In} = \frac{\Delta G^{\circ}_{In_2O_3}}{2RT} - \frac{3}{4} \cdot \ln p''_{O_2} .$$

(7)

Fig.4 shows the relative partial excess free energy of Sn in the concentration range of the terminal solid solution at 800, 900 and 1000°C.

The oxygen pressures in the presence of two palladium-rich alloys were also measured directly in a high vacuum system using a McLeod gauge. Table II shows the results from the direct and the emf method for comparison.

TABLE II Results from the direct and the emf method

x_{Sn}	T	direct method		emf method	
		p_{O_2}	$\Delta \bar{G}^E_{Sn}$	p_{O_2}	$\Delta \bar{G}^E_{Sn}$
	[°C]	[mbar]	[kJ/mol]	[mbar]	[kJ/mol]
0.015	900	0.113	−217.7	0.106	−216.8
	1000	1.303	−212.5	1.40	−213.6
0.029	1000	0.851	−215.0	0.94	−217.1

Considering the absolute errors of the $\Delta \bar{G}^E_{Sn}$ values which are 0.4 kJ/mol (direct method) and 1.5 kJ/mol (emf method) the results agree within the limits of error.

The thermodynamic properties of Pd-Sn alloys were previously determined with the help of a thoria electrolyte which is unsuitable to use at oxygen pressures above 10^{-5} mbar (see chapter 1) [14]. For palladium-rich alloys in particular, this study yielded distinctly smaller deviations from ideality ($\Delta \bar{G}^E_{Sn} = -130$ kJ/mol at $x_{Sn} = 0$).

The best method to check the reliablity of activity data is to calculate enthalpies of mixing using the temperature dependence of the activity and to compare the results with calorimetric data. The integral properties of α-Pd-alloys at 900°C obtained from our emf results by Gibbs-Duhem integration are presented in Fig.5 by the solid lines. The ΔH^M-curve lies within the two sets of calorimetric data determined by Darby et al.[31] and by Bryant et al.[32], respectively. Consequently, emf measurements with a zirconia electrolyte appear to be a reliable method not only to obtain free energies but also enthalpies of mixing.

The integral quantities calculated from the emf data of Bryant et al. are shown in Fig.5 as dashed lines. Com-

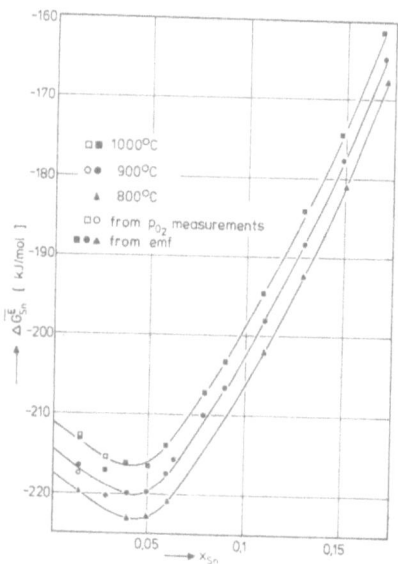

Fig.4 Relative partial excess Gibbs energy of Sn in Pd-Sn
alloys [29] [Reference State: Sn solid]

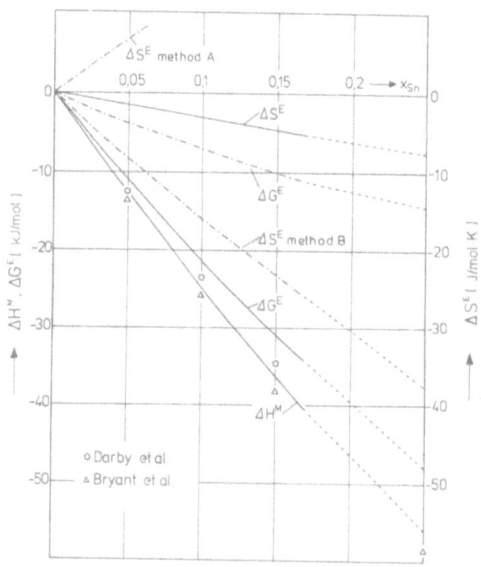

Fig.5 Integral thermodynamic properties of Pd-Sn alloys at
900°C. Schaller and Brodowsky [29]:————, Bryant
et al. [14]: —·—·—   [Reference state: Sn solid]

pared to our own results, the excess Gibbs energies are less negative by a factor of 3.

The excess entropies calculated from the temperature dependence of the emf (method A) assume unusually large positive values (e.g. $\Delta S^E = 7$ J/mol·K at $x_{Sn} = 0.05$). Combining the excess Gibbs energies at 900°C with calorimetric data, however, yields extremely negative excess entropies (method B). These discrepancies in the paper of Bryant et al. illustrate the influence of mixed conduction in the electrolyte on the result of emf measurements.

The relative partial excess free energy of In and the integral properties of the Pd-In system are shown in Fig.6 and Fig.7, respectively. The integral heats of mixing calculated from the emf results (solid line) are in good agreement with the data points from calorimetric measurements [31,32].

The solution behavior of In and Sn in Pd exhibits similar characteristics: strongly negative deviations from ideality which are more pronounced in the case of Pd-Sn alloys, and a minimum in the $\Delta \overline{G}^E_i$ curve in the terminal solid solution range. Two opposing effects are considered to account for these properties [29,30]. An elastic effect is associated with the lattice distortion brought about by the dissolved particles. The lattice strain is reduced if two isolated particles enter adjacent sites. The resulting

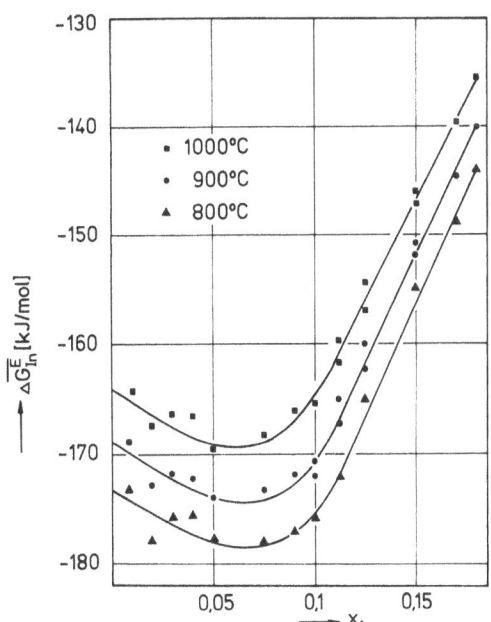

Fig.6 Relative partial excess Gibbs energy of In in Pd-In alloys [30] [Reference state: In solid]

pair interaction can be used to calculate the elastic or
dilatational (superscript: "d") contribution $\Delta \bar{G}_i^{E\,d}$ to the
excess function by means of the quasichemical method.
Assuming an energy of attraction, the quasichemical
approach predicts a negative gradient of $\Delta \bar{G}_i^{E\,d}$.

An electronic effect is based on the observation that
a number of metals behave as electron donors with respect
to Pd [33] and Pt [34] and donate their valance electrons
to the electron gas of the alloy. The charge transfer gives
rise to a bonding or electronic (superscript: "e") contribution $\Delta \bar{G}_i^{E\,e}$ to the excess function. Neglecting an inter-
relation of these effects, the excess function can be
separated in electronic and elastic parts:

$$\Delta \bar{G}_i^E = \Delta \bar{G}_i^{E\,e} + \Delta \bar{G}_i^{E\,d} \,. \qquad (8)$$

The large negative values of $\Delta \bar{G}_i^E$ in Fig.4 and Fig.6 indicate the dominating influence of the bonding contribution
$\Delta \bar{G}_i^{E\,e}$ on the solution behavior of these alloys. The deviations from ideality are more pronounced in the case of
Pd-Sn alloys; this finding indicates an influence of the
valency of the donor metal on the thermodynamics of Pd alloys.

The occurance of the minima in Fig.4 and Fig.6, however, reflects the influences of the two opposing effects.
Adding an electron donor metal, the increase of the elec-

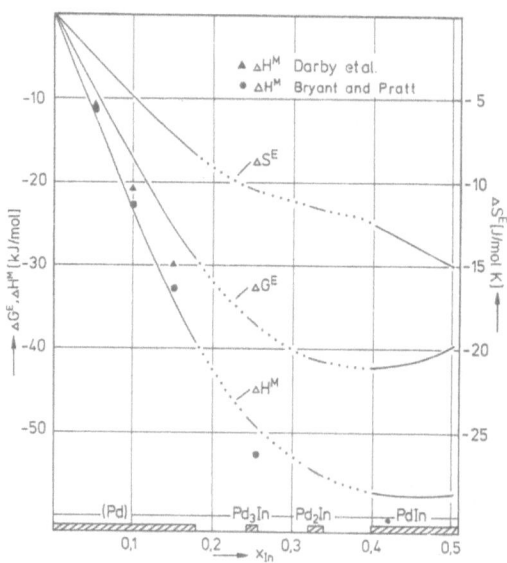

Fig.7 Integral thermodynamic properties of Pd-In alloys at
900°C [30] [Reference state: In solid]

tron concentration leads to a rise of the Fermi energy, which corresponds to a rise in $\Delta \bar{G}_i^{E\bullet}$. Because of the characteristic band structure of the transition metals Pd and Pt (high and low density of states in the d- and s-band, respectively), the Fermi energy shows a moderate rise upon filling d states and a steep rise upon filling s states. Therefore, the slope of the elastic contribution dominates in the d-band regime and brings about a negative gradient of the total function $\Delta \bar{G}_i^E$ on the palladium-rich side. The reversal of the slope is caused by the step rise of the Fermi energy upon filling s-states.

A detailed analysis of the Pd-Sn and Pd-In alloys based on these ideas [29, 30] yielded the following results, which refer to zero solute concentration ($x_{Sn}= 0$, $x_{In}= 0$) and 800°C:

$$\Delta \bar{G}^{Ed}_{Sn} = 58\pm12 \text{ kJ/mol}, \quad \Delta \bar{G}^{E\bullet}_{Sn} = -275\pm14 \text{ kJ/mol},$$

$$\Delta \bar{G}^{Ed}_{In} = 62\pm16 \text{ kJ/mol}, \quad \Delta \bar{G}^{E\bullet}_{In} = -235\pm18 \text{ kJ/mol}. \tag{9}$$

3.2. Pt-Sn, Pt-In

Analogous cells were used to determine the activity of Sn and In in Pt [35]. The results are shown in Fig.8 and Fig.9. As in the case of the corresponding Pd alloys, the reference state of the solute atoms were the hypothetically solid metals Sn and In at the elevated temperatures of measurement. Compared to the corresponding palladium alloys, the deviations from ideality are less pronounced. Moreover, the excess functions do not exhibit a minimum but a strong negative gradient across the terminal solid solution range. These findings point to a stronger influence of the lattice distortion on the solution behavior of these alloys.

According to elastic theories [36-39], the misfit energy is proportional to the bulk modules B of the host metal. Upon introducing solutes with larger or smaller atomic volumes, palladium (B = $1.85\cdot10^5$ J/cm³)is more easily expanded or compressed than platinum (B = $2.98\cdot10^5$ J/cm³).

A separation of the excess functions in elastic and electronic parts yielded the following values which refer to zero solute concentration and 800°C [35]:

$$\Delta \bar{G}^{Ed}_{Sn} = 93\pm15 \text{ kJ/mol}, \quad \Delta \bar{G}^{E\bullet}_{Sn} = -209\pm17 \text{ kJ/mol},$$

$$\Delta \bar{G}^{Ed}_{In} = 89\pm15 \text{ kJ/mol}, \quad \Delta \bar{G}^{E\bullet}_{In} = -194\pm17 \text{ kJ/mol}. \tag{10}$$

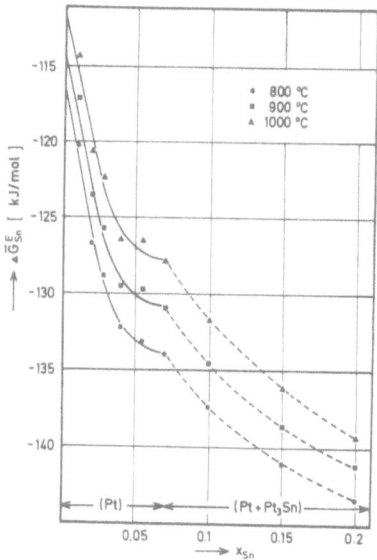

Fig.8 Relative partial excess Gibbs energy of Sn in Pt-Sn
alloys [35] [Reference state: Sn solid]

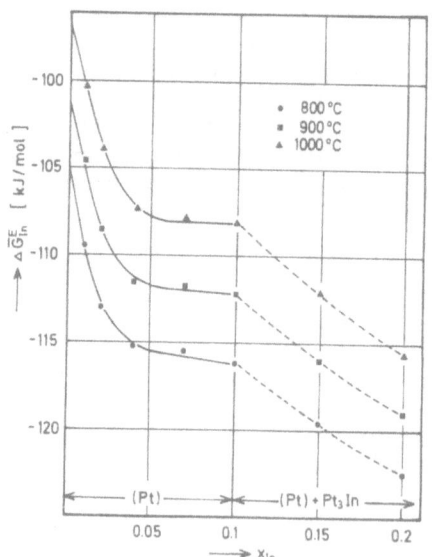

Fig.9 Relative partial excess Gibbs energy of In in Pt-In
alloys [35] [Reference state: In solid]

3.3. Pd-Al, Pt-Al

The activity of Al in Pd and Pt was determined between 700
and 800° C by emf measurements on fluorine concentration
cells using monocrystalline CaF_2 as the electrolyte [40].
The following type of cell was operated:

$$C|Al,AlF_3|CaF_2|X-Al,AlF_3|C, \quad X=Pd \text{ or } Pt. \qquad (VII)$$

Because of the volatility of Al and AlF_3 at the temperatures of measurements, the electrodes were encapsuled in carbon cups. The experimental setup for cell (VII) was similar
to that in Fig. 3 except that two carbon cups were used.
The spring pushed the polished edges of the carbon cups
onto the surfaces of the CaF_2 single crystal and formed an
effective seal. As in the previous examples, the measurements were performed under a protective atmosphere of carefully purified argon. Usually a time of less than 8h was
needed for the emf to reach a steady value.

The emf of cell (VII) is given by the expression

$$E = - \frac{RT}{3F} \cdot \ln a_{Al}. \qquad (11)$$

The excess quantity $\Delta \bar{G}^E_{Al}$ of Al in Pd at 700, 750, and 800° C
calculated from the emf results is shown in Fig.10.

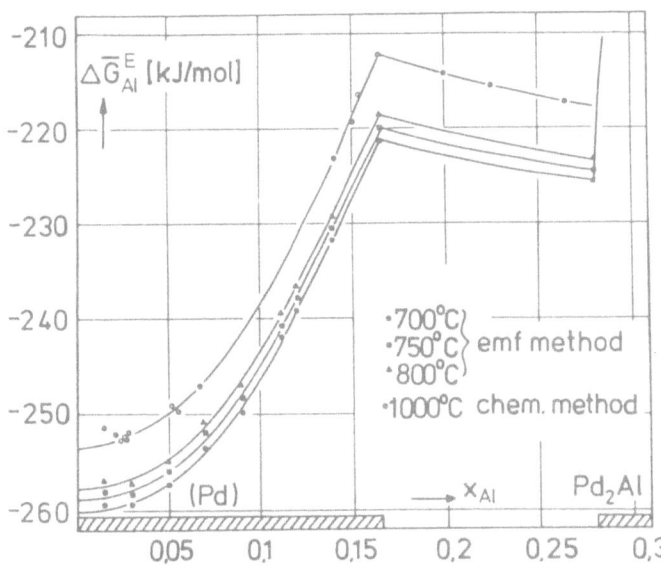

Fig.10 Relative partial ecxess Gibbs energy of Al in Pd-Al
alloys [40] [Reference state: Al solid]

In comparison with the preceding alloy systems, the deviations from ideality are even more pronounced. Because of extremely low activity coefficients of Al in palladium-rich alloys, a chemical method may be applied to examine these alloys. The isotherm at 1000°C in Fig.10 was determined by reducing Al_2O_3 in the presence of Pd by controlled H_2-H_2O mixtures and measuring the aluminium concentration of the alloy. The isotherm determined by this method is in excellent agreement with results of the emf study.

Fig.11 shows the activity of Al at 1000°C on a logarithmic scale over the entire composition range. The solid line which extends up to the stochiometric phase PdAl represents the emf results. The broken curve was determined by Ettenberg, Komarek and Miller [41] by using an isopiestic technique. The activities obtained by the two methods in the two-phase region between Pd_2Al and PdAl agree well with each other.

The activity curve is characterized by strongly negative values in the terminal solid solution (Pd) and by steep rises of about 5 and 4 orders of magnitude across the phase fields of Pd_2Al and PdAl, respectively. This behavior is typical of highly ordered phases. Following Wagner's theory of substitutional disorder [42], the intrinsic degree of disorder α was calculated from the activity change to be $\alpha = 2,7 \cdot 10^{-6}$ for Pd_2Al [40]. The corresponding value for PdAl is $\alpha = 2,5 \cdot 10^{-4}$ [41].

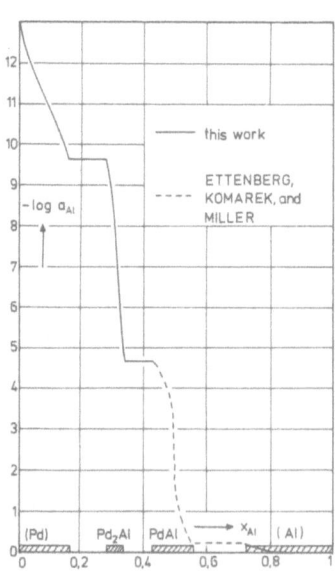

Fig.11 Activity of Al in Pd-Al alloys at 1000°C.
Schaller [40]: ———, Ettenberg et al. [41]: - - -
[Reference state: Al liquid]

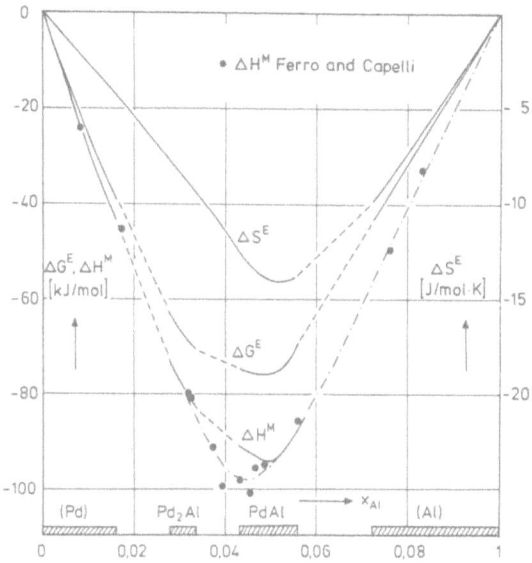

Fig.12 **Integral thermodynamic properties of Pd-Al alloys.
Schaller [40]: ———, Ferro and Capelli [43]: —·—·—·—
[Reference state: Al solid]**

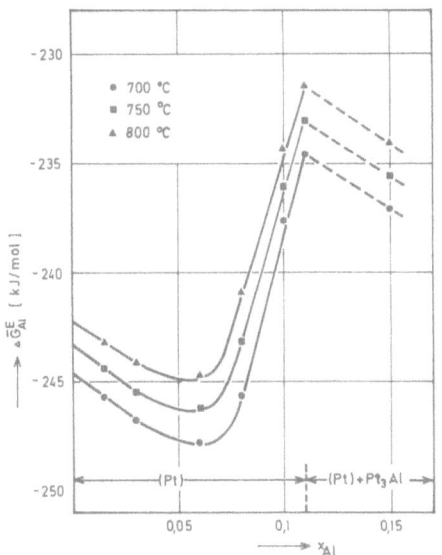

Fig.13 **Relative partial excess Gibbs energy of Al in Pt-Al
alloys [35] [Reference state: Al solid]**

The integral properties of the system at 1000°C calculated from the activities and their temperature dependence are plotted in Fig.12. Enthalpies of formation obtained calorimetrically and recalculated for 100°C [43] are shown for comparison. Considering the difference in the temperatures of measurement and the scattering of the data, the agreement between calorimetric and emf results is rather good. Consequently, emf measurements with CaF_2 as an electrolyte may also be regarded as a reliable method to determine enthalpies of formation.

The relative partial excess free energy of Al in Pt-Al alloys [35] is shown in Fig.13. The deviations from ideality are of the same order of magnitude as in the Pd-Al alloys ($\Delta\bar{G}^E_{Al}$ = -245 kJ/mol and $\Delta\bar{G}^E_{Al}$ = -260 kJ/mol for Pt and Pd alloys, respectively, at infinite dilution and 700°C). A separation of the excess function into electronic and elastic parts yielded the following values [35] which refer to infinite dilution (x_{Al} = 0) and 700°C:

Pd-Al: $\Delta\bar{G}^E_{Al}{}^d$ = 8±3 kJ/mol, $\Delta\bar{G}^E_{Al}{}^e$ = -268±6 kJ/mol,

$$(12)$$

Pt-Al: $\Delta\bar{G}^E_{Al}{}^d$ = 27±4 kJ/mol, $\Delta\bar{G}^E_{Al}{}^e$ = -272±8 kJ/mol.

Compared to other solutes, adding of Al to Pd or Pt causes significantly smaller volume changes. Since the misfit energy is proportional to the square of volume change in first approximation [39], the elastic effect exerts a minor influence on the energetics of these alloys. Therefore, the relative partial excess Gibbs energy in Fig.10 does not exhibit a minimum which is a characteristic of a number of other Pd alloys.

3.4. Pd-Gd, Pd-Y, Pd-Th

Emf measurements with CaF_2 as a solid electrolyte may also be applied to investigate strongly electropositive metals such as Gd, Y, and Th. The activities of these components in Pd alloys were determined on cells of the following type [44,45]:

Ta|X,XF_z|CaF_2|Pd-X,XF_z|Ta   X=Gd,Y (z=3), Th (z=4)   (IX)

The activity of Y relative to the pure solid metal is shown in Fig.14.

The activity coefficients of Y in palladium-rich alloys assume extremely small values. The activity change of more than 10 orders of magnitude within the narrow phase field of Pd₃Y indicates a high degree of order. The phase diagram of the Pd-Y system is well established [46]. The activity data are in accordance with the existence of the phases included in the diagram.

The relative partial excess free energies of Gd and Th at 800°C are shown in Fig.15 and Fig.16, respectively. The systems are characterized by extremely negative deviations from ideality, e.a. up to $\Delta \overline{G}^E_{Th}$ = -415 kJ/mol for Pd-Th alloys. At present, no other system is known which approaches this outstanding stability. As in the preceding examples, the electron transfer is supposed to be the factor responsible for these extreme deviations from ideality. Though the electronic effect dominates, the pronounced minima in Fig.15 and Fig. 16.point to the influence of the size effect.

Following a continuum elastic approximation [39], the elastic contribution was calculated independently of the thermodynamic data. The separation of the excess function into elastic and electronic parts yielded the following result [44, 45] which refers to infinite dilution:

Pd-Y: $\Delta \overline{G}^{E,d}_Y$ = 92±30 kJ/mol, $\Delta \overline{G}^{E,e}_Y$ = -412±33 kJ/mol,

Pd-Gd: $\Delta \overline{G}^{E,d}_{Gd}$ = 108±35 kJ/mol, $\Delta \overline{G}^{E,e}_{Gd}$ = -414±38 kJ/mol,   (13)

Pd-Th: $\Delta \overline{G}^{E,d}_{Th}$ = 180±50 kJ/mol, $\Delta \overline{G}^{E,e}_{Th}$ = -585±54 kJ/mol.

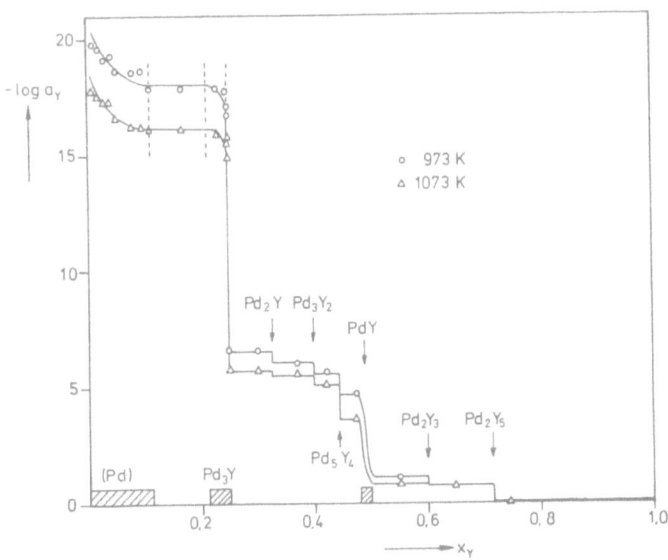

Fig.14 Activity of Y in Pd-Y alloys [44]

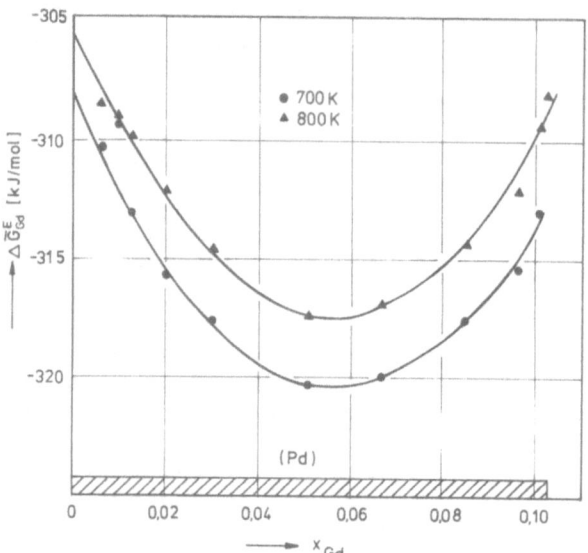

Fig.15 Relative partial excess Gibbs energy of Gd in Pd-Gd
alloys [44]

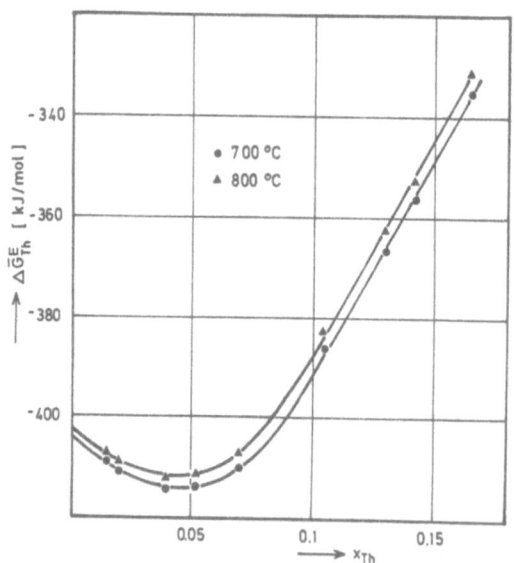

Fig.16 Relative partial excess Gibbs energy of Th in
Pd-Th alloys [45]

3.5. Pd-Ce

The thermodynamic properties for solid Pd-Ce alloys were determined by emf measurement on the following type of cells [26, 47]:

$$\text{Fe} \,|\, \text{Bi-Ce} \,|\, \text{CeF}_3 \,|\, \text{Pd-Ce} \,|\, \text{Re}, \qquad (X)$$
$$x_{Ce} = 0.15$$

$$\text{Re} \,|\, \text{Ce} \,|\, \text{CeF}_3 \,|\, \text{Bi-Ce} \,|\, \text{Fe}. \qquad (XI)$$
$$x_{Ce} = 0.15$$

A CeF₃ single crystal was used as a solid electrolyte in order to avoid unwanted side reactions. CeF₃ exhibits pure ionic conductivity up to a_{Ce} = 1 (see chapter 1). A liquid two phase Bi-Ce alloy ($x_{Ce} \cong 0.15$) encapsuled in iron cups was used as a reference electrode for experimental reasons. Using pure Ce as a reference electrode was not practical because of its high affinity to oxygen. The reduced activity of Ce in the Bi-Ce alloy helps to avoid oxygen contamination.

In cell (X) thin films of Pd (0.3-0.5 μm) were sputtered onto the CeF₃ single crystal. Re foils turned out to be suitable as an electrical contact for the counter-electrodes at all cerium activities. The Pd-Ce electrode in

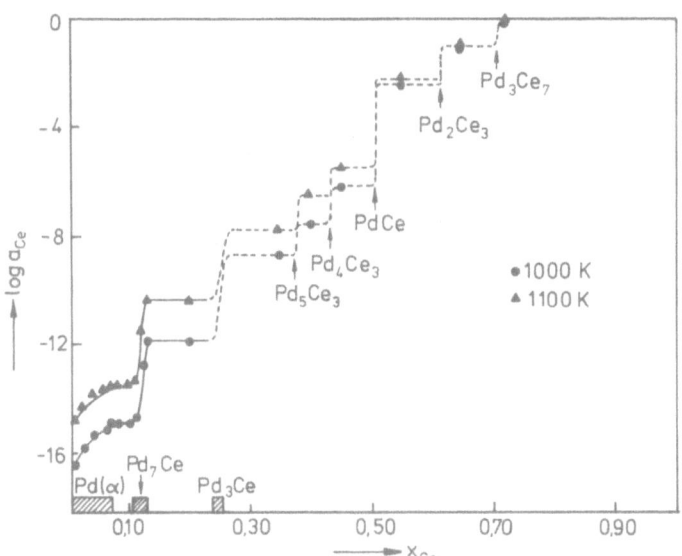

Fig.17 Activity of Ce in Pd-Ce alloys [26, 47]
[Reference state: Ce solid]

cell (X) was produced in situ by passing controlled currents through the cell. This also helped to avoid oxygen
contamination since any direct handling of cerium was
eliminated. Moreover, the concentration of the alloy can be
controlled with high accuracy. As indicated by the
monitored emf, equilibrium in the thin Pd film was attained
within 2 to 6h after each titration. Due to the low chemical diffusion coefficients, thin films have to be used in
order to obtain equilibrium in reasonable times. In order
to avoid an influence of surface effects on the Gibbs
energy, however, the thickness of the films should not be
reduced below a critical value of about 0.2 μm [48]. The
pure Ce counterelectrode in cell (XI) was also produced by
cathodic deposition. A similar technique has previously
been described by Eagan [49].

The activity of Ce in Pd-Ce alloys is given by

$$\ln a_{Ce} = - \frac{3F}{RT} \cdot (E_1 - E_2) \tag{14}$$

where E_1 and E_2 are the emfs measured on cell (X) and (XI),
respectively.

Fig.17 shows the activity of Ce in Pd on a logarithmic
scale at 1000 and 1100K. The measurements were focused on
palladium-rich alloys in order to clarify the phase relation up to x_{Ce} = 0.25. There are a number of indications
that the published phase diagram [46] has to be revised.
According to this diagram, the α-phase is supposed to extend up to x_{Ce} = 0.13 and to be in equilibrium with Pd_3Ce.

As indicated by the pronounced change of the activity, an
additional phase with the stochiometry Pd_7Ce was shown to
exist. This phase was also identified by X-ray and DTA investigations. The revised phase diagram, will be published
elsewhere [50]. The dashed line (x_{Ce} > 0.25) was constructed in accordance with the existence of a number of phases
reported in the literature.

In conclusion it can be seen that emf measurements in
combination with coulometric titrations are a valuable tool
in the study of thermodynamic properties and phase diagrams.

4. ELECTRONIC EFFECTS IN Pd AND Pt ALLOYS

The results of the investigations on Pd and Pt alloys are
compiled for comparison in Fig.18 and Fig.19, respectively.
Included are relative partial Gibbs energies of Ag [51],
Cd [52], and Zr [53] in Pd and of Au [54] and Zr [53] in
Pt, which were determined by other than emf methods.

Considering the excess functions of Ag, Cd, In, and Sn in Fig.18,the influence of the valency on the solution behavior is evident. With increasing valency of the solute $\Delta \bar{G}_i^E$ decreases from about zero for Ag down to about - 200 kJ/mol for Sn. The influence of the Fermi energy on the solution behavior may be illustrated by considering solutes with equal valency such as Sn, Zr and Th. The more electropositive the solutes, the more pronounced are the deviations from ideality.

The positive values of $\Delta \bar{G}_{Au}^E$ for Pt-Au alloys in Fig.19 indicate a dominating influence of the elastic effect. Upon alloying Pt with polyvalent solutes, however, the electronic effect dominates and $\Delta \bar{G}_i^E$ assumes negative values down to -380 kJ/mol for Pt-Zr alloys.

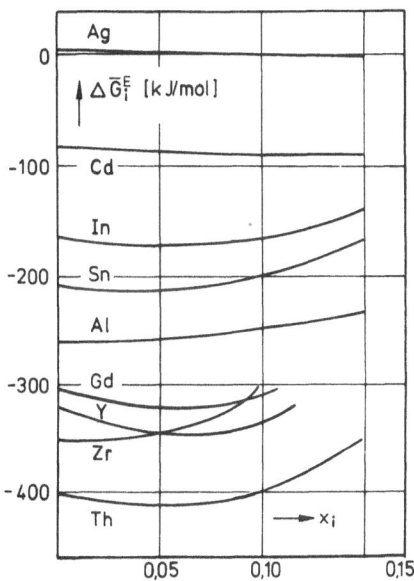

Fig.18 Relative partial excess Gibbs energy of various
    solutes in Pd at 800° C [39]

Although the electronic effects are quite noticeable in the experimental excess functions of most of the systems, a quantitative descripton calls for a separation of $\Delta \bar{G}_i^E$ into electronic and elastic parts, as outlined in the preceding chapter.

At infinite dilution, electrons are transfered from the Fermi level of the donor metal to the Fermi level of Pd and Pt, respectively. Therefore, the electronic part of the

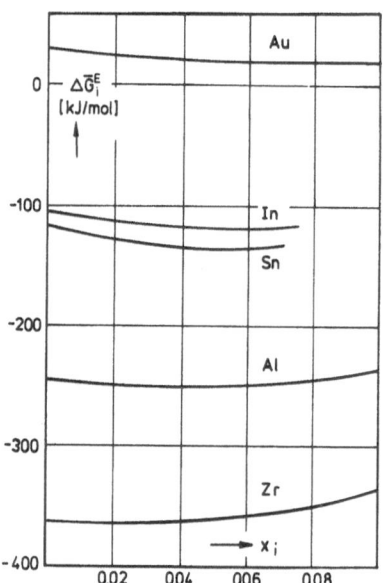

Fig.19 Relative partial excess Gibbs energy of various
solutes in Pt [39]

excess function is supposed to be determined by

$$\Delta \overline{G}_i^{E\bullet} \ (x_i = 0) = z_i \cdot (\mu_i^F - \mu_{Pd}^F) \tag{15}$$

where z_i is the valency of the solute i and μ^F the Fermi
energy of the components. Using the work function Φ as a
measure of the Fermi energy, the electronic part may be
estimated independed of thermodynamic data:

$$\Delta \overline{G}_i^{E\bullet} \ (x_i = 0) = z_i \cdot (\Phi_i - \Phi_{Pd}). \tag{16}$$

The electronic contribution obtained by splitting the excess function is plotted versus the difference of the work
function times valency (Eq.16) for Pd and Pt alloys in
Fig.20 and Fig.21, respectively. The values scatter around
the predicted lines with slope 1. The observed deviations
may be attributed to the approximate nature of the description on one hand and the inaccuracy of the data involved on
the other hand.

The results represented in Fig.20 and Fig.21 lead to
the conclusion, that the electronic mixing effects of Pd
and Pt alloys are determined by the position of the Fermi
levels and the valency of the solutes. Apparently the work

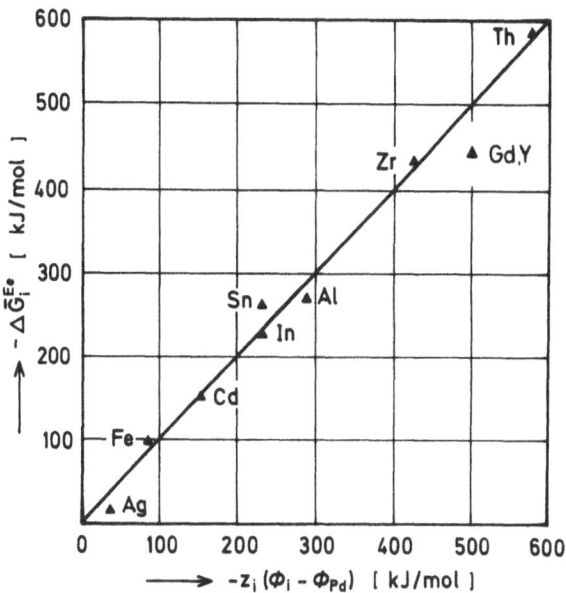

Fig.20 Bonding contribution $\Delta\bar{G}_i^{Ee}$ ($x_i=0$) of various solutes vs $z_i(\Phi_i - \Phi_{Pd})$ for Pd alloys

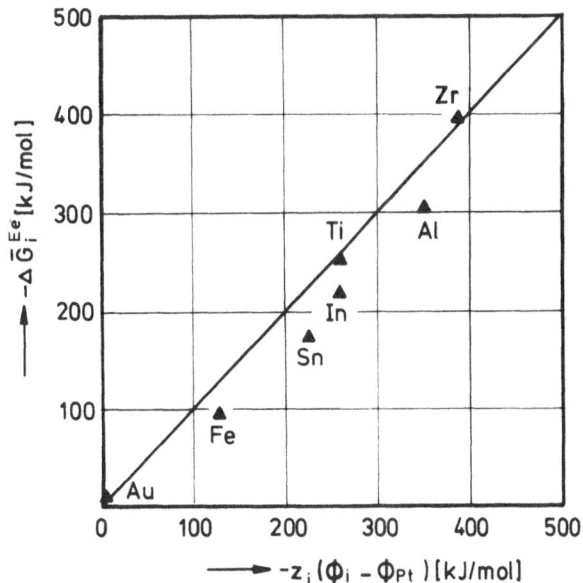

Fig.21 Bonding contribution $\Delta\bar{G}_i^{Ee}$ ($x_i=0$) of various solutes vs $z_i(\Phi_i - \Phi_{Pt})$ for Pt alloys

356

function is quite a good measure of the Fermi energy and
Eq.16 may be used to predict the bonding contribution
independent of thermodynamic data.

Acknowledgements

The author is pleased to acknowledge the valuable experimental contributions by his coworkers Drs. S. Paasch and
Th. Bretschneider. The research was supported by Deutsche
Forschungsgemeinschaft and Fonds der Chemie. For his helpful suggestions and assistance in preparing the manuscript,
the author is indebted to Prof. H. Brodowsky.

5. REFERENCES

1. K.Kiukkola and C.Wagner,
   J. Electrochem. Soc. **104**, 308 (1957).
2. K.Kiukkola and C.Wagner,
   J. Electrochem. Soc. **104**, 379 (1957).
3. R.A.Rapp and F.Maak, Acta Met. **10**, 63 (1962).
4. R.A.Rapp and D.A.Shores, in *"Physicochemical
   Measurements in Metals Research"* Part 2, ed. by
   R.A.Rapp, Wiley-Interscience, New York, 1970
   pp. 123-192.
5. W.A.Fischer and D.Janke, *"Metallurgische Elektrochemie"*
   Stahleisenverlag, Düsseldorf, 1975
6. S.Geller (ed.), *"Solid Electrolytes"* Springer Verlag,
   Berlin, 1977.
7. E.C.Subbarao (ed.), *"Solid Electrolytes and Their
   Applications"* Plenum Press, New York, 1980.
8. H.Rickert, *"Electrochemistry of Solids"*, Springer
   Verlag, Berlin, 1982.
9. C.Wagner, Z. Phys. Chem. B **21**, 25 (1933).
10. H. Schmalzried, Z.Phys.Chem. **25**, 178 (1960)
11. J.W.Patterson, J. Electrochem. Soc. **118**, 1033 (1971).
12. J.W.Patterson, E.C.Bogren, and R.A.Rapp,
    J. Electrochem. Soc. **114**, 752 (1967).
13. I.Barin and O.Knacke, *"Thermochemical Properties of
    Inorganic Substances"*, Springer, Berlin 1973.
14. A.W.Bryant, W.G.Bugden, and J.N.Pratt,
    Acta Met. **18**, 101 (1970).
15. R.W.Ure, Jr., J. Chem. Phys. **26**, 1363 (1957).
16. C.Wagner, J. Electrochem. Soc. **115**, 933 (1968).
17. J.W.Hinze and J.W.Patterson,
    J. Electrochem. Soc. **120**, 96 (1973).
18. J.Delcet, R.J.Heus, and J.J.Egan,
    J. Electrochem. Soc. **125**, 775 (1978).
19. E.Mollwo, Nachr. Ges. Wiss. Göttingen, Math.-Phys. Kl.
    Fachgruppe II, **NF** 1, 19 (1934).

20. C.Wagner, Z. Elektrochem. **60**, 4 (1956).
21. R.Benz and C.Wagner, J. Phys. Chem. **65**, 1308 (1961).
22. J.J.Egan, J. Phys. Chem. **68**, 978 (1964).
23. H.Kleykamp, Ber. Bunsenges. Phys. Chem. **87**, 777 (1983).
24. L.E.Nagel and M.O'Keeffe, in *"Fast Ion Transport in Solids"*, ed. by W.van Gool, North-Holland Publ. Co, Amsterdam, London, 1970, p. 165.
25. T.Takahaski, H.Iwahara, and T.Ishikawa, J. Electrochem. Soc. **124**, 280 (1977).
26. Th.Bretschneider, Ph.D.thesis, Univ. Kiel, 1988.
27. O.Greis and J.M.Haschke, in *"Handbook on the Chemistry and Physics of Rare Earths"* ed. by K.A.Gschneider jr. and L.Eyring, North-Holland Publ. Co., Amsterdam/ New York, Vol V, p. 387, 1982.
28. L.I.Kholokhonova and T.N.Rezukhina, Russ. J. Phys. Chem. **50**, 451 (1976).
29. H.-J.Schaller and H.Brodowsky, Z. Metallkde. **69**, 87 (1978).
30 H.-J.Schaller and H.Brodowsky, Ber. Bunsenges. Phys. Chem. **82**, 773 (1978).
31. J.B.Darby, K.M.Myles, and J.N.Pratt, Acta Met. **19**, 7 (1971).
32. A.W.Bryant and J.N.Pratt, Proc. Colloque Int. du C.N.R.S. **201**, 241 (1972).
33. D.Gerstenberg, Ann. Phys. **2**, 236 (1958).
34. H.-J.Schaller, Z. Metallkde. **70**, 318 (1979).
35. H.-J.Schaller, Z. Phys. Chem. **NF 112**, 85 (1978).
36. A.W.Lawson, J. Chem. Phys. **15**, 831 (1947).
37. T.Heumann, Ber. Bunsenges. Phys. Chem. **57**, 724 (1953).
38. J.D.Eshelby, J. Appl. Phys. **25**, 225 (1954).
39. H.-J.Schaller, Ber. Bunsenges. Phys. Chem. **87**, 734 (1983).
40. H.-J.Schaller, Ber. Bunsenges. Phys. Chem. **82**, 365 (1978).
41. M.Ettenberg, K.L.Komarek, and E.Miller, Met. Trans. **2**, 1173 (1970).
42. C.Wagner, *"Thermodynamics of Alloys"*, Addison Wesly Co., Reading, London, 1952.
43. R.Ferro and R.Capelli, Atti Accad. naz. Lincei, Rend., Cl. Sci. fisiche mat. natur. **34**, 659 (1963).
44. S.Paasch and H.-J.Schaller, Ber. Bunsenges. Phys. Chem. **87**, 812 (1983).
45. H.-J.Schaller, Z. Naturforsch. **34**, 464 (1979).
46. T.B.Massalski (ed), *"Binary Alloy Phase Diagrams"*, Vol. 1, American Soc. for Metals, Metals Park, Ohio, 1986.
47. Th.Bretschneider and H.-J.Schaller, submitted to Ber. Bunsenges. Phys. Chem.
48. J.R. Sambles, J. Phys. Chem. Solids **46**, 525 (1985)

358

49. R. Alqasmi and J.J. Egan, Ber. Bunsenges. Phys. Chem.
    87, 815 (1983).
50. Th. Bretschneider and H.-J. Schaller, submitted to
    Z. Metallkde.
51. N.G. Schmahl and W. Schneider, Z. Phys. Chem. NF 57,
    218 (1968)
52. 51 H.Brodowsky, Y.S.Oei, and H.-J.Schaller,
    Z. Metallkde. 71, 593 (1980).
53. H.-J.Schaller,
    Ber. Bunsenges. Phys. Chem. 80, 999 (1976)
54. H.-J.Schaller, Z. Metallkde. 70, 354 (1979).

Thermodynamic Properties of Solid Copper-Aluminium and Copper-Germanium Alloys

H.-J.Schaller, G.Fickel, and A.Maaz
Institut für Physikalische Chemie
der Universität Kiel
Olshausenstr. 40
2300 Kiel 1
F.R. Germany

ABSTRACT. The activity of Al in Cu–Al alloys (0–34.9 at.% Al) was determined between 1000 and 1100K by emf measurements on fluorine concentration cells with CaF_2 as the solid electrolyte. Oxygen concentration
cells with doped ZrO_2 as the electrolyte were used to determine the
thermodynamic properties of Cu–Ge alloys (0–9.6 at.% Ge) between 900
and 1100K. The systems exhibit pronounced negative deviations from
ideality. The relative partial excess Gibbs energies of Al and Ge are
−130 and −53 kJ/mol, respectively, at zero solute concentration and
1000K. The thermodynamic properties of Cu-base alloys are interpreted in
terms of the following effects: The charge transfer upon alloying Cu with
electron donors, the lattice distortion resulting from the size misfit of
the components, and the energy change upon transforming the structure
of the solutes into the fcc structure of Cu. The influence of these effects
on the solution behaviour is illustrated by splitting the partial excess
Gibbs energies into electronic and elastic parts.

1. INTRODUCTION

According to previous studies on Pd–Al [1], Pt–Al [2], and Co–Al [3] alloys, the thermodynamic behaviour of these systems is characterized by
pronounced negative deviations from ideality. The interpretation of these
properties was based on the observation that Al behaves as an electron
donor with respect to the noble host metals [1,2]. The transfer of valence
electrons from the Fermi level of Al to the Fermi level of the alloy gives
rise to a bonding contribution to the thermodynamic excess quantities.
Considering the lattice distortion brought about by the size misfit of the
components as an additional effect, the mixing behaviour of these
systems could be well described. These ideas which have contributed to
correlate the thermodynamics of a number of Pd and Pt alloys should also
be applicable to solid solutions with Cu as host metal.

The thermodynamic behaviour of liquid Cu–Al alloys at elevated
temperatures has been studied extensively [4–8] whereas only limited information is available concerning solid alloys. The activity of Al in solid
Cu–Al alloys was determined by Ali et al. [9] in the ranges 660–760°C

359

H. Brodowsky and H.-J. Schaller (eds.), Thermochemistry of Alloys, 359–370.
© 1989 by Kluwer Academic Publishers.

and 0-35 at.% Al, by Ali and Geiderikh [10] in the ranges 400-500°C and 37-60 at.% Al, and by Hair and Downie [8] at 550°C in the whole composition range. The two published sets of activity data in the copper-rich terminal solid solution (0-18 at.% Al) are in poor agreement and, moreover, limited to three or four compositions each . This situation makes it desirable to provide more detailed activity data for solid alloys.

In previous papers, emf measurements with CaF_2 as a solid electrolyte turned out to be a suitable method to determine the activity of Al at elevated temperatures [1,2]. Experimental problems associated with the volatility of the electrode components were overcome by encapsulating the electrodes. This procedure also helped to extend the temperature range of the measurements.

Solid Cu-Ge alloys have been studied by Predel and Schallner [11], using oxygen concentration cells with doped ThO_2 as the solid electrolyte. In this study analogous cells with a $ZrO_2(+CaO)$ electrolyte were employed.

2. EXPERIMENTAL

2.1. Method

The activity of Al in Cu was determined by emf measurements on cells of the type

$$C \mid Al,AlF_3 \mid CaF_2 \mid Al(Cu),AlF_3 \mid C.$$

A CaF_2 single crystal served as the solid electrolyte.

The emf E is determined by the activity of Al in Cu, a_{Al}, and given by the equation

$$E = (RT/3F) \cdot \ln a_{Al} . \qquad (1a)$$

Experimental problems associated with liquid Al and the volatility of AlF_3 at elevated temperatures were overcome by encapsulating the electrodes in graphite crucibles. The arrangement of the alloy electrode in Figure 1 shows that direct contact between the alloy pellet and the electrolyte was brought about by a rhenium spring. The surfaces of the crucible and the single crystal were polished to give a good seal. The cells were operated in an protective argon atmosphere. Before entering the cell assembly, bottled argon was passed through dry $Mg(ClO_4)_2$ at room temperature and over Zr turnings at 1000°C to eliminate traces of water and oxygen, respectively. The emf was measured with a sensitive electrometer (input resistance $2 \cdot 10^{14} \Omega$). Further details of the apparatus have been given in previous papers [1,12].

The thermodynamic properties of Cu-Ge alloys were determined on oxygen concentration cells of the type

$$Ge,GeO_2 \mid ZrO_2(+CaO) \mid Ge(Cu),GeO_2.$$

The emf of the cells is given by

$$E = (RT/4F) \cdot \ln a_{Ge} \quad .$$

(1b)

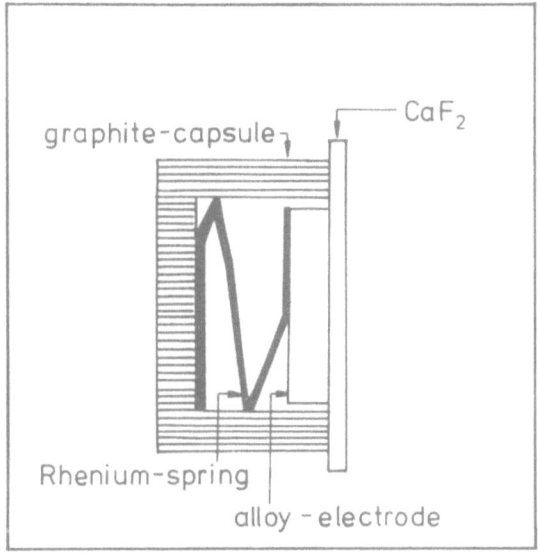

Figure 1.   Encapsulation of the alloy electrode

2.2. Preparation of Alloys

The alloys were prepared by melting accurately weighed amounts of Cu (99.9998%, Preussag) and Al (99.999%, Koch–Light) or Ge (99.999%, Koch–Light) in a high-frequency induction furnace under pure argon. Since weight losses before and after melting were insignificant, the alloy concentrations were directly calculated from the weighed charges.

The alloy ingots were annealed for three days at 1100 K in an argon atmosphere. The powder for the alloy electrodes was prepared from the ingots by milling with a tungsten carbide drill. The alloy powder was then mixed with AlF₃ or GeO₂, respectively, and pressed into pellets.

3.RESULTS

After starting a run, sufficient time (up to 12 h) for the cells to reach equilibrium was allowed to pass. On subsequent changes of the temperature, the new equilibrium value of the emf was attained in less than 4 h. The recorded emf had the same value irrespective of whether the temperature was approached from above or below. Passing a small current through the cell, the observed voltage returned to the original value. These criteria indicate the reversibility of the cells.

3.1. Cu-Al

The plot of the Al activity versus composition at 1050K is shown in Figure 2. The results of Ali et al. [9] and of Hair and Downie [8] have been extrapolated to 1050K and referred to hypotheticallly solid Al as standard state. The phase field limits shown in Figure 2 were taken from the published phase diagram [13].

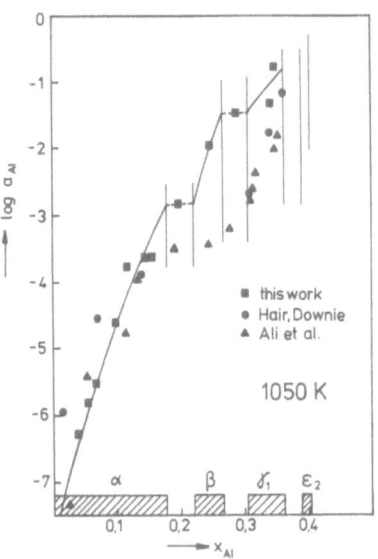

Figure 2. Activity of Al in Cu-Al alloys at 1050K [Reference state: hypothetically solid Al]

Figure 2 shows that there is only fair to poor agreement between the three activity studies in the α-phase region. Compared to the data published by the other authors, the present data set shows less scattering and permits a more precise fixing of the activity versus composition variation. The agreement in the β- and γ₁-phase fields is poor: Ali et al. stated that their activity data in the α-β and β-γ₁ two phase regions showed a lack of reproducibility; moreover, their results turned out to be inconsistent with the well established phase diagram.

Figure 3 shows a plot of the relative partial excess free energy of Al, $\Delta \bar{G}_{Al}^E$, versus mole fraction in the α-phase region. The excess function exhibits a positive slope and assumes a value of -112 kJ/mol at 1050K and zero solute concentration.

3.2. Cu-Ge

The activity of Ge in α-phase alloys obtained in the present work at 900, 1000, and 1100K is shown in Figure 4 along with the results of Predel and Schallner [11] at 1000K. Considering the concentration dependence, the two isotherms at 1000 K show a similar tendency; the data of the present study, however, assume distinctly smaller values.

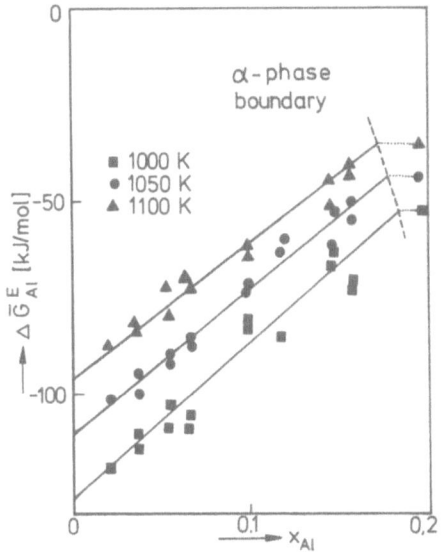

Figure 3.   Relative partial excess free energy of Al in Cu–Al alloys [Reference state: hypothetically solid Al]

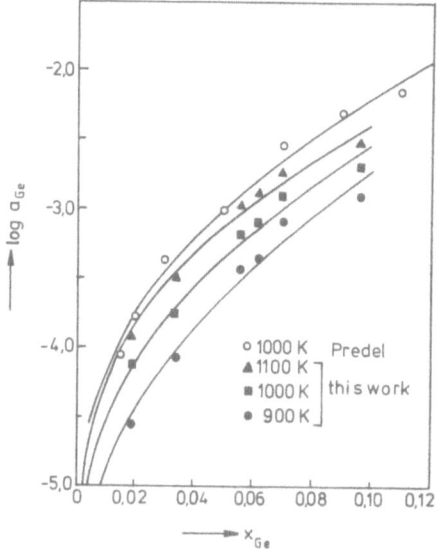

Figure 4.   Activity of Ge in Cu–Ge alloys [Reference state: solid Ge]

4. DISCUSSION

Figure 5 shows the relative partial excess free energy of Al, Ge, and Sb [18] in α-phase Cu alloys at 1000K for comparison. Cu-Al alloys in particular exhibit pronounced negative deviations from ideal mixing behaviour whereas α-phase Cu-Sb alloys are close to ideality. Moreover, the slope of the excess functions is positive in the case of Al and Ge and slightly negative in the case of Sb.

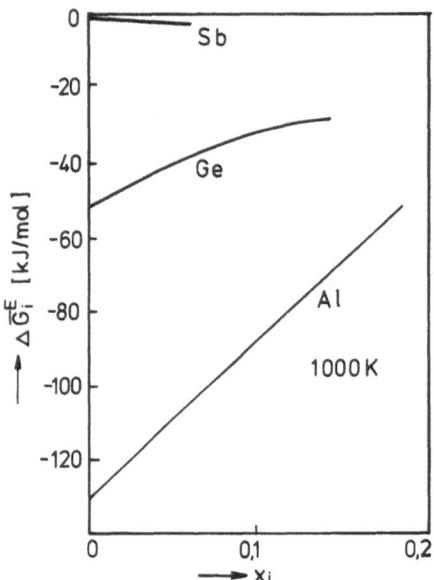

Figure 5. Relative partial excess free energies of Sb, Ge, and Al in α-phase Cu alloys at 1000K

The thermodynamic behaviour of the Cu base alloys is interpreted by considering an electronic and an elastic effect on the energetics of alloy formation.

As in the case of the analogous Pd and Pt alloys, the electronic effect is based on the assumption that the solutes Al, Ge, and Sb behave as electron donors with respect to Cu. This charge transfer gives rise to a bonding contribution $\Delta \bar{G}_i^{Ee}$ to the relative partial excess free energy.

The elastic effect is caused by the expansion of the Cu lattice upon alloying Cu with Al, Ge, or Sb. The elastic energy associated with the lattice distortion reduces the solubility.

Due to the different structures of Cu and Ge, an additional effect has to be taken into account in the case of Cu-Ge alloys: the transformation of Ge from the equilibrium diamond structure into the hypothetical fcc structure [11]. The free enthalpy of transformation, $\Delta \bar{G}_{Ge}(trans)$, has been estimated to be 42 kJ/mol [14]. Assessments made for a number of metals [15] have shown that the energy difference between the rhombohedral and the fcc structure is distinctly smaller. Since no reliable information is available, the transformation energy of Sb is not taken into account in the case of Cu-Sb alloys.

The described effects may be interrelated, and to discuss them separately is a first approximation. With this caveat in mind, a separation into an electronic (superscript e) and an elastic or dilatational (superscript d) part can be accomplished if independent information on at least one of these effects is available or by model considerations.

There is a straightforward relation between the separate parts of the integral excess Gibbs energy ΔG^E and its derivatives $\Delta \overline{G}_i{}^E$, $\Delta \overline{G}_{cu}{}^E$, and $(\Delta \overline{G}_i{}^E - \Delta \overline{G}_{cu}{}^E)$. For substitutional alloys it is most practical to select the following function [16,17]:

$$(\partial \Delta G^E / \partial x_i) = (\Delta \overline{G}_i{}^E - \Delta \overline{G}_{cu}{}^E) \qquad (2)$$

$$= (\Delta \overline{G}_i{}^{Ee} - \Delta \overline{G}_{cu}{}^{Ee}) + (\Delta \overline{G}_i{}^{Ed} - \Delta \overline{G}_{cu}{}^{Ed}) + \Delta \overline{G}_i(trans)$$

(i: Al, Ge, or Sb).

One advantage is that the term $(\Delta \overline{G}_i{}^{Ee} - \Delta \overline{G}_{cu}{}^{Ee})$ may be identified with the rise of the Fermi energy upon alloying Cu with electron donors. The difference of the elastic parts of the excess functions in Eq.(2) can be directly described by means of the quasichemical method:

$$(\Delta \overline{G}_i{}^{Ed} - \Delta \overline{G}_{cu}{}^{Ed}) = -6w + 12RT\ln \frac{2 - 2x_i}{1 + \beta + 2x_i} \qquad (3)$$

$$\beta = [1 - 4x_i(1 - x_i)(1 - e^{-w/RT})]^{1/2}.$$

The pair interaction parameter w is assumed to be an energy of attraction. An explanation for the attraction between like particles is offered in terms of the lattice strain which is caused by the misfit of the components. The lattice strain is reduced if two isolated solute atoms enter adjacent sites. Assuming an energy of attraction, the quasichemical approach yields a negative gradient of the elastic contribution to the excess function. At zero solute concentration, Eq.(3) is reduced to

$$\Delta \overline{G}_i{}^{Ed}(x_i=0) = -6w. \qquad (4)$$

In previous publications, the following relation based on an elastic lattice model was applied in order to estimate $\Delta \overline{G}_i{}^{Ed}(x_i=0)$ independent of thermodynamic data [3, 18]:

$$\Delta \overline{G}_i{}^{Ed}(x_i=0) = \frac{B^{\circ}_{cu}}{b_{cu}} (\overline{V}_i - V^{\circ}_{cu}) +$$

$$+ -\frac{B^{\circ}_{cu}V^{\circ}_{cu}}{b_{cu}(1-b_{cu})} [1-(\overline{V}_i/V^{\circ}_{cu})] \quad {}^{(1-b_{cu})}. \qquad (5)$$

$\overline{V}_i$ is the partial molar volume of the solute at infinite dilution, V°_{cu} the molar volume of copper, B°_{cu} the bulk modulus at $p = 1$ bar, and b_{cu} its pressure coefficient $(\partial B_{cu}/\partial p)_T$.

The difference $(\overline{V}_i - V^{\circ}_{cu})$ can be calculated from the linear lattice spacing $a=a_{cu}+bx_1$ of the solid solution alloys by means of the following equation (N_A denotes Avogadro's number):

$$(\overline{V}_1 - V^{\bullet}_{cu}) = (3/4)\ N_A a^2_{cu}b. \qquad (6)$$

According to Figure 6 which shows the lattice spacings of the α-phase Cu alloys for comparison, the misfit of Sb in Cu is distinctly larger than the misfit of Al or Ge. Using the values $\overline{V}_{sb} = 13.6$ cm³/mol and $V^{\bullet}_{cu}=7.11$ cm³/mol, the relative volume difference between Sb and Cu is as large as 90%. According to the 2nd Hume-Rothery rule, the extent of primary solid solutions is seriously hindered whenever the disparity in atomic radii exceeds 15% which corresponds to a volume difference of about 45%. Nevertheless, the solid solution extends up to 6 at.% Sb. The exception to Hume-Rothery's size rule indicates that an opposing effect has to take an active part in balancing the influence of the misfit.

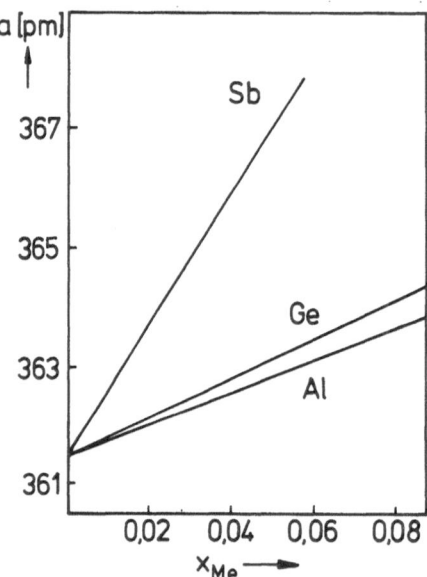

Figure 6.   Lattice spacings of α-phase Cu alloys

The misfit energies calculated with the literature data $b_{cu} = 5.1$ and $B^{\bullet}_{cu} = 1.35 \cdot 10^5$ J/cm³ [19] are listed in Table 1 along with the bonding contribution. All values refer to zero solute concentration.

The relative partial excess free energy of Cu in Eq.(2) was determined by integrating the Gibbs-Duhem equation. The quasichemical approximation (Eq.(3)) was applied to extrapolate the elastic part of the excess function from zero solute concentration over the solid solution range. Finally, the difference of the electronic excess functions was obtained using Eq.(2).

The separation of the excess functions for Cu-Sb, Cu-Al, and Cu-Ge alloys is presented in the Figures 7, 8, and 9.

TABLE 1. Thermodynamic excess properties of Al, Ge, and Sb in Cu alloys with elastic and electronic contributions. The values refer to zero solute concentration ($x_1 = 0$) and 1000K.

System	$\Delta\bar{G}_1^E$ [kJ/mol]	$\Delta\bar{G}_1^{Ed}$ [kJ/mol]	$\Delta\bar{G}_1^{Ee}$ [kJ/mol]	$\Delta\bar{G}_1(trans)$ [kJ/mol]
Cu–Al	−130	17	−147	−
Cu–Ge	− 51	27	−120	42
Cu–Sb[18]	0	131	−131	−

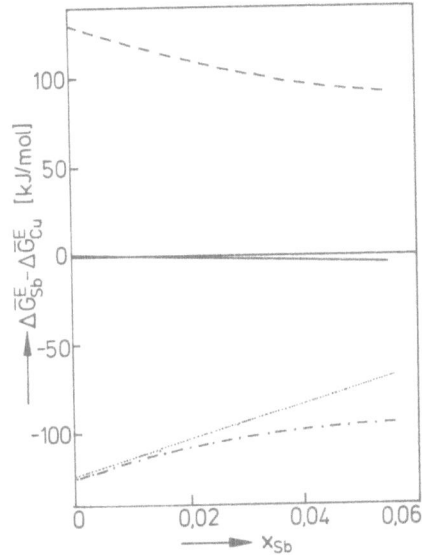

Figure 7. Separation of the excess functions for Cu–Sb alloys (1000K); ——— ($\Delta\bar{G}_{Sb}^E - \Delta\bar{G}_{Cu}^E$), - - - elastic part, −·−·− electronic part, ········· rise of the Fermi energy according to the rigid band model

    Considering the results for the Cu–Sb system shown in Table 1 and Figure 7, the small deviations from ideality do not result from moderate elastic and electronic effects as might be expected: on the contrary, the separation of the excess function indicates the existence of two large opposing effects. Since elastic and electronic contribution almost balance each other, the mixing behaviour of α–phase Cu–Sb alloys is close to ideality.

    Due to the large misfit of Sb in Cu, the attraction parameter w of the quasichemical approximation assumes a large absolute value (w = −21.8 kJ/mol). This causes a marked negative gradient in the elastic part of the excess function which overcompensates the rise of the Fermi energy. Consequently, the total excess function shows a negative gradient.

In the case of Cu–Al and Cu–Ge alloys, the results of the thermo-dynamic analysis in the Figures 8 and 9 indicate moderate elastic and strong electronic influences on the solution behaviour. In both systems the rise of the Fermi energy overcompensates the negative slope of the elastic contribution. Therefore, the total excess function exhibits a positive gradient.

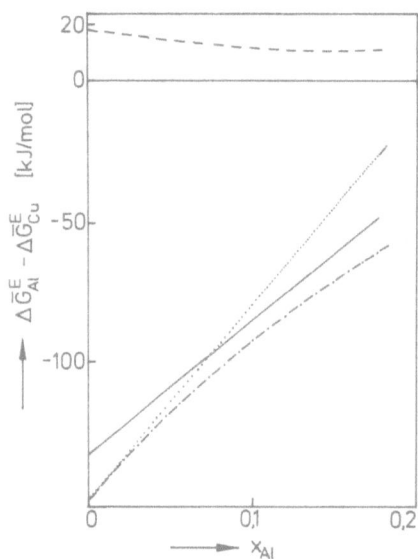

Figure 8. Separation of the excess functions for Cu–Al alloys (1000K); ———— ($\Delta \bar{G}_{Al}^E - \Delta \bar{G}_{Cu}^E$), – – – elastic part, –·–·– electronic part, ········· rise of the Fermi energy according to the rigid band model

At infinite dilution, the bonding contribution of Cu–Al alloys assumes a value of $\Delta \bar{G}_{Al}^{Ee} = -147$ kJ/mol (Table 1). An analogous analysis of Pd–Al and Pt–Al alloys yielded the values of $\Delta \bar{G}_{Al}^{Ee} = -274$ kJ/mol and $\Delta \bar{G}_{Al}^{Ee} = -272$ kJ/mol, respectively, which refer to zero solute concentration [1,2].

For Al and a number of other solutes in the host metals Pd or Pt, the bonding contribution was correlated with the Fermi energy μ_i^F of the components and the valency z_i of the solute by means of the relation

$$\Delta \bar{G}_i^{Ee} = z_i(\mu_i^F - \mu_M^F) \tag{7}$$

M: host metal,
i: solute.

Using the work function Φ_i as a measure of the Fermi energy, the $\Delta \bar{G}_i^{Ee}$ values were predicted in good approximation independently of thermodynamic data.

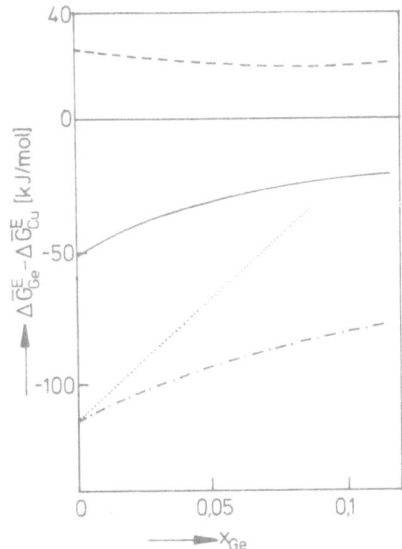

Figure 9. Separation of the excess functions for Cu-Ge alloys (1000K);
——— $(\Delta \overline{G}_{Ge}^E - \Delta \overline{G}_{Cu}^E)$, - - - elastic part, -·-·- electronic part,
········· rise of the Fermi energy according to the rigid band model

Inserting the literature data $\Phi_{Al} = 4.2$ eV and $\Phi_{Cu} = 4.72$ eV [20] in Eq.(7), the electronic contribution at zero Al concentration, $\Delta \overline{G}_{Al}^{Ee}$, is calculated to be -141 kJ/mol. This prediction is in good accordance with $\Delta \overline{G}_{Al}^{Ee} = -147$ kJ/mol, the result obtained from the experimental excess Gibbs energy (-130 kJ/mol) and the misfit energy ($+17$ kJ/mol).

Within the scope of the rigid band model, the Fermi energy rise may be calculated from the electronic specific heat of Cu. The increase of the valence electron concentration upon exchanging a host atom with a solute atom is determined by the valence difference Δz of the components [14]. The rise of the Fermi energy calculated with $\Delta z=4$ for Cu-Sb, $\Delta z=3$ for Cu-Ge and $\Delta z=2$ for Cu-Al is plotted in Figures 7, 8, and 9 for comparison.

Only in the case of very dilute Cu-Al and Cu-Sb alloys conforms the rise of the Fermi energy approximately to the rise predicted by the rigid band model. Apart from this, the rigid band approximation yields distinctly steeper slopes. The deviations observed point to an increase of the density of states upon alloying Cu with electron donors.

5. Literature

[1]   H.-J. Schaller, Ber. Bunsenges. Phys. Chem. **82**, 365 (1978).
[2]   H.-J. Schaller, Z. Physik. Chem. **NF** 112, 85 (1978).
[3]   H.-J. Schaller and Th. Bretschneider, Z. Metallkde. **76**, 143 (1985).
[4]   G. Grube and P. Hantelmann, Z. Elektrochem. **48**, 399 (1942).

[5]   D.I. Layner, L.M. Ostrovskaya, and O.S. Serebryannikova,
        Russ. Metall. 1, 15 (1976).
[6]   M. Hiroyasu and N. Hiroshi, Nippon Kinzoku Gakkaishi 33, 344
        (1969).
[7]   T.C. Wilder, Trans. Met. Soc. AIME 233, 202 (1965).
[8]   J. Hair and D.B. Downie, Faraday Symposia Chem. Soc. 8, 56 (1973).
[9]   S.A. Ali, V.V. Samokhval, V.A. Geiderikh, and A.A. Vecher,
        Russ. J. Phys. Chem. 46(1), 139 (1972).
[10]   S.A. Ali and V.A. Geiderikh, Russ. J. Phys. Chem. 47, 22 (1973).
[11]   B. Predel and U. Schallner, Mater. Sci. Eng. 10, 249 (1972).
[12]   H.-J. Schaller and H. Brodowsky, Z.Metallkde. 69, 87 (1978).
[13]   T.B. Massalski, J.L. Murray, L.H. Bennett, and H. Baker, "Binary
        Alloy Phase Diagrams", American Society for Metals, Metals Park,
        Ohio (1986).
[14]   B. Predel and D.W. Stein, Z. Naturforsch. 26a, 722 (1971).
[15]   A.P. Midownik in: L.H. Bennett (ed.),"Computer Modeling of Phase
        Diagrams", The Metallurgical Society, Inc., Warrendale, Pennsylvania
        (1986).
[16]   H. Brodowsky, Z. Naturforsch. 22a, 130 (1967).
[17]   H.-J. Schaller and H. Brodowsky, Ber. Bunsenges. Phys. Chem.
        82, 773 (1978).
[18]   H. Brodowsky, A. Fruma, H. Sagunski, and H.-J. Schaller,
        Z. Metallkde. 73, 354 (1982).
[19]   Landolt-Börnstein, "Tabellen und Funktionen", Vol.2, Part 1,
        Springer Verlag, Berlin, New York (1979).
[20]   J. Hölzl and F.K. Schulte, "Solid Surface Physics", Springer Verlag,
        Berlin, New York (1979).

CaF2 SOLID ELECTROLYTES USED TO STUDY THERMODYNAMICS OF ALLOYS

J. J. Egan
Brookhaven National Laboratory
Department of Applied Science
Upton, NY 11973, U.S.A.

ABSTRACT. Galvanic cells using CaF2 as a solid electrolyte are
described for obtaining thermodynamic and phase equilibria data on
alloys both in the liquid and solid state. The method of coulometric
titration, which controls the composition of alloy electrodes very
accurately, is also explained. Results yield information on the Gibbs
energies of mixing for liquid phases. In the solid state the method
identifies what intermetallic compounds are present in a binary system
as well as their range of homogeneity and Gibbs energy of formation.
This information combined with appropriate atomic models gives the
type and degree of atomic or electronic disorder present in the
intermediate phase. Examples will be presented for binary systems
involving alkali and alkaline earth metals and alkali metals dissolved
in their alkali halides.

1. INTRODUCTION

Solid electrolytes generally are ionic conductors but possess a
component of electronic conductivity under some conditions. Before
they can be used in galvanic cells to obtain thermodynamic data on
alloys, it is necessary to determine the transference number of ions
or electrons as a function of the activity of the component metal
e.g. the activity of Ca in CaF_2. Several methods are available to
determine the fraction of electronic conduction in solid electrolytes,
but one of the most accurate ways is the Wagner polarization
technique. This method has been applied to CaF_2 and the experiments
and results will be discussed in section 2.

Section 3 describes the use of the coulometric titration method
illustrated here by the study of solid Ca-Au alloys. This technique
is usually applied to systems with high diffusion coefficients of one
of the components e.g. Ag_2S, Cu_2S, Ag_2Se and several Li alloys. It is
used in this work to study a system with slower diffusion and this is
accomplished by using thin film electrodes (2 to 25 microns).

CaF_2 has also been used to obtain thermodynamic data on some
liquid Ca alloys under conditions where the electrolyte is a partial

371

H. Brodowsky and H.-J. Schaller (eds.), Thermochemistry of Alloys, 371–398.

electronic conductor. Results obtained on Ca-Zn alloys compared with literature values obtained by vapor pressure measurements show excellent agreement indicating validity of this type of measurement. These findings are detailed in section 4.

In the next section (5) ionic alloys composed of alkali metals and semimetals Bi and Sb are studied. A specially designed sealed cell arrangement allows the study of these alloys at temperatures near the boiling point of the alkali metal. These alloys are semiconductors in the liquid state and results are important for models considering the effect of electron and electron hole concentration changes on the thermodynamic properties.

Finally the use of CaF_2 to study the thermodynamics of mixing between Na and NaBr will be reviewed. The galvanic cell can also be used to generate Na metal and control the composition of mixtures contained in a conductivity cell.

2. POLARIZATION MEASUREMENTS TO DETERMINE THE FRACTION OF ELECTRONIC CONDUCTIVITY IN CaF_2

Several authors have treated the problem of electronic conduction in CaF_2. Wagner [1] has calculated the transference number of electrons as a function of Ca activity using the measurements of Mollwo [2] on the number of color centers in CaF_2 and their mobility, Hinze and Patterson [3-4] have measured the total electrical conductivity as a function of the partial pressure of fluorine. They concluded that CaF_2 exhibits negligible electronic conductivity even when equilibrated with Ca metal. This is in disagreement with the conclusions of Delcet and Egan [5] and Wagner [1]. Baukal [6] has measured the electronic conduction of CaF_2 doped with NaF at temperatures between 490°C and 550°C using the polarization technique.

In the present work, which is a review of that reported by Delcet and Egan [5], the polarization method is applied to pure CaF_2 single crystals between 800°C and 950°C.

2.1 General Approach and Equations

The theory and equations of the Wagner polarization method have been explained in detail by Wagner [7] and also reviewed by Kröger [8] so that only an outline will be presented here.

The CaF_2 forms part of a galvanic cell with two electrodes. One electrode possesses a known activity of Ca and acts as the reference electrode, the other electrode is an inert electronic conductor. By applying a dc potential between the two electrodes a known activity of Ca is impressed on the CaF_2 in contact with the inert electrode. Using only potentials below the decomposition potential of CaF_2, ionic currents are suppressed and only electronic currents are measured. Measuring the current at several applied potentials (differing Ca activities) and knowing the ionic conductivity which does not change with Ca activity, the transference number of electrons as a function of Ca activity or chemical potential is obtained. In this method at steady state there is no potential gradient across the electrolyte since no ions are moving, thus the electrons are moving only in a concentration gradient.

To study CaF_2 the following cell was employed:

$$Ca-Pb(1)|CaF_2(s)|Fe(s)$$
$$x_{Ca} = 0.90$$

(I)

where the reference electrode is the Ca-Pb alloy. The current at various applied voltages is given by the expression [7,8]

$$i = \frac{RT}{FG} \sigma_e' \left[1 - \exp\left(- \frac{EF}{RT}\right) \right]$$

(1)

The Ca-Pb alloy is negative with respect to the iron electrode. Here G is the cell constant (thickness of CaF_2 crystal divided by the area of the electrodes), σ_e' is the electronic conductivity of CaF_2 equilibrated with the reference electrode and E is the potential applied to cell (I). A Ca-Pb alloy was used instead of pure Ca metal for experimental reasons. Ca-Pb ($x_{Ca} = 0.90$) is liquid at 800°C and could be contained in iron cups which fit closely onto the CaF_2 crystals. Electrodes of solid Ca formed a conducting film on the CaF_2 surface and interfered with the measurements. The activity of Ca in the Ca-Pb alloy was calculated from the heat of fusion of Ca and the Ca-Pb phase diagram.

Results of current-potential curves on cell(I) yield values of σ_e'. Knowing σ_e' one obtains the electronic conduction at other activities using the equation:

$$\sigma_e = \sigma_e' \left(\frac{a_{Ca}}{a_{Ca}'}\right)^{\frac{1}{2}}$$

(2)

where a'_{Ca} is the activity of the reference electrode. This equation assumes that the electronic conductivity is proportional to the square root of the Ca activity in CaF_2. This assumption was confirmed by operating cells where the reference electrode has a much smaller activity of Ca and were made positive with respect to the iron electrode.

$$Fe(s)|CaF_2(s)|Ca-Bi(1)$$
$$x_{Ca} = 0.25$$

(II)

Under reducing conditions, the steady-state current in cell(II) is given by:

$$i = \frac{RT}{FG} \sigma_e'' \left[\exp\left(\frac{EF}{RT}\right) - 1 \right]$$

(3)

where σ_e'' is the electronic conductivity of CaF_2 whose Ca activity is the same as that for Ca-Bi ($x_{Ca} = 0.25$).

The activity of Ca in the Ca-Bi alloy is obtained from emf measurements on the auxiliary cell

$$\text{Ca-Pb(1)}\,|\,\text{CaF}_2\text{(s)}\,|\,\text{Ca-Bi(1)} \qquad\qquad (III)$$
$$x_{Ca} = 0.90 \qquad x_{Ca} = 0.25$$

In this cell the electrolyte has partial electronic conduction but the activity of Ca in Bi can still be obtained from an equation derived by Wagner [9] in 1933, namely;

$$E = \frac{1}{2F}\int_{\mu''_{Ca}}^{\mu'_{Ca}} t_{ion}\,d\mu_{Ca} \qquad\qquad (4)$$

where E is the emf of cell(III). Since a small current is running through this cell at all times it is important that the electrodes are liquid to minimize polarization effects. In equation (4)

$$t_{ion} = \frac{\sigma_{ion}}{\sigma_{ion}+\sigma_e} = \frac{\sigma_{ion}}{\sigma_e^\circ a_{Ca}^{\frac{1}{2}} + \sigma_{ion}} \qquad\qquad (5)$$

and σ_e° is the electronic conductivity of CaF_2 with $a_{Ca} = 1$.
Equation (4) may then be expressed as:

$$E = \frac{RT}{2F}\int_{a''_{Ca}}^{a'_{Ca}} \frac{\sigma_{ion}}{a_{Ca}(\sigma_e^\circ\, a_{Ca}^{\frac{1}{2}} + \sigma_{ion})}\,da_{Ca} \qquad\qquad (6)$$

where a'_{Ca} is the activity at the left hand electrode and a''_{Ca} is that of the right hand electrode. Upon integration

$$E = \frac{RT}{2F}\ln\frac{a'_{Ca}}{a''_{Ca}} + \frac{RT}{F}\ln\frac{\sigma_e'' + \sigma_{ion}}{\sigma_e' + \sigma_{ion}} \qquad\qquad (7)$$

The quantities σ_e' and σ_e'' are related by

$$\sigma_e' = \sigma_e''\left(\frac{a_{Ca}'}{a''_{Ca}}\right)^{\frac{1}{2}} \qquad\qquad (8)$$

Equation (7) may be solved explicitly for a''_{Ca}, the activity of Ca in the Ca-Bi ($x_{Ca} = 0.25$) alloy yielding

$$a_{Ca}''^{\frac{1}{2}} = \frac{\exp\left(-\dfrac{nF}{RT}\right) - \sigma_e'' a_{Ca}'^{\frac{1}{2}}}{\sigma_{ion}} \qquad\qquad (9)$$

where

$$\eta = E - \frac{RT}{2F}\ln a'_{Ca} - \frac{RT}{2F}\ln(\sigma_{ion} + \sigma_e'') \qquad\qquad (10)$$

Thus knowing a'_{Ca}, σ_{ion}, and σ''_e which is obtained from steady state measurements on cell(II) using equation (3) as well as the emf of cell(III) one obtains the activity of Ca in the Ca-Bi alloy. Knowing σ''_e from experiments on cell(II) one may obtain conductivity value at other activities by use of the equation

$$\sigma_e = \sigma''_e \left(\frac{a_{Ca}}{a_{Ca''}} \right)^{\frac{1}{2}} \tag{11}$$

2.2 Experimental Details

The experimental setup for both cell(I) and cell(II) is shown in figure 1. The iron cup is filled with either a Ca-Pb alloy or a Ca-Bi alloy. The CaF_2 crystals are about 0.2 cm thick and the diameter of the inert iron electrode is 1.0 cm for cell(I) and 1.6 cm for cell(II). In order to minimize the edge effect caused by the CaF_2 extending beyond the edges of the iron rod, the ratio of the crystal thickness to the electrode area was kept small in accordance with Barrer, Barrie, and Rogers [10]. The cell arrangement was designed to compress the electrodes onto the CaF_2 single crystal which softens somewhat at around 600°C. Since the iron rod has a higher coefficient of expansion than the four tantalum rods the center part of the cell is compressed upon heating forming tight seals at the electrode - CaF_2 interface. The entire cell is contained in a vacuum tight Vycor

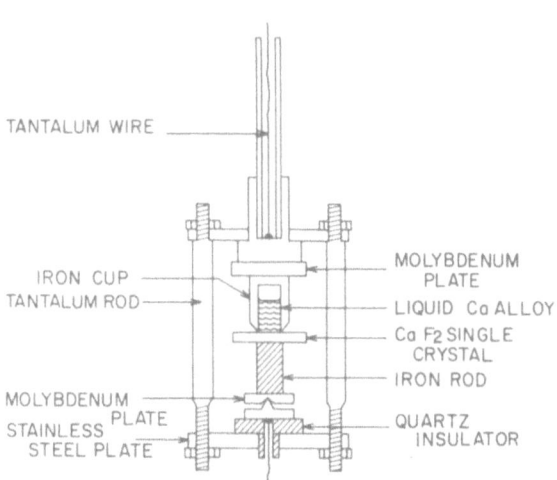

Fig. 1 Experimental arrangement for cell(I) and (II).

jacket lined with tantalum foil which is filled with purified argon. Cell(III) used essentially the same arrangement as shown in figure 1 except that two iron cups were used and the cell operated in a horizontal position.

2.3 Results and Discussion

Steady-state currents obtained at various applied voltages for cell(I) operated at 800°C are shown in Table I. Steady values were obtained only after several days of cell operation at a given voltage. Values of σ_e' were calculated from equation (4) and values of σ_e° from equation (2) with $a_{Ca} = 1$. Only currents measured on the plateau of the current voltage curve yielded consistent value of t_{ion}. Attempts to operate cell(I) at voltages lower than 0.30V proved unsuccessful. It is believed that the greater solubility of Ca in the crystal at lower voltages caused a density change and broke the seal between the alloy cup and the crystal thus shorting out the cell.

Results of cell(II) and (III) are shown in Table II at temperatures between 800°C and 950°C. Rearranging equation (3) and taking the logorithm of both sides gives

Table I

			Results from Cells of Type (I)			
E	i(μa)	G	σ'_e	σ_e°	t_{ion}°	a'_{Ca}
1.000	250	.302	8.17×10^{-4}	8.6×10^{-4}	.635	.90
.500	242	.302	7.94×10^{-4}	8.4×10^{-4}	.641	.90
1.000	344	.222	8.27×10^{-4}	8.7×10^{-4}	.633	.90
.500	330	.222	7.95×10^{-4}	8.4×10^{-4}	.641	.90

Table II

			Results from Experiments on Cells of Type (II) and (III)			
$T^{\circ}C$	E_{III}(volts)	σ_{ion}	σ''_e	a''_{Ca}	σ_e°	t_{ion}°
800	.7436	1.5×10^{-3}	1.72×10^{-7}	3.89×10^{-8}	8.72×10^{-4}	.632
825	.7391	2.45×10^{-3}	3.55×10^{-7}	5.95×10^{-8}	1.46×10^{-3}	.627
850	.7345	4.0×10^{-3}	7.1×10^{-7}	9.47×10^{-8}	2.31×10^{-3}	.634
875	.7300	6.1×10^{-3}	1.36×10^{-6}	1.42×10^{-7}	$3.61 \times 1-^{-3}$	.628
900	.7256	9.3×10^{-3}	2.5×10^{-6}	2.16×10^{-7}	5.38×10^{-3}	.634
950	(.7164)	2.1×10^{-2}	8.4×10^{-6}	4.59×10^{-7}	1.24×10^{-2}	.629

Results of cell(II) and (III) are shown in Table II at temperatures between 800°C and 950°C. Rearranging equation (3) and taking the logorithm of both sides gives

$$\log\left(i\frac{FG}{RT}\right) = \log \sigma_e'' + \log\left[\exp\left(\frac{EF}{RT}\right) - 1\right] \tag{12}$$

Therefore a plot of $\log i[\frac{FG}{RT}]$ versus $\log[\exp\frac{EF}{RT} - 1]$ gives a straight

line with a slope of one and an ordinate intercept at $\log \sigma_e''$. Such a plot is shown in figure 2. Values of σ_e'' were taken from the ordinate intercept. Cell(III) was operated between 800°C and 900°C and the values at 950°C represent an extrapolation. In Table II the values of E_{III} represent results on cell(III). Values of σ_{ion} are taken from the work of Ure [11], values of a_{Ca}'' are from equation (10) and (11) and σ_e° is calculated from equation (2) with $a_{Ca} = 1$. The values of σ_e° obtained from cell(I) and cell(II) agree quite well indicating that the electronic conductivity is proportional to $a_{Ca}^{\frac{1}{2}}$ over a large range of activities.

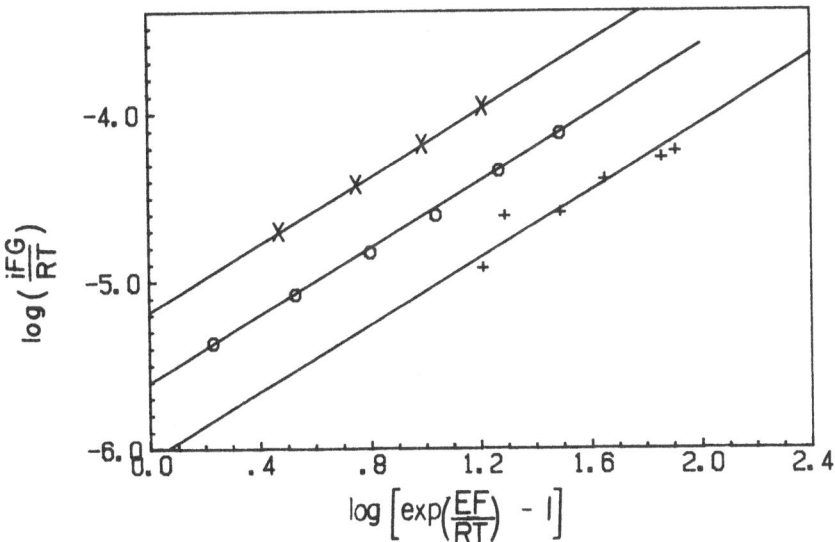

Fig. 2   Results of polarization measurements on cell(II)

Values of t_{ion} were calculated using equation (5) in combination with σ_{ion} from results of Ure. Figure 3 shows the transference number of ions in CaF_2 at various activities of Ca. It can be seen that CaF_2 can be used as a valuable electrolyte in galvanic cells where the activity of Ca is less than 10^{-3}. Thus it can be used for measurements on electropositive metals such as U, Th, Na, and Mg in contrast to solid oxide electrolytes.

378

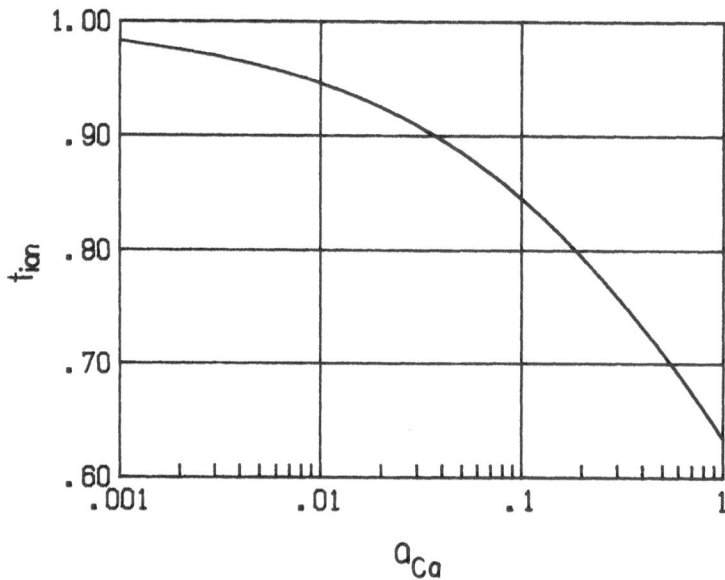

Fig. 3 The transference number of ions in CaF$_2$ at various Ca activities

3. Ca-Au ALLOYS

3.1 General Approach

The thermodynamic properties of solid Ca-Au alloys were studied by an emf technique using galvanic cells of the following type at 700°C

$$Ca-Sn(1)\,|\,CaF_2(s)\,|\,Ca-Au(s) \qquad\qquad\qquad (IV)$$
$$x_{Ca}=0.10$$

In cell (IV) the relative partial molar Gibbs energy of Ca in Ca-Au is given by:

$$G_{Ca}^M = G_{Ca}^M(ref.) - 2FE_{IV} \qquad\qquad\qquad (13)$$

where G_{Ca}^M (ref.) is the partial molar Gibbs energy of Ca in the Ca-Sn reference electrode. This has been determined in a separate study [12] and is equal to -132 kJ/mole.

The composition of the alloy phase was controlled by the method of coulometric titration first employed by Wagner [13] to study solid Ag_2S. In this method current is passed thru the cell under constant potential or constant current conditions and the coulombs measured with a coulometer. Using Faraday's law the exact amount of Ca entering the Au is obtained.

Thin film electrodes (10 to 25 microns) are used to study solid alloy phases so that the electropositive species diffuses thru the host metal in a reasonable time (generally less than one half hour).

When several intermediate phases (intermetallic compounds) are present in the phase diagram, the emf vs. mol fraction curve obtained will be similar to that shown in figure 5. Several plateaus are separated by steep drops in the emf. The plateaus are two phase regions and the region of emf drop represents the intermetallic compound. Thus one is able to identify the compounds present in the binary system.

The Gibbs energy of formation may be obtained by using equation (1) in conjunction with equations (2) and (3).

$$G_{Ca}^E = G_{Ca}^M - RT \ln x_{Ca} \tag{2}$$

where G_{Ca}^E is the excess partial molar Gibbs energy of Ca and x_{Ca} is the mol fraction. Also

$$G^M = (1-x_{Ca}) \int_0^{x_{Na}} \frac{G_{Ca}^E}{(1-x_{Ca})^2} dx_{Ca} + RT \left[x_{Ca} \ln x_{Ca} + (1-x_{Ca}) \ln(1-x_{Ca}) \right] \tag{15}$$

G^M is the integral molar Gibbs energy and represents the Gibbs energy change for the reaction

$$xCa + (1-x) Au = Ca_x Au_{1-x}$$

where x is the mol fraction x_{Ca}. This is the Gibbs energy of formation of the compound per gram atom.

The range of homogeneity or deviation from exact stoichiometric composition of each compound may also be obtained. Since the coulometer is so accurate a measure of Ca added to the compound one obtains curves similar to those shown in figure 6 which clearly show the range of homogeneity. Further by applying models such as that of Wagner-Schottky [14] model to the data, reasonable speculations may be deduced concerning the type and degree of disorder in each compound. This subject will be treated in section 3.4.

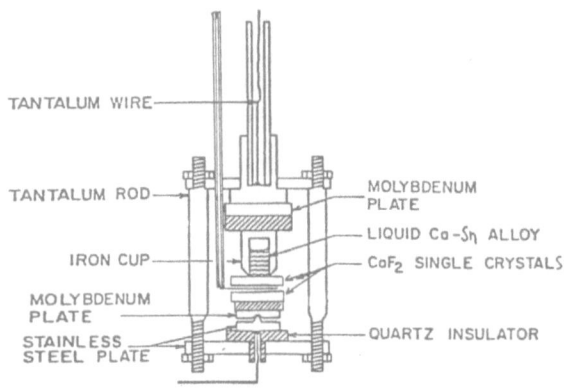

TANTALUM WIRE

TANTALUM ROD

IRON CUP

MOLYBDENUM
PLATE

STAINLESS
STEEL PLATE

MOLYBDENUM
PLATE

LIQUID Ca-Sn ALLOY

CaF$_2$ SINGLE CRYSTALS

QUARTZ INSULATOR

Fig. 4 Experimental arrangement for Ca-Au alloy cells.

3.2 Experimental Details

The experimental arrangement of the cell to study Ca-Au alloys is
shown in figure 4. Single crystals of CaF$_2$ (Harshaw) approximately
0.5 cm thick and 2.5 cm diameter were used. The Ca-Sn reference
electrode (x_{Ca} = 0.10) was contained in a molybdenum cup. A thin
gold foil (.0025 cm) was sandwitched between the CaF$_2$ electrolyte and
another CaF$_2$ crystal. A small molybdenum electrode lead contacted the
gold foil. The seals formed between the Ca-Sn alloy and the crystal
were obtained in the same way as the cell described in section 2.

The coulometric titrations were carried out with the help of a
Princeton Applied Research digital coulometer and potentiostat. This
allowed Ca to be added or removed from the alloy sample under constant
current or constant potential conditions. Usually emf vs. mol
fraction curves were obtained with addition of Ca metal but the same
results were obtained by removal of Ca.

Equilibrium was obtained between additions of Ca in less than one
half hour when only a single phase was present. When two phases were
present and phase changes were occurring, the equilibrium times were
sometimes considerably longer.

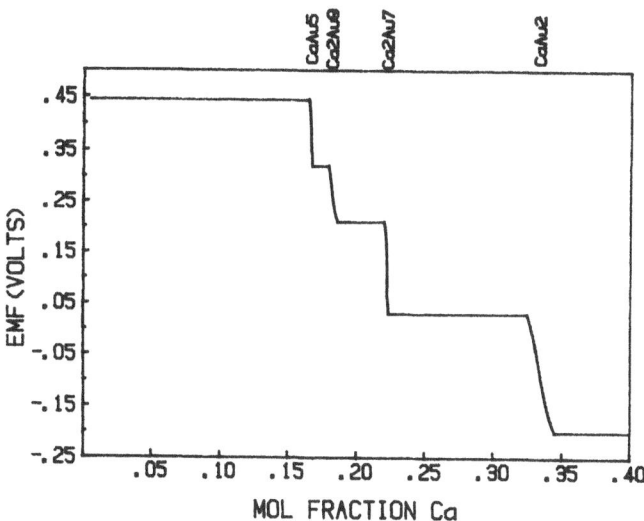

Fig. 5 Titration curve of solid Ca-Au alloys.

3.3 Results

The Ca-Au system was studied between x_{Ca} = 0 and x_{Ca} = .40.
Higher compositions of Ca were inaccessable due to electronic
conduction in the electrolyte. Results are shown in figure 5 and
associate thermodynamic quantities listed in Table III using equations
(13), (14), and (15). Four compounds were found, namely; $CaAu_5$,
Ca_2Au_9, Ca_2Au_7 and $CaAu_2$. These findings differ from the only known
phase diagram published by Weibke [15]. He lists three compounds at
$CaAu_4$, $CaAu_3$, and $CaAu2$. The titration curves for the four compounds
found in this work are shown in figure 6. The range of homogeneity is
shown in Table IV.

Table III

Thermodynamic Properties of Ca-Au Intermetallic
Compounds at 700°C

Compound	x_{Ca}	G_{Ca}^M, kJ/g · atom	G^M, kJ/g · atom
$CaAu_5$	0.1667	−224.0	−37.11
Ca_2Au_9	0.1818	−200.3	−40.08
Ca_2Au_7	0.2222	−179.5	−46.94
$CaAu_2$	0.3333	−146.4	−60.79

Therefore the coulometric titration gives the compounds present in the phase diagram, their thermodynamic properties, and their range of homogeneity.

Table IV

Range of Homogeneity for Ca-Au Compounds

Compound	Δx_{Ca}	$\Delta \delta$
CaAu$_5$	.1655 to .1673	-.008 to .0048
Ca$_2$Au$_9$	.1809 to .1835	-.012 to .022
Ca$_2$Au$_7$	.2214 to .2226	-.010 to .004
CaAu$_2$	.3277 to .3410	-.025 to .035

Table V

Type and Degree of Disorder in Ca-Au Compounds

Compound	Type of Disorder	Degree of Disorder
CaAu$_5$	Substitutional	$x^{o}_{Ca_{Au}} = .0011$
Ca$_2$Au$_9$	Substitutional	$x^{o}_{Ca_{Au}} = .0035$
Ca$_2$Au$_7$	Substitutional	$x^{o}_{Ca_{Au}} = .0004$
CaAu$_2$	Au-Interstitial	$x^{o}_{Au_i} = .0053$

3.4 Discussion and Application of Wagner-Schottky Equations

The shapes of the curves shown in figure 6 may be analyzed using the equations of Wagner and Schottky [14] concerning the type and degree of disorder in intermediate phases. These authors treat three types of disorder, namely; substitutional with Ca atoms on Au sites and Au atoms on Ca sites, Ca atoms on interstitial sites and vacancies formed on the Ca sublattice, and lastly Au atoms on interstitial sites and vacancies formed on the Au sublattice. The formulas for these three types of disorder are given here. The equation for substitutional disorder is:

$$\delta = 2x^{\circ}_{Ca_{Au}} \left(\frac{\nu_{Ca}+\nu_{Au}}{\nu_{Au}} \right) \sinh \left[\frac{\nu_{Ca}+\nu_{Au}}{\nu_{Au}} (\varepsilon^{\circ}-\varepsilon) \right] \qquad (16)$$

where δ has been defined in the previous section, ν_i is the stoichiometric number and $\varepsilon = 2FE/RT$ with E being the emf. $x^{\circ}_{Ca_{Au}}$

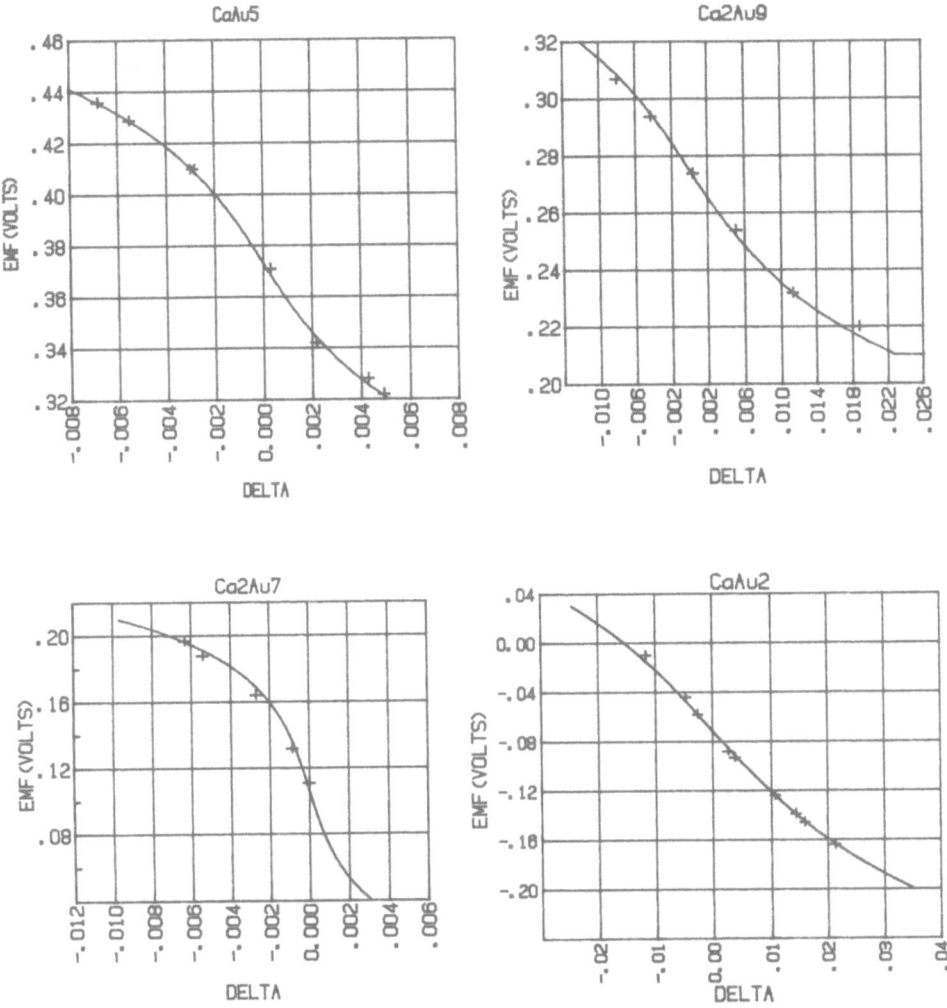

Fig. 6 Titration curves for intermetallic compounds of Ca and Au.

degree of disorder or number of Ca atoms on Au sites per total number
of sites at exact stoichiometry. The equation for Ca on interstitial
sites and vacancies in Ca sublattice is

$$\delta = 2x^{\circ}_{Ca_i} \sinh (\varepsilon^{\circ}-\varepsilon) \tag{17}$$

where $x^{\circ}_{Ca_i}$ is the degree of disorder i.e. the number of Ca atoms
on interstitial sites per total number of sites at exact stoichiometry. For Au atoms on interstitial sites and vacancies in the Au
sublattice the equation is

$$\delta = \left[\frac{\nu_{Ca}}{\nu_{Au}}\right] 2x_{Au_i}^{\bullet} \; \sinh\left[\frac{\nu_{Ca}}{\nu_{Au}} \; (\varepsilon^{\circ}-\varepsilon)\right] \tag{18}$$

when x_{Au_i} is the degree of disorder i.e. the number of Au atoms on interstitial sites or number of Au vacancies per total number of sites at exact stoichiometry. The equations for these three types of disorder have been put in a simpler form from the original published equations by the use of hyperbolic sines.

The results shown in figure 6 were fit to the equations either by methods described by Wagner[16] or by a general least squares computer technique [17]. Since the three equations each yielded a different shape curve, the best fit indicated the type and degree of disorder. The curves drawn through the points in figure 6 are given by the appropriate equation with the appropriate constants.

These conclusions on the type of disorder are only meant to be suggestive. Further corroborative evidence is needed. Also electronic disorder was not considered since conductivity measurements were not made as a function of Ca content of the intermediate phases. Certainly further measurements are necessary to be sure of the type of disorder in these compounds.

4. THE STUDY OF SOME LIQUID Ca ALLOYS UNDER CONDITIONS WHERE THE ELECTROLYTE EXHIBITS PARTIAL ELECTRONIC CONDUCTION

The Gibbs energies of mixing of liquid Ca-Zn alloys were determined at 800°C using cells of the following type

$$\text{Ca-Zn(1)} \,|\, \text{CaF}_2(s) \,|\, \text{Ca-Bi(1)} \tag{V}$$
$$x_{Ca} = 0.10$$

where the Ca-Bi ($x_{Ca}=0.10$) was used as a reference electrode of known activity. This is similar to cell(III) described in the polarization measurement section. The emf is given by equations (4-7) since some Ca-Zn alloys have an activity greater than 10^{-3} and the electrolyte has some electronic conductivity. Equation 7 may be solved for a'_{Ca}, the activity of the left hand electrode. Since the reference electrode has a very low activity of Ca (see following) $\sigma''_e = 0$. Solving equation (7) for a'_{Ca} one obtains

$$a_{Ca}^{'\,\frac{1}{2}} = \frac{\sigma_{ion} \, \exp\left(\frac{nF}{RT}\right)}{1-\sigma_e^{\bullet} \, \exp\left(\frac{nF}{RT}\right)} \tag{19}$$

where

$$\eta = E - \frac{RT}{F}\ln \sigma_{ion} + \frac{RT}{2F}\ln a''_{Ca} \tag{20}$$

Here a''_{Ca} is the activity of Ca in the right hand electrode.

The Ca activity in the reference electrode was determined by operating the cell

$$\text{Ca-Bi(1)} \mid \text{CaF}_2\text{(s)} \mid \text{Ca-Bi(1)} \qquad\qquad \text{(VI)}$$
$$x_{Ca}=0.25 \qquad\qquad x_{Ca}=0.10$$

which gave E=0.082 at 800°C. The value of the activity of Ca in Ca-Bi (x_{Ca}=0.25) was determined in conjunction with polarization experiments and is described in section 2. From Table II the value is $a_{Ca}=3.89\times10^{-8}$. At such low activities of Ca, electronic conduction in cell(VI) is neglible so according to the Nernst formula

$$a_{Ca}(x_{Ca}=0.10)=6.6\times10^{-9}$$

The experimental arrangement for cell(V) and (VI) was similar to those in figures 1 and 4 except that two tantalum cups were used to hold the liquid alloys and the cells were operated horizontally.

Results are listed in Table (VI) and equations (19), (20), as well as equations (14) and (15) were used to obtain the thermodynamic quantities.

The thermodynamic values were compared with measurements of Chiotti and Hecht [18] who studied the Ca-Zn system by vapor pressure

Table VI.

Some Thermodynamic Properties of Ca-Zn Alloys at 800°C

x_{Ca}	E[v]	η	a_{Ca}	$-G^M_{Ca}$,J	$-G^E_{Ca}$,J	$-G^M_{Zn}$,J
0.10	0.480	0.210	2.15×10^{-4}	75,380	57,580	1860
0.20	0.585	0.315	2.16×10^{-3}	54,720	40,370	5140
0.30	0.672	0.402	1.54×10^{-2}	37,210	26,455	9665
0.40	0.736	0.466	7.17×10^{-2}	23,500	15,340	14,790
0.50	0.764	0.494	1.47×10^{-1}	17,900	10,900	19,470
0.60	0.786	0.516	2.72×10^{-1}	11,620	7060	24,300
0.70	0.803	0.533	4.39×10^{-1}	7340	4160	29,540
0.80	0.816	0.546	6.56×10^{-1}	3760	1750	36,640
0.90	0.824	0.554	8.50×10^{-1}	1340	420	44,100

$\sigma_{ion}=1.5\times10^{-3}$
$\sigma°_e=8.72\times10^{-4}$
$a_{Ca}(\text{Ca-Bi})=6.6\times10^{-9}$

measurements. These are shown in figure 7. The comparison was very good so that it was concluded that reliable measurements could be made using CaF$_2$ with partial electronic conduction present.

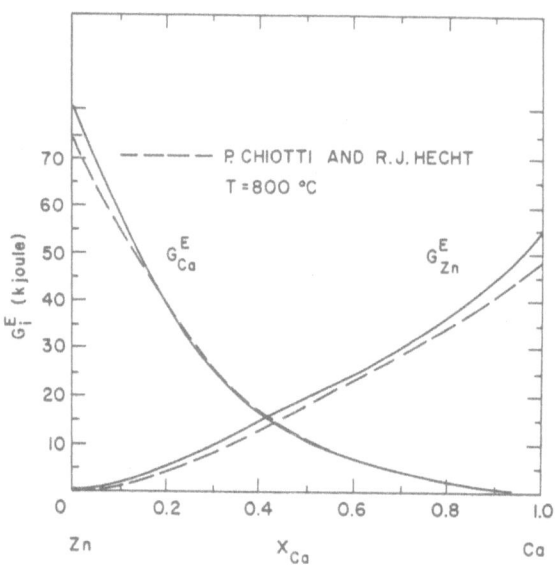

Fig. 7 Thermodynamic results for liquid Ca-Zn alloys.

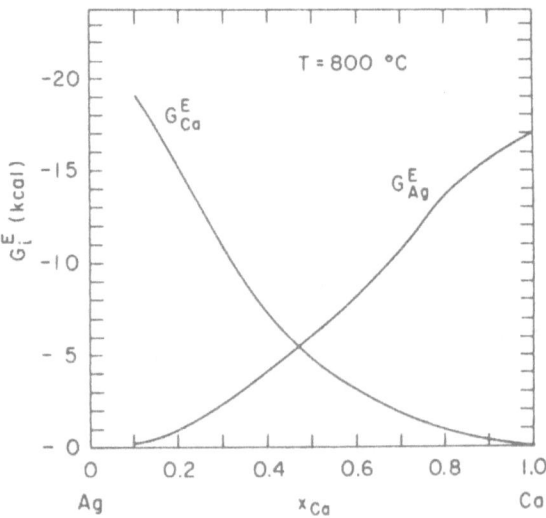

Fig. 8 Thermodynamic results for liquid Ca-Ag alloys

Since the technique was successful with Ca-Zn alloys whose thermodynamic properties were known, two other unknown systems Ca-Ag and Ca-In were studied in the liquid state. Results are presented in figures 8 and 9. This section is a review of work previously performed by Delect and Egan [19-20].

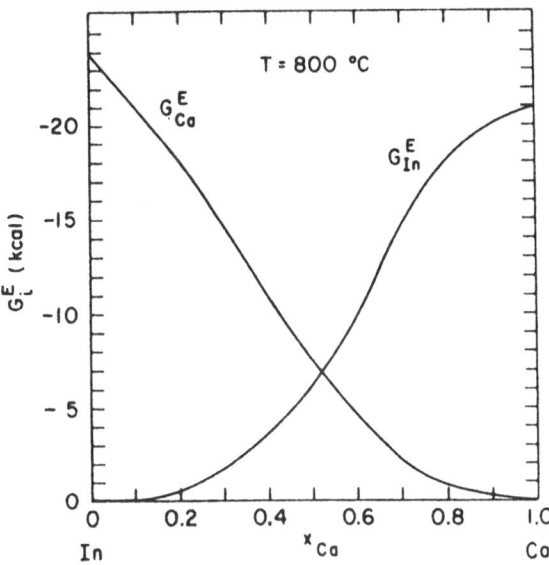

Fig. 9   Thermodynamic results for liquid Ca-In alloys.

5.   IONIC ALLOYS

5.1   General Approach

The coulometric titration method was used to study liquid alloys of Na-Bi and K-Bi and solid alloys of Na-Bi and Na-Sb. Alloys of Na-Bi and Na-Sb were studied using cells of the following type:

$$Ca\text{-}Sn(1)\,|\,CaF_2(s)\,|\,NaF,\ Na\text{-}Bi(1\ or\ s) \qquad\qquad (VII)$$
$$x_{Ca}=0.10$$

$$Ca\text{-}Sn(1)\,|\,CaF_2(s)\,|\,NaF(s),Na(1) \qquad\qquad\qquad (VIII)$$
$$x_{Ca}=0.10$$

where the relative partial molar Gibbs energy is given by

$$G_{Na}^{M} = -F(E_{VII} - E_{VIII}) \qquad\qquad\qquad (21)$$

Operating two cells instead of one was found useful for experimental reasons since it was difficult to use Na electrodes as references. In cell (VIII) the Na was generated in situ by passing current through the cell. Alloys of K-Bi were studied using the following cells:

$$Ca\text{-}Sn(1)\,|\,CaF_2(s)\,|\,CaF_2\cdot KF,\ K\text{-}Bi(1) \qquad\qquad (IX)$$
$$X_{Ca}=0.10$$

$$Ca\text{-}Sn(1)\left|CaF_2(s)\right|CaF_2 \cdot KF, \; K(1) \tag{X}$$
$$X_{Ca}=0.10$$

In the CaF_2-KF system a double salt is formed that is not present in the CaF_2-NaF system. This necessitated the use of the double salt in these cells. Precautions were taken to ensure that the activity of KF did not change during the course of the titration. Again

$$G_K^M = -F(E_{IX} - E_X) \tag{22}$$

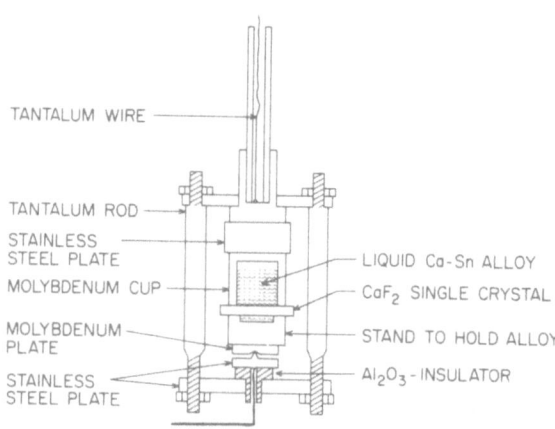

Fig. 10 Experimental arrangement of cells to study ionic alloys.

5.2 Experimental Details

The experimental set-up for studying sodium and potassium alloys is shown in figure 10. The Ca-Sn reference electrode (10 at % Ca) was contained in a molybdenum cup. The alloy under study was contained in the small compartment of molybdenum as shown. This compartment also contained a small amount of appropriate fluoride salt. In the case of sodium alloys this was NaF and for potassium alloys the double salt $CaF_2 \cdot KF$ was used. The quantity of material composing the alloy (5-10 mg) was very much smaller than that contained in the reference electrode so that during the course of a coulometric titration the composition of the reference electrode did not change. Thin films of Bi and Sb were evaporated onto the molybdenum alloy holder in preparation for experiments.

Table VII
Thermodynamic Properties of Na-Sb Alloys at 800°C

x_{Na}	emf, V	G_{Na}^M kJ/mol	G_{Na}^E kJ/mol	a_{Na}	γ_{Na}
.300	.657	-63.39	-53.15	.00082	.00273
.400	.585	-58.44	-48.65	.00179	.00447
.500	.504	-48.63	-42.73	.00429	.00858
.600	.435	-41.97	-37.63	.00905	.01509
.681	.380	-36.66	-33.40	.016	.024
.748	.352	-33.96	-31.49	.022	.030
.750	.290	-27.98	-25.53	.043	.058
.752	.156	-15.05	-12.63	.185	.246
.754	.080	-7.72	-5.32	.421	.558
.755	.039	-3.76	-1.37	.658	.869

5.3 Results

Na-Sb alloys were studied at 800°C by coulometric titration. The
phase diagram shows that only compositions near $x_{Na}=0.75$ are solid
at this temperature. The thermodynamic properties are shown in Table
VII with results for $x_{Na}<.70$ representing liquid alloys. The emf
values listed are for $E_{VII} - E_{VIII}$ where $E_{VIII}=0.110V$. The
range of homogeneity extends from $x_{Na}=0.748$ to $x_{Na}=0.755$. The
Gibbs energy of formation for 1 gm. atm. of Na_3Sb was calculated to be
-32kJ using equation (15).

Na-Bi alloys were also studied at 800°C where Na_3Bi is solid.
Thermodynamic value are shown in Table (VIII) where values for
$x_{Na}<0.70$ are liquid alloys. Figure 12 shows the titration curve for
the solid. The Gibbs energy of formation is -28kJ per gm. atm. A
titration curve from Pfeifer et al. [26] for the liquid at 850°C is
shown in figure 13 for comparison.

Liquid K-Bi alloys were studied at 750°C and results are shown in
Table (IX) and figure 14.

5.4 Discussion

The results presented here for these ionic alloys can be explained
quantitatively by Wagner's model of electronic disorder. This model
has been presented in some detail in the literature [16, 21-25] so
that only pertinent equations will be given here. The general idea is
that the thermodynamics is influenced and often controlled by changes
in the concentration of electrons and electron holes with
concentration. Equations may be derived for the change in activity

with compositon of the alloy and for changes in conductivity with composition.

Table VIII
Thermodynamic Properties of Na-Bi alloys at 800°C

x_{Na}	emf, V	G_{Na}^M kJ/mol	G_{Na}^E kJ/mol	a_{Na}	γ_{Na}
.100	.771	-74.39	-54.80	.00024	.00239
.200	.670	-64.64	-50.96	.00071	.00356
.300	.589	-56.83	-46.59	.00171	.00571
.400	.514	-49.59	-41.80	.00385	.00963
.500	.444	-42.84	-36.94	.008	.016
.600	.377	-36.37	-32.03	.017	.028
.700	.309	-29.81	-26.79	.035	.051
.720	.289	-27.88	-25.09	.044	.061
.740	.260	-25.09	-22.52	.060	.081
.750	.196	-18.91	-16.46	.120	.160
.760	.115	-11.10	-8.76	.288	.379
.770	.058	-5.60	-3.37	.534	.694
.780	.038	-3.67	-1.55	.663	.850

Table IX
Thermodynamic Properties of Liquid K-Bi Alloys at 750°C

x_{Ka}	emf, V	G_K^M kJ/mol	G_K^E kJ/mol	a_K	γ_K
.300	.679	-65.51	-55.27	.00045	.00151
.400	.560	-54.03	-46.24	.00174	.00436
.500	.455	-43.90	-38.00	.00573	.01147
.600	.347	-33.48	-29.13	.01952	.03253
.680	.267	-25.76	-22.48	.048	.071
.700	.245	-23.64	-20.60	.062	.089
.710	.231	-22.29	-19.37	.079	.102
.720	.215	-20.74	-17.95	.087	.121
.730	.196	-18.91	-16.23	.108	.148
.740	.174	-16.79	-14.23	.139	.188
.751	.142	-13.70	-11.26	.200	.286
.760	.117	-11.29	-8.95	.265	.349

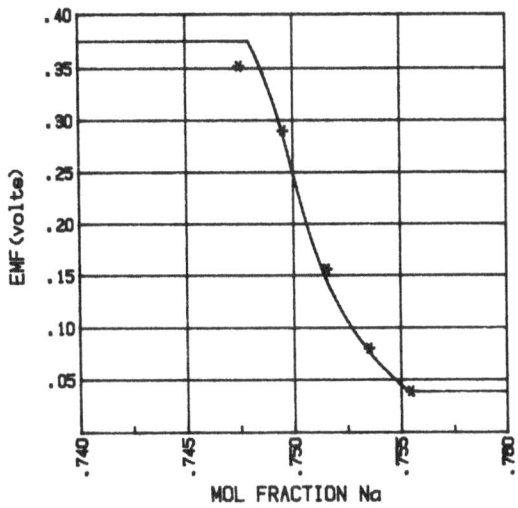

Fig. 11   Titration curve for solid Na₃Sb at 800°C.

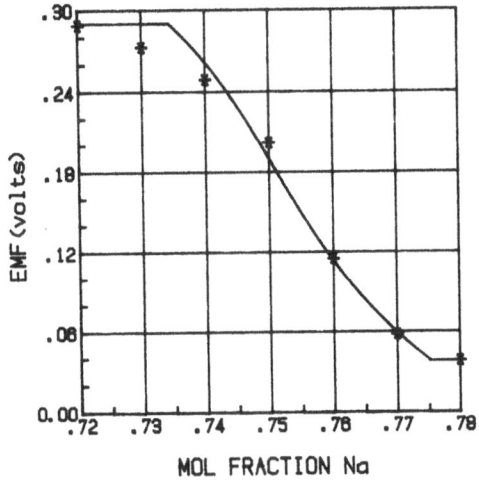

Fig. 12.   Titration curve for solid Na₃Bi at 800°C.

According to Wagner, for solid compounds such as Na_3Bi and Na_3Sb when electrons and electron holes are the minority defects

$$\delta = 2x_e^\circ \sinh\,[\ln a_{Na}/a_{Na}^\circ] = 2x_e^\circ \sinh\,[(E^\circ-E)F/RT] \tag{23}$$

where x_e° is the ratio of the number of electrons to the number of Bi or Sb atoms at exact stoichiometry, a_{Na}° is the activity of Na at exact stoichiometry, and E° the emf at $x_{Na}=0.75$. Thus the emf is given as a function of compositon. Methods to test the data for conformity to equation (23) and determine x_e° are described by Wagner [16].

Comparisons are made in figures 11 and 12. The agreement is very good except for a point on the Na_3Bi curve which is believed to caused by slow equilibration. The value for x_e° in Na_3Sb is 0.009 and for Na_3Bi it is 0.09. the model likewise holds for liquids as seen in figures 13 and 14. Here the values for x_e° are 0.175 for $Na_3Bi(l)$ and 0.255 for $K_3Bi(l)$.

The model also predicts the shape of the electrical conductivity curve with varying composition. The equations are

$$\sigma = \sigma_{min}\,\cosh[(E_{min}-E)F/RT] \tag{24}$$

where

$$\sigma_{min} = 2x_e^\circ(u_e u_h)^{\frac{1}{2}}F/V_m \tag{25}$$

and

$$u_e/u_h = \exp[(E_{min}-E^\circ)2F/RT] \tag{26}$$

σ_{min} is the minimum conductivity, E_{min} is the potential of minimum conductivity, u_e and u_h are the mobilities of electrons and electron holes which are assumed to be independent of concentration.

The conductivities of liquid Na-Bi [26] and K-Bi [27] alloys have been measured and are shown in figure (15) and (16) compared with calculations according to equation (24). Again the comparisons are excellent showing the validity of this model for these liquid semiconductors. No conductivity data was available for solid Na_3Sb and Na_3Bi.

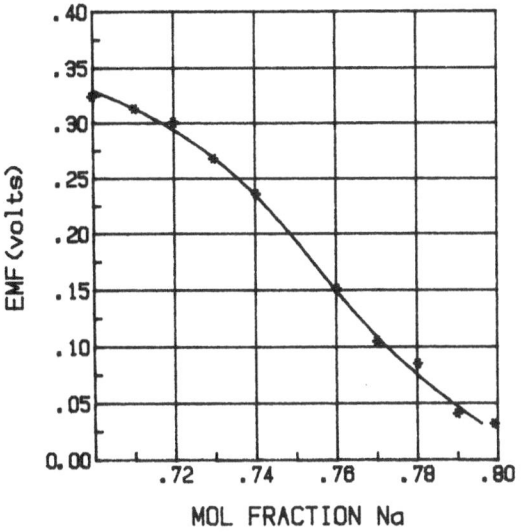

Fig. 13 Titration curve for liquid Na-Bi alloys at 850°C.

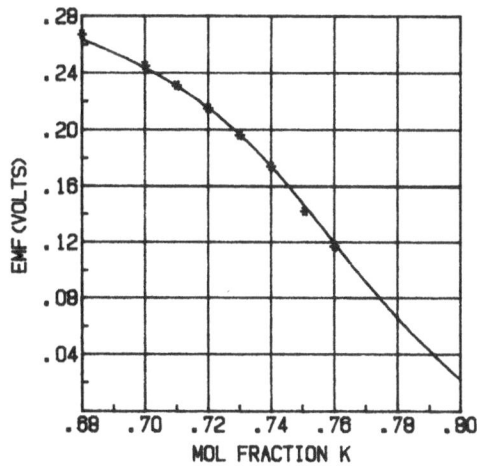

Fig. 14. Titration cruve for liquid K-Bi alloys at 750°C.

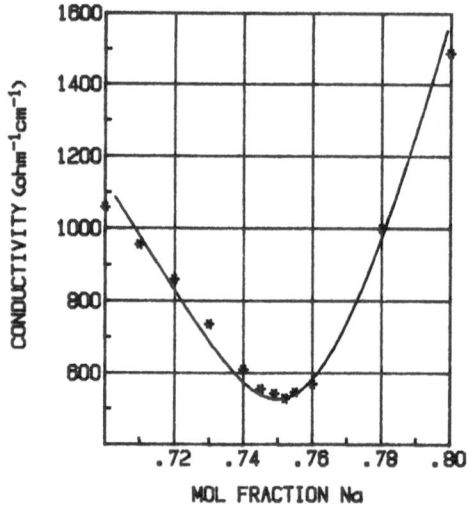

Fig. 15 Electrical conductivity of liquid Na-Bi alloys at 850°C.

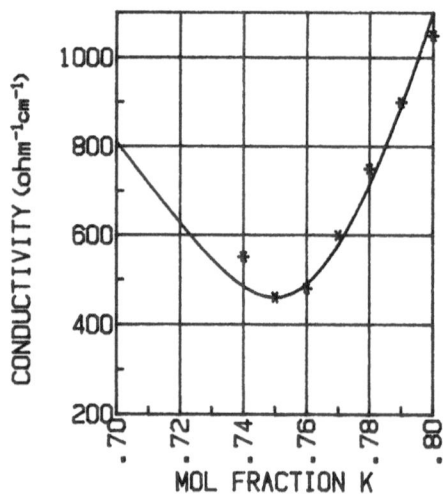

Fig. 16 Electrical conductivty of liquid K-Bi alloys.

6. THE STUDY OF SODIUM DISSOLVED IN MOLTEN NaBr

The thermodynamics of mixtures of alkali metals in their molten halides may be studied with emf cells using CaF_2 solid electrolytes. The following cells were employed to study Na-NaBr at 800°C:

$$Ca-Sn(1)|CaF_2(s)|NaF(s)-NaBr(1) \qquad (XI)$$
$$x_{Ca}=0.10$$

$$Ca-Sn(1)|CaF_2(s)|NaF(s), Na(1) \qquad (XII)$$
$$x_{Ca}=0.10$$

where the NaF and NaBr are in separate compartments joined together through the gas phase. Again

$$G^M_{Na} = -F(e_{XI} - E_{XII}) \qquad (27)$$

The composition of the Na-NaBr mixture may be controlled by coulometric titration. Sodium is formed from NaF and in turn is transferred through the gas phase into the NaBr. Activity and activity coefficients are obtained from the usual equations e.g.

$$G^M_{Na} = RT\ln a_{Na} = RT\ln x_{Na}\gamma_{Na} \qquad (28)$$

where a_{Na} is the activity coefficient and x_{Na} is the mol fraction of Na in NaBr. The experimental arrangement is shown in figure 17 which shows how the NaF and molten NaBr are separated. The seal formed between CaF_2 and the holder for NaF and NaBr is very tight so that no Na vapor escapes from the system.

Results on the activity of Na in NaBr as a function of deviations from exact stoichiometry are shown in figure 18. Also shown on the curve are vapor pressure results of Smirnov et al. [28]. The agreement is very good showing the validity of the method. Further details concerning mixture of molten Na-NaBr are given by Egan and Freyland [29].

Since Na vapor may be generated in known quantities by coulometric titrations this technique was used to study the electrical conductivity of Na-NaBr and Na-NaI mixtures [30]. The cell employed for these measurements is shown in figure 19.

ACKNOWLEDGMENT

The author would like to thank the following people with whom he had the pleasure to have worked with on these problems. R. J. Heus, J. Delcet, H.-P. Pfeifer, S. Rowley, H. Heyer, and W. Freyland. The financial support for this work was supplied by the Division of Chemical Sciences, U.S. Department of Energy, Washington DC, under contract No. DE-AC02-76CH00016.

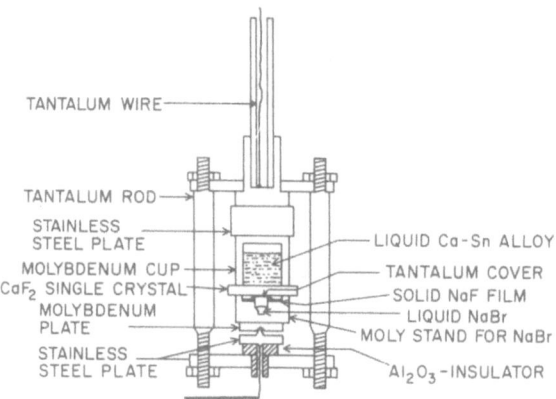

Fig. 17 Experimental arrangement for cells to study Na-NaBr.

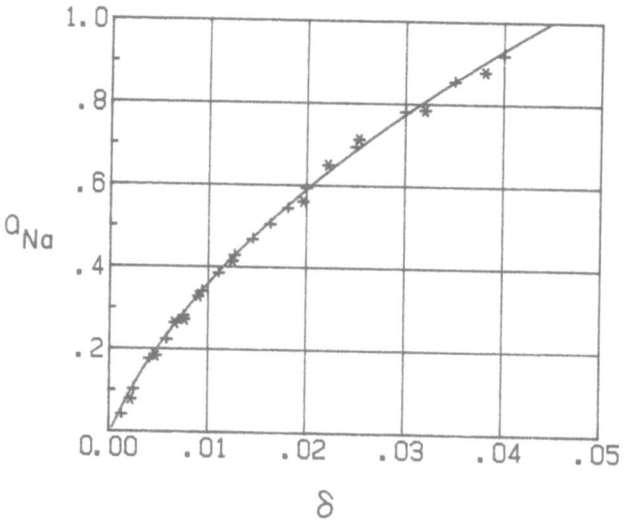

Fig. 18 Activity of sodium in NaBr.

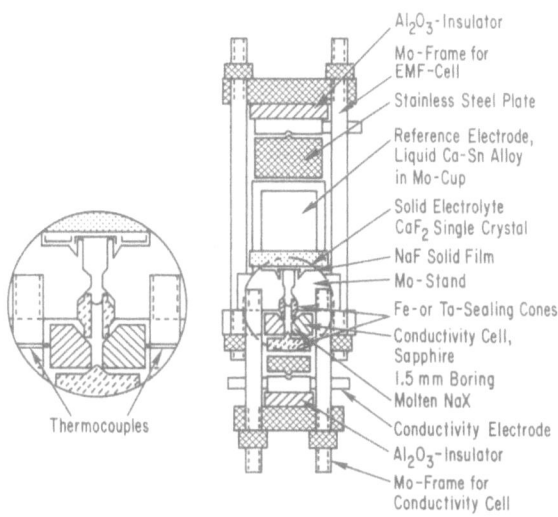

Fig. 19 Experimental arrangement to study electrical conductivity in
Na-NaBr mixtures.

References

1. C. Wagner, J. Electrochem. Soc. 115, 933 (1968).
2. E. Mollwo, Nachr, Gesellsch. Wissennack. Göttingen Matk. Physik.
    Kl. N.F., 6, 79 (1934).
3. J. W. Hinze and J. W. Patterson, J. Electrochem. Soc., 120, 96
    (1973).
4. J. W. Hinze and J. W. Patterson, J. Electrochem. Soc., 120, 1792
    (1973).
5. J. Delcet, R. J. Heus, and J. J. Egan, J. Electrochem. Soc.,
    125, 755 (1978).
6. W. Baukal, Ber. Sunsenges. Phys. Chem., 79, 1148 (1975).
7. C. Wagner, Proc. CITCE, 7, 361 (1957).
8. F. A. Kröger, The Chemistry of Imperfect Crystals, Vol. 2, 2nd
    edition (North Holland Publ. Comp., Amsterdam) 1974.
9. C. Wagner, Z. Phys. Chem., B21, 25 (1933).
10. R. M. Barrer, J. A. Barrie, and M. G. Rogers, Trans. Faraday
    Soc., 58, 2473 (1962).
11. R. W. Ure Jr., J. Chem. Phys., 26, 1363 (1957).
12. J. Delcet, A. Delgado-Brune, and J. J. Egan, in Calculation of
    Phase Diagrams and Thermochemistry of Alloy Phases, The
    Metallurgical Society of AIME, Warrendale, PA, 1979, p. 339.
13. C. Wagner, J. Chem. Phys., 21, 1819 (1953).
14. C. Wagner and W. Schottky, Z. Phys. Chem., B11, 163 (1931).

15. F. Weibke and W. Bartels, Z. Anorgan. Chem., 218, 241 (1934).
16. C. Wagner, in 'Progress in Solid State Chemistry,' Vol. 6, H. Reiss and J. McCaldin, eds., Pergamon, New York, 1971, p. 1-15.
17. D. W. Marquardt, J. Soc. Indus. Appl. Math., 2, 431 (1963).
18. P. Chiotti and R. J. Hecht, Trans. TMS-AIME, 239, 536 (1967).
19. J. Delcet and J. J. Egan, Metall. Trans. B, 9B, 728 (1978).
20. J. Delcet and J. J. Egan, J. Less-Common Met., 59, 299 (1978).
21. J. B. Wagner and C. Wagner, J. Chem. Phys., 26, 1602 (1957).
22. V. Wehfritz, Z. Physik. Chem. N.F., 26, 339(1960).
23. N. Valverdi, Z. Physik. Chem. N.F. 61, 92 (1968).
24. N. Valverdi, Z. Physik. Chem. N.F. 70, 113 (1970).
25. H. Rickert, 'Electrochemie fester Stoffe, Springer, Berlin 1973.
26. H.-P. Pfeifer, R. J. Heus, and J. J. Egan, in <u>Chemical Metallurgy-A Tribute to Carl Wagner</u>, N. Goken, Ed., The Metallurgical Society of AIME, Warrendate, PA, 1981 p. 339.
27. W. Freyland and H.-P. Pfeifer, private communication.
28. M. V. Smirnov, V. V. Chebykin, and L. A. Tsiovkina, Electrochem. Acta, 26, 1275 (1981).
29. J. Egan and W. Freyland, Ber. Bunsengesell. Phys. Chem., 89, 381 (1985).
30. G. M. Haarberg, K. Osen, J. J. Egan, H. Heyer, and W. Freyland, to be published in Ber. Bunsengesell. Phys. Chem.

THERMODYNAMIC PROPERTIES OF LIQUID As - Cd - Zn ALLOYS

A. Mikula and K.L. Komarek
Institut für Anorganische Chemie, Universität Wien
Währingerstraße 42
A-1090 Vienna
Austria

ABSTRACT. Using an emf method with a liquid $KCl-LiCl-ZnCl_2$ electrolyte the activity of Zn in the ternary As-Cd-Zn system was determined at four cross sections with a constant Cd to Zn ratio of 9:1, 3:1, 2:1 and 1:1, resp. The activities show a crossover point in all four sections at approximately 30 at% As.

On some alloys the liquidus temperature was measured by DTA and the results agree well with the discontinous change in slope of the emf vs. temperature curves. In the measured temperature range up to 1123 K these curves in the liquid range are straight lines with constant slopes.

1. INTRODUCTION

During the last twenty years more and more evidence has been obtained, pointing to the existence of some sort of ordering in certain liquid alloys (1). The thermodynamic properties of such alloys exhibit a characteristic concentration and temperature dependence which can be attributed to the formation of so called clusters or associations.

Komarek and coworkers investigated the binary alloys of the group II B and V A elements using emf or calorimetric methods. The following systems were investigated: Bi-Zn (2), Bi-Cd (3), Cd-Sbs (4,5), Sb-Zn (6) and As-Cd (7).

In the Sb-Zn system (6) several effects on the thermodynamic properties due to the formation of clusters are listed e.g.

a_i - composition curves with a cross - over point,

H_i - composition curves of S-type with an inflection point, and

S^M - composition curves with a minimum.

H. Brodowsky and H.-J. Schaller (eds.), Thermochemistry of Alloys, 399–408.

These effects vary with temperature and they disappear at higher temperatures when the clusters dissociate.

Fig. 1 shows the entropy of mixing for the investigated alloys. No abnomal behavior of the entropy is found in the Bi-Cd and Bi-Zn systems but some tendency for compound formation can be observed in the Cd-Sb system and this effect is more pronounced in the Sb-Zn and As-Cd system. In this series of investigations the As-Zn system is missing which should show an even greater abnormal behavior. Because of the high melting point of the intermetallic compound As_2Zn_3 of 1288 K (8), and the high vapor pressure of both components it is very difficult to carry out thermodynamic measurements in the liquid state. Therefore no thermodynamic data of the liquid As-Zn system are reported in the literature.

Since emf measurements on the liquid As-Cd system were carried out at our institute we thought it would be of interest to investigate the ternary As-Cd-Zn system if such anomalies could also be detected in this system. An emf technique was applied to determine the thermodynamic properties of liquid zinc at five cross sections with a constant cadmium to zinc ratio.

2. EXPERIMENTAL PROCEDURE

The ternary alloys were prepared from high purity 5N metals (As from Johnson Matthey Chemicals Limited, UK and Cd and Zn from Ventron, Karlsruhe, FRG). The oxide layer of Cd and Zn was removed by melting the metals in vacuum and filtration of the melt through quartz wool. The As was used without further treatment.

The metals were weighed on an semi-micro balance and sealed into quartz capsules under vacuum, melted together at 1300 K and homogenized at this temperature for three days. The alloys were quenched in ice water, broken in small pieces, and approximately 3g of the samples were used for the emf measurements.

As electrolyte the eutectic mixture of KCl-LiCl with 0.5 mol% of $ZnCl_2$ was used. Since the measurements had to be carried out at temperatures above 1100 K great care was taken in the preparation of the electrolyte. For LiCl, KCl and $ZnCl_2$ p.A. Chemicals of the Fa. E. Merck (Darmstadt, FRG) were used. The electrolyte was purified by passing clorine gas through the molten salt until it had a yellow-greenish colour. The chlorine gas was purged by a highly purified argon gas, which was gettered over a Ti sponge, until the melt was again colorless. The liquid electrolyte was filled into glass ampules which were sealed under vauum.

For the measurements the following cell arrangement was used.

$$Zn(1)/Zn^{++}(KCl- LiCl)(1)/Alloy(1)$$

In Fig. 2 the cell for the emf measurements is shown. The cell, the electrode containers, the capillaries and the thermocouple protection tube were made of quartz. The tungsten electrode wire was sealed into the capillaries and the electrode containers were also sealed on to the capillaries.

Zinc for the pure electrode and the alloy samples were filled into the quartz containers which were sealed to minimize the evaporation of the metals. A small hole in these containers, just above the surface of the melt, allowed for contact with the electrolyte. In each experiment four electrodes, one of pure zinc were used. The electrodes could be raised or lowered through the O-ring seals in the brass head which sealed the top of the emf cell.

Before each experiment the cell was evacuated and flushed several times with pure argon and finally filled with argon at reduced pressure so that the argon pressure in the cell at the highest operating temperature was approximately two atmospheres. The cell was placed into a larger metal block in order to reach a uniform temperature in the electrolyte. This metal block was heated by a resistance furnace and the temperature was measured with a Ni/NiCr thermocouple.

At the beginning of each experiment the electrodes were pulled out of the furnace. After the electrolyte was molten and had reached a temperature of 900 K the electrodes were quickly submerged into the electrolyte to prevent evaporation. The surface of the electrolyte was approximately 8 cm above the electrodes.

Measurements of the emf were carried out on heating and cooling, the heating and cooling rate was 8 to 10 degrees per hour and the emf and the temperature was automaticaly recorded every five minutes. Every fifty degrees the temperature was kept constant for a longer period of time to check the stability of the emf. For each run two heating and two cooling curves were recorded and the agreement between two consecutive curves was good for alloys with an arsenic content less than 40 at% and at temperatures below 1073 K. At higher arsenic contents and higher temperatures the difference between two curves became as large as 5 to 10 mV.

Ten to fifteen hours after the start of the experiement a thin mirror-like metallic deposit began to appear on the cooler part of the emf cell. This deposit was zinc which evaporated from the pure electrode. At temperatures above 1123 K and at or high arsenic content evaporation of arsenic occurred too.

On several samples DTA measurements were carried out to determine the liquidus temperature. These measurements were performed with a commercial thermal analyzer (DTA 404S/3, Netzsch, FRG) using samples of about 0.5g sealed under vacuum in special quartz containers. The heating rate was 2 K/min.

3. EXPERIMENTAL RESULTS AND DISCUSSION

In the liquid ternary As-Cd-Zn alloy the emf vs. temperature
curves in the measured temperature range are straight lines with
constant slopes. In Fig. 3 some typical curves are shown. The
emf plotted in these graphs and also used for the calculation of
the activity of liquid zinc was taken in all cases from the
first cooling curve of all experiments.

Since the emf vs. temperature curves show a distinct break
at the liquidus temperature this temperature could be obtained
from the emf measurements. To check the accuracy of this
liquidus temperature DTA measurements on the same samples were
carried out and as it is shown in the graphs both temperatures
agree very well.

From the emf data the activity of liquid zinc was
calculated for two different temperatures. At all four cross
sections (Cd-Zn = 9:1, 3:1, 2:1, 2:1 and 1:1) which have already
been measured the activity curves exhibit a cross over point at
approximately 30 at% As. This means that the activity of liquid
zinc is higher at lower temperature at a low arsenic content and
lower at a higher arsenic content (Fig. 4a-d) The scatter of
the emf data in the ternary As-Cd-Zn system is much larger than
in the As-Cd system (7); due to the higher temperature at which
the measurements had to be carried out the evaporation of the
metals becomes larger. For certain alloys a non linear
temperatur dependence was found in the As-Cd system (7). This
could not be detected in the As-Cd-Zn system even at the cross
section Cd:Zn = 9:1. This might be explained, that in the As-Cd
system the emf of cadmium was measured whereas in the ternary
system the emf of zinc was determined. This might also explain
why the crossover points in both systems do not exactly
correlate. In order to draw some final conclusions of the
thermodynamic data and of the formation of cluster in this
system it is necessary to try to estimate some thermodynamic
values of the binary liquid As-Zn phase and then calculate the
integral quantities of the ternary phase.

4. REFERENCES

(1) K.L. Komarek, Ber. Bunsenges. 81 (1977) 936.

(2) E. Hayer, K.L. Komarek, and A. Mikula, Monatsh. Chem. 107 (1976) 1437.

(3) K.L. Komarek, and G. Stummerer, Monatsh.Chem. 104 (1973) 32.

(4) G. Schick and K.L. Komarek, Z. Metallk. 65 (1974) 112.

(5) R. Geffken, K.L. Komarek, and E. Miller, Trans.Met.Soc. AIME 239 (1967) 1151.

(6) I.B. Rubin, K.L. Komarek, and E. Miller, Z. Metallk. 65 (1974) 191.

(7) K.L. Komarek, A. Mikula, and E. Hayer, Ber.Bunsenges. 80 (1976) 765.

(8) M. Hansen and K. Anderko, Constitution of Binary Alloys. New York, Mc Graw-Hill 1958.

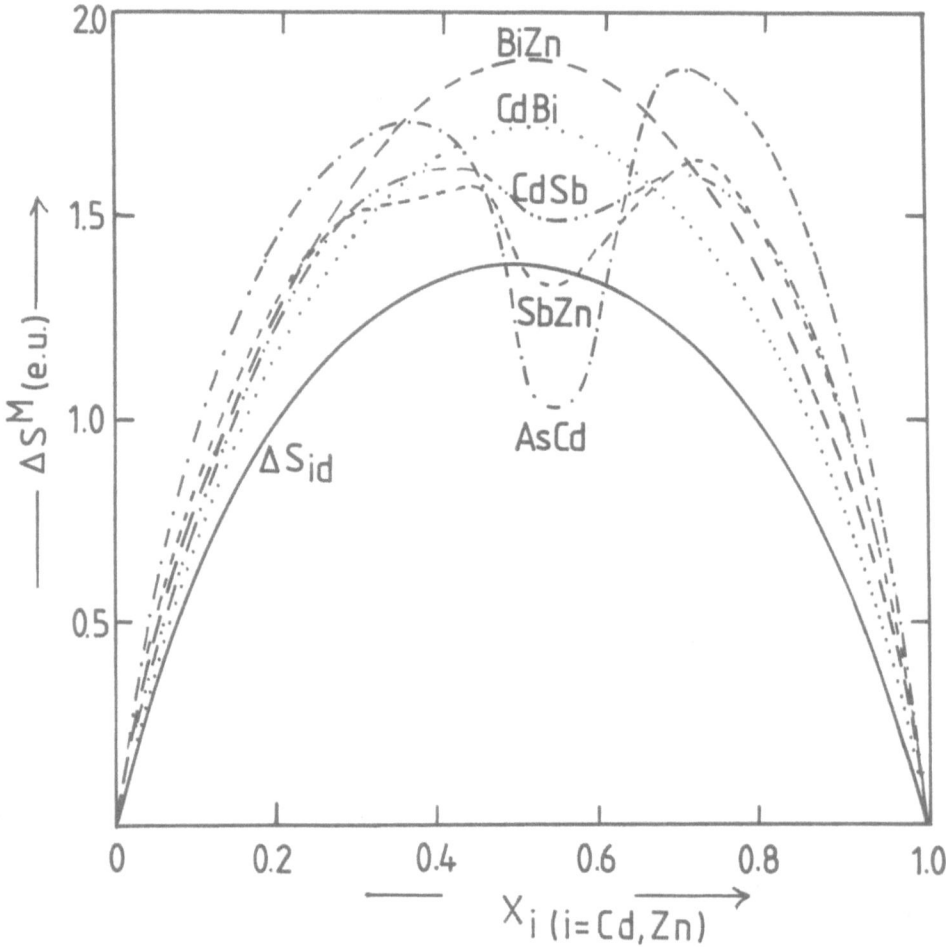

Fig. 1. Integral entropy of mixing of liquid alloys in the
systems As-Cd, Bi-Cd, Bi-Zn, Cd-Sb and Sb-Zn.

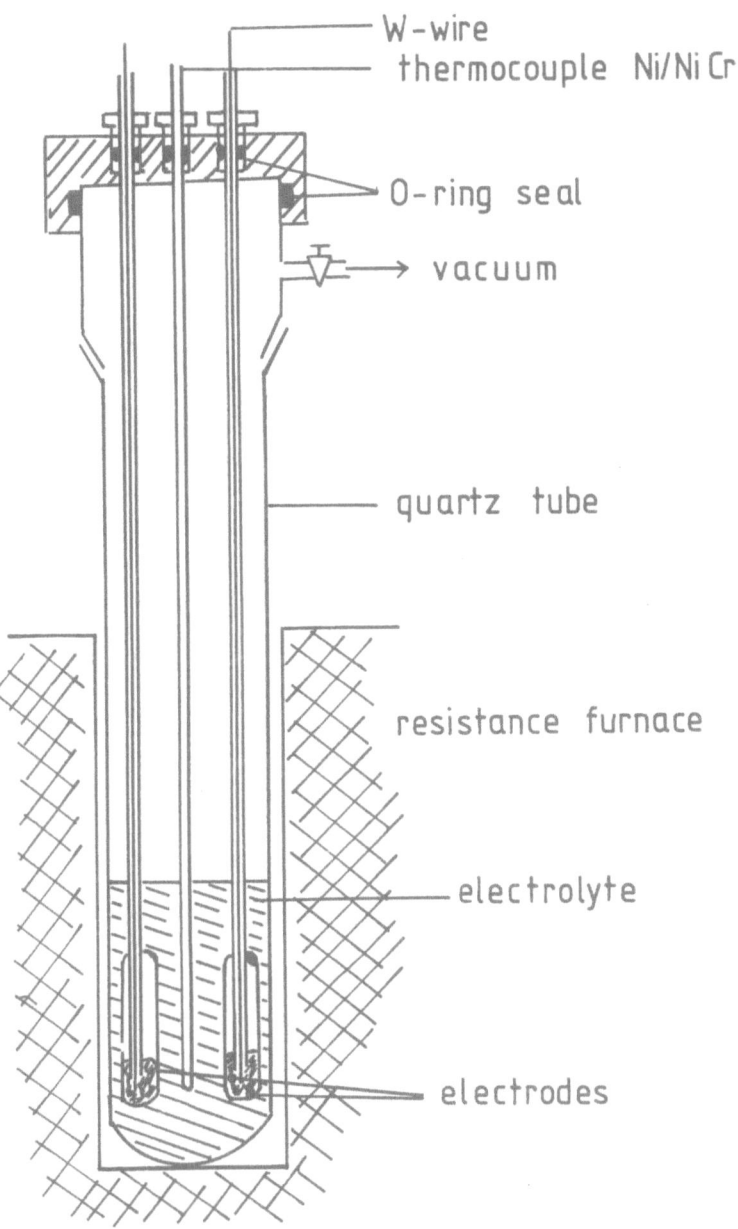

Fig. 2. Arrangement of the emf cell.

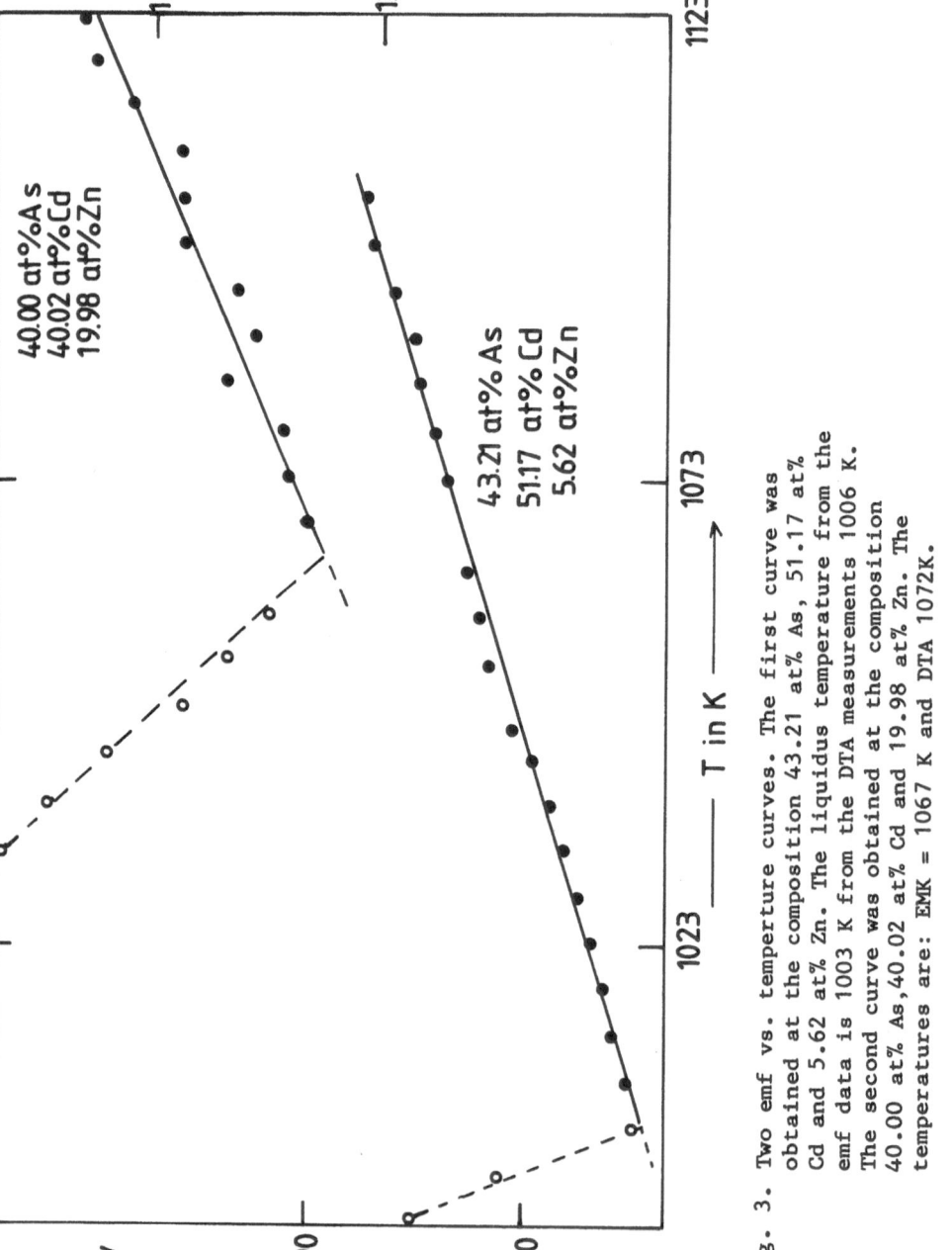

Fig. 3. Two emf vs. temperture curves. The first curve was obtained at the composition 43.21 at% As, 51.17 at% Cd and 5.62 at% Zn. The liquidus temperature from the emf data is 1003 K from the DTA measurements 1006 K. The second curve was obtained at the composition 40.00 at% As, 40.02 at% Cd and 19.98 at% Zn. The temperatures are: EMK = 1067 K and DTA 1072K.

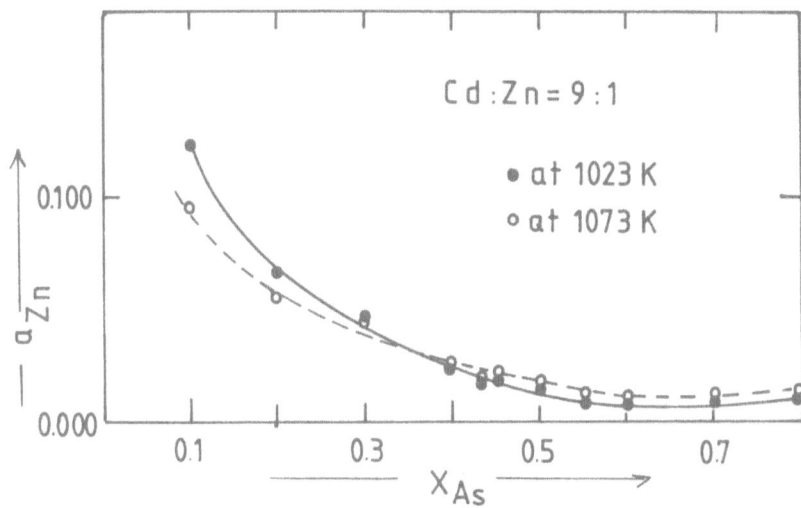

Fig. 4. Activity curves of Zn at different cross sections and
at two temperatures.

a) Cd:Zn = 9:1   ● at 1023 K
                  O at 1073 K

b) Cd:Zn = 3:1 ⎤
               ⎥   ● at 1073 K
c) Cd:Zn = 2:1 ⎬
               ⎥   O at 1123 K
d) Cd:Zn = 1:1 ⎦

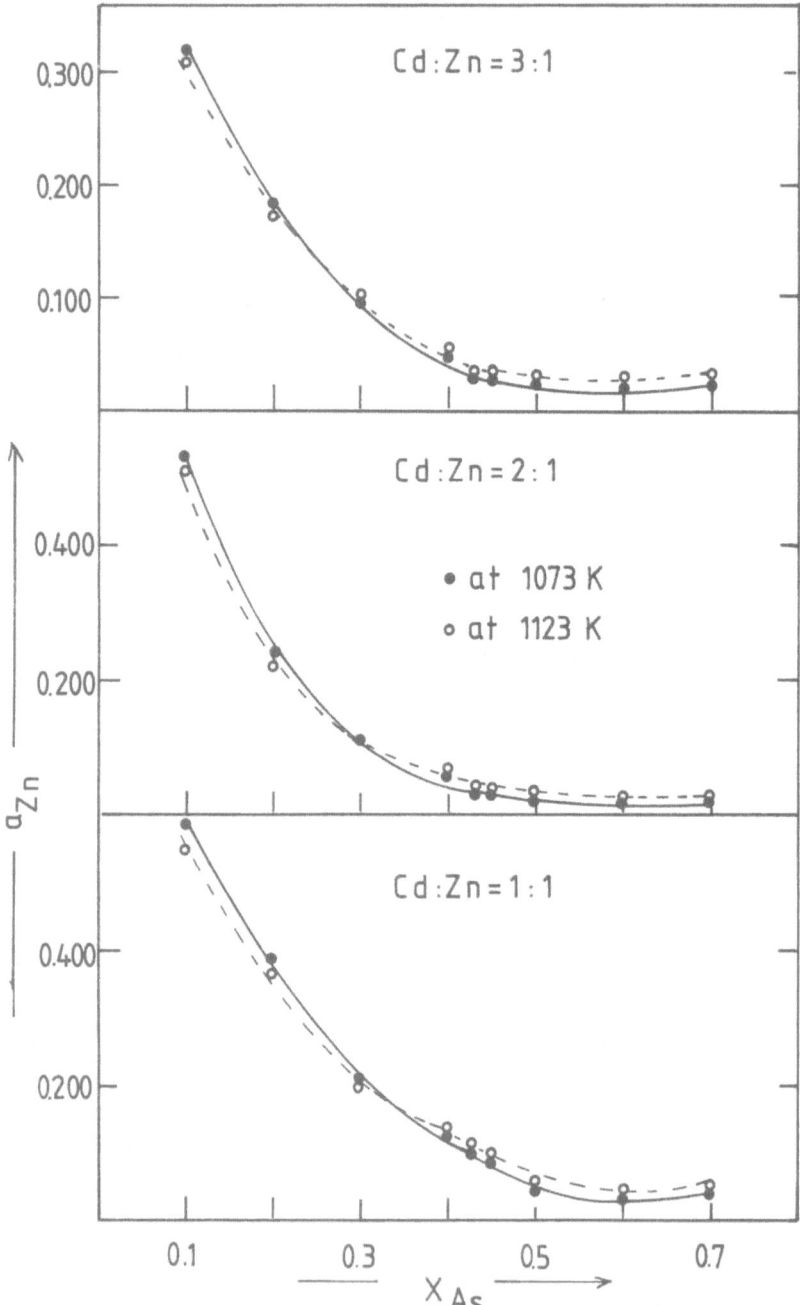

EMF MEASUREMENTS OF HYDROGEN ACTIVITY IN METALS

F.Jaggy, W.Kieninger, T.Mütschele and R.Kirchheim
Max-Planck-Institut für Metallforschung
Institut für Werkstoffwissenschaften
Seestrasse 92
D-7000 Stuttgart 1
FR-Germany

ABSTRACT. The activity of hydrogen which is dissolved in metals can be
determined by emf measurements in aqueous electrolytes, if the
conditions for a dynamic equilibrium between dissolved hydrogen and
protons are fulfilled. Therefore the surface conditions of the metal
must allow a rapid transfer of hydrogen from the metal into the
electrolyte and vice versa. This is the case for palladium and most of
its alloys. Other metals usually have an oxide layer on the surface
which acts as a diffusion barrier for hydrogen. The oxide can be removed
by evaporation at elevated temperatures or by sputtering at room
temperature in an ultra high vacuum apparatus. The plating of a thin
film of palladium onto the clean metal surface allows emf measurements
again. From changes of the emf after applying an external stress the
partial molar volume of hydrogen can be calculated. The sensitivity and
reliability of the emf method and examples for pure palladium,
nano-crystalline palladium and amorphous Pd-Si-alloys will be discussed.
The results of the measurements for these materials are interpreted in
the framework of Statistical Thermodynamics.

1. INTRODUCTION

The high mobility of hydrogen dissolved in metals allows the hydrogen
activity to be measured at room temperature by either direct partial
pressure measurements or emf measurement, which was mainly used for pure
palladium /1/. Other metals are mostly covered with an oxide film, which
acts as a diffusion barrier for hydrogen. However, this film can be
replaced by a palladium layer (i.e. by evaporation /2/), which
facilitates the attainment of equilibrium between dissolved hydrogen and
either gaseous hydrogen or protons. By using two electrolytic
compartments the emf technique can be also applied to the determination
of hydrogen diffusivities in metals /3/. One of the major advantages of
the electrochemical measurements is the possibility of precise doping
samples with hydrogen by coulometric titration /4/.

H. Brodowsky and H.-J. Schaller (eds.), Thermochemistry of Alloys, 409–418.
© 1989 by Kluwer Academic Publishers.

This paper deals mainly with the used emf technique and results obtained for the hydrogen activity in different disordered materials. We also report on changes of H-activity by externally applied stresses.

2. EXPERIMENTAL METHOD

The schematic diagram of the electrochemical cell is shown in Figure 1. The sample (usually a foil) separates two compartments which are filled with an electrolyte. Some characteristic features of the cell are as follows:
- the temperature of the whole system can be controlled.
- if necessary argon can be bubbled through each side of the double cell to remove residual gases.
- Haber-Luggin-capillaries located at each side of the sample are used to measure the electrode potentials against a saturated calomel electrode (SCE).
- a viscous electrolyte, which is a mixture of two parts of glycerine and one part phosphoric acid, is used. In this electrolyte the emf is more stable than in sulfuric acid.

There is also a platinum counter electrode, from which a current can flow to the sample (the working electrode), in order to load the sample with hydrogen by coulometric titration. The two main requirements

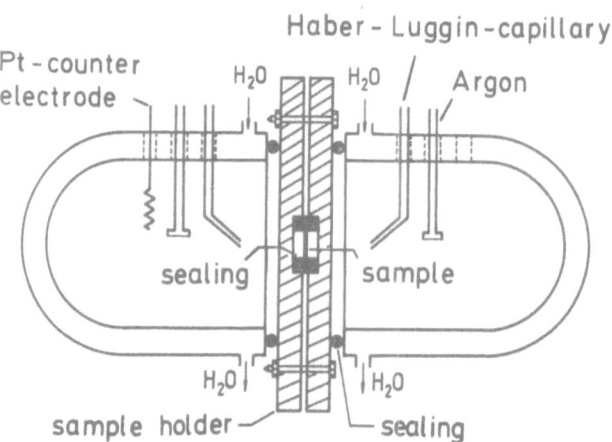

Figure 1. Schematic figure of the electrochemical double cell as proposed in Ref. 3 and as used in this study.

for this process are: First a surface of the sample which has to be permeable in order to attain a dynamic equilibrium. This is fulfilled for palladium and palladium alloys, because there is no oxide on the surface. If an oxide existed, it was removed by sputtering with argon ions in an UHV-apparatus. After that procedure a thin palladium film was sputtered onto the clean surface. This was done for instance in the case of niobium and amorphous NiTi- and NiZr-alloys.

The second condition is a current yield of approximately 100%. There is evidence that this condition was fulfilled. If hydrogen is removed from the sample by anodic oxidation the same charge can be measured as in the doping procedure. Another piece of evidence is the fact that the resistivity changes and length changes, which were measured during the doping of a Pd-sample with hydrogen are in good agreement with literature data /5/. Two examples are shown in Figures 2 and 3. In Figure 2 the results of resistivity measurements are plotted versus the hydrogen concentration. After the sample was doped to a given hydrogen concentration 1000 atppm were withdrawn by anodic polarisation (800 mV/SCE) several times and subsequently added again by cathodic charging. The results show that reversible resistivity changes occur. The reproducibility of the resistivity change is better than 5%.

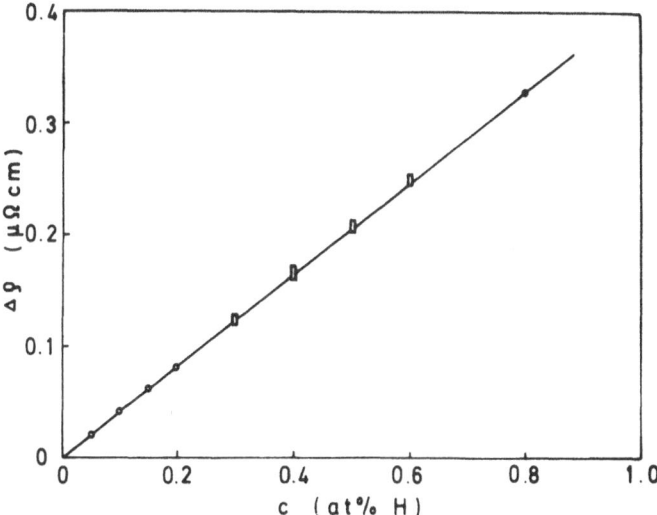

Figure 2. Resistivity increase of palladium at 298 K as a function of the hydrogen content of the sample. Between 0.3 and 0.6 at.% H the hydrogen concentration within the sample was increased or decreased several times by cathodic or anodic polarization, respectively. The changes in concentration corresponded to 0.1 at.% and reproduced the resistivity results within the error bars given by the size of the rectangular symbols.

Dilatometric measurements were performed in a apparatus schematically shown in Figure 4. From measurements of length changes of a metal ribbon during the doping procedure with hydrogen a value for the partial molar volume of hydrogen can be calculated. The determined value of 1.6 cm^3/mol H for polycrystalline Pd is in very good agreement with literature data /6,7/. In Figure 3 the sensitivity and reliability of this method is shown. 100 atppm of hydrogen can be added to the sample and also withdrawn causing the same length change of the sample.

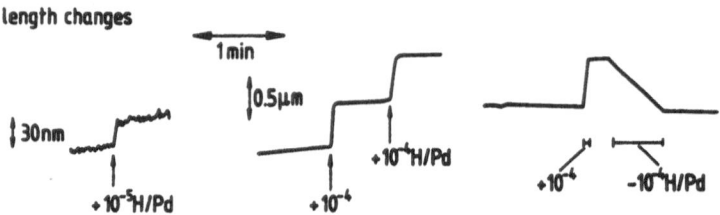

Figure 3. Length changes of a palladium wire due to doping with hydrogen (a chart recorder plot of the output voltage of device No. 13 in Figure 4). The changes are reversible as seen in the right part of the figure.

The current yield is less than 100% dissolved hydrogen at high hydrogen activities (hydrogen partial pressure larger than 1 atm), because the formation of hydrogen gas bubbles and, therefore, hydrogen losses occur. At very small hydrogen activities other electrochemical reactions can determine the electrode potential (i.e. reducing oxygen dissolved in the electrolyte). But nevertheless, reliable measurements of hydrogen activities over ten orders of magnitude are possible (cf. Figure 5).

By incorporation of a stress gauge into the apparatus shown in Figure 4 we were able to measure emf-change caused by externally applied stresses.

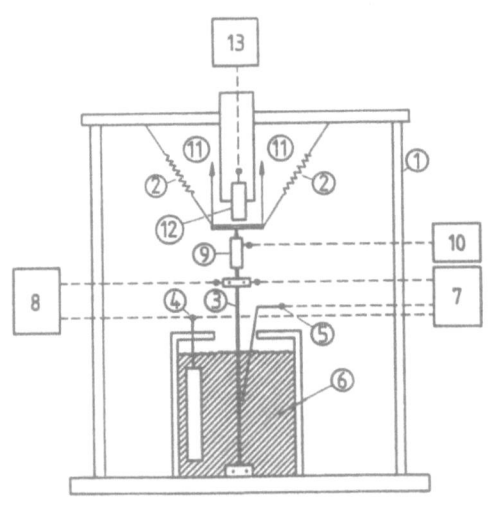

Figure 4. Schematic figure of the apparatus for measurements of strain, emf and stress. 1. Solid frame, 2. spring, 3. sample, 4. counter electrode, 5. reference electrode, 6. electrolyte, 7.+8. current source and emf-measurements, 9.+10. stress gauge, 11. variable force, 12.+13. srain gauge

3. RESULTS AND DISCUSSION

For an ideal dilute solution of hydrogen in the electrode (sample) the emf is given by Nernst's Law

$$E = E^o + \frac{RT}{F} \ln c = E^{(p)} + \frac{RT}{2F} \ln p \tag{1}$$

where E^o ($E^{(p)}$) is the emf for a concentration of $c = 1$ (for a partial pressure of hydrogen of $p=1$ atm) and R, T and F have the usual meaning. Concentrations in this study are ratios of the number of H to metal atoms. At high pressures the fugacity has to be used in Eq. 1 instead of p. Figure 5 shows that for lower hydrogen concentrations in single crystalline palladium and polycrystalline niobium a straight line is obtained at room temperature in a plot of E vs. log c with a slope of RT/F ln 10 according to Eq. 1. From the intercept at log c = 0 the free energy of dissolution of hydrogen can be calculated. At high H-concentrations an emf plateau appears, which corresponds to a two phase region (solid solution and hydride) of the metal/hydrogen system. The plateau value of the emf yields the free energy of hydride formation.

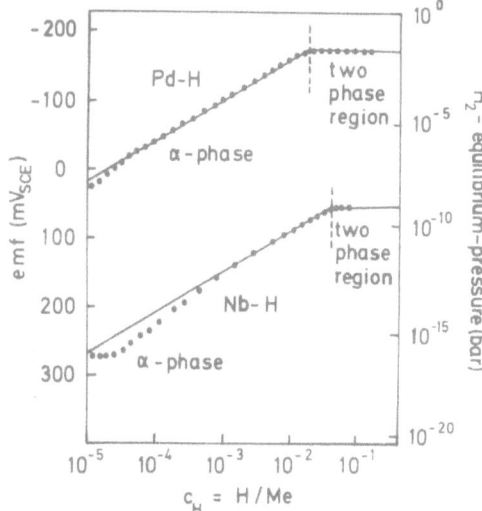

Figure 5. Emf and partial pressure of hydrogen at 295 K as a function of H-concentration for Nb and Pd. The straight lines have a slope of RT/F ln 10 (cf. Eq. 1.)

The ideal dilute behavior of hydrogen in single crystalline palladium and in polycrystalline niobium corresponds to a random distribution of hydrogen atoms among equal interstitial sites in the metallic matrix. In other materials, as nano-crystalline palladium or amorphous Pd-Si alloys, different interstitial sites exist with different site energies, which can be occupied by hydrogen. The distribution of hydrogen atoms among the different sites is described by Fermi-Dirac-statistics

$$p(E) = \frac{1}{1 + \exp((E-\mu)/RT)} \qquad (2)$$

where $p(E)$ is the probability that a site of energy E at temperature T and chemical potential μ is occupied by hydrogen. Multiplying this expression with the distribution function of the site energies $n(E)$ and intergrating over all site energies gives the total concentration, c, as a function of the chemical potential /8/.

$$c = \int_{-\infty}^{\infty} \frac{n(E)\ dE}{1 + \exp(\frac{E-\mu}{RT})} \qquad (3)$$

By measuring the emf ($= \mu/F$) at a given concentration allows to obtain information on the energy distribution of sites which is related to structural or compositional disorder in a given material.

3.1 Amorphous Pd-Si-alloys

If a Gaussian distribution of site energies is assumed for an amorphous metal it was shown /8/ that the chemical potential of hydrogen can be related to the hydrogen concentration by the following equation

$$\mu = G^{o} + \sigma\ \mathrm{erf}^{-1}(1-2c) \qquad (4)$$

where σ is the width and G^{o} the average energy of the Gaussian distribution. Thus a plot of the chemical potential versus the inverse

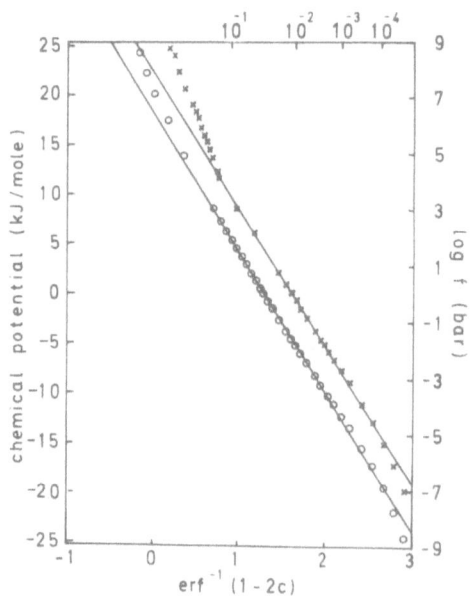

Figure 6. Emf and fugacity of hydrogen at 295 K for two amorphous Pd-Si alloys (circles = 20 at.% Si, crosses = 15 at.% Si) as a function of the inverse error function of 1-2c according to Eq. 4. Deviations from the linear behavior occur at high fugacities due to filling of available sites or H-H interaction /9/.

error-function of 1-2c should result in a linear behavior. This has been confirmed in a recent study, where emf- and high pressure results have been combined /9/. The validity of Eq. 3 was demonstrated over many orders of magnitude in hydrogen concentration and pressure. An example is shown in Figure 6. The samples were produced by melt-spinning and had a σ-value of about 11 kJ/Mol H.

3.2 Nano-crystalline Palladium

The nano-crystalline palladium was produced by the solidification of metal vapor in a high-purity helium atmosphere. In this process small crystallites with a grain diameter between 8 and 12 nm can be obtained. These are compacted to pellets of about 5 mm in diameter and thicknesses between 80 and 120 µm. Therefore, the volume fraction of the distorted material belonging to the grain boundaries is 0.2 to 0.5, if a thickness of the grain boundary between 0.5 and 1.5 nm is assumed.

In a first order approximation a nano-crystalline metal can be considered of containing two different regions. The first are the single-crystalline regions within the grains and the second are the distorted regions within the grain boundaries. So it appears to be reasonable to assume a continuous spectrum of site energies for the boundary regions of the nano-crystalline metal (Figure 7). This distribution can be determined for hydrogen from emf results and can be compared with that of an amorphous alloy.

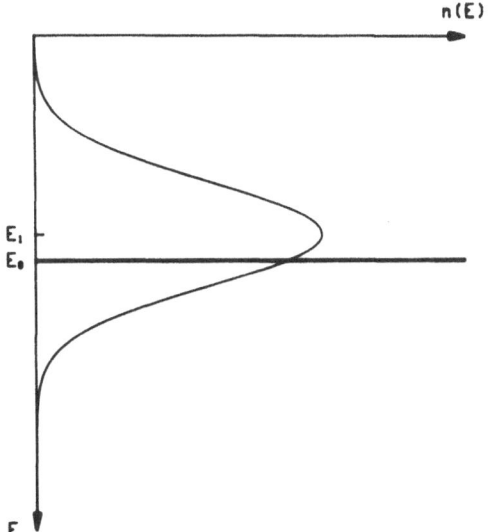

Figure 7. Proposed distributions of site energies for hydrogen in nano-crystalline palladium. The Gaussian distribution around E_1 corresponds to sites in the boundary whereas the line at E_0 represents sites within the grains. The experimental results of this study can be explained with the following values E_0 = 3.9 kJ/mol H (as for single crystalline Pd), E_1 = 9.2 kJ/mol H with respect to 1 atm of H_2 and σ (width of the Gaussian) = 15 kJ/mol H.

416

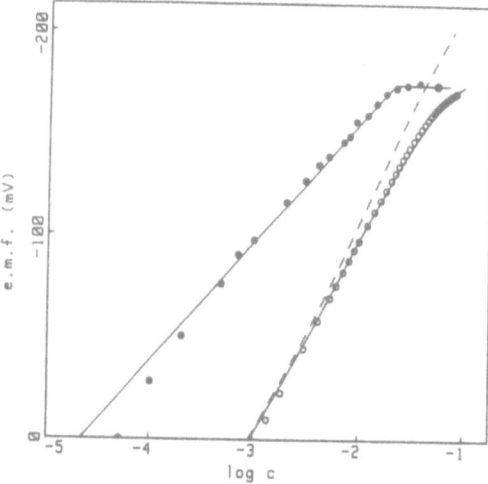

Figure 8. Emf of hydrogen in a single crystal of Pd (closed circles) and in nano-crystalline Pd (open circles) as a function of hydrogen concentration c (ratio of H to Pd atoms) at 293 K. The straight line through the data points of the single crystal has the theoretical slope (cf. Eq. 1). The lines through the points of the nano-crystalline Pd are calculated by assuming a distribution of site energies as shown in Figure 7 (dashed line: without H-H interaction).

The emf results for nano-crystalline palladium are shown in Figure 8 in comparision with the results for single crystalline palladium. There is a remarkable difference in the solution behaviour between the single crystal and the nano-crystalline Pd, where the hydrogen concentrations are larger by about one to two orders of magnitude, because hydrogen segregates at the grain boundaries. For a given equilibrium pressure of hydrogen the local concentration of hydrogen has to be the same in the grains as in the single crystal, because the sites within the grains were assumed to be the same as in the single crystal. Therefore, the quantity of hydrogen segregated at the grain boundaries can be calculated from the difference between the total concentration in the nano-crystalline sample and the one in the single crystalline Pd. Assuming for the spectrum of site energies in the grain boundary regions a Gaussian distribution with a width of 15 kJ/Mol H the solubility results can be described well only for lower hydrogen concentrations. At higher concentrations the local concentration at the grain boundaries becomes so high (>10%) that H-H-interaction must be included. This can be done by the quasichemical approach using the interaction parameter of polycrystalline palladium /10/. Now an excellent agreement between calculated and measured solubility can be obtained. This is also shown in Figure 8.

3.3. Emf as a function of external stress

Direct measurements of the change of the chemical potential of hydrogen by external stresses were made for Pd-Ag alloys by Kussner /11/ and R.A. Oriani /12/. In both studies it was shown that the chemical potential increases proportionally with the applied tensile stress. This can be confirmed by measurements made in this study with an amorphous ribbon of Pd-Si, which are presented in Figure 9. However, in this case the slope of the straight lines, which is proportional to the partial molar volume of hydrogen /7/ depends on H-concentration. At smaller H-concentrations the partial molar volume of hydrogen is smaller in agreement with the concept of a distribution of interstitial sites being present in the amorphous matrix, where the larger sites are occupied first /8/.

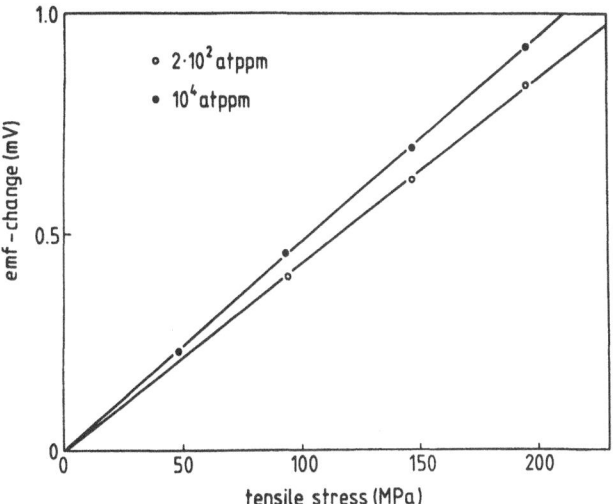

Figure 9. Emf-changes of amorphous $Pd_{80}Si_{20}$ for two different concentrations of hydrogen due to an external tensile stresses.

It can be also shown that the magnitude of the emf change depends on the stress state of the sample /7/, where in the case of torsional stresses and a cubic symmetry of the hydrogen distortion field no changes are observed at all.

Acknowledgements: One of the authors (R.K.) is grateful for financial support from the Deutsche Forschungsgemeinschaft.

REFERENCES

/1/ F.A. Lewis, "The Palladium-Hydrogen System", Academic Press, London, New York (1967)

/2/ N. Boes and H. Züchner, Ber. Bunsenges. physik. Chem., **80** (1976) 22

/3/ M.A. V. Devanathan and Z. Stachurski, Proc. Roy. Soc. Lond. A **270** (1962) 90

/4/ R. Kirchheim and R.B. McLellan, J. Electrochem. Soc. **127** (1980) 2418

418

/5/ A.Szökefalvi-Nagy, X.Y. Huang and R. Kirchheim, J. Phys. F: Met. Phys. **17** (1987) 427

/6/ H. Peisl in 'Hydrogen in Metals I', G. Alefeld and J. Volkl (ed.), Springer, Berlin (1978), p. 52

/7/ R. Kirchheim, Acta metall. **34** (1986) 34

/8/ R. Kirchheim, Acta metall. **30** (1982) 1069

/9/ A. Szökefalvi, S. Filipek and R. Kirchheim, J. Phys. Chem. Solids **48** (1987) 613

/10/ E. Wicke and J. Blaurock, Ber. Bunsenges. physik. Chem., **85** (1980) 1091

/11/ A. Küssner, Z. Naturf. **21a** (1966) 515

/12/ H.A. Wriedt and R.A. Oriani, Acta metall. **18** (1970) 753

THERMODYNAMIC BEHAVIOUR OF OXYGEN IN MOLTEN METALLIC ALLOYS

Sabri Anik and Martin G. Frohberg
Institut für Metallurgie - Allgemeine Metallurgie
Technische Universität Berlin
Joachimstaler Straße 31-32;
D-1000 Berlin 15

ABSTRACT. A statistical solution model is introduced to explain the thermodynamic behaviour of oxygen in liquid binary metallic solvents. The model permits the prediction of thermodynamic data of oxygen in the whole concentration range of homogeneous alloy melts only from the properties of the limiting binary systems. The approach is capable to evaluate the self interaction parameter and therefore to describe the thermodynamics of oxygen not only in its dilute but also in higher concentration range up to saturation. By means of emf-measurements and quenching experiments the activities and solubilities of oxygen in the liquid systems Cu-O-Bi, Cu-O-Pb, Bi-O-Pb, Bi-O-Sb and Pb-O-Sb were determined. The theoretical approach is checked by the results.

1. INTRODUCTION

Considerable theoretical work has been done on the prediction and explanation of thermodynamic properties of interstitial solutes such as nitrogen, carbon and oxygen in binary and higher component alloy solvents /1-9,13/. All of these approaches have been restricted to the infinitely dilute solution range and have been based, predominantly on the quasi-chemical theory. Block /2/ and Jacob and Alcock /4/ have developed expressions for the calculation of thermodynamic properties of nitrogen and oxygen respectively, under the assumptions that the energy and entropy of an interstitial atom is a linear function of surrounding atoms and that this atom effects the bond energies between the surrounding atoms. These models yield good results in case of nitrogen, but they succeed only in few cases for the explanation of the behaviour of oxygen. Further the effect of the presence of interstitial atoms on bond energies is empirically adjusted to arrive at best fit curves with experimental values. Wagner /3/, in the development of a model for the behaviour of oxygen in binary regular solvents, has shown that the energy of an interstitial atom can be described by a quadratic function of the types of the surrounding solvent atoms. This model was further modified empirically by Chiang and Chang /7/and applied quite successfully for the explanation of the behaviour of oxygen.

H. Brodowsky and H.-J. Schaller (eds.), Thermochemistry of Alloys, 419–428.

In the treatment of interstitial solutes in binary solvents we have stressed upon the need to further improve of Wagner's model. The aim of the present work, therefore, is to present a unified approach to explain the behaviour of interstitial components in binary solvents not only in dilute solutions but also at higher concentration range of the solved element and to predict its self interaction coefficient.

2. EXPERIMENTAL

In order to check the theoretical considerations it is necessary to investigate not only the region of the pure metals but also the total concentration range of the alloy up to the saturation of the solved element. Here was a lack of experimental data. Therefore we have determined the oxygen activities in the systems Cu-O-Bi, Cu-O-Pb at 1473 K and Bi-O-Pb,Bi-O-Sb and Pb-O-Sb at 1173 K electrochemically using a concentration cell with magnesia stabilised zirconia as solid electrolyte. The saturation concentration of oxygen was additionally determined by quenching experiments.

3. THEORETICAL CONSIDERATIONS

3.1. Assumptions of the statistical model

In earlier works /14-15/ we introduced a statistically derived approach which leads to a modified Wagner equation. Oxygen is known to occupy interstitial sites in the solvent. In order to derive an expression for its behaviour, one can proceed with the construction of a semi-grand partition function for the system containing the metallic components A and B and solved oxygen. To evaluate this partition function we made following assumptions.

a) The solvent atoms form a three dimensional lattice in which the interstitial sites are partly occupied by oxygen atoms. No vacancies exist in the system at substitutional sites.

b) Each oxygen atom is surrounded by Z_s solvent atoms in its first substitutional coordination shell and Z_i sites in its first interstitial coordination shell. Each solute oxygen atom can thus be distinguished from the other by the number of A-atoms in its substitutional coordination shell and also by the number of oxygen atoms in its interstitial coordination shell. In the case of infinitely dilute solution of oxygen, it is assumed that no oxygen atoms will be found in the interstitial coordination shell of a solved oxygen atom. At higher concentration of oxygen it is supposed, that there is one oxygen atom additionally placed in the interstitial coordination shell.

c) The energy of a central oxygen atom is considered to be a function of the type of atoms in its coordination shell. Atom constellations only in the first coordination shells will be taken into account.

d) Exchange of oxygen atoms by the solution with the gas phase does not cause a rearrangement of the solvent atoms.

3.2. Main equations of the model

The mathematical evaluation requires the application of the maximization method, which leads to energetical expressions for oxygen in the most probable distribution of the atoms. For the mole fraction of oxygen in its most probable distribution we get the basic expression

$$N_0 = \sum_{i=0}^{Z_s} \left\{ \binom{Z_s}{i} N_A^i \, N_B^{Z_s-i} \, N_V^{Z_i} \, \exp\left(-\frac{g_0^{i0}}{kT}\right) \left[1 - Z_i \frac{N_0}{N_V} \exp\left(-\frac{\delta g_0^{i1}}{kT}\right) \right] \right\} \exp\left(\frac{\mu_0}{NkT}\right) \qquad (1)$$

N_A, N_B, N_V : Mole fractions of A, B and vacancies
$\quad$ k : Boltzmann's constant
$\quad$ N : Avogadro's number
$\quad$ T : Absolute temperature
$\quad Z_s$: Coordination number of the substitutional shell
$\quad Z_i$: Coordination number of the interstitial shell
$\quad$ i : Number of A-atoms in the substitutional shell
$\quad \mu_0$: Chemical potential of oxygen in the gas phase
$\quad g_0^{i0}$: Energy term of the substitutional shell
$\quad \delta g_0^{i1}$: Energy term of the interstitial shell

In infinite dilute solutions of oxygen in a binary solvent, the above equation yields for $N_0 \to 0$ a modified Wagner-formula in which for the explanation of the energy term g_0^{i0} the two parameters g_A and g_B are introduced.

$$\frac{1}{\gamma_{0(A+B)}^0} = \sum_{i=0}^{Z_s} \left\{ \binom{Z_s}{i} \left(\frac{N_A}{\gamma_{0(A)}^{01/z_s}}\right)^i \left(\frac{N_B}{\gamma_{0(B)}^{01/z_s}}\right)^{Z_s-i} \exp\left(\frac{i(Z_s-i)}{2RT} \frac{(2Z_s-3)g_B+(Z_s-3)g_A+i(g_A-g_B)}{3(Z_s-2)}\right) \right\}$$

$$(2)$$

$\gamma_{0(A)}^0$, $\gamma_{0(B)}^0$, $\gamma_{0(A+B)}^0$: Activity coefficient of oxygen in A, B and A+B in infinite dilute solution respectively
$\quad\quad\quad\quad\quad\quad$ R : Gas constant

The system specific terms g_A and g_B express the energies in the substitutional coordination shell for the case of $N_B \to 0$ and $N_A \to 0$ respectively. Therefore both terms are directly correlated to the interaction parameters of oxygen in A at $N_B \to 0$ and in B at $N_A \to 0$ and can be calculated mathematically from these interaction parameters. The application of equation (2) is illustrated on various metal-oxygen -metal systems in the figures 1 - 5, where the points are experimentally determined by us. The curves are calculated by equation (2).

3.3. The self interaction parameter

In homogeneous molten metallic alloys the activity coefficient of oxygen γ_0 can be expressed in a simplified form of the Taylor-series:

$$\ln \gamma_0 = \ln \gamma_0^o + \varepsilon_0^0 N_0 \tag{3}$$

For all measurements of solved oxygen in alloys, this equation has been proved to be valid up to the saturation limit of oxygen /16-18/. Therefore by the prediction of the self interaction parameter ε_0^0 it is possible to describe the thermodynamic behaviour of oxygen also in higher concentrated regions. The basic equation (1) enables to evaluate further equations to describe and calculate this self interaction parameter. The energy term of the interstitial coordination shell $\delta g_0^{i^1}$ contains the energy change when one oxygen atom is built into the vacant interstitial coordination shell of a central oxygen atom. At higher oxygen concentrations the energy of a central oxygen atom therefore will not be specificated alone by the constellation of the metal atoms in the first substitutional shell but also by the influence of the neighbouring oxygen atom. This influence is introduced by the system specific energy term $\delta g_0^{i^1}$. In a realistic view of the problem this energy term is dependent from the place of the neighbouring oxygen atom in the interstitial shell for which Z_i probabilities exists. In our approach we assumed a system dependent constant value for $\delta g_0^{i^1}$ which is denoted with g_0.

Equation (1) leads to expressions for the prediction of the self interaction parameter ε_0^0 which was the subject of an earlier publication /15/. In many cases these expressions are only applicable if there is sufficient experimental data existent. Under the assumption that the interactions between the oxygen atoms in the melt are small, we get the simple equation

$$\varepsilon_{0(A+B)}^0 = N_A \, \varepsilon_{0(A)}^0 + N_B \, \varepsilon_{0(B)}^0 - \frac{Z_s(Z_s-1)}{2RT} \, g_0 \, N_A \, N_B \tag{4}$$

$\varepsilon_{0(A)}^0$, $\varepsilon_{0(B)}^0$, $\varepsilon_{0(A+B)}^0$: Self interaction parameters in A, B and A+B

4. DISCUSSION

As we measured the self interaction parameters of oxygen in various molten metallic alloys for the first time we could check equation (4), where we calculated the energy term g_0 by regression analysis of the experimental data. The figures 6-10 show that by application of the model (curves) the experimental values (points) can be fairly described except in the system Cu-O-Bi. This discrepancy can be explained by the high oxygen solubilities of molten Cu-Bi alloys, where the simplifying assumptions of the model are not realistic. Therefore the aim of our actual investigations is to find a physical interpretation for the energy term g_0 in order to reduce the number of simplifying assumptions and also to reduce the experimental effort to a minimum.

Starting point of our considerations was the typical curve-run of the self interaction parameters in the system Cu-O-Bi. The same run can be observed in the coordination number curve of the system Cu-Bi measured by Zaiss and Steeb /19,20/ independently. We find out that in the case of predicting the self interaction parameter it is necessary to build in the real coordination number of the respective alloy. Furthermore the energy term g_0 is dependent of geometrical structur factors like the radius of the viewed central cell and the percentage of the occupied places in the cell.

The investigations treated in this report have been made with funds of "Deutsche Forschungsgemeinschaft" to which we would express our thanks.

5. REFERENCES

1) J.F. Elliott and J. Chipman, *Chemical Metallurgy of Iron and Steel*, Iron and Steel Inst., London (1973),348.
2) U. Block, Met. Trans., **1** (1970),2018.
3) C. Wagner, Acta Met., **21** (1973),1297.
4) K.T. Jacob and C.B. Alcock, Acta Met., **20** (1972),221.
5) M.L. Kapoor, Tans. JIM, **18** (1977),125.
6) M.L. Kapoor, Scripta Met., **10** (1976),323.
7) T. Chiang and Y.A. Chang, Met. Trans., **7B** (1976),453.
8) S.H. Kuo and Y.A. Chang, Met. Trans., **9B** (1978),154.
9) Y.A. Chang and D.C. Hu, Met. Trans., **10B** (1979),43.
10) M.L. Kapoor, Int. Met. Reviews, **20** (1975),150.
11) I. Prigogine, *The Molecular Theory of Solutions*, North-Holland Pub.Co., Amsterdam (1957).
12) R.H. Fowler and E.A. Guggenheim, *Statistical Thermodynamics*, University Press, Cambridge (1960).
13) M. Blander, M.L. Saboungi and P. Cerisier, Met. Trans., **10B** (1979),613.
14) S. Anik, M.L. Kapoor and M.G. Frohberg, Z. Metallkde., **74** (1983),53.
15) S. Anik, M.L. Kapoor and M.G. Frohberg, Z. Metallkde., **74** (1983),372.
16) W. Stichel, Dissertation TU Berlin 1967.
17) B. Isecke, Dissertation TU Berlin 1977.
18) S. Anik and M.G. Frohberg, Z.Metallkde., **74** (1983),530.
19) W. Zaiss und S. Steeb, Phys. Chem. Liq., **6** (1976),1.
20) W. Zaiss und S. Steeb, Phys. Chem. Liq., **6** (1976),43.

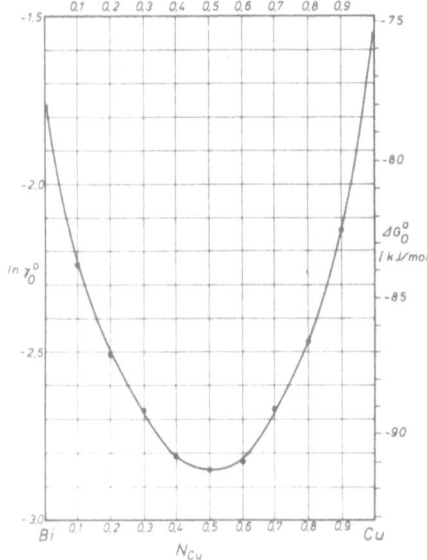

Figure 1.   Activity coefficient of oxygen in liquid Bi–Cu alloys at 1473 K in infinite dilution.

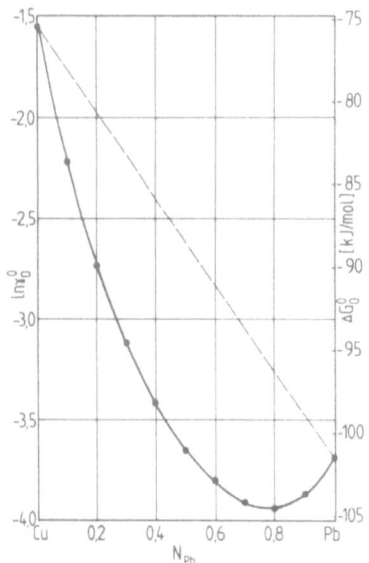

Figure 2.   Activity coefficient of oxygen in liquid Cu–Pb alloys at 1473 K in infinite dilution.

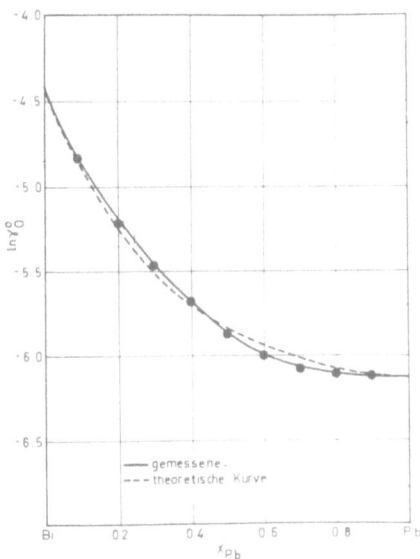

Figure 3. Activity coefficient of oxygen in liquid Bi–Pb alloys at 1173 K in infinite dilution.

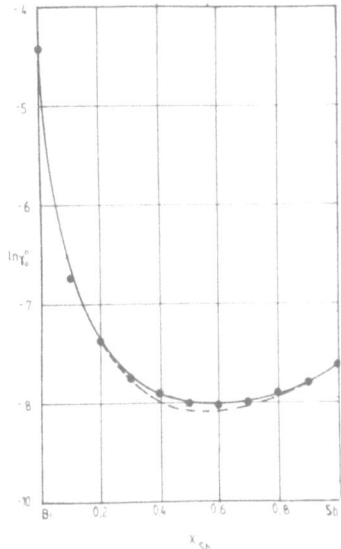

Figure 4. Activity coefficient of oxygen in liquid Bi–Sb alloys at 1173 K in infinite dilution.

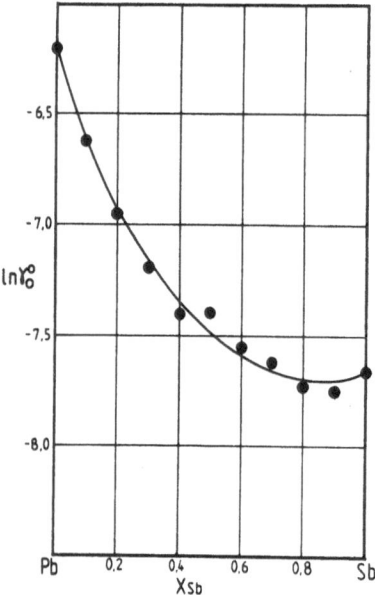

Figure 5. Activity coefficient of oxygen in liquid Pb-Sb
alloys at 1173 K in infinite dilution.

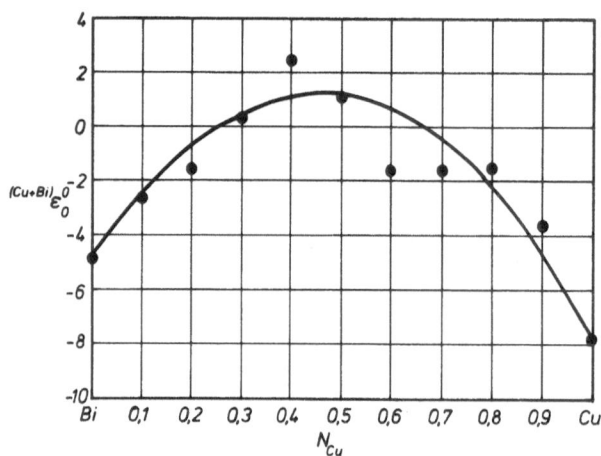

Figure 6. Self interaction parameter of oxygen in liquid
Bi-Cu alloys at 1473 K.

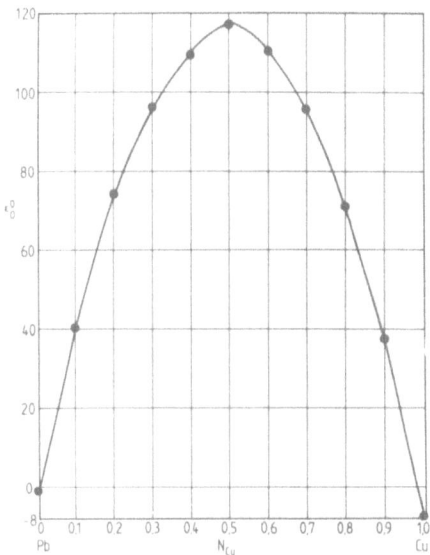

Figure 7. Self interaction parameter of oxygen in liquid Pb-Cu alloys at 1473 K.

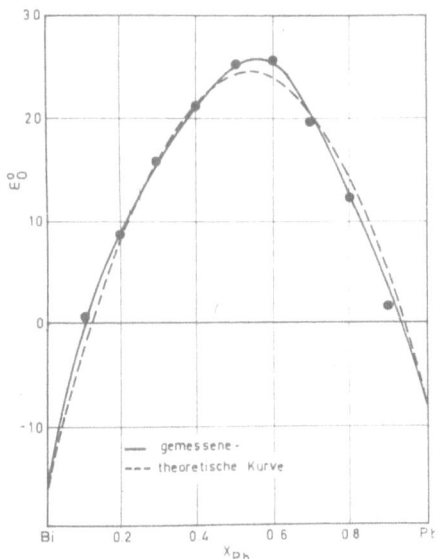

Figure 8. Self interaction parameter of oxygen in liquid Bi-Pb alloys at 1173 K.

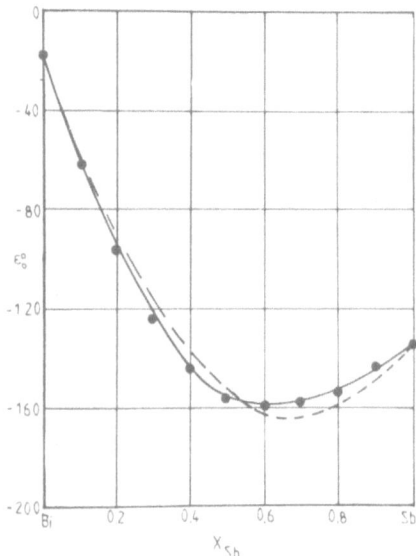

Figure 9.   Self interaction parameter of oxygen in liquid
            Bi-Sb alloys at 1173 K.

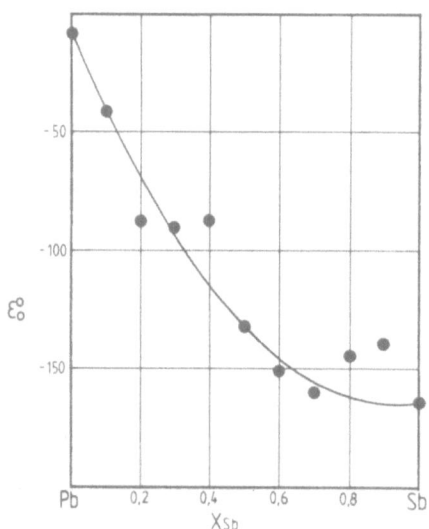

Figure 10. Self interaction parameter of oxygen in liquid
           Pb-Sb alloys at 1173 K.

THERMODYNAMICS OF LIQUID ALLOYS

B. Predel
MPI f. Metallforschung, Inst. f. Werkstoffwissenschaften
und Institut f. Metallkunde der Universität Stuttgart
Seestraße 75
D-7000 Stuttgart 1, FRG

ABSTRACT. In an introduction, some methods for the experimental
determination of important thermodynamic quantities are explained.
After a general survey of experimental results, structural as well
as energetic conditions in liquid alloys with a tendency to compound
formation are more closely inspected. The possibility of a description
of short range ordering phenomena in such alloy melts on the basis of
association equilibria is discussed. In connection herewith the Pd-Si,
Ag-Te, and Ag-Ge systems are more closely regarded and some causes for
the formation of metastable solid crystalline phases are indicated.

1. INTRODUCTION

As is well known, as results of the solidification of liquid alloys not
only stable crystalline phases can occur, but also metastable crystalline phases or metallic glasses may be formed. This depends mainly on
the conditions of solidification but also on the atomic structure and
the energetic conditions of the liquid alloys. Within certain limits,
the cooling conditions can be chosen by technical measures. Necessary
informations about the atomistic-structural and energetic factors have
to be disclosed by controlled experiments. The most important methods
are, on the one hand, X-ray and neutron diffraction experiments and, on
the other one, thermodynamic investigations. Subsequently the energetics
of the formation of liquid alloys shall be entered into, in the course
of which structural considerations are essential for the comprehension
of the observed phenomena.

2. EXPERIMENTAL METHODS

For the disclosure of thermodynamic properties of liquid alloys, on the
one hand, equilibrium methods for the determination of partial Gibbs
free energies and, on the other one, calorimetric methods for the
determination of the integral mixing enthalpy have been developed.
In order to enable a complete thermodynamic description of a system,

H. Brodowsky and H.-J. Schaller (eds.), Thermochemistry of Alloys, 429–450.
© 1989 by Kluwer Academic Publishers.

in the simplest case the knowledge of the mixing enthalpy ΔH and of the Gibbs free energy ΔG is necessary. According to the Gibbs-Helmholtz equation

$$\Delta G = \Delta H - T\Delta S \tag{1}$$

then, also the integral mixing entropy ΔS is accessible. The equilibrium measurements, as a rule, are yielding values for the partial Gibbs free enthalpy of mixing $\overline{\Delta G_i}$ of only one component. Applying the Gibbs-Duhem equation, also the partial Gibbs free enthalpy of the second component, $\overline{\Delta G_j}$ can be disclosed. Then, there follows:

$$\Delta G = x_i\overline{\Delta G_i} + x_j\overline{\Delta G_j} \ . \tag{2}$$

Thus, in order to get an optimal access to the energetics of a system, in general it is practical to bring up in addition to calorimetric measurements, a suitable equilibrium method. Here, some experimental methods shall be only briefly described.

2.1 E.M.F. Method

This method is, among others, applicable if the alloy components are markedly differring in regard to their electrochemical nobility. So, e.g. for the investigation of the Ag-Zn system the following galvanic chain can be used:

$$Zn\,|\,ZnCl_2 + (KCl + LiCl)_{eut.}\,|\,Ag\text{-}Zn\,| \tag{3}$$

The electrolyte can consist of a molten eutectic mixture of KCl and LiCl with 5 mole-% $ZnCl_2$. A possible experimental arrangement is represented in Fig. 1 (1). Pure zinc and the alloy, respectively are put into alumina crucibles and are covered by the liquid electrolyte. The partial Gibbs free energy of zinc is connected with the E.M.F. E of the galvanic cell in the following way:

$$\overline{\Delta G_{Zn}} = - z \cdot F \cdot E \tag{4}$$

z is the charge number of the Zn ions in the elextrolyte (z = 2) and F, the Faraday constant. Numerous variations, also with solid electrolytes, have been developed.

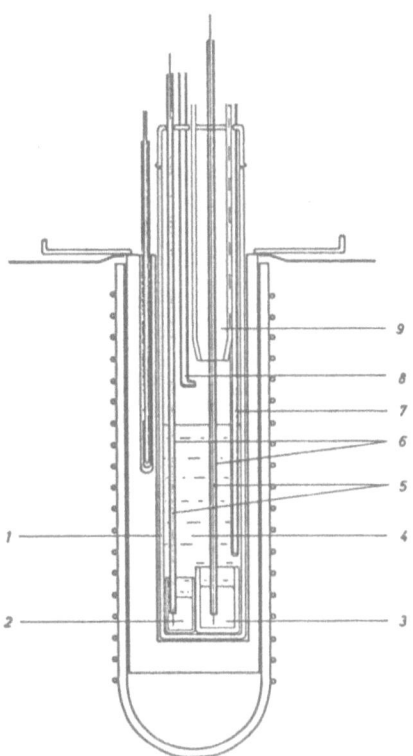

Figure 1. Experimental arrangement for the determination of partial
Gibbs free energies of mixing of liquid Ag–Zn alloys (1).
1 Quartz glass vessel
2 Alumina crucible with pure zinc
3 Alumina crucible with a silver-zinc alloy
4 Electrolyte
5 Tungsten wires
6 Quartz glass protective tube
7 Glass tube
8 Tube for the introduction of zinc.

2.2 Partial Vapour Pressure Measurements

The partial Gibbs free enthalpy of mixing of the A component of an alloy
is connected with the thermodynamic activity according to:

$$\overline{\Delta G}_A = RT \ln a_A \tag{5}$$

Furthermore, there is:

$$a_A = \frac{p_A}{p_A^o} \tag{6}$$

Here, p_A means the partial vapour pressure of A over the A–B alloy and p_A^o the vapour pressure of the pure A component at the same temperature T.

Many methods for the measurement of partial vapour pressures have been developed. Here, a method presented by O. Ruff and B. Bergdahl (2) for the determination of the total vapour pressure of inorganic substances ought to be mentioned. F. Sommer (3) has developed further this method for the experimental determination of the activities a_A and a_B in an alloy whose components exhibit similar vapour pressures. The method has been applied, e.g. for the determination of the partial vapour pressures in alloys both components of which are earth alkali metals.

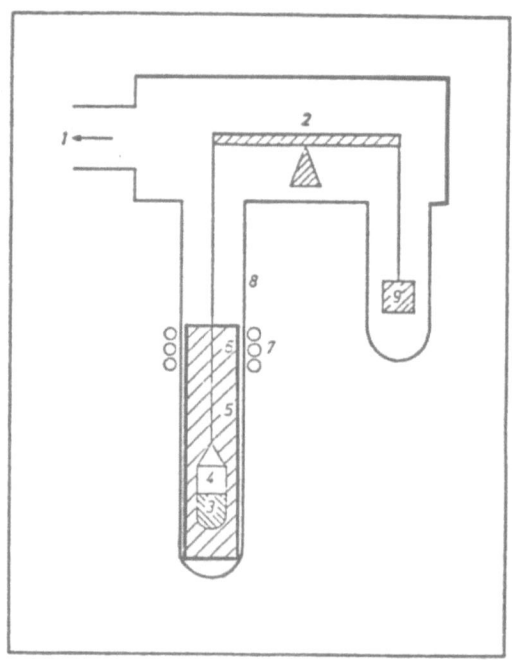

Figure 2. Experimental arrangement for the determination of partial vapour pressures of an alloy according to F. Sommer (3).

1) to vacuum pump   6) condensate
2) balance   7) water cooling
3) substance to be investigated 8) quartz tube
4) crucible   9) counterbalance
5) iron tube

The alloy resting in an iron crucible is attached to a sensitive electronic balance. Earth alkali metals don't react with iron. The crucible is, as can be seen from Fig. 2, surrounded by an iron tube. This arrangement is placed in a vessel filled with argon. The crucible is held at a constant temperature. The argon pressure is varied and the

weight loss per unit of time is recorded. If the argon pressure is
lower than the total pressure of the alloy, the evaporation rate rapidly increases. In a plot of the evaporation rate as a function of the
argon pressure, in the resulting curve a break point is appearing at
an argon pressure corresponding with the total pressure of the alloy
(see Fig. 3).

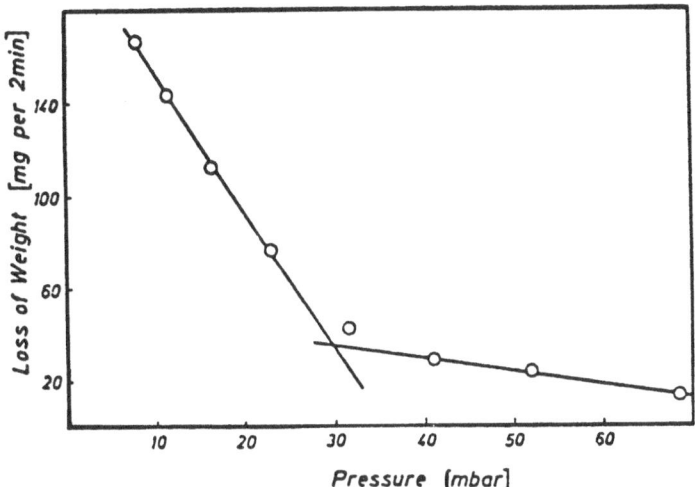

Figure 3. Weight loss of a pure magnesium sample as a function of
the argon pressure in the experimental arrangement at 1010 ± 2 K
according to F. Sommer (3).

The metal vapour is condensing at the iron tube which is cooled
at the top. The analytically determined concentration ratio $x_A : x_B$
which is present in the condensate permits, on the basis of the simultaneously determined total vapour pressure, the determination of the
partial vapour pressure and, herewith of the thermodynamic activities
a_A and a_B.

2.3 Calorimetry

With particular exactness the integral mixing enthalpy can be determined if both liquid components are combined in the calorimeter to form
the liquid alloy. As most metals are melting at elevated temperatures
according high temperature calorimeters are necessary. A simple calorimeter which can be used up to about 1300 K is represented in Fig. 4
(4). By lifting the plug in the upper crucible the one liquid component
is flowing into the lower crucible in which the second liquid component
is present. By a stirrer a quick concentration equalization is guaran-
teed. The temperature change of the alloy caused by the mixing enthalpy is recorded as a function of time, with a sensitive thermopile.
Calibration is done by dropping a pure metal into the calibration tube.
The melting enthalpy introduced and the heat content of the calibration substance for the temperature interval between room temperature

434

and the calorimeter temperature are yielding the calibration effect in the temperature-time diagram.

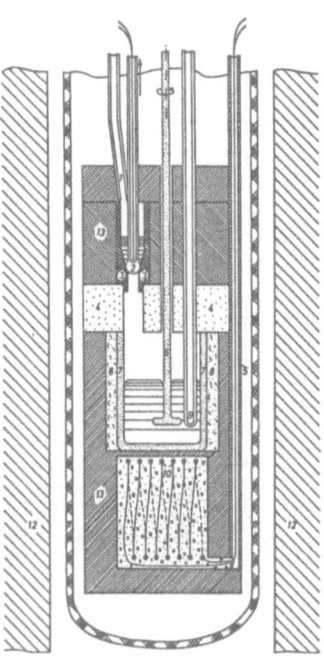

Figure 4. Principle of a high temperature calorimeter for the determination of mixing enthalpies up to 1300 K (4).

1) Filling tube
2) Plug
3) Melting crucible
4) Ergan ring
5) Protective tube for the thermocouple wires
6) Stirrer
7) Reaction crucible
8) Thermal insulation
9) Calibration tube
10) Thermopile
11) outer tube
12) Thermax blocks

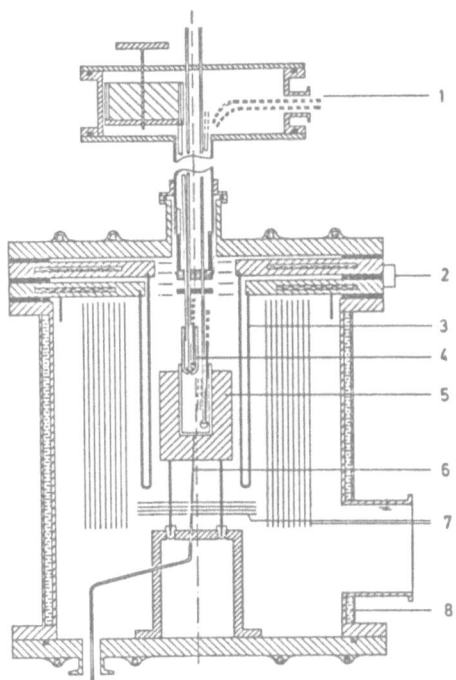

Figure 5. High temperature calorimeter for the determination of mixing
enthalpies up to temperatures of 2000 K after (5).
1) Charge vessel
2) Powder leads
3) Molybdenum short circuit heater
4) Charge crucible with stopper
5) Massive molybdenum block with crucible, stirrer and thermocouple
6) Reference thermocouple
7) Molybdenum radiation shields
8) Vacuum tank with water cooling

3. ATOMIC ARRANGEMENT AND ENERGETICS OF LIQUID ALLOYS

3.1 Survey of Some Experimental Results

For the description of thermodynamic properties of metals J.H. Hildebrand (6) starts with the assumption that between the atoms central interaction forces are effective and that the different kinds of atoms
are arranged according to a random distribution. On the basis of this
regular solution model the concentration dependence of the mixing enthalpy is given by:

$$\Delta H = \Delta H^{o} \cdot x_{A} \cdot x_{B} \tag{7}$$

ΔH^{o} is a constant. Thus, the ΔH-x_{A} curve is a parabola of the 2nd degree.

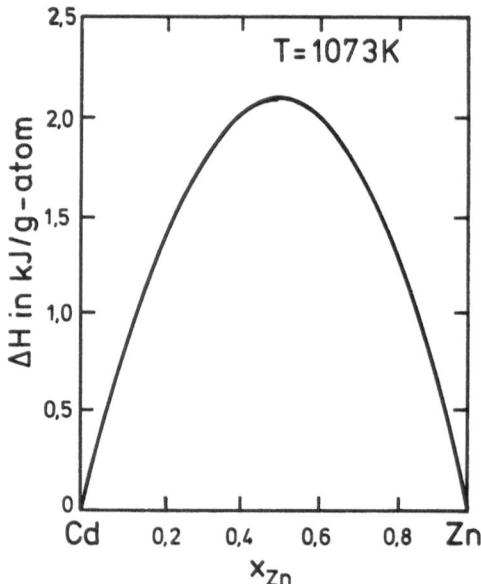

Figure 6. Mixing enthalpy of liquid Cd-Zn alloys as a function of the concentration according to (7).

This model is well fulfilled in some cases. Fig. 6 shows the mixing enthalpy of liquid Cd-Zn alloys as a function of the atomic fraction of zinc. The concentration dependence indeed is complying with a parabola of the 2nd degree.

To be sure also distinct deviations from this shape have been found. An example is the ΔH-x curve in the Li-Tl system (8) which is approximately parabolic, but not symmetrical to x = 0.5 (see Fig. 7). In the Cs-Au system the ΔH-x curve isn't even parabolic any more but almost triangular (9). Finally, the mixing enthalpies in the Ag-Sb system are having, in the Ag rich region, another sign as in the Sb rich one (see Fig. 9).

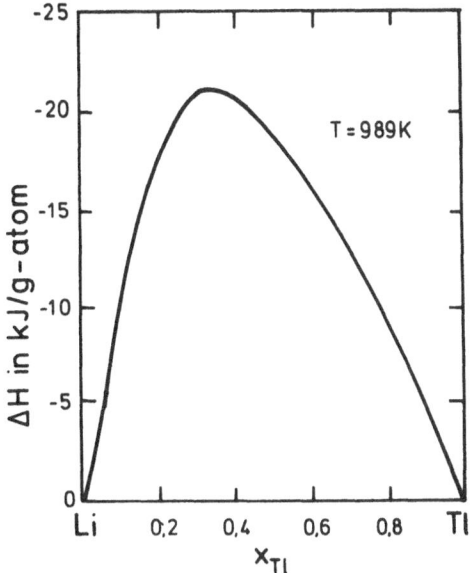

Figure 7. Mixing enthalpy of liquid Li-Tl alloys according to (8).

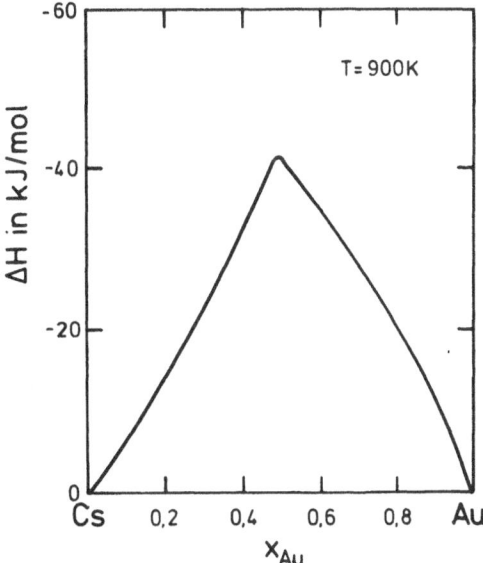

Figure 8. Mixing enthalpy of liquid Cs-Au alloys according to (9).

438

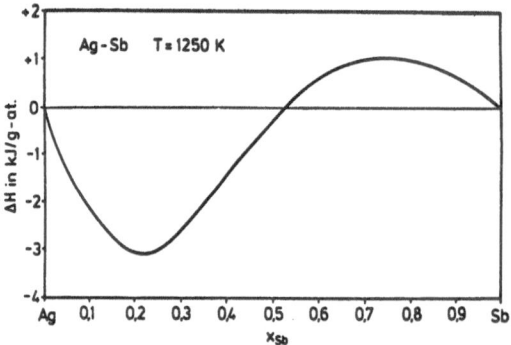

Figure 9. Mixing enthalpy of liquid Ag-Sb alloys according to (7).

3.2 Description of the Deviation from the Random Distribution of the Atoms by an Association Equilibrium

Naturally, the entire variety of thermodynamic manners of behaviour cannot be described by the regular solution model obviously. In many cases the presumption of a random distribution stipulated there is not fulfilled. Such a distribution, at all, can be exactly expected only in the case if, for the interatomic interactions there is:

$$\omega_{AB} = \frac{\omega_{AA} + \omega_{BB}}{2} \tag{8}$$

Here, ω_{AB} is the interactive force between the atoms denoted ij, respectively. For the case in equation (7), naturally $\Delta H = 0$. At least, the either positive or negative ΔH values have to be small in order to still enable the application of the regular solution model.

If, between A and B atoms interaction forces are effective which are considerably stronger than it would be in accordance with the mean of the interactions between like atoms it is to be expected that in the surroundings of an A atom there are more B atoms than in a random distribution of the kinds of atoms. There is a short range order as can also be affirmed by X-ray small angle diffraction. Such a short range order also can be understood in this way that in a limited measure molekule-like species e.g. A_iB_j are present which are existing in a dynamic equilibrium with the also present monatomic species A_1 and B_1:

$$i\ A_1 \cdot j\ B_1 \rightleftharpoons A_iB_j \tag{9}$$

We are following the considerations of F. Sommer (10). The numbers of moles of the monatomic species are:

$$n_{A_1} = n_A - i \ n_{A_i B_j} \tag{10}$$

$$n_{B_1} = n_B - j \ n_{A_i B_j} \tag{11}$$

Here, n_A and n_B are denoting the nominal numbers of moles of the A and B components, respectively. Naturally, the entire number of moles is:

$$n = n_{A_1} + n_{B_1} + n_{A_i B_j} \tag{12}$$

For the equilibrium, in a homogeneous phase (eq. 9), there is:

$$\left(\frac{\partial \Delta G}{\partial n_{A_i B_j}} \right)_{T,p,n_A} = 0 \tag{13}$$

From this, the mass action law is following:

$$\frac{(a_{A_1})^i \cdot (a_{B_1})^j}{(a_{A_i B_j})} = \frac{1}{K_{A_i B_j}} = \frac{1}{\exp\left[-\dfrac{\Delta G^o_{A_i B_j}}{RT} \right]} \tag{14}$$

a are denoting the activities of the respective species; $K_{A_i B_j}$ is the mass action constant, R is the General Gas Constant, T the temperature, and $\Delta G^o_{A_i B_j}$ is the change of the Gibbs free energy on the formation of one mole of associates from the according amounts of the monatomic species. This quantity can be split up according to the Helmholtz-Gibbs equation:

$$\Delta G^o_{A_i B_j} = \Delta H^o_{A_i B_j} - T \cdot \Delta S^o_{A_i B_j} \tag{15}$$

$\Delta H^o_{A_i B_j}$ is the enthalpy change on the formation of 1 mole of associates.

$\Delta S^o_{A_i B_j}$ is the entropy change on the formation of 1 mole of associates.

At first, the integral mixing enthalpy of a liquid alloy containing associates shall be considered. As is already known, it is the result of the change of the interatomic interactions on the formation of an alloy from the pure components. Simplifyingly it shall be assumed that in the liquid alloy all species – A_1, B_1 and $A_i B_j$ – are statistically distributed. The interactions which are effective between these species are resulting in the following change of the enthalpy ΔH^{reg}:

$$\Delta H^{reg} = \frac{n_{A_1} \cdot n_{B_1}}{n} \cdot C^{reg}_{A_1,B_1} + \frac{n_{A_1} \cdot n_{A_iB_j}}{n} \cdot C^{reg}_{A_1,A_iB_j} +$$

$$+ \frac{n_{B_1} \cdot n_{A_iB_j}}{n} \cdot C^{reg}_{B_1,A_iB_j} \qquad (16)$$

Here the C quantities are meaning interaction constants for the inter-action parameters denoted, respectively.

In addition to the interactions given in equ. (16), there is an-other contribution to the enthalpy change on alloy formation stemming from the interaction between the atoms within the associates A_iB_j. Nat-urally, this is proportional to the number of moles of the associates:

$$\Delta H^{as}_{A_iB_j} = n_{A_iB_j} \cdot \Delta H^{o}_{A_iB_j} \qquad (17)$$

The integral mixing enthalpy of the liquid alloy is given by:

$$\Delta H = \Delta H^{reg} + \Delta H^{as}_{A_iB_j} \qquad (18)$$

Also, for the integral mixing entropy an anlogous consideration can be carried out. From this follows:

$$\Delta S = - R \left[n_A \cdot \ln x_{A_1} + n_{B_1} \cdot \ln x_{B_1} + n_{A_iB_j} \cdot \ln x_{A_iB_j} \right] +$$

$$+ n_{A_iB_j} \Delta S^{o}_{A_iB_j} \qquad (19)$$

Meanwhile, a larger number of systems have been considered under this aspect of an associate formation. It is apparent that in many cases the interaction between the monatomic species A_1 and B_1 and the asso-ciates, respectively, is considerably smaller than the interaction bet-ween A_1 and B_1 as well as between the atoms within the associates.

4. LIQUID Pd-Si ALLOYS AS AN EXAMPLE FOR STRONG ASSOCIATE FORMATION

Often the interaction within the associations is particularly strongly prevailing, that is, the mixing enthalpy is essentially determined by $\Delta H^{as}_{A_iB_j}$. As an example, the Pd-Si system shall be mentioned.

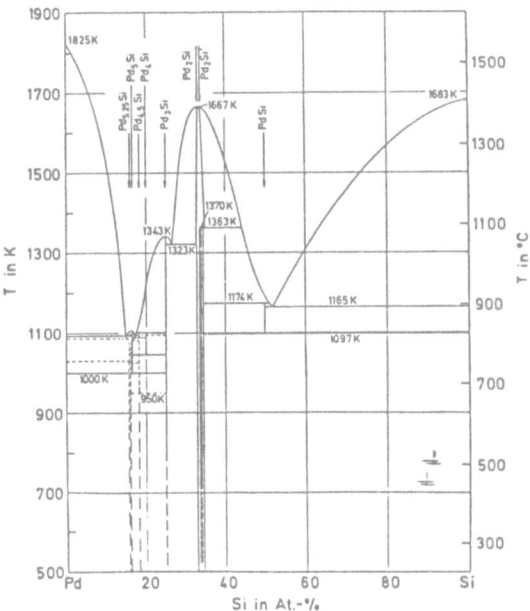

Figure 10. The Pd-Si phase diagram according to (11).

In the phase diagram (see Fig. 10) the intermetallic compounds PdSi, Pd_2Si, Pd_3Si, Pd_4Si, and Pd_5Si are occurring. The Pd_2Si phase has the highest melting point and therefore is the most stable of all inter-metallic compounds in this system. It ought to be expected that, in the liquid phase, the tendency to the formation of associates with the stoi-chiometric composition of the strongest solid sompound is prevailing. This is indeed the case, as Fig. 11 shows (12). Experimentally, not the parabolic ΔH-x curve is found as is to be expected in systems which can be described with the aid of the regular solution model. Here, on the contrary, the interaction between the atoms within the associates is dominating. There is an almost triangular course of the ΔH-x curve. The apex of the triangle coincides with the stoichiometry of the asso-ciates, which is complying in this case, to the formula Pd_2Si. At the Si-rich side there is an entirely linear course of the ΔH-x curve. There are no noticeable interactions between the Si_1 and Pd_2Si species. At the Pd-rich side the pertinent part of the ΔH-x dependence is slightly cur-ved. This is an indication of the presence of noticeable interactions between the Pd_1 and Pd_2Si species.

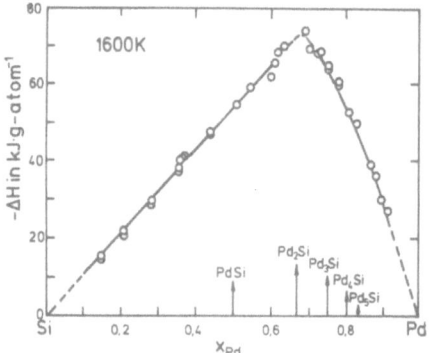

Figure 11. Mixing enthalpies of liquid Pd-Si alloys at 1600 K according
to (12); — calculated ΔH-x curve assuming Pd$_2$Si associates; 0 = measured points.

The ΔH-x curve can be quantitatively described by the association
model. The line drawn in Fig. 11 exactly fits the experimentally obtained points. Thus, the assumption of associates can well reproduce
reality.

5. THE MISCIBILITY GAP AS AN EFFECT OF STRONG ASSOCIATION

The mixing enthalpies of liquid Ag-Te alloys have been calorimetrically
determined by R. Castanet and C. Bergmann (10), (13). As can be seen
from Fig. 12, obviously an associate formation takes place. The stoichiometric composition of the associates lies, as the apex of the nearly
triangular ΔH-x curve is showing, at Ag$_2$Te. On the Te-rich side of the
'triangle' the dependence of the mixing enthalpy on the concentration
is linear. There are no perceptible interactions between Ag$_2$Te and Te.
On the Ag-rich side of the 'triangle' the curve is deviating from a
straight line, that is, in the direction towards diminished amounts of
negative values. This means a demixing tendency between associates and
the Ag$_1$ species which are not contained within the associates. This
interaction is described by the second term in equation (16). This
equation has to be rewritten for the case of liquid Ag-Te alloys, as:

$$\Delta H = \frac{n_{Ag} \cdot n_{Ag_2Te}}{n} \cdot C^{reg}_{Ag,Ag_2Te} + n_{Ag_2}Te \cdot \Delta H^o_{Ag_2Te} \qquad (20)$$

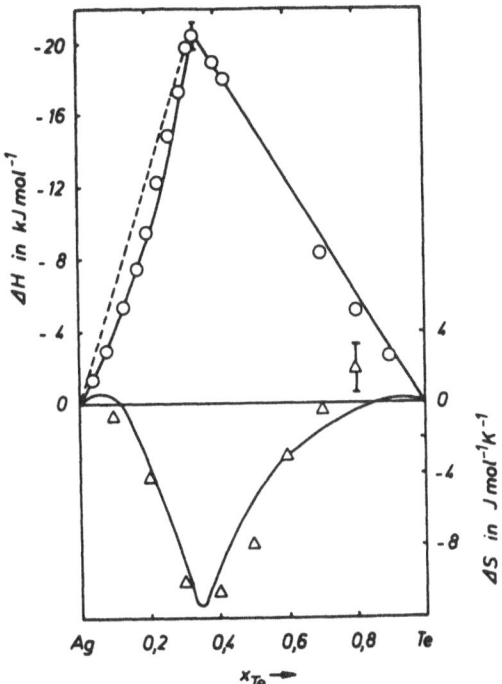

Figure 12. Mixing enthalpies and mixing entropies of liquid Ag-Te alloys according to (10), (13).

In the molten Ag_2Te, obviously strongly heteropolar bonding fractions are present as a consequence of a charge transfer from Ag to Te. This is analogous to the circumstances prevailing in a molten mixture of metallic potassium and molten potassium chloride. A complete miscibility of the both differently bonded and structured liquids is possible only at high temperatures. At lower temperatures a miscibility gap is existing in the liquid state.

Due to the strong association which is caused by a strong bonding between the unlike kinds of atoms, the molar heat is considerable changed. At the same time, the configuration part of the mixing entropy is small. This results in a considerable amount of the negative mixing entropy. The highest negative value of ΔS lies, as can be expected, at the stoichiometric composition Ag_2Te (13) (see Fig. 12). With these ΔS values and with the mixing enthalpies from Fig. 12, F. Sommer (10) has calculated the Gibbs free energies. ΔG, as a function of x_{Te} beneath a critical temperature shows a fold indicating a miscibility gap. The miscibility gap calculated in this way is in good agreement - as Fig. 13 shows - with the phase diagram (10).

444

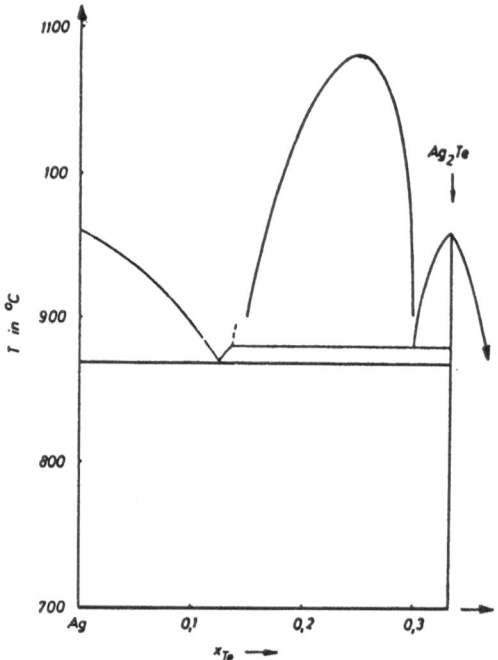

Figure 13.   The Ag-Te phase diagram (10).

6. DISCLOSURE OF SHORT RANGE ORDERING PHENOMENA FROM MELTING
    EQUILIBRIA REGARDING THE Ag-Ge SYSTEM

On the basis of the rules of Hume-Rothery, in the Ag-Ge system Hume-
Rothery compounds ought to occur. Namely, this system comprises a
monovalent noble metal and a higher-valent main group element. But,
as the phase diagram presented in Fig. 14 shows, it is just a simple
eutectic system (14). The mixing enthalpies determined by Ehrlich (15),
however, are indicating that there is, in the liquid alloys, a tendency
to form Hume-Rothery phases. In the concentration range around x_{Ge} = 0.2
in which the according VEC values for the formation of Hume-Rothery
phases are present, the mixing enthalpy exhibits a maximum of its
negative values (15) (see Fig. 15). A comparison with the analogous
Ag-Sn system is showing (see Fig. 16) that this minimum value just at
the appropriate VEC is hinting at the tendency towards the formation
of solid Hume-Rothery compounds. The ∆H-x curve for the liquid Ag-Sn
alloys is presented in Fig. 17. Also here, a minimum of the ∆H values
is present in a region in which, in the solid state, Hume-Rothery
phases are existing. The cause for the phenomenon that the liquid
alloys in the Ag-Sn and Ag-Ge systems are similar in regard to their
thermodynamic properties, their phase diagrams, however, are largely
differing is that the higher-valent components Sn and Ge have different
structures in their stable elementary modifications. These higher-valent

components are positioned, e.g. in the ε-Hume-Rothery phase at sites of
the hexagonally close packed structure. In their stable elementary structures, however, they are present in the tetragonal lattice of the white
tin or, respectively, in the diamond lattice (Ge). If an ε-Hume-Rothery
phase with Ag and another closely packed component (e.g. Cd or Al) is
formed, the Cd or the Al, respectively, is not subject to a considerable
structural change. Such Hume-Rothery phases are rather stable. If the
enthalpies of formation of the ε-phases of the Ag-Sn and Ag-Ge systems
shall be compared with those of the Ag-Cd or Ag-Al in regard to the change
of the bonding conditions, this is only correct if both Ag-Sn and Ag-Ge
systems are starting from an analogous initial state on compound formation, that is, from a hypothetical most closely packed structure of tin
or germanium, respectively. Before the formation of compounds a transformation of Sn and Ge into such a hypothetical modification is necessary. The transformation enthalpies for this amount to ΔH_T^{Sn} = 5.4 kJ/g-
at. and ΔH_T^{Ge} = 54 kJ/g-at., respectively. The energy gain on the change
of the bonding conditions can still surmount this transformation enthalpy which is entering into the whole energy balance of the compound
formation, in the case of Ag-Sn; in the case of the formation of the
ε-phase in the Ag-Ge system this wouldn't be possible any more. Therefore the ε-Hume-Rothery phase in the Ag-Ge system is not able to exist
as stable phase in competition with the solid solutions.

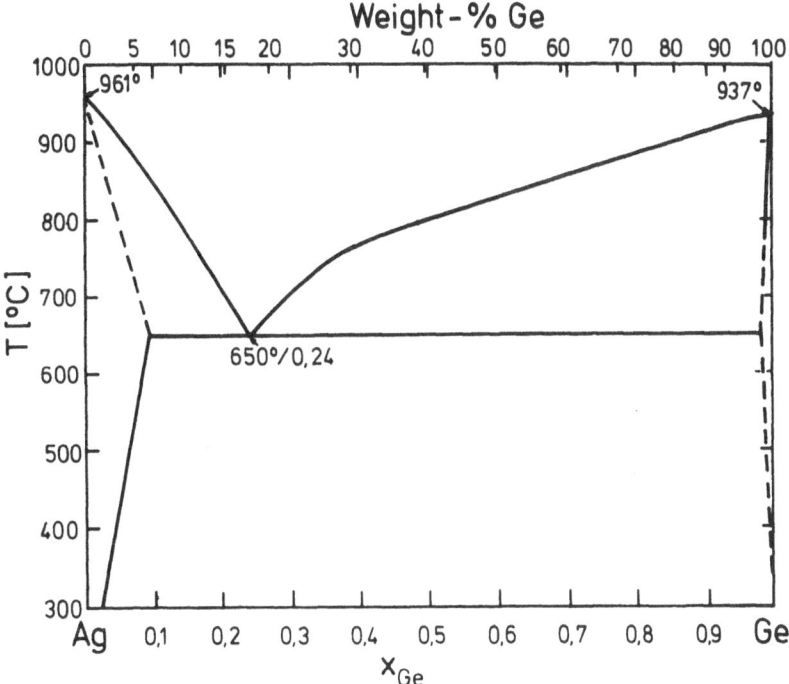

Figure 14.   The Ag-Ge phase diagram (14).

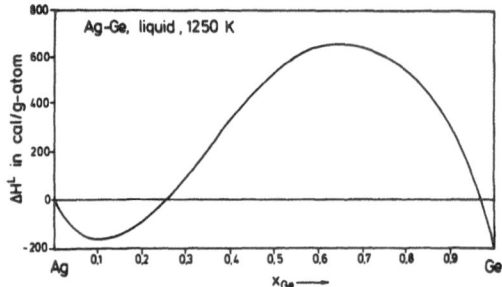

Figure 15. Mixing enthalpies of liquid Ag-Ge alloys according to (15), (7).

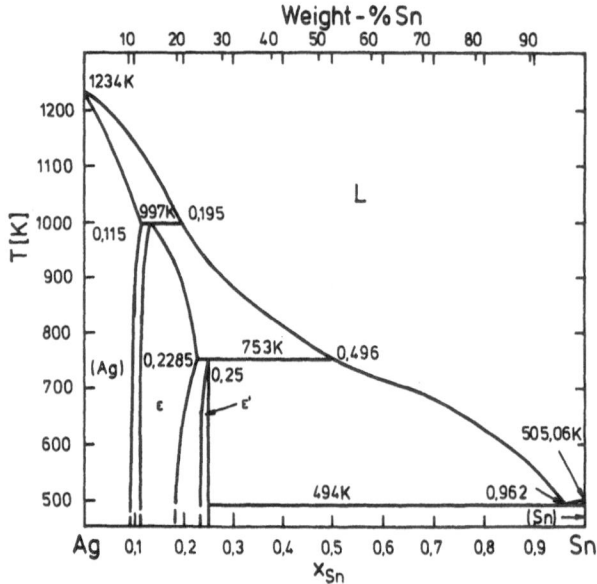

Figure 16. The Ag-Sn phase diagram according to (7).

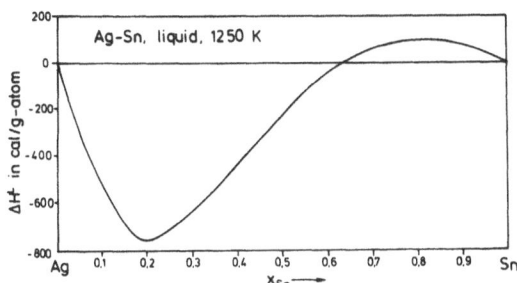

Figure 17. Mixing enthalpies of liquid Ag-Sn alloys according to (7).

Obviously, such an influence of structural differences between the initial and final states of alloy formation is not present in the liquid state. The metallic melts are, as a rule, of similar atomic structure. So, the inclination, according to the bonding, to the formation of Hume-Rothery phases can express itself in the thermodynamic data of the according liquid alloys.

Naturally, such peculiarities of liquid alloys have to become manifest in the solid-liquid equilibria of a phase diagram. Therefore, it is possible to disclose these peculiarities also by an analysis of the melting equilibria. So, it is possible in a relatively simple way to determine the integral excess Gibbs free energy of mixing, in the Ag-Ge system from the position of the phase equilibria. There is:

$$\Delta G^{ex} = \Delta G - \Delta G^{ideal} \tag{21}$$

The result obtained for liquid Ag-Ge alloys is represented in Fig. 18 (14). Near $x_{Ge} = 0.2$ the VEC values are corresponding with those required for the formation of solid Hume-Rothery phases. In this concentration region lies the minimal value of ΔG^{ex}. Obviously, the phase diagram yields more information than is evident on the basis of a simple qualitative consideration.

448

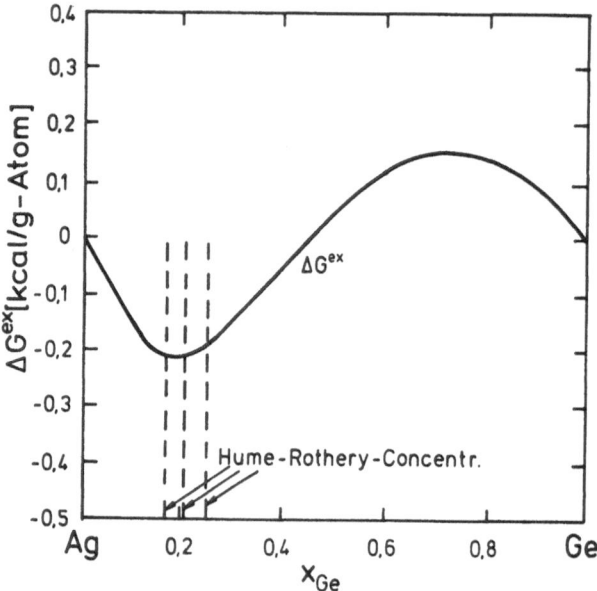

Figure 18. Integral excess Gibbs free energies of liquid Ag-Ge alloys which have been disclosed from the phase equilibria (Fig. 14) (14).

Now, it is not surprising that it is possible to obtain, by very rapid cooling of Ag-Ge melts with a VEC = 1.70, the ε-Hume-Rothery phase as a metastable compound. Its formation enthalpy has a positive sign. For the energy balance, the following values shall be compiled:

$$\Delta H^F = + 7.32 \text{ kJ/g-at.}$$

$$\Delta H_T^{Ge} = + 54 \text{ kJ/g-at.}$$

$$\Delta H_B = - 5.14 \text{ kJ/g-at.}$$

Here, ΔH_B is the fraction of the enthalpy which has to be attributed solely to the change of the bonding conditions on alloy formation. There is:

$$\Delta H^F = \Delta H_B + x_{Ge} \cdot \Delta H_T^{Ge} \qquad (22)$$

or, in numbers:

$$7.32 = - 5.14 + 12.46 \text{ kJ/g-at.}$$

Now, the bonding fraction ΔH_B composed from the formation enthalpy ΔH^F and the transformation term $x_{Ge} \cdot \Delta H_T^{Ge}$ can be compared with the bonding fraction of other Hume-Rothery phases. This is performed in Fig. 19, in which ΔH_B is plotted as a function of the square of the electronegativity difference.

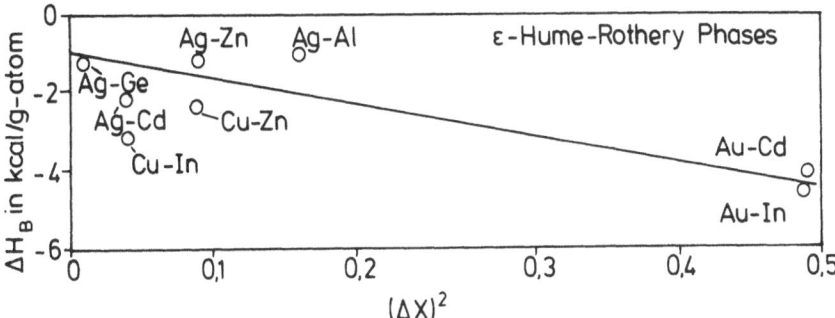

Figure 19. Bonding fractions of hexagonal Hume-Rothery phases as a function of the square of the electronegativity difference of the components.

$(\Delta x)^2$ is a measure for the heteropolarity of the bondings of a phase. The value of ΔH_B for the metastable Ag-Ge phase is well fitting into the course of the other ΔH_B values. As this value, however, is rather small, it cannot compensate the transformation term ΔH_T^{Ge}, as we have also rather roughly disclosed above.

7. CONCLUDING REMARKS

The conception of the molecule-like associates in liquid alloys with compound formation tendency has proven its usefulness in a series of systems investigated. It is remarkable that the Cowley ordering parameters which can be disclosed from X-ray small angle diffraction experiments and which can be used for the quantitative description of short range ordering phenomena in liquid alloys can be immediately converted into the degree of association as resulting from the treatment of thermodynamic data according to the association model. This has been demmonstrated by, e.g. F. Sommer et al. (16) taking for an example liquid alloys of the Cu-Ti system. An advantage of this model is the fact that under physically reasonable conditions, it is able to comprehend with a relatively small mathematical expense, the whole of the thermodynamic state functions of liquid alloys as dependent on the concentration and the temperature. This is useful among other facts, for the calculation of phase equilibria.

It ought to be mentioned that there are some variations of the associate model, e.g. where it is tried to take into account the size of the associates (17). Furthermore, association models, e.g. of P. Ramachandrarao et al. (18) and of A.S. Jordan (19) are to be called to mind. Also, other conceptions have been used, e.g. a quasilattice model of

450

M. Hillert et al. (20) and a model of H. Hoch and I. Arpshofen (21). All of these conceptions are trying to give an as realistic as possible description as well as a practicable comprehension of the peculiar situation in alloy melts which can't be easily described and which is lying between the entirely random distribution of a regular solution and the strict order of the crystal lattice of an intermetallic phase.

Literature

(1) U. Gerling, M.J. Pool and B. Predel, Z. Electrochem., 70 (1979) 224.

(2) O. Ruff and B. Bergdahl, Z. Anorg. Chem., 106 (1919) 76.

(3) F. Sommer, Z. Metallkde., 70 (1979) 545.

(4) G. Oehme and B. Predel, Thermochim. Acta, 22 (1978) 267.

(5) R. Lück and B. Predel, Z. Metallkde., 76 (1985) 684.

(6) J.H. Hildebrand and R.L. Scott, Solubility of Nonelectrolytes, A.C.S. Monograph, Vol. 17 (1950).

(7) R. Hultgren, P.D. Desai, D.T. Hawkins, M. Gleiser and K.K. Kelley, Selected Values of Thermodynamic Properties of Binary Alloys, Amer. Soc. for Metals, Metals Park, Ohio, 1973.

(8) B. Predel and G. Oehme, Z. Metallkde., 70 (1979) 618.

(9) F. Sommer, D. Eschenweck, B. Predel and R.W. Schmutzler, Conf. Proceedings 'Chemical Metallurgy - A Tribute to Carl Wagner' - N.A. Gokcen (Editor), Metallurg. Soc. AIME, Chicago (1981) 19.

(10) F. Sommer, Z. Metallkde., 73 (1982) 72,77.

(11) H. Langer and E. Wachtel, Z. Metallkde., 74 (1983) 535.

(12) I. Arpshofen, M.J. Pool, U. Gerling, F. Sommer, E. Schultheiß and B. Predel, Z. Metallkde., 72 (1981) 776.

(13) R. Castanet and C. Bergman, J. Chem. Thermodyn., 11 (1979) 83.

(14) B. Predel and H.P. Bankstahl, J. Less-Common Met., 43 (1975) 191.

(15) K. Ehrlich, Inaugural-Dissertation, Ludwig-Maximilian-Universität, München, Germany (1965).

(16) F. Sommer, K.H. Klappert, I. Arpshofen and B. Predel, Z. Metallkde., 73 (1982) 581.

(17) R. Lück, U. Gerling and B. Predel, Z. Metallkde., in press.

(18) S. Lele and P. Ramachandrarao, Met. Trans. B, 12B (1981) 659.

(19) A.S. Jordan, Met. Trans. B, 7B (1976) 191.

(20) M. Hillert, B. Jansson, B. Sundman and J. Agren, Met. Trans. A, 16A (1985) 261.

(21) M. Hoch and I. Arpshofen, Z. Metallkde., 75 (1984) 23.

METALLIC GLASSES

B. Predel
MPI f. Metallforschung, Inst. f. Werkstoffwissenschaften
und Institut f. Metallkunde der Universität Stuttgart
Seestraße 75
D-7000 Stuttgart 1
FRG

ABSTRACT. A general survey of the production, the properties, and the application of metallic glasses, is presented. Especially, the nature of the glass transformation as well as the preconditions for a high glass formation ability are entered into. Furthermore, relaxation processes are mentioned and crystallization processes of the metastable amorphous alloy are dealt with.

1. INTRODUCTION

Only towards the end of the last century a systematic development of metallic materials set in. Until then, essentially the same metallic materials were available for the construction of machinery and heavy mobile apparatus as for the old Romans, that is: iron, brass, copper, and bronce. With the introduction of new methods for the extraction of metals from oxides difficult to reduce, such as e.g. of aluminium and after the development of conceptions for systematically and purposefully influencing the properties by alloying additions, the unfolding of metallic materials commenced. Among the most recent and by no means fully developed metallic materials are the amorphous alloys or metallic glasses.

Metallic glasses have been discovered in 1960 by W. Klement, R.H. Willens and P. Duwez (1). Their peculiarity in regard to materials technology is that they are possessing combinations of properties which are very useful for some applications. They may have a high mechanical strength at extremely soft magnetic behaviour. Also, they may be outstanding as to high corrosion and wear resistance. Amorphous alloys can be produced rather easily. In addition, they can be obtained from inexpensive and - in sufficient amounts - available raw materials. The interest aroused by these numerous advantages has led to a rapid research on the conditions under which metallic glasses can be produced. Research on production conditions, however, not only concerns the technical realisation of production processes but also very essentially fundamental atomic facts. It is not possible to produce, under given technical conditions, a metallic glass in any arbitrary system. Also,

H. Brodowsky and H.-J. Schaller (eds.), Thermochemistry of Alloys, 451–469.

it is not possible to accomplish this even in systems in which glass
formation is occurring, for any arbitrary composition. Finally, it
ought to be reminded that solid alloys, as a general principle, are
thermodynamically stable only as crystalline materials. Non-crystalline
solid bodies are metastable. For their practical use, the question of
their durability, their applicability at certain temperatures and for
sufficiently long times, is of importance. Subsequently, the questions
of glass formation ability, the possibilities of production and their
stability shall be treated.

2. THE GLASS TRANSFORMATION PROCESS

The usual glasses, for instance window glass, are silicates. In the
simplest case of the soda-lime silicates, on the cooling-down of the
melts not crystallization but solidification to a glass, is taking
place. The easy supercooling ability is connected with the atomic
structure of the liquid silicates. Its structural units are SiO_4^{-4}
tetrahedra which are arranging themselves to form a net or lattice-
like structure. In the cavities, the Na^+ and Ca^{++} cations are placed.
The netting causes a high viscosity of the melt. The mobility of the
structural units even at the liquidus temperature, is so low that the
formation of the long range order which is present in crystals can
take place only slowly. The melt doesn't crystallize just on the trans-
gression below the liquidus temperature. As the mobility of the structural units decreases with decreasing temperature, accordingly crystal
nucleation becomes less probable. Eventually, the viscosity of the
silicate is so high that the material has the consistency of a solid
body - it is now present as a glass.

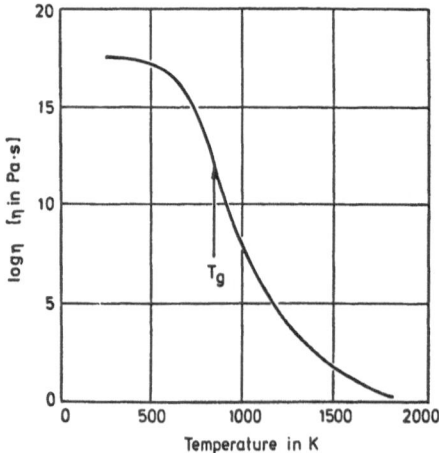

Figure 1. The logarithm of the viscosity as a function of the temperature for a soda-lime glass (2).

Fig. 1 shows the dependence of the logarithm of the viscosity η of a soda lime silicate melt on temperature. In the interesting temperature range between about 1800 K and room temperature (~ 300 K) the viscosity changes by about 17 orders of magnitude. The consistency of a solid body is reached when a viscosity of $\eta \approx 10^{12}$ Pa $\cdot$ s is present. This viscosity is occurring at a temperature T_g corresponding to the inflection point of the log η against T curve. This is the so-called glass point or the glass transformation temperature.

As can be seen from Fig. 1, on the glass transformation the viscosity does not change discontinuously, as this is the case for a crystallization process. Also, other properties are changing continuously at T_g. Therefore, the transition melt $\rightarrow$ glass is no transformation of the first order, as this is the case for crystallization or vaporisation. Nevertheless, the glass is different from the undercooled melt. Above T_g, at a given cooling rate the structural units can arrange themselves without delay in a way so that the entire arrangement of the atoms at any given temperature is complying with a minimum of the Gibbs free energy of the system. Below T_g, the mobility of the structural units is too small to render this possible. Thus, T_g represents the limit between both possibilities. Necessarily in this process T_g, at which just the limiting mobility is reached depends upon the cooling rate. On slow cooling, T_g is low, at high cooling rates, it is high. Thus, T_g is no thermodynamic equilibrium temperature as the fusion or the liquidus temperatures but a kinetically determined limiting temperature.

From these facts it has to be concluded that in the glass the atomic arrangement is the same one as it is present in the supercooled melt at the glass point T_g. With respect to the crystalline equilibrium phase, the supercooled melt as well as the glass are metastable. The supercooled melt, however, above T_g exists in an internal equilibrium, whilst the glass at a temperature $T < T_g$ is in a state of internal imbalance. If, at a temperature $T < T_g$, the kinetic possibility to change its structure slowly is given to a glass, that is, over a long period, then this change will always be in such a direction that an atomic arrangement is approached as it would be present in the supercooled melt at T on attaining the internal equilibrium. Such a change, below T_g, is called a relaxation process. During it, still no crystallization is occurring, but nevertheless considerable changes of certain properties can be connected with this process.

On the formation of metallic glasses analogous conditions are existing. The viscosity of metallic melts above the liquidus point, however, is considerably lower than that of most silicates which is connected with the different atomic structure of liquid alloys. Accordingly, on undercooling below the liquidus point crystallization may easily set in. In order to prevent this, an extraordinately high cooling rate has to be applied. Only by that means the melt can be cooled down to sufficiently low temperatures in such a short time that T_g is reached before a crystal nucleus has been formed. Experience shows that for alloys inclining towards glass formation, the necessary cooling velocities are above 10^4 K $\cdot$ s^{-1}.

3. PREPARATION METHODS

Throwing small alloy droplets into water or streaming round small a-
mounts of melt with a cold gas yields cooling rates up to about
10^4 K $\cdot$ s^{-1}. Only in a few favourable cases, metallic glasses can be
obtained in this way. As heat transfer from the melt by convection
doesn't yield a perceptibly higher cooling rate, for the production
of glasses usually the heat is carried off from the liquid alloy by
conduction. In order to achieve a high effectivity the dimensions of
the melt at least in one direction have to be small and the heat conductivity of the substrate into which the heat is carried off, has to
be large. In the simplest case this is achieved by rapidly spreading
out the melt on a massive metallic base plate. Some of the possibilities which have been used successfully are shown in Figs. 2 and 3.

The particular methods are differring only in the kind of mechanism by which the melt is spread out to a thin film and is pressed
against the substrate. Using the hammer-and-anvil method, a droplet
falls between a resting and a movable metallic block with smoothly
polished planes. Triggered by the passing of the droplet through a
light-ray bar above the blocks, the movable block is shot against the
resting one, hitting it just when the falling droplet is between the
blocks.

At the centrifugal method, a crucible with the melt is fastened to
a rotating arm. When a sufficiently high rotation velocity is reached,
a small hole in the wall of the crucible is opened. The melt, flying
out tangentially with high velocity hits a cylindrical cooled substrate (copper) and here, on hitting, is spread out to a thin foil.

In the shock wave tube according to Duwez and co-workers (5) at
the lower end of a crucible above a small hole there rests a melt droplet which cannot fall through the hole due to the surface tension. The
upper part of the tube is separated from the lower one by a thin foil.
If the pressure of an inert gas in the upper part is increased, eventually the foil will burst. The resulting shock wave flings the droplet
downward with supersonic velocity, where it is sliding along on a cur-
ved substrate and spread out to a thin film.

A further discontinuous method is schematically depicted in Fig.
2d. A droplet will be squeezed through a hole in the bottom of the crucible by a certain pressure thrust of the inert gas above it. Falling,
it is caught by a quickly rotating (10.000 rev./min) wing-like substrate and, by the impact is spread out to a thin film, which is, in addition, pressed quite forcefully against the well heat-conducting substrate.

Considerable amounts of thin ribbons are obtained by the methods
sketched in Figs. 3a and 3b. In Fig. 3a melt is continuously cast between two cooled rollers, in Fig. 3b it is poured upon the surface of
a rotating drum. According to 3c a glass tube in which the melt is
included is drawn to a thin filament. In doing so, the melt can solidify as a glass. Finally, as indicated in Fig. 3d, the melt sprayed upon
a substrate with a highly heated-up plasma jet can solidify to a glass
foil.

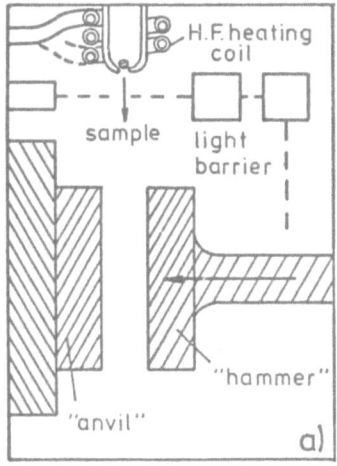

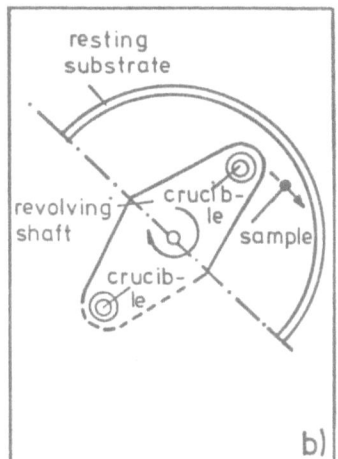

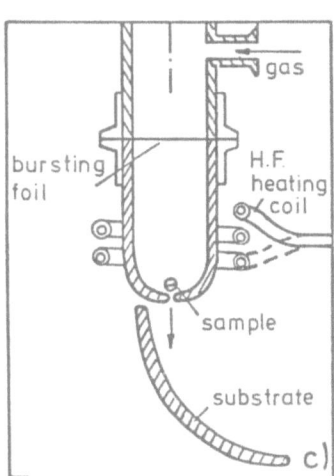

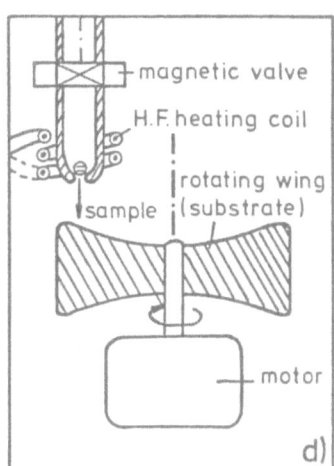

Figure 2. Discontinuous methods for the production of metallic glasses.
a) Hammer-and-anvil method (3)   c) Shock wave tube (5)
b) Centrifugal method (4)   d) Rotating wing method (6)

456

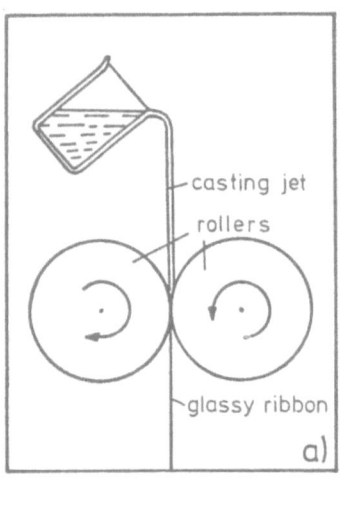

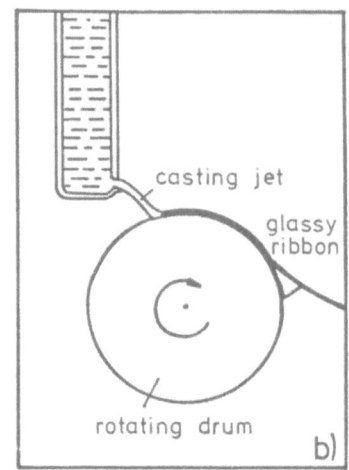

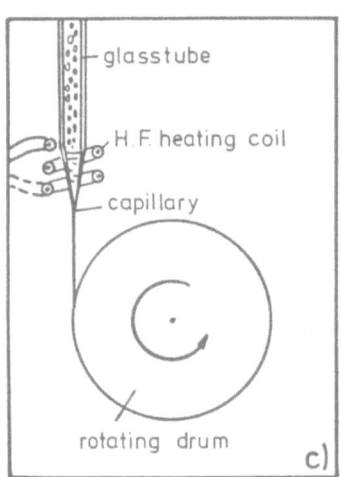

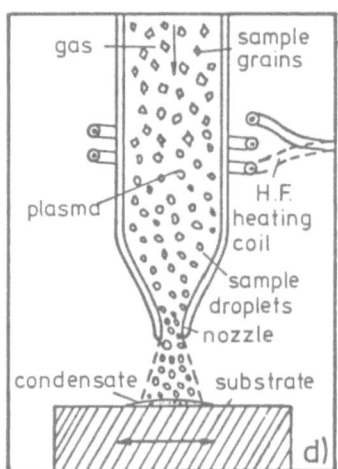

Figure 3. Continuous methods for the production of metallic glasses.
a) Roller method (7)   c) Taylor (melt spinning method) (9)
b) Rotating drum method (8)   d) Plasma sputtering (10)

By means of Discontinuous methods cooling rates up to 10^{10} K · s^{-1} can be reached. To be sure, the amount of glass obtained in this way is small. Applying continuous methods, within short time relatively big amounts of metallic glass in the form of ribbons can be obtained, as they are suitable for technological applications. In doing so, however, the cooling rates are relatively small so that it is necessary to employ alloys suitable for glass formation in a systematic way in order to get a metallic glass which, in addition, ought to have the desired special properties. These facts have given reason for a thorough research on the pre-conditions for a good glass formation ability of alloys. As can be expected from this short consideration of glass formation in silicate melts energetic-structural factors in metallic systems should play a role. But also kinetical conditions have to be presumed as crucial, so the glass temperature T_g is, as already stressed, no thermodynamic equilibrium temperature but is determined, on the one hand, by the mobility of the structural units, and, on the other one, by the cooling rate.

4. THERMODYNAMIC PREREQUISITES FOR THE GLASS FORMING ABILITY

4.1 Atomic Structure of Liquid Alloys

The net or lattice of SiO_4^{-4} tetrahedra in silicate melts is the result of directional bonding between silicon and oxygen. In metallic melts a similar atomic arrangement is not possible as the metallic bonding is not directional and there is a tendency to stack around a given atom as many other atoms as there is room for them. This tendency towards a most closely packed structure is also prevailing in metallic melts. If, however, the interaction between the unlike sorts of atoms A and B in a binary A-B alloy is bigger than between like kinds of atoms then, as is well-known, more A atoms are placed around a B atom, and vice versa, than would be in accordance with a random statistical atomic distribution. A so-called short range order is resulting which will be so more distincly marked as the interaction between A and B atoms is bigger than the mean value of all possible interactions; that is, A-B, A-A, and B-B. Then, a tendency for compound formation is present which is, as a rule, documenting itself in the solid state by the occurrence of one or more intermetallic compounds.

Such a short range order results in a lower mobility of the atoms as compared with an imaginative random distribution in the alloy melts, at the same concentration and temperature. The structural units, however, are not so unwieldy as in the silicates, so that the viscosity at the liquidus temperature is still very low, that is, comparable in its order of magnitude to that of water at room temperature. Therefore, as has already been mentioned, the supercoolability of alloy melts, as a rule, is poor. In order to avoid crystal nucleation before T_g – where $\eta \approx 10^{12}$ Pa · s^{-1} – is reached, extremely high cooling rates are required.

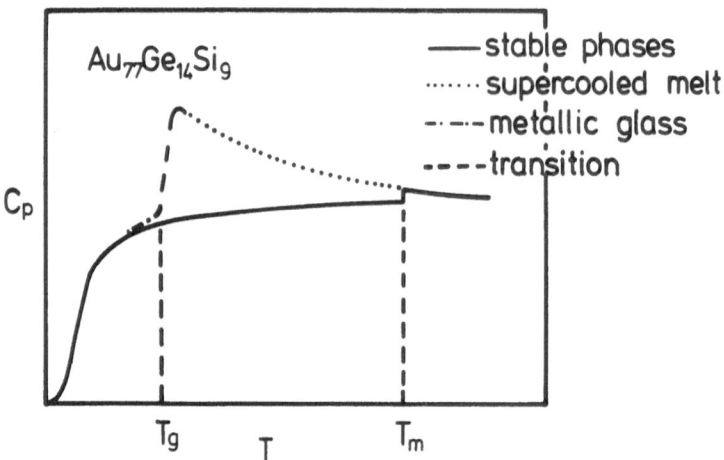

Figure 4. Molar heat of an $Au_{77}Ge_{14}Si_9$ alloy as a function of the temperature (4) (after (11); schematically).

The glass transformation in metallic alloys with regard to the transition from the internal equilibrium in the melt to the frozen-in state in the glass can easily be followed by means of the molar heat. In Fig. 4, as an example the molar heat of the $Au_{77}Ge_{14}Si_9$ alloy as a function of temperature for the equilibrium phase, the undercooled melt, and for the glass is represented schematically. As is well known, the molar heat is defined as the number of Joules required to heat 1g atom of the alloy by 1 K. If the melt exhibits a short range order with preferred neighbouring of unlike atoms, the degree of ordering is decreasing with increasing temperature. Therefore, for this case of a short range ordered melt energy has to be expended, among others, also to reduce the short range order on an increase of the temperature. This part is increasing with decreasing temperature and, herewith, with increasing degree of ordering, also in the region of supercooling (see Fig. 4). Below T_g, the adaptation to the equilibrium isn't possible any more due to kinetic reasons. Therefore, in this temperature range C_p is rapidly decreasing, namely to a value lying only slightly above the C_p value of the crystalline equilibrium phase.

It is evident that the glass temperature T_g and, herewith, the glass forming ability are actually connected with the state of the short range order in a melt. Indeed, as yet no metallic glasses have been found in systems in which there is no tendency for a short range order with preferred neighbouring of unlike kinds of atoms.

4.2 Influence of the Energetics of Crystal Nucleation

At first, it seems to be rather odd that in systems which are inclined to the formation of glasses due to their compound formation tendency, there are only narrow concentration regions in which, with the cooling rates usual today, glass formation actually takes place. This is connected with the kind of the solid phase emerging on crystallization and, in particular, with the solid-liquid phase equilibria, that is, with the phase diagram. Indeed, the reaction competing with glass formation is crystal nucleation.

At first it ought to be mentioned that, in general, the formation of a short range order in the melt causes a diminution of the Gibbs free energy of the liquid phase. This means a stabilization of the melt as against a given solid body of equal composition, and, herewith, a reduction of the probability of crystal nucleation. Obviously this has, in general, as a consequence an enhancement of the glass forming ability of the melt. As mentioned above, all hitherto found glass forming metallic systems are distinguished by a short range order in the melt. This general condition for the glass formation ability, however, will be modified by individual peculiarities.

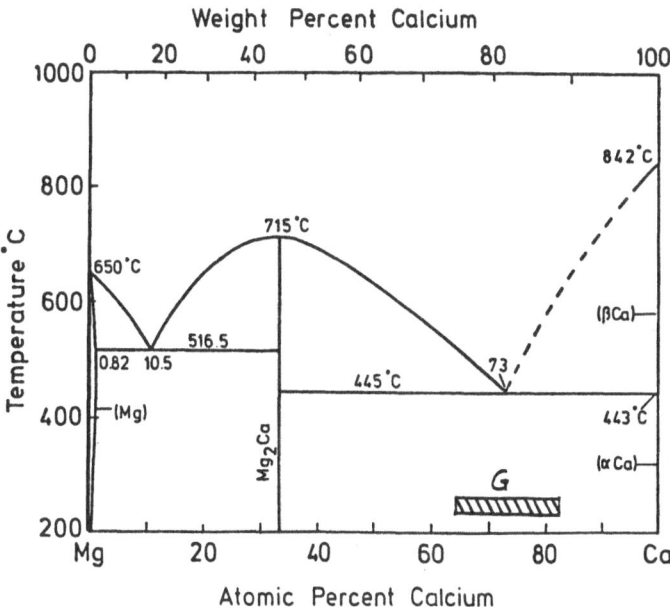

Figure 5. The Ca-Mg phase diagram. Drawn-in is also the concentration range of glass formation at a cooling rate of 10^6 K · s^{-1}.

The energetic situation of nucleation shall be demonstrated taking for an example the Ca-Mg system. In Fig. 5, the phase diagram is reproduced. Also, the concentration region is marked in which at a cooling rate of 10^6 K · s^{-1} glass formation is found. Fig. 6 reproduces the Gibbs free energy of the liquid alloys as well as of the solid equilibrium phases at the glass temperature, namely at $T_g \approx 400$ K. The three partial pictures are showing the energetic conditions for crystal nucleation at different alloy compositions. The diminution of the Gibbs free energy ΔG_n on crystal nucleation results from the construction of the respective tangents.

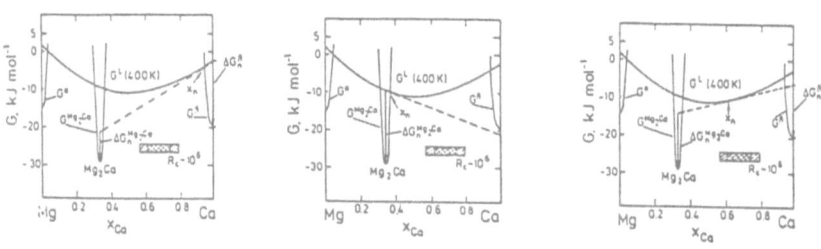

Figure 6. On the energetics of nucleation in the Ca-Mg system.

Case a) The driving force for the nucleation of the β-phase (Ca-rich solid solution), ΔG_n, is large, the force for the nucleation of the intermetallic Mg$_2$Ca phase is small. For the crystallization of β, only a small diffusion transport is necessary. In order to attain the Mg$_2$Ca structure, a considerable rearrangement of the atoms in the melt would have to take place, as the structures of the melt and of Mg$_2$Ca are not similar. This would take time. Therefore, the β phase crystallizes without considerable delay. There is no glass formation, as crystallization takes place before T_g is reached. This complies with the experimental findings.

Case b) If the molar fraction of the liquid alloy is near the stoichiometric composition Mg$_2$Ca, then $\Delta G_n^{Mg_2Ca}$ is large and ΔG_n^{β} is small. The probability for the formation of Mg$_2$Ca crystal nuclei is large. Again, no glass formation occurs, that is, this time because nucleation of the intermetallic phase sets in before T_g is reached.

Case c) If x_n lies about in the middle between the molar fraction for Mg$_2$Ca and for the β phase, then $\Delta G_n^{Mg_2Ca}$ and ΔG_n^{β} are both smaller than in the cases for the probable nucleation of Mg$_2$Ca (case b) or of a probable crystal nucleation of β (case a). Therefore, for none of the crystalline phases in question easy nucleation is to be expected. On the cooling of the liquid alloy T_g can be reached before crystal nucleation commences. So this is the region in which easy glass formation can be expected and is indeed found.

4.3 Kinetic Factors of Glass Formation

In so far as the short range order in the melt is analogous to the long range order present in the solid equilibrium phase a rapid crystal nucleation suggests itself, even then, if the gain in Gibbs free energy is not important. The diffusion steps required for the formation of the long range order are easily to perform. Then, the glass forming ability – in spite of a distinct short range order in the melt – is poor.

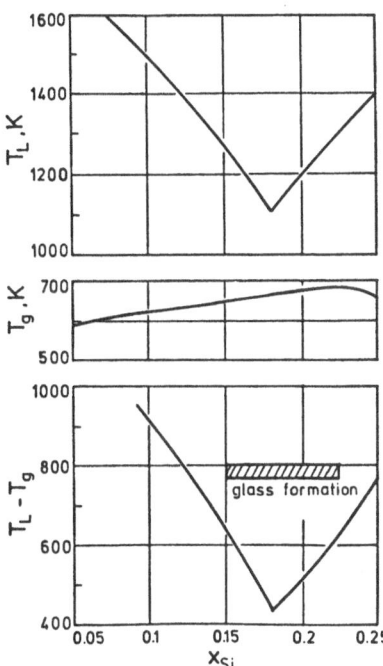

Figure 7. On the influence of the difference between the liquidus and solidus temperatures on the glass forming ability in the Pd-Si system according to (14) to (17).

Conspicuous is the fact that glass formation often occurs at concentrations near an eutectic. This is also the case in the Ca-Mg system and shall be more closely expounded for the Pd-Si system. As can be seen from Fig. 7, the liquidus temperature in the surrounding of the eutectic which is positioned at about 18 at.% Si is strongly dependent of the concentration whilst this is not true for the glass temperature T_g. Therefore, the difference $T_L - T_g$ strongly varies with the composition. This temperature interval now represents the range in which there is the danger of crystal nucleation. Above T_L, no thermodynamic driving force for crystal nucleation is present. Below T_g, crystal nucleation can not take place any more due to kinetic causes. At the eutectic concentration this interval which has to be surmounted on cooling without crystal nucleation, is minimal. Therefore, obviously also the time is minimal at a given cooling rate, which is required for the passing through this temperature interval $T_L - T_g$ at the eutectic concentration. This minimal time interval corresponds with a minimal probability of the occurrence of crystal nucleation. So it is not surprising that in the neighbourhood of an eutectic often glass formation is found, especially in the cases when there is a deeply cut-in eutectic.

In the surroundings of an eutectic a further factor favouring glass formation is coming up: As already demonstrated, the degree of ordering in the melt is increasing with decreasing temperature. At the eutectic concentration it is possible to reach rather low temperatures without surpassing T_L. In doing so, the viscosity assumes much higher values than would be the case in an equally composed but not short range ordered melt. Therefore, when reaching the eutectic temperature, the atomic mobility is relatively low and accordingly, the probability for supercooling is higher than for liquidus temperatures at other concentrations. An easily occurring supercooling causes a further lowering of the mobility, a lowering of the mobility a reduction of the nucleation rate, and so on.

According to the findings pointed out above, high glass formation ability can be expected if T_g is high. A high T_g can be expected if, as already mentioned, already a trifling below T_g the mobility of the atoms is so small that a "freezing-in" of the structure takes place, that means, that already at a small undercooling the mobility of the structural units isn't any more sufficient to achieve immediately an internal equilibrium at the given cooling rate. This is the case when the interatomic interactions are strong. As a measure of the strungth of these interactions the enthalpy of vaporization can be taken, which complies with the energy of cohesion. As an approximation for it the ideal vaporization enthalpy ΔH_i^V can be used, which is additively composed of the weighted vaporization enthalpies of the pure components.

As Fig. 8 is showing, the simple relationship between T_g and ΔH_i^V which is to be expected is indeed approximately given. Among the totality of the metallic glasses three groups can be distinguished. With increasing ΔH_i^V, the strongest increase of the glass temperature is observed in glasses on the basis of earth alkali metals. If the glass, on the other hand consists of a noble metal (e.g. Au) and of a tetravalent element (e.g. Ge, Si), then T_g only slightly increases with increasing ΔH_i^V. Between these extreme kinds of behaviour the ΔH_i^V-T_g value pairs for glasses with transition metals can be found.

Doubtlessly, these differences are connected with the structural and bonding conditions in the liquid alloys. Au alloys belong to systems having a less strongly distinct compound formation tendency whilst in the earth alkali metal systems the compound formation tendency is very pronounced. Also, it ought to be mentioned that the ternary alloys containing transition metals at a given ΔH_i^V always exhibit higher T_g values than glasses of binary alloys with transition metals. This obviously hints at further factors which possibly are connected with diffusion problems in the adaptation of the system to the state of internal equilibrium.

In connection with this it still ought to be mentioned that in general the glass formation ability, as experience shows, is increasing with increasing number of alloy components. It is assumed that this is connected with diffusion phenomena on crystal nucleation. The more components are present and the more equilibrium phases have to be formed on the crystallization of a eutectic mixture, the more difficult will be the distribution of the components which are homogenously dispersed in the melt to the locally separated solid phases. This fact is occasionally summarized by the word "confusion principle".

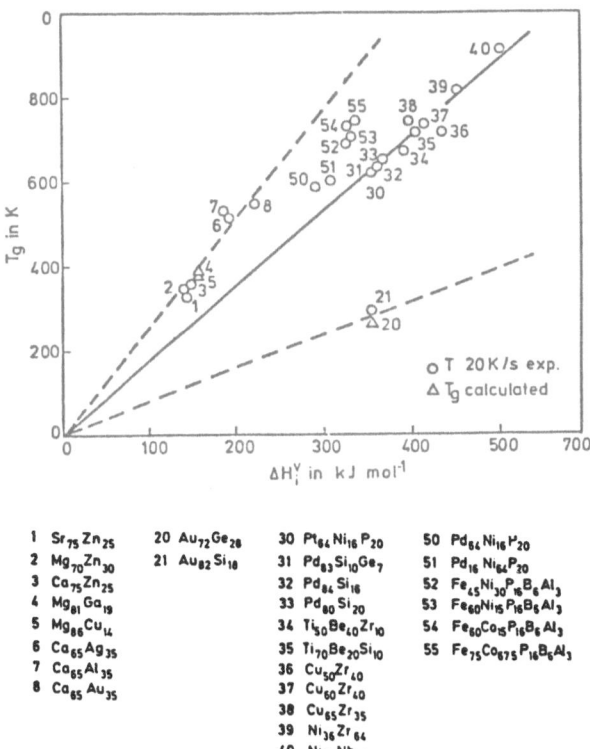

1 $Sr_{75}Zn_{25}$	20 $Au_{72}Ge_{28}$	30 $Pt_{64}Ni_{16}P_{20}$	50 $Pd_{64}Ni_{16}P_{20}$
2 $Mg_{70}Zn_{30}$	21 $Au_{82}Si_{18}$	31 $Pd_{83}Si_{10}Ge_7$	51 $Pd_{16}Ni_{64}P_{20}$
3 $Ca_{75}Zn_{25}$		32 $Pd_{84}Si_{16}$	52 $Fe_{45}Ni_{30}P_{16}B_6Al_3$
4 $Mg_{81}Ga_{19}$		33 $Pd_{80}Si_{20}$	53 $Fe_{60}Ni_{15}P_{16}B_6Al_3$
5 $Mg_{86}Cu_{14}$		34 $Ti_{50}Be_{40}Zr_{10}$	54 $Fe_{60}Co_{15}P_{16}B_6Al_3$
6 $Ca_{65}Ag_{35}$		35 $Ti_{70}Be_{20}Si_{10}$	55 $Fe_{75}Co_{8.75}P_{16}B_6Al_3$
7 $Ca_{65}Al_{35}$		36 $Cu_{50}Zr_{40}$	
8 $Ca_{65}Au_{35}$		37 $Cu_{60}Zr_{40}$	
		38 $Cu_{65}Zr_{35}$	
		39 $Ni_{36}Zr_{64}$	
		40 $Ni_{60}Nb_{40}$	

Figure 8. Glass temperature T_g as dependent on the ideal vaporization enthalpy ΔH_i^V of some glass forming alloys according to Sommer (12).

5. RELAXATION AND CRYSTALLIZATION OF METALLIC GLASSES

As the glass temperature T_g depends on the cooling rate and, therefore, at various cooling rates different atomic arrangements with different degrees of ordering are "frozen-in", the metallic glasses obtained at various cooling rates are, as already indicated at the beginning, not entirely equal even at the same chemical composition. They are differring in their atomic structures and consequently, also in their Gibbs free energies. To be sure, the mobility underneath of T_g is so small that the structure cannot change so rapidly as it would be necessary to change the atomic arrangement in such a manner that at any temperature passed the atomic configuration complying with internal equilibrium would be present. But, if the glass is annealed, in particular at a temperature slightly underneath T_g, a change of the atomic structure is taking place, as already mentioned, which is naturally connected with a reduction of the Gibbs free energy. On this relaxation, some properties can be changed considerably.

At sufficiently high temperatures and annealing times the relaxation process is followed by crystallization. On this occasion, the properties are showing peculiarly drastic changes. Also during the crystallization process of the glass, thermodynamic as well as kinetic influences are of importance. Primarily, the crystallization reaction can yield stable as well as metastable solid phases. According to the Ostwald rule, metastable intermediate steps may occur which only after a longer period at elevated temperatures are transitioning into the equilibrium phases. The respective course of the reaction will be determined by the interplay of kinetic and thermodynamic influences.

As an example, the crystallization processes in metastable iron-boron glasses shall be investigated here. They have been disclosed in detail by Köster and Herold (19). A schematic survey of the Gibbs free energies as a function of the concentration for the interesting phases is reported in Fig. 9.

In the case when the B concentration of the glass is small (arrow (1)) or is approximately in accordance with the stoichiometry of an intermetallic phase (arrow (4)), only one phase is crystallizing: α-Fe-B solid solution in reaction (1) or, respectively, the metastable intermetallic Fe_3B compound in reaction (4). In these reactions, the maximally possible Gibbs free energy is not gained. Only in a later reaction the α-Fe-B solid solution can disintegrate into a stable mixture of an α-Fe-B solid solution with a lower B concentration and of the stable intermetallic Fe_2B. Also, the primarily emerging metastable Fe_2B compound has to change into a mixture of the same stable phases.

If the B contents of the glass phase is increasing, the gain in Gibbs free energy due to the reaction (1) is sinking. As competing reaction, eventually the precipitation of α-Fe solid solution according to reaction (2) is coming into play. This precipitation results in the metastable equilibrium between the α-Fe-B solid solution and a glass matrix with the residual B concentration determined by the tangent t_2. Thus, B is enriched in the glass matrix; the B concentration in the α-Fe-B solid solution is lower than that of the initial matrix. The both phases of this metastable equilibrium can pass into equilibrium phases in a further reaction.

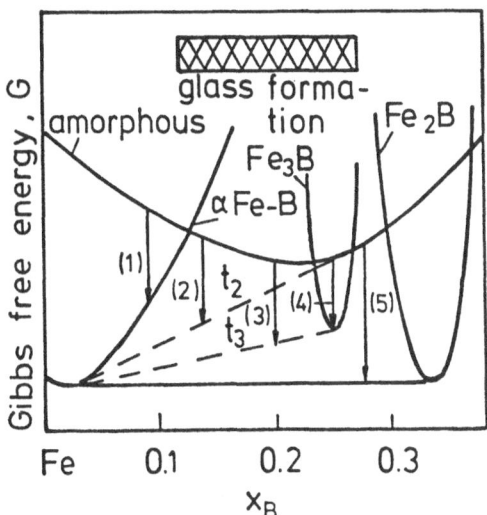

Figure 9. The gibbs free energy as a function of the concentration for some phases in the Fe-B system according to U. Köster and Herold (19).

If the composition of the glass is near the Fe_3B stoichiometry, then in reaction (3), in addition to α-Fe-B solid solution the metastable Fe3B compound is developing according to the tangent t_3. The spatial distribution of these both phases will be similar to that one resulting from an eutectoid reaction. Lamellar precipitation colonies are emerging. If the initial glass has a B concentration which is near the stoichiometry of the stable Fe_2B phase, in the reaction (5) immediately the stable phases α-Fe-B solid solution and Fe_2B are resulting. Also all intermediate steps which are primarily formed in other reactions are eventually transformed, as already indicated, into this stable two-phase equilibrium.

The choice of the respective reaction course shall be dealt with only briefly here. The transformation of the glass into a unique phase - either the α-Fe-B solid solution or the metastable Fe_3B phase - proceeds with the highest velocity, as a redistribution of the single alloy components to different phases by diffusion is not necessary. The formation of the crystal nuclei preferably takes place at the surface of the glass.

The cooling rate at the glass production is without influence on the formation of crystal nuclei. Namely, before crystallization sets in, all structural peculiarities which could be possible centres for nucleation are eliminated by relaxation processes.

As the crystallization temperature T_c, the temperature is defined at which the crystallization reaction sets in which most rapidly forms nuclei. The knowledge of T_c is of importance for technically applicable metallic glasses as with the onset of crystallization considerably property changes are connected, as already stated. Above all, it is the intention to produce glasses with high T_c. In Table 1, some examples of T_c values are reported.

TABLE I Crystallization temperature T_c in K of metallic glasses according to (25).

met. glass	T_c in K	Literature
$Ta_{55.5}Ir_{44.5}$	> 1223	(20)
$Ni_{60}Nb_{40}$	923	(21)
$(Fe_{40}Ni_{60})_{75}P_{16}B_6Al_3$	714	(22)
$Pd_{82.4}Si_{17.6}$	639	(23)
$Mg_{86}Cu_{14}$	380	(24)

6. PROPERTIES AND POSSIBILITIES FOR THE APPLICATION OF METALLIC GLASSES

Due to their outstanding properties, metallic glasses can be used as materials for specific applications. Table 2 gives a short, comparative survey of the most important properties of glasses as well as of crystalline metallic materials. Of special importance there are magnetic, electric, mechanical, and corrosion properties.

It is quite possible to produce crystalline, ferromagnetic materials which are similarly soft-magnetic as ferromagnetic metallic glasses. The production of such soft-magnetic materials, however, is connected with a rather lengthy and costly heat treatment, whereas the metallic glass can be obtained as a ribbon, which essentially needs no further processing. In many cases, the combination of the soft-magnetic behaviour with a high hardness is desirable, which cannot easily be achieved in crystalline materials. Just how advantageous, under this aspect, the ferromagnetic metallic glasses are can be seen from Fig. 10, wich gives a prospect on some commercial soft-magnetic materials.

TABLE II Properties of metallic glasses as compared with crystalline metals and silicate glasses after Güntherodt (27).

Property	Crystalline Metals	Metallic Glasses	Silicate Glasses
Plasticity	good; ductile	good; ductile	bad; brittle
Hardness	small	high	high
Tensile strength	high	high	low
Optical properties	opaque	opaque	transparent
Electrical and thermal conductance	good	good	bad
Corrosion resistance	low	high	high
magnetism	various phenomena	various phenomena	non-magnetic

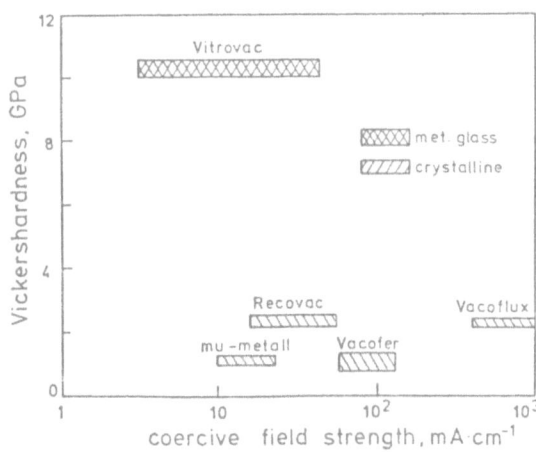

Figure 10. Comparison of mechanical and magnetic propertis of some commercial soft-magnetic materials (Vakuumschmelze Hanau (26)). Simple hatching: crystalline material; crosshatched: metallic glass.

Also the special combination of magnetic and electrical properties proves to be useful. The specific electrical resistance of metallic glasses is higher by a factor of about 2 to 3 than that of a crystalline alloy of the same composition. This is very suitable for the construction of transformers with small energy losses. On the one hand, in transformers whose ferromagnetic part consists of a metallic glass the hysteretic losses due to the reversion of the magnetization are small, on the other hand, due to the natural geometry of the thin foils in combination with still satisfactory electrical properties, the eddy current losses are also very low.

Although metallic glasses are metastable, they are considerably more resistant against a corrosive attack than crystalline alloys of the same composition. Due to energetic causes, the corrosion reaction ought to proceed much quicker in a glass than on the attack of an e-qually composed crystalline material. This correlation, which is to be expected is indeed there, but also of importance is the probability of the onset of corrosion. The inhibition of the corrosion of metallic glasses takes place right at its initial stage. The initial reaction of the attack by a corrosive agent in a crystalline material usually begins at lattice defects, such as dislocation line penetration points and grain boundary intersection lines, at the surface of the metal. Such lattice defects which could be good starting places for the corrosion reaction, are not present in a metallic glass. Metallic glasses containing chromium, are spontaneously forming a passive layer. In the case of crystalline materials a passive layer, at the surface penetration points of dislocation lines, shows irregulatities. In the case of metallic glasses there is an extensively smooth passive layer which is much more resistant than a disturbed one. There are much less special concentration conditions but, on the contrary, lattice defects which are also decisive for the corrosion behaviour (28).

References

(1) W. Klement, R.H. Willens and P. Duwez, Nature, 187 (1960) 869.

(2) H. Schulze, 'Glas', Vieweg u. Sohn, Braunschweig (1965).

(3) P. Pietrokowski, Rev. Sci. Instr., 34 (1963) 445.

(4) B. Predel, and K. Hülse, J. Less-Common Met., 63 (1979) 45.

(5) P. Duwez, R.H. Willens, and W. Klement, J. Appl. Phys., 31 (1960) 1136.

(6) B. Predel, and G. Duddek, Z. Metallkde., 69 (1978) 773.

(7) H.S. Chen, and C.E. Miller, Rev. Sci. Instr., 41 (1970) 1237.

(8) R. Pond jr., and R. Maddin, Trans. Metall. Soc. AIME, 245 (1969) 247.

(9) J. Nixdorf, Draht, 53 (1967) 696.

(10) M. Moss, D.L. Smith, and R.A. Lefevre,
Appl. Phys. Lett., $\underline{5}$ (1964) 120.

(11) H.S. Chen , and D. Turnbull, Appl. Phys. Lett., $\underline{10}$ (1967) 284.

(12) F. Sommer, Z. Metallkde., $\underline{72}$ (1981) 219.

(13) F. Sommer, Z. Metallkde., $\underline{73}$ (1982) 72, 77.

(14) B.G. Lewis, and H.A. Davies, in P.H. Gaskell (Editor):
'The Structure of Non-Crystalline Materials', Taylor
and Francis, London (1977).

(15) P. Duwez, Trans. Amer. Soc. Met., $\underline{60}$ (1967) 607.

(16) R.W. Cahn, Contemp. Phys., $\underline{21}$ (1980) 43.

(17) H.S. Chen, and K.A. Jackson, in 'Metallic Glasses', Papers
presented at Sem. of Mater., Soc. Div. Amer. Soc. Met.,
(1976), Amer. Soc. Met., Metals Park (Ohio) (1978).

(18) H.J. Güntherodt, and H. Beck (Eds.): 'Glassy Metals I' in
Topics in Applied Physics, Vol. $\underline{46}$; Springer Verlag, Berlin
(1981).

(19) K. Köster, and K. Herold, in H.J. Güntherodt and H. Beck (Eds.):
'Glassy Metals I', Topics in Applied Physics, Vol. $\underline{46}$;
Springer-Verlag, Berlin (1981).

(20) M. Fischer, D.W. Polk, and B.C. Giessen, in R. Mehrahian,
B.H. Kear and M. Cohen (Eds.): Proc. Conf. on Rapid Solidification
Processing, Claitors Publ. Div., Baton Rouge (1978).

(21) N.A. Pratten,and H.G. Scott, Scr. Metall., $\underline{12}$ (1978) 137.

(22) H.S. Chen, Acta Metall., $\underline{24}$ (1976) 153.

(23) H.S. Chen, and B.K. Park, Acta Metall., $\underline{21}$ (1973) 395.

(24) F. Sommer, G. Bucher, and B. Predel, J. Phys.,
$\underline{C8}$ (1980) 563.

(25) B. Predel, Nachr. Chem. Techn. Lab., $\underline{31}$ (1983) 168.

(26) H. Warlimont, Inst. Phys., (1980) 29.

(27) H.J. Güntherodt, Metall, $\underline{33}$ (1979) 723.

ASPECTS OF HIGH TEMPERATURE CORROSION OF METALS BY GASES

Per Kofstad
Department of Chemistry
University of Oslo
P.B. 1033, 0315 Oslo 3,
Norway

ABSTRACT. The paper surveys high temperature corrosion phenomena involving diffusion-controlled growth of continuous scales. Lattice diffusion is generally rate-determining only for scales with relatively high concentrations of point defects and at high temperatures, e.g. oxidation of Co to CoO above about 1000 °C. At more reduced temperatures and for scales with relatively low point defect concentrations scale growth predominantly takes place by grain boundary diffusion, e.g. growth of NiO, Cr_2O_3, and Al_2O_3. The common use metals may corrode rapidly in many sulfur-containing atmospheres due to the formation of sulfide phases which are highly nonstoichiometric and exhibit high rates of diffusion. This is illustrated for corrosion of metal and dilute nickel alloys in O_2+SO_2/SO_3 atmospheres. It is emphasized that the distribution of the sulfide phases in the scales is important for the reaction mechanisms.

1. INTRODUCTION

Metals are thermodynamically unstable in most gaseous environments. Depending on the applications, the ambient atmospheres may consist of oxygen, air, combustion gases, special gas mixtures in high temperature chemical and metallurgical processes, gasification of coal, operation of nuclear reactors, etc. In dry, gaseous environments at room or low temperatures corrosion rates are often low and of little practical consequence. But reaction rates increase rapidly with increasing temperatures, and at high temperatures gaseous corrosion and our ability to protect metals is of great importance.

Gaseous corrosion involves a number of different part-reactions and processes depending on the metal, gaseous environment, temperature, time of reaction, etc. (1). If one exposes a clean metal to reacting gases, the initial process is adsorption of gas on the metal surface. This is followed by nucleation and growth of nuclei of the reaction products and they generally grow laterally to cover the metal surface as a film. The reactants, the metal and the reacting gas species, are then separated and the reaction proceeds through solid state transport

471

H. Brodowsky and H.-J. Schaller (eds.), Thermochemistry of Alloys, 471–480.

through the film. At high temperatures the films grow in thickness to thicker scales. In order to achieve good corrosion resistance, the solid state transport through the growing, protective scale must be as low as possible.

But additional complications arise. Growth stresses generally develop in the scales and the underlying metal. Stresses and strains may be relieved through various processes. The scales may crack and they then loose at least part of the protective properties and the reacting gas may penetrate into the scale or to the metal/scale interface. Alternatively, the scales may plastically deform through high temperature creep. But the process may also, in turn, produce microcracks or channels in the scales and these may also serve as transport paths of reacting gas molecules into the inner part of the scales.

For some systems reaction products may partially be liquid, and for still others relatively volatile products may be formed. These features may significantly affect the reaction behaviour. In many practical applications, particularly exposures to combustion gases, salts (e.g. alkali chlorides, sulphates and vanadates) may be deposited on the corroding metal surfaces. These may often form liquid phases on the surface and induce rapid corrosion and breakdown of the materials.

All in all, high temperature corrosion may involve complicated and interrelated processes and where the overall reaction can be extremely complex. This presentation focuses on fundamental aspects and processes of growth of continuous scales. This will be exemplified through reactions of metals with single reactants such as oxygen and sulphur and with mixed reactants containing both oxygen and sulphur, i.e. mixtures of SO_2+O_2.

2. REACTIONS CONTROLLED BY SOLID STATE DIFFUSION THROUGH SCALES

Solid state diffusion may take place through lattice, grain boundary, and dislocation diffusion. Grain boundary and dislocation diffusion are generally orders of magnitude faster than lattice diffusion and are also often termed short-circuit diffusion. Oxide scales may grow by all types of solid state diffusion; in the following it is convenient to consider these separately.

2.1. Lattice diffusion-controlled scale growth

The basic theory for this type of scale growth was developed by C. Wagner (2,3). He assumed that lattice transport of ions of the reactants (e.g. metal and/or oxygen ions in metal-oxygen reactions) is the rate-determining process in the growth of scales which are electron-conductors; alternatively electron-transport is rate-determining for growth of scales which are ionic conductors. Since lattice transport is rate-determining, he further assumed that interface reactions are rapid and that thermodynamic equilibria are established at the metal/scale and scale/gas interfaces.

On plane surfaces this type of reaction behaviour leads to a scale growth that is parabolic with time. This may be described by the equations

$$\frac{dx}{dt} = k_p' \frac{1}{x} \tag{1a}$$

or

$$x^2 = 2k_p't + C = k_pt + C \tag{1b}$$

where x is the oxide thickness, t is time, k_p' and k_p parabolic rate constant ($k_p = 2k_p'$), and C the integration constant.

In his theory Wagner expressed the parabolic rate constant either in terms of the electrical conductivity and transport numbers of electrons and ions or in terms of the self-diffusion coefficients of the reacting ions through the scale. In the latter case and when the scale thickness is expressed in cm, the parabolic rate constant for growth of a scale of the oxide M_aO_b is given by

$$k_p' = \frac{1}{2} \int_{p_{O_2}^i}^{p_{O_2}^o} \frac{z_c}{|z_a|}(D_M+D_O)d\ln p_{O_2} \tag{2}$$

D_M and D_O represent the self-diffusion coefficients of the cations and anions, respectively; z_c and z_a are the valences of the corresponding ions, and $p_{O_2}^o$ and $p_{O_2}^i$ are the partial pressures of oxygen at the scale (gas/scale and metal/scale interfaces, respectively. In order to evaluate k_p' from Eq. 2 the oxygen pressure dependence of D_M and D_O must be known.

In detailed analyses of the defect structures of oxide scales, the oxygen pressure dependencies of D_M and D_O may for many oxides over their entire existence range be complicated expressions, and the integrations in Eq. 2 may then be difficult to perform. However, simplified expressions may be obtained for some ideal, limiting conditions. By way of example, let us assume that i) $D_M \gg D_O$ and that the same defect structure situation prevails through the entire scale. Let us further assume that metal vacancies with an effective charge α are the predominant lattice defects and that the ambient oxygen pressure, $p_{O_2}^o$, is much larger than that at the metal/scale interface, i.e. $p_{O_2}^o \gg p_{O_2}^i$. For this limiting condition, k_p' becomes

$$k_p' = (\alpha+1)D_M^o \tag{3}$$

where D_M^o is the self-diffusion coefficient of the metal ions at the scale/gas interface (1).

In order to test the validity of the Wagner theory, comparisons have been made between directly measured values of k_p' and values calculated from experimental values of self-diffusion coefficients or electrical conductivity. For some systems good agreement has been obtained confirming the validity of the theory for selected systems (1). For metal-oxygen reactions there are few systems for which such comparisons can be made due to the lack of appropriate data on self-diffusion data in oxides. However, oxidation of cobalt to CoO is an example where data are available. CoO is a metal-deficient oxide where the predominant defects are cobalt vacancies with one negative effective charge - at least at near-atmospheric oxygen pressures (4). Following Eq. 3 the oxygen pressure dependence of k_p' and D_{Co} (the random self-diffusion coefficient of cobalt in CoO) should be equal. This is also found experimentally at temperatures above about 1000 °C. Furthermore, the ratio of k_p' to D_{Co} is expected to be 2; similarly, and taking into account the correlation effects of tracer diffusion which for CoO amounts to f = 0.79, the ratio of k_p' to D_{Co}^{tr} (the cobalt tracer self-diffusion coefficient in CoO) is expected to be close to 2.5. This has also been confirmed experimentally, and this strongly indicates that the growth of CoO-scales predominantly takes place by cobalt lattice diffusion (1).

All in all, it is to be expected that lattice diffusion is the predominant mode of transport in growth of scales at high temperatures and for oxides with relatively large concentrations of lattice defects.

2.2. Scale growth controlled by grain boundary diffusion

In view of the similarities in properties of CoO and NiO it is of interest to make a corresponding comparison between the parabolic rate constant for oxidation of nickel and the nickel self-diffusion coefficient in NiO. This oxide is also metal-deficient and the predominant defects constitute nickel vacancies. However, the nickel vacancy concentration in NiO is more than an order of magnitude smaller than the cobalt vacancy concentration in CoO under the same conditions (4). In NiO the nickel self-diffusion coefficient is 2-3 orders of magnitude higher than that of oxygen diffusion (1). Figure 1 shows such a comparison of k_p' and D_{Ni}^T (the nickel tracer lattice self-diffusion coefficient in NiO) (5-11). Below about 1200 °C the ratio k_p'/D_{Ni}^T gradually increases and at 500-600 °C the observed parabolic rate constants are several orders of magnitude higher than values expected from Ni lattice diffusion in NiO.

The reasons for this difference have been the subject of numerous investigations in recent years. A number of aspects have been studied: microstructures of scales, tracer distributions in growing scales, self-diffusion of nickel in oxide scales, effects of metal orientation and surface preparation prior to the oxidation, etc., and it can be unequivocally concluded that the increased rate of oxidation with decreasing temperature is due to a corresponding increase in the importance of grain boundary diffusion (1,10,12). At reduced temperatures (500-800 °C) grain boundary diffusion predominates; lattice diffusion is of minor importance. Dislocation diffusion has also been shown to be

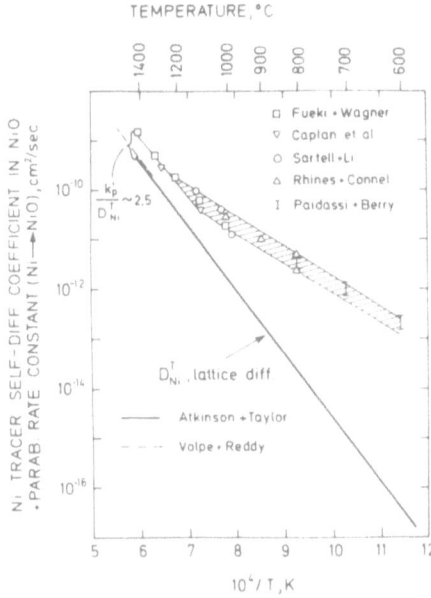

Fig. 1. Comparison of the parabolic rate constants for oxidation of high purity nickel and the nickel tracer lattice diffusion coefficient in NiO. Results of oxidation studies after Refs. 5-9 and of diffusion studies after Refs. 10 and 11.

of little importance compared to grain boundary diffusion (10).

Using the values of the grain boundary diffusion coefficient for nickel and knowledge of the microstructure of the scales, an oxidation model may be set up and values of the rate constant may be calculated. In this model the diffusion coefficient is expressed as an effective diffusion coefficient which constitutes a weighted sum of the diffusion coefficients for lattice and grain boundary diffusion (13).

$$D_{eff} = D_{latt.}(1-f) + D_{gb}f \qquad (4)$$

where f denotes the fraction of the total member of diffusion sites that are in the boundaries. Calculated and directly measured values of the oxidation rate constants for nickel have been in good agreement (14). Grain boundary diffusion also predominates in other systems (1). Thus for chromia scales growth rates are generally orders of magnitude faster than values of the chromium lattice diffusion coefficient in Cr_2O_3. Growth of alumina scales at high temperatures are also in all probability governed by grain boundary diffusion. Thus scales that are of great technological importance in providing corrosion resistance to numerous high temperature alloys predominantly grow by grain boundary diffusion.

3. STRESSES AND STRAINS IN GROWING SCALES; FORMATION OF VOIDS AND MICROCHANNELS:

So far the scales have been considered to be dense and continuous. But as described in the introduction scales may crack or plastically deform

476

and voids and microchannels may develop in the scales. In order to describe these phenomena and their effects on the reaction behaviour, it
is convenient to consider stresses that develop in the scales and the
underlying metal.

The stresses that develop in the scales and the underlying metal
depend on a number of factors: the ratio of the volume of the oxide to
the volume of the metal from which it is formed, properties of the
oxides (e.g. high temperature creep), mechanisms of solid state diffusion through the scales, growth mechanisms, a.o. For high temperature
metals and alloys compressive stresses generally develop in the scales
under isothermal conditions. As a result of this, oxide scales often
develop columnar grains at sufficiently high temperatures. In other
cases oxide scales may buckle and partially detach from the metal substrate; this is often observed for chromia and alumina scales. An example of this is illustrated for a chromia scale formed on unalloyed
chromium at 1200 $^{\circ}$C (1). It has been proposed that the large stresses

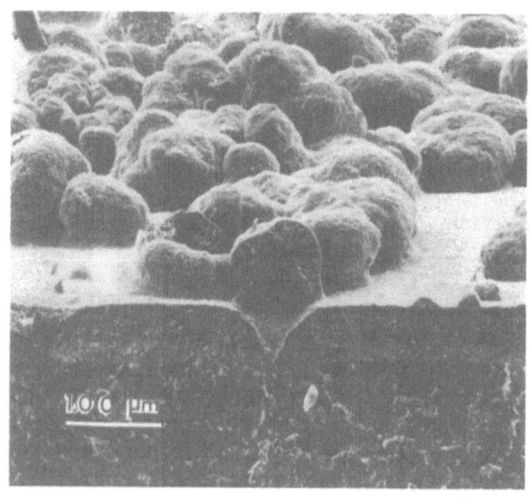

Fig. 2. Scanning electron
micrograph of Cr_2O_3 scale
formed on unalloyed chromium in 1 atm. O_2 at
1200 $^{\circ}$C (1).

that develop in growth of such chromia scales are due to the growth
mechanism that involves counter-current diffusion of chromium and oxygen along grain boundaries and a resultant growth of new oxide crystallites at the grain boundaries. In this way the oxide scale not only grows
normal to the metal surface, but also sideways. This creates large
stresses which may result in cracking, buckling or extensive plastic deformation of the scales depending upon the reaction conditions. Chromia
scales have an increasing tendency to deform plastically with decreasing
oxygen pressure, and under highly reduced oxygen pressures the chromia
scales may grow detached from the metal substrate as balloon-like oxide
shells (2).

As a result of the deformation through the scales and that this
varies with the variation in oxygen activity through the scale, it

has also been suggested that microcracks, pores and microchannels develop in the scales (1).

4. HIGH TEMPERATURE CORROSION IN SULFUR AND SULFUR-CONTAINING ATMOSPHERES:

Metals that constitute the base metals and important alloying additions in high temperature alloys (e.g. nickel, cobalt, chromium, iron) corrode rapidly in sulfur gas and in many sulfur-containing gases. Reactions may be several orders of magnitude faster than in oxygen or under conditions where only oxide scales are formed (1).

The reason for this is that many sulfide phases that are formed as scales or part of the scales are highly nonstoichiometric and defective (e.g. Co_8S_7, Ni_3S_2, Cr_2S_3) and as a result exhibit very rapid lattice diffusion rates.

The reaction behaviour of metals in gases containing sulfur oxides (SO_2, SO_3) may under many conditions prove extremely complex. Due to the presence of two basic reactants (sulfur and oxygen) the reaction products may comprise oxides, sulfides and/or sulfates. Let us give a brief illustration of this by the reaction of high purity nickel with gas mixtures of SO_2+O_2 (9,10).

Extensive studies of this reaction have shown that nickel may at 600-900 °C corrode several orders of magnitude faster in such gas mixtures than in oxygen gas. Furthermore, the critical parameter for this rapid reaction is the partial pressure of SO_3 at the surface of the metal specimen. Such rapid reactions take place when nickel sulfate may be formed on the surface of the scale through the reaction

$$NiO + SO_3 = NiSO_4 \hspace{4cm} (5)$$

Accordingly the mechanism is often termed the sulfate mechanism.

The overall reaction can be described as follows: During the very initial reaction of nickel with SO_3 the scale consists of a mixture of $Ni_{3\pm x}S_2$ and NiO. The sulfide phase "wets" the oxide crystallites and forms a continuous network through the scale. Nickel diffuses rapidly outward through the scale to the scale surface where additional reaction products are formed. When $NiSO_4$ may be formed on the surface through Eq. 5, the outwardly diffusing nickel reacts with $NiSO_4$ to form a mixture of nickel sulfide and oxide:

$$9Ni + 2NiSO_4 \rightarrow Ni_{3\pm x}S_2 + 8NiO \hspace{3cm} (6)$$

As a result the scale consists of the sulfide/oxide mixture and rapid reaction continues for extended periods.

If the SO_3-pressure is not sufficiently high to form $NiSO_4$, the scale gradually builds up to consist predominantly of NiO, and the reaction rate is then of the same order as that for the reaction of nickel in oxygen gas.

478

It is concluded from these studies that a very important feature of the reaction mechanism is that nickel sulfide $(Ni_{3\pm x}S_2)$ is able to wet the nickel oxide grains and in that way forms a three-dimensional network through the scale and that serves as a rapid diffusion path for the nickel. Thus the composition and the properties of the grain boundary regions in the scales are extremely important for the reaction behaviour.

This aspect may be further illustrated by similar studies on commercial nickel (termed Wiggin Ni-200 and Ni-201) with a purity of about 99.6%. Under similar reaction conditions these commercial qualities react considerably slower than the high purity nickel as illustrated in Fig. 3 (17-19). The important metal impurities in the

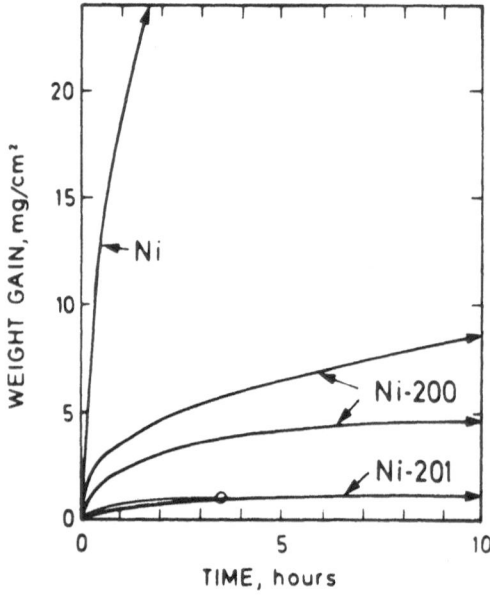

Fig. 3. Comparison of corrosion rates of high purity nickel (Ni) and commercial types of nickel (Ni 200 and Ni 201) at 700 °C in $O_2+4\%SO_2$ at a total pressure of 1 atm. (19).

commercial qualities are iron, silicon, manganese and copper. It was hypothesized that these impurities may affect the reaction behaviour and accordingly a series of dilute nickel alloys with these alloying elements were prepared and studied (17-19). Small additions of iron, silicon and manganese proved to have beneficial effects on the corrosion behaviour, and even further improvements and synergistic effects were obtained for dilute ternary and quaternary alloys. The composition of the scales on these dilute alloys proved to be similar to that on high purity nickel: an outer sulfate layer (consisting of a mixture of nickel sulfate and the sulfate of the alloying element, e.g. $MnSO_4$) and an inner layer consisting of oxide (NiO) and sulfide. The silicon was furthermore evenly distributed through the scales. Judging from the composition of the scales the reaction behaviour thus seemed to be

similar to that for high purity nickel, but still the reaction rates
are much slower for the dilute alloys. The provisional interpretation
of this is that the sulfide in the scale on the dilute alloys does not
wet the oxide grains and accordingly does not form a three-dimensional
network through the scale. Rather, the sulfide is seemingly formed as
particles in the scales and as such does not serve as rapid diffusion
paths through the scale. This, in turn, reflects a change in the grain
boundary properties in the scale, and this is proposed to be due to an
enrichment or accumulation of the alloying elements at the surfaces of
the oxide grains, e.g. as silicates for the silicon-dontaining alloys
(17-19).

High temperature corrosion of metals and alloys comprises numerous
additional phenomena which can not be covered in a brief presentation.
However, the examples discussed above emphasize one feature that has
become increasingly recognized in the interpretation of mechanisms of
high temperature corrosion in more recent years: the importance of properties of and transport processes along grain boundaries. The author
is convinced that further fundamental knowledge of these aspects will
contribute significantly to an improved understanding and elucidation
of high temperature corrosion.

REFERENCES

1) P. Kofstad, 'High Temperature Corrosion', Elsevier Applied
    Science, London, 1988
2) C. Wagner, Z. Phys. Chem., B21 (1933) 25
3) C. Wagner, Progr. Solid State Chem., 10 (1975) 3
4) P. Kofstad, 'Nonstoichiometry, Diffusion, and Electrical
    Conductivity in Binary Metal Oxides', Wiley-Interscience,
    New York, 1972
5) K. Fueki and J.B. Wagner, Jr., J. Electrochem. Soc., 112
    (1965) 384
6) D. Caplan, M.J. Graham and M. Cohen, J. Electrochem. Soc., 119
    (1972) 1295
7) J.H. Sartell and C.H. Li, J. Inst. Metals, 90 (1961-62) 92
8) F.N. Rhines and R.G. Connel, Jr., J. Electrochem. Soc., 124
    (1977) 1122
9) J. Paidassi and L. Berry, Compt. rend. Acad. Sci. Paris, 262
    (1966) 1553
10) A. Atkinson and R.I. Taylor, J. Mat. Science, 13 (1978) 427;
    Phil. Mag., A39 (1979) 51; 43 (1981) 979
11) M.L. Volpe and J. Reddy, J. Chem. Phys., 53 (1970) 1117
12) A. Atkinson, R.I. Taylor, and P.D. Goode, Oxidation of Metals, 13
    (1979) 519
13) W.W. Smeltzer, R.R. Haering, and J.S. Kirkaldy, Acta Metall., 9
14) A. Atkinson, M.L. Dwyer and R.I. Taylor, J. Mater. Sci., 18
    (1983) 2371

480

15) B. Haflan and P. Kofstad, Corrosion Science, $\underline{23}$ (1983) 1333
16) K.P. Lillerud, B. Haflan, and P. Kofstad, Oxidation of Metals $\underline{21}$ (1984) 119
17) B. Haflan and P. Kofstad, Oxidation of Metals, $\underline{25}$ (1986) 217
18) B. Haflan, K.P. Lillerud, and P. Kofstad, in "Reactivity of Solids", ed. by P. Barret and L.-C. Dufour, Elsevier Science Publishers B.V., Amsterdam, 1985, p. 67.
19) A. Andersen, B. Haflan, P. Kofstad, and K.P. Lillerud, Materials Science and Engineering, $\underline{87}$ (1987) 45.

THE HOT CORROSION OF METALS BY FUSED SALTS

Robert A. Rapp
Department of Metallurgical Engineering
The Ohio State University
116 West 19th Avenue
Columbus, Ohio 43210

ABSTRACT. The corrosion of metals by electrically conducting fused salts is a subject of scientific interest and engineering importance. Accelerated oxidation of a metal beneath a thin fused alkali sulfate film degrades turbine alloys and coatings, as one example. Understanding the fundamental steps involved in this "hot corrosion" requires a knowledge of phase stabilities, acid-base behavior of the salt electrolyte, solubilities of the protective oxides in the salt, and details of the electrochemical reactions for the metals and solutes. Experimental testing of mechanistic models is best made using electrochemical methods. The state of knowledge for these topics is reviewed in this presentation.

INTRODUCTION. In many of the traditional (smokestack) industries of the world, such as electrical power generation or refuse incineration, as well as in many high-tech systems such as gas turbines, high temperature fuel cells, or solar heat transfer systems, etc., alloys come into contact with corrosive fused salts at elevated temperatures. Because salts such as alkali sulfates, carbonates, nitrates and vanadates are generally electrolytic conductors, corrosion processes must be understood in terms of electrochemical corrosion mechanisms, analogous to the aqueous corrosion of metals near room temperature, a subject which has received detailed investigation for a century. While the electrochemical principles are common for both types of corrosion, almost every detail about the metal/electrolyte interaction is quite different. Rahmel (1) has reviewed the accepted mixed potential theory of electrochemical corrosion by fused salts. One particularly important type of fused salt corrosion occurs when only a thin film of electrolyte separates the protective oxide on an alloy from the oxidizing gas phase. This "hot corrosion", which occurs as alkali sulfate vapors condense or particles impact onto gas turbine hardware at high temperatures, enjoys a geometric similarity to atmospheric corrosion by ambient water films on metals, but the specific details for the two processes are quite different. The author (2,3) has recently reviewed the chemistry and electrochemistry of hot corrosion.

H. Brodowsky and H.-J. Schaller (eds.), Thermochemistry of Alloys, 481–494.

482

THERMODYNAMIC CONSIDERATIONS. An important initial aspect of hot corrosion
is the consideration of phase stability, which, for the equilibria between
pure metals and pure oxyanion salts, requires only a knowledge of the
standard Gibbs formation energies of the compounds in the system. The
introduction of activity coefficients for alloys and salt solutions is
obvious. Figure 1 is a phase stability diagram for the Na-S-O system at
1200K (4). Fused Na2SO4 is frequently a major constituent of corrosive
deposits because of its high thermodynamic stability coupled with the
general presence of Na and S in most combustion systems, particularily
near seawater. The solid lines on Fig. 1 indicate the conditions for the
equilibrium coexistence of phases having equal activities (usually unity).
Over the huge regime of Na2SO4 existence, the abscissa may be expressed
redundantly by either log PSO3 or -log a Na2O. The dashed lines within
the Na2SO4 field are predominance lines, identifying the principal minority
solutes in the system (assuming an ideal solution). The predominance
fields prove to be very useful in the interpretation of chemical or
electrochemical reactions, whereby the principle of forming stable species
from unstable species needs to be respected. The light sloping lines are
isopotential lines for a specific Na^+-ion reference electrode with a scale
indicated at the lower right of the diagram.

For such an oxyanion salt system, an acid-base behavior is important,
and the quantity $+log$ a_{Na2O} can be defined as the salt basicity, with pure
solid Na2O as the standard state. This definition avoids the need to
treat the activity of any ionic species. Later, electrochemical probes
are described which permit the quantitative measurement of melt basicity
and oxidizing potential, the two coordinates of Fig. 1.

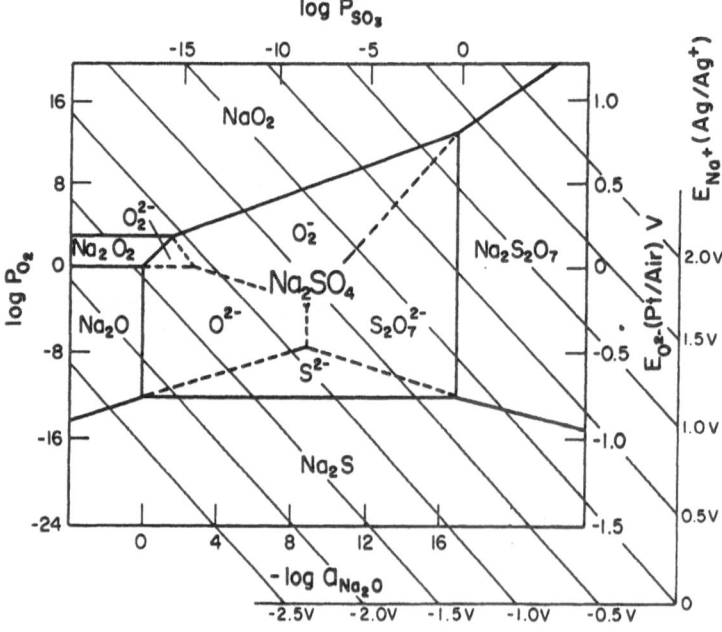

Fig. 1 Phase Stability Diagram for Na-S-O System at 1200K.

Metals and alloys are protected from accelerated hot corrosion by
the formation of an adherent, slow growing oxide film, generally
comprised of Cr_2O_3, Al_2O_3, or SiO_2. However, such oxides exhibit a
limited regime of phase stability in contact with the fused salt. For
example, Fig. 2 illustrates the superposition of the stability diagram
for the Na-Cr-S-O system with the solvent Na-S-O system. Within the
Na_2SO_4 stability field, Cr_2O_3 may react to form any of four corrosion
products; $Cr_2(SO_4)_3$, CrS, Na_2CrO_4, or $NaCrO_2$, depending upon melt
basicity. For oxidizing basic conditions, Na_2CrO_4 is stable, and under
very reducing environments CrS should form, etc. Notice that the metal
Cr is not stable under any conditions in contact with Na_2SO_4, so any
crack or fissure through a protective Cr_2O_3 film on an alloy or coating
would result in a reaction to form CrS, or perhaps $NaCrO_2$, depending
upon the melt basicity. Even within the Cr_2O_3 field of Fig. 2, the
oxide exhibits a certain solubility which depends upon melt basicity.

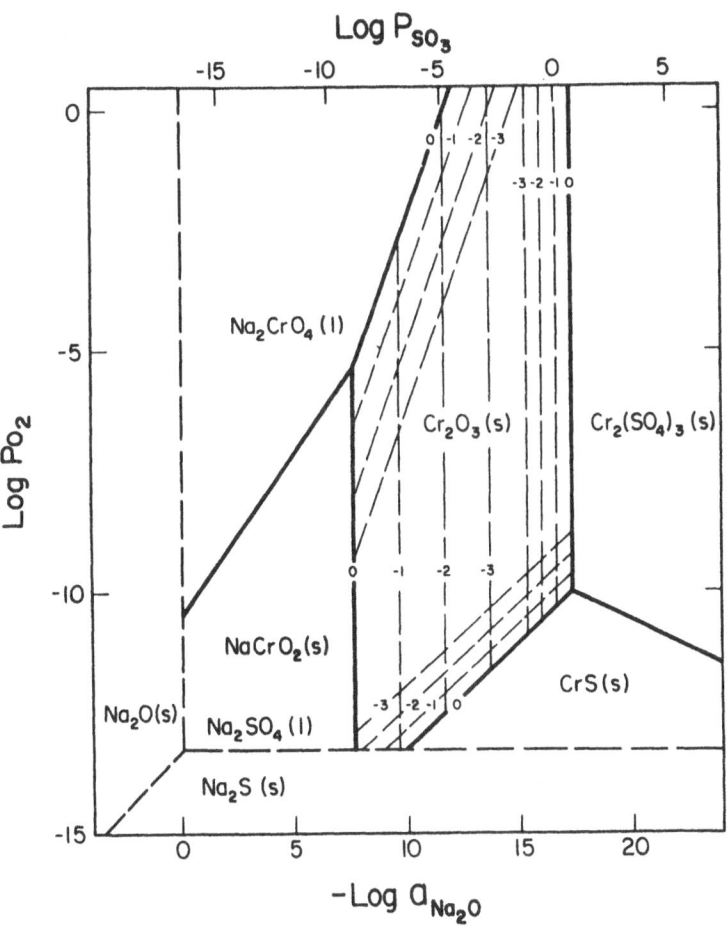

Fig. 2   Phase Stability Diagram for Na-Cr-S-O System at 1200K.

Isoactivity lines for the four possible solute compounds are indicated
in Fig. 2 by calculated dotted lines for solute activities, 10-1, 10-2,
etc., assuming ideal solutions. This construction is exactly analogous
to the well known Pourbaix diagrams, plots of E vs pH for aqueous
solutions, except that the sense for the acid-base abscissa has been
reversed.

In experimental studies, the measurement of the coordinates of Figs.
1 and 2, i.e., the oxidizing potential and the basicity of the Na2SO4
melt are achieved by the use of two solid electrolyte reference electrodes which contact the fused salt. These two reference electrodes
are shown schematically in Fig. 3 (5). The most precise explanation
and interpretation of the cell voltages are given as follows (4):

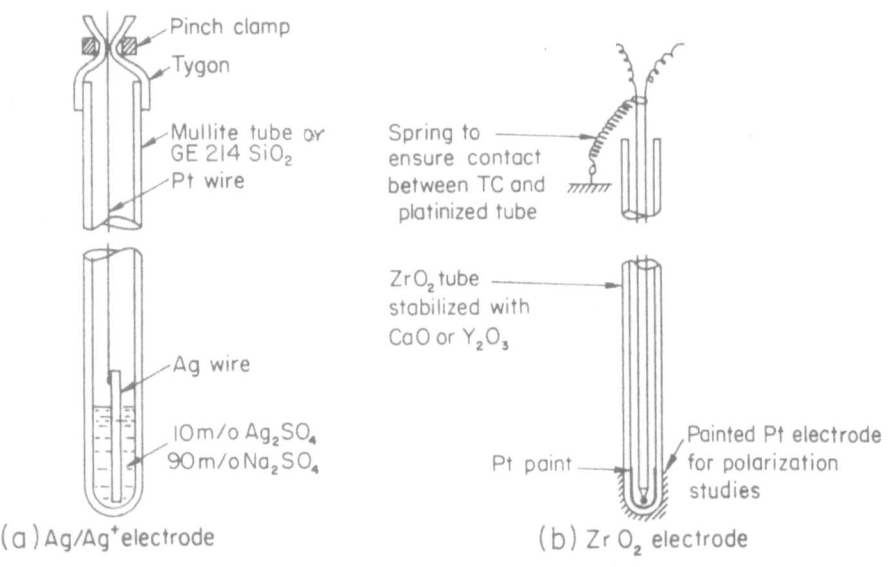

Fig. 3   Solid Electrolyte Reference Electrodes for Studies in Fused
Na2SO4 Salt.

A CaO - stabilized zirconia electrode is used to measure the oxygen
activity of the Pt WE in the melt.

$$(-)Pt, O_2(air)/ZrO_2/Na_2SO_4, SO_3, SO_2, O_2, Pt(+) \qquad [I]$$

Half-cell reactions for cell I are as follows:
$$O^{2-}(ZrO_2/Pt) = 1/2 \ O_2(air) + 2e- \qquad [1]$$

$$O^{2-}(ZrO_2/Pt) = O^{2-}(ZrO_2/Na_2SO_4) \qquad [2]$$

For local equilibrium between Na_2O, $Na+$, and O^{2-} at the ZrO_2/Na_2SO_4 interface

$$Na_2O(ZrO_2/Na_2SO_4) = 2Na+(ZrO_2/Na_2SO_4) + O^{2-}(ZrO_2/Na_2SO_4) \quad [3]$$

$$2Na+(WE) = 2Na+(ZrO_2/Na_2SO_4) \quad [4]$$

For local equilibrium at the Pt WE

$$2Na+(WE) + O^{2-}(WE) = Na_2O(WE) \quad [5]$$

$$1/2\ O_2(WE) + 2e- = O^{2-}(WE) \quad [6]$$

The total virtual cell reaction is obtained by summing Eqs. [1] - [6]

$$Na_2O(ZrO_2/Na_2SO_4) + 1/2\ O_2(WE) = Na_2O(WE) + 1/2\ O_2(air) \quad [7]$$

The Nernst equation for reaction [7] describing cell [I] is
$$E_1 = \frac{RT}{2F} \ln(P_{O_2}^{1/2}(air)\ a_{Na_2O}(WE)/P_{O_2}^{1/2}(WE)\ a_{Na_2O}(ZrO_2/Na_2SO_4)) \quad [8]$$

For melt in internal equilibrium, i.e., without concentration gradients in the melt, $a_{Na_2O}(WE) = a_{Na_2O}(ZrO_2/Na_2SO_4)$.

$$E_1(equilibrium) = -\frac{RT}{4F} \ln(P_{O_2}(air)/P_{O_2}(WE)) \quad [9]$$

For 900°C
$$E_1(equilibrium) = 0.0394 + 0.0582 \log P_{O_2}(WE) \quad [10]$$

A second closed-end electrolyte tube

$$(-)Ag,\ 10m/o\ Ag_2SO_4 - Na_2SO_4/\underset{mullite}{(Na+)}/Na_2SO_4,SO_3,_{WE}SO_2,O_2,\ Pt(+) \quad [II]$$

Since mullite provides exclusive sodium ion conduction:

$$E_{II} = \frac{RT}{F} \ln\ (a_{Na}(ref.)/a_{Na}(WE)) \quad [11]$$

For 900°C, Shores and John (8) evaluated the a_{Na}(ref.) to give

$$E_{II} = -2.748 - 0.233 \log a_{Na}(WE) \quad [12]$$

or using the equilibrium condition relating Na_2O, Na, and O_2

$$E_{II} = -1.427 - 0.116 \log(a_{Na_2O}/P_{O_2}^{1/2})_{WE} \qquad [13]$$

From Eq. [12], the EMF measured from cell [II] is decided only by the sodium activity at the WE.

The basicity of the melt, defined as $\log a_{Na_2O}$, can be measured electrochemically by combining cell [I] and cell [II]. The voltage E_I measured between the zirconia RE and the WE may be subtracted from E_{II} measured between the mullite RE and the WE to obtain a_{Na_2O} at 900°C. Since the potential of the WE is clearly canceled by this subtraction, the cell voltage E_{III}, measured directly between the two reference electrodes, gives the same result

$$E_{III} = E_{II} - E_I = -1.466 - 0.116 \log a_{Na_2O} \ (ZrO_2/Na_2SO_4) \qquad [14]$$

where E_{III} corresponds to the voltage between the zirconia and mullite electrodes, and indicates the activity of sodium oxide locally at the ZrO_2/Na_2SO_4 interface.

SOLUBILITY MEASUREMENTS. The measurement of oxide solubilities (as well as electrochemical studies) is carried out in an experimental arrangement shown in Fig. 4. Within an enclosed mullite reaction tube, an Al_2O_3 or silica crucible with 2 cm depth of fused Na_2SO_4 is saturated with respect to excess oxide powder. The two reference electrodes of Fig. 3 are immersed in the melt to establish the P_{O_2} and a_{Na_2O} values. After long equilibration times, a cold Pt wire is used to withdraw a sample of the saturated salt, which is then weighed, dissolved in water, and analyzed for solute content by atomic absorption spectroscopy. For a fixed P_{O_2}, the basicity of the melt is then changed to a new value, and further equilibrium samples are taken and analyzed. As an example study, the excellent results of Zhang (6) for solubility measurements of Cr_2O_3 are reviewed.

Figure 5 presents the measured solubility of Cr_2O_3 in fused Na_2SO_4 at 1atm O_2 and 1200K(6). The dissolution reactions resulting in the three solutes Na_2SO_4, $NaCrO_2$, and $Cr_2(SO_4)_3$ are written in Fig. 5. Clearly, the slopes predicted for the three different solutes are derived solely from the stoichiometric coefficients for the dissolution reactions. The analysis assumes Henrian behavior, i.e., a constant activity coefficient for each solute for sufficiently dilute solutions. There is excellent agreement between the predicted and the measured slopes.

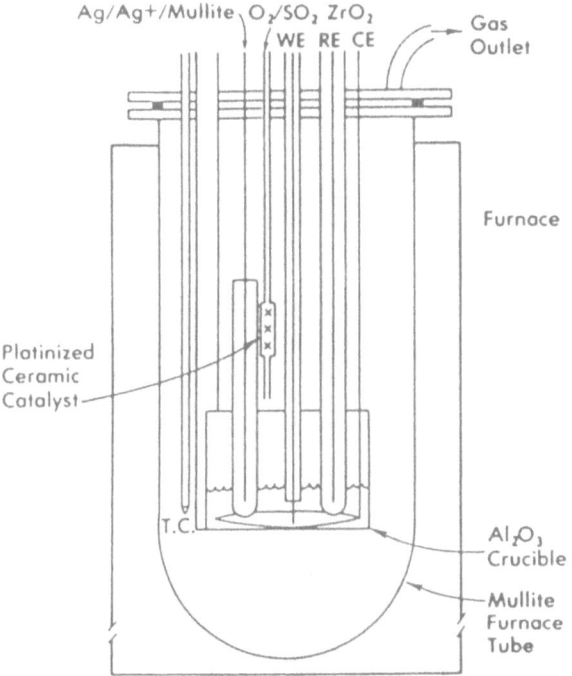

Fig. 4   Schematic Illustration of Experimental Appartus for Solubility and Electrochemical Studies in Fused Na2SO4.

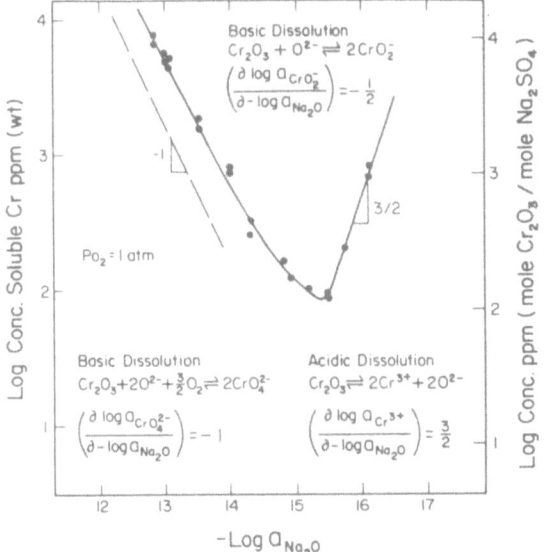

Fig. 5   Solubility of Cr2O3 in Fused Na2SO4 at 1atm O2 and 1200K.

488

From the experimental solubility values for known activities of the individual solutes (as plotted onto Fig. 2.), the Raoultian activity coefficients at infinite dilution can be calculated. Figure 6 presents measurements for the solubility of Cr_2O_3 as $NaCrO_2$ and CrS at $PO_2 = 10^{-11.5}$ atm (Ni-NiO eq'm) at 1200K. The measured slopes are in exact agreement with those expected for the solutes of Fig.2, again permitting the calculation of activity coefficients.

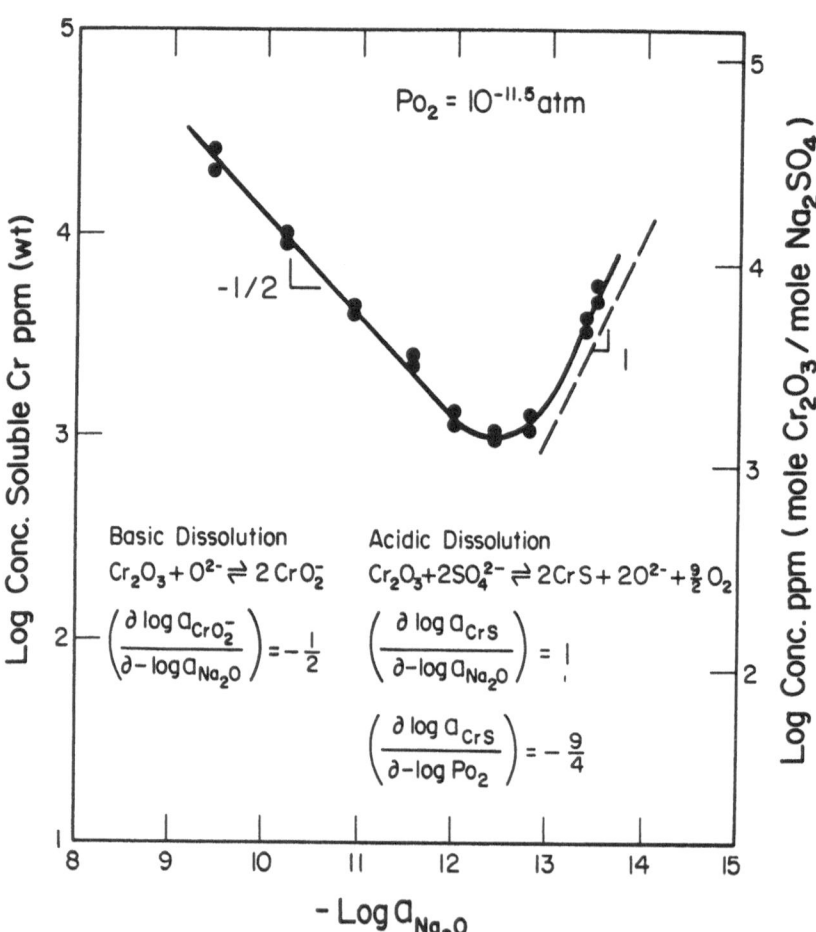

Fig. 6   Solubility of Cr2O3 in Fused Na2SO4 at PO2 = $10^{-11.5}$ and 1200K.

Figure 7 presents a masterplot for the solubility of Cr2O3 in fused Na2SO4 for any combination of conditions of PO2 and melt basicity; the plot is calculated by using the four activity coefficients resulting from the solubility measurements.

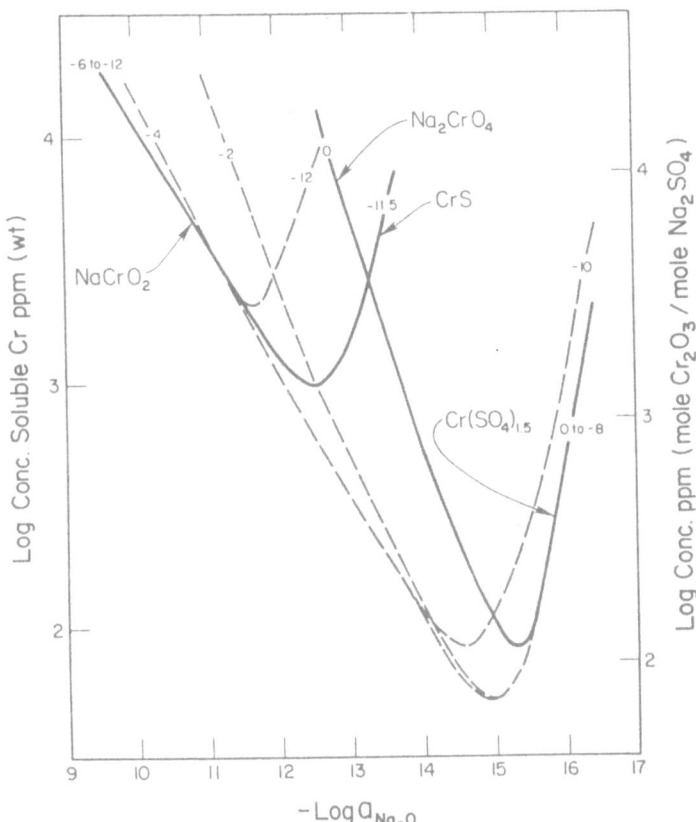

Fig. 7   Measured and Calculated Solubilities of Cr2O3 in Fused Na2SO4 at 1200K for Several Oxygen Activities.

In a manner again analogous to the Pourbaix treatment for aqueous solutions, Fig. 8 is a solute preponderance diagram, in which the regime of Cr2O3/Na2SO4 coexistence is divided into fields which identify the particular majority solute for the given conditions. A similar detailed solubility study for iron oxides (Fe3O4 and Fe2O3) in fused Na2SO4 at 1200K was reported by Zhang and Rapp(7).

Figure 9 presents a compilation (2) of recently measured solubility plots in fused Na2SO4 at 1200K and 1atm O2 for the oxides NiO(8), Co3O4(8), Fe2O3(7), Cr2O3(6), Al2O3(9), and SiO2(10). The six orders of magnitude span in Na2O activity between the minima for these oxides emphasizes the importance of acid-base behavior in deciding the

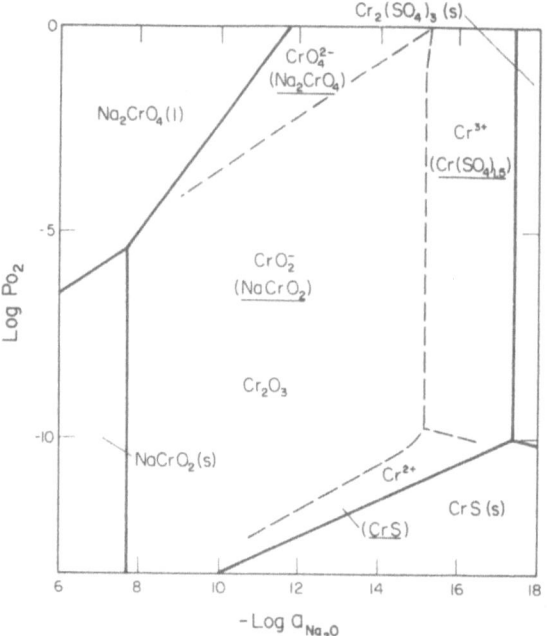

Fig. 8   Solute Preponderance Diagram for Solutes of Cr2O3 in Fused
         Na2SO4 at 1200K.

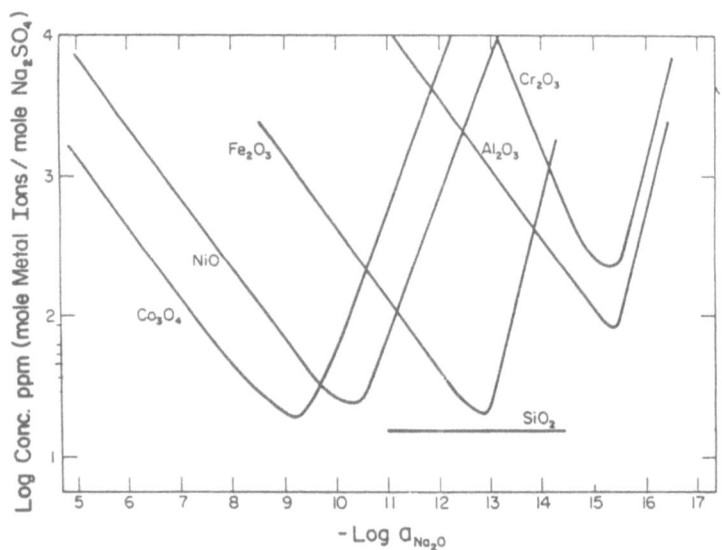

Fig. 9   Compilation of Measured Oxide Solubilities in Fused Na2SO4 at
         1atm O2 and 1200K.

stability of oxides and oxide-protected metals in contact with fused
Na_2SO_4. The obvious first rule in material selection for given service
conditions is to choose a protective oxide with the least solubility
(closest to its minimum). The solubility for SiO_2 (at 900°C) in Fig. 8
lacks any dependence on melt basicity, because for the experimental
range studied, SiO_2 does not measurably form any ionized acidic or
basic solutes, only a molecular solute. Not only is the solubility of
SiO_2 in acidic fused Na_2SO_4 very low, but the absence of any dependence
on melt basicity represents an immunity of SiO_2 to dissolution/
reprecipitaion reactions in a salt film which may support basicity and
therefore solubility gradients. However, it is well known that SiO_2
is subject to dissolution in basic melt conditions.

In the engineering practice of combustion systems, the presence of a
strong acid , such as V_2O_5, and its salts ($NaVO_3$, etc.) are known to
accelerate corrosive attack. While a number of chemical and electrochemical factors are probably involved, the effect of the $NaVO_3$ solute in
Na_2SO_4 on the solubilities of the oxides HfO_2, CeO_2, and Y_2O_3 were
recently measured (11). Without presenting the detailed results, each
of these three oxides were found to exhibit greatly enhanced acidic
solubilities in a Na_2SO_4-$0.3NaVO_3$ solvent as the metavanadate (VO_3^-)
anion complexed with product oxygen anions to form orthovanadate (VO_4^{3-})
anions, according to a reaction such as

$$3CeO_2 + 4NaVO_3 \longrightarrow 2Na_2O + Ce_3(VO_4)_4 \qquad [15]$$

For comparison the solubility of CeO_2 in pure Na_2SO_4 was also measured.
Figure 10 presents the comparison of CeO_2 solubilities in pure Na_2SO_4
and in Na_2SO_4-$0.3NaVO_3$ in 1atm O_2 at 1200K. The very significant rise
acidic solubility (up to nearly 3 orders of magnitude) and the shift in
the solubility minimum (more than 2 orders of magnitude in Na_2O activity)
help to explain the severity of Na_2SO_4-$NaVO_3$ condensates. Because the
cerium ion is only involved incidentally in Eq. (15), while the important
reaction is the conversion of metavanadate to orthovanadate anions, this
observed augmentation in acidic solubility must be valid equally for any
oxide, not just CeO_2.

ELECTRODE POLARIZATION. For any electrochemical corrosion process, a
knowledge of the prevailing cathodic reduction reaction, and the local
chemical conditions existing at an electrode surface during polarization,
are matters of critical importance. Indeed, the fast reaction of non-
preoxidized metals and alloys upon exposure to fused Na_2SO_4 at 1200K,
generally with the profuse formation of sulfides, discourages the direct
study of reactive electrodes, at least for the purpose to learn about
fundamental reaction step possibilities for the electrolyte/environment
combination. For this reason, the cathodic reduction processes have
been studied initially on Pt electrodes (4,12). In particular, as
explained by Eqs. [7] and [8], if the working electrode in a 3-electrode
cell is porous Pt painted onto the external surface of a ZrO_2 reference
electrode, then both the local oxygen activity and the local Na_2O
activity at the Pt electrode can be monitored during electrode polarization by a Na^+-ion conducting electrode(4). Figure 11 presents the
trace of local P_{O_2} and Na_2O activities as a porous Pt electrode was
polarized from the open-circuit condition in a cyclic voltammetry

492

experiment.

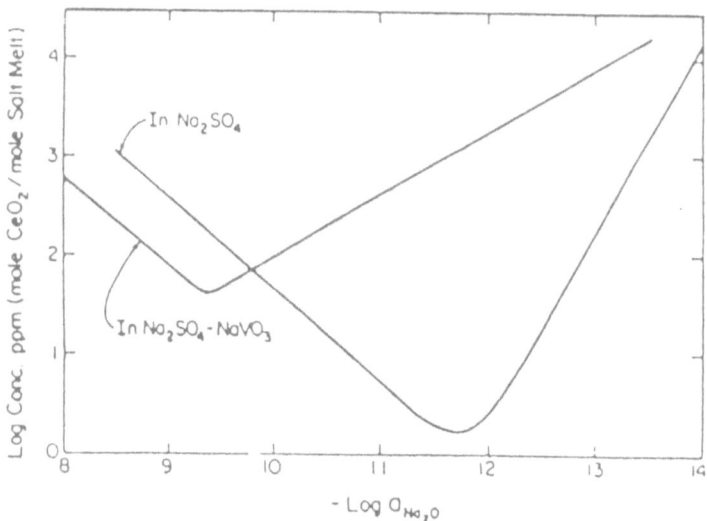

Fig. 10   Comparison of CeO2 Solubilities in pure Na2SO4 and in Na2SO4
-30m/o NaVO3 at 900°C in 1atm O2.

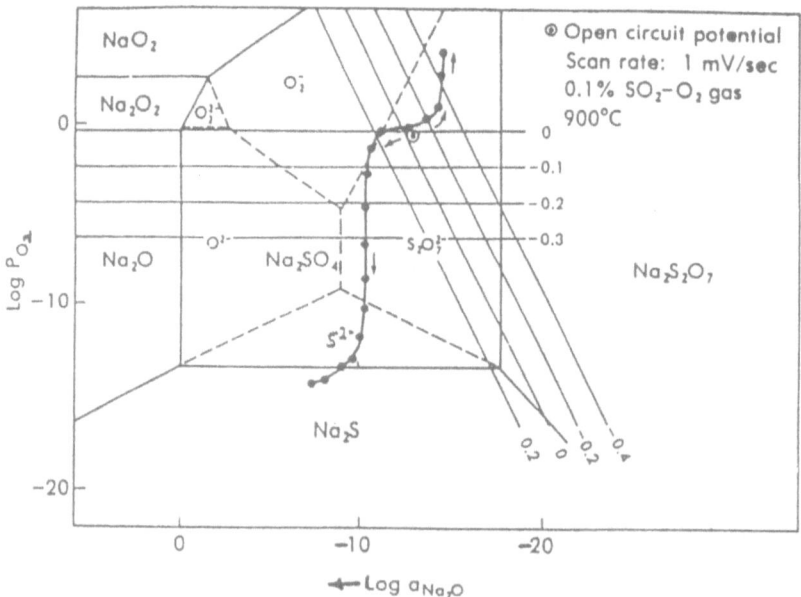

Fig. 11   Trace of basicity and oxygen activity measurements on a porous
Pt working electrode during cyclic voltammetry measurement.

Upon cathodic polarization, the melt became locally less acidic at nearly constant P_{O_2}, presumably while $S_2O_7^{2-}$ ions were reduced (probably to form O_2^- product ions). Then further cathodic polarization generated a steep drop in local P_{O_2} for nearly constant basicity, presumably as the very dilute molecular oxygen dissolved in Na_2SO_4 was reduced. Finally, upon entering a field of ionic preponderance for S^{2-}, the local electrode chemistry experienced combined reductions in melt oxidation state and melt acidity, presumably as SO_4^{2-} ions were reduced. The anodic polarization behavior is of much less interest. For the cathodic polarization trace of Fig. 11, the correlation of changes in behavior with crossing of solute preponderance lines provides a logical basis to interpret the gross reaction steps of cathodic polarization in terms of thermodynamic stabilities of the species involved. Recently, the technique has been adapted to study the salt chemistry changes during hot corrosion of preoxidized metal or alloy coupons covered by a thin Na_2SO_4 salt film.

CONCLUDING REMARKS. Some of the important applications of thermodynamics to the corrosive attack of metals by the fused oxyanion salt Na_2SO_4 have been discussed. The more direct application of these considerations to models of the corrosion mechanism are treated elsewhere (2,3). Current studies are examining the specific electrochemical action of the oxyanion solutes VO_3^-, CrO_4^{2-} and MoO_4^{2-} in fused Na_2SO_4. Clearly analogous considerations are relevant to corrosion by other fused oxyanion salts: carbonates, nitrates and hydroxides.

ACKNOWLEDGEMENTS. This research was sponsored by the National Science Foundation, Metallurgy Program of the Division of Materials Research, under grant DMR-8620311, with additional support from The Ohio State University.

REFERENCES

1. A. Rahmel, _Mater. Sci. Eng._, __87__ (1987).

2. R.A. Rapp, _Corrosion_, __42__ (1986) 568.

3. R.A. Rapp, _Mater. Sci. Eng._, __87__ (1987) 319.

4. C.O. Park and R.A. Rapp, _J. Electrochem. Soc._, __133__ (1986) 1636.

5. R.A. Rapp, Proc. JIMIS-3. _Trans. Jap. Inst. Metals, Suppl._, p. 23 (1983).

6. Y.S. Zhang, _J. Electrochem. Soc._, __133__ (1986) 655.

7. Y.S. Zhang and R.A. Rapp, _J. Electrochem. Soc._, __132__ (1985) 734 and 2498.

8. D.K. Gupta and R.A. Rapp, _J. Electrochem. Soc._ , __127__ (1980) 2194 and 2656.

9. P.D. Jose, D.K. Gupta, and R.A. Rapp, J. Electrochem. Soc., 132 (1985) 735.

10. D.Z. Shi and R.A. Rapp, J. Electrochem. Soc., 133 (1986) 849.

11. Y.S. Zhang and R.A. Rapp, Corrosion, 43 (1987) 348.

12. W.C. Fang and R.A. Rapp, J. Electrochem., Soc., 130 (1983) 2335.

MIXED OXIDATION OF MATERIALS IN HIGH TEMPERATURE ENVIRONMENTS CONTAINING CHLORINE

Michael McNallan
CEMM Department, m/c 246
University of Illinois at Chicago
P.O. Box 4348, Chicago, IL 60680

ABSTRACT. The kinetics of high temperature corrosion by chlorine differ from those of high temperature corrosion of most other oxidizing
gases because of the volatility of many chloride reaction products.
The economic importance of high temperature corrosion in chlorine con-
taminated environments is increasing because of the presence of
chlorine in many municipal and industrial wastes which must be disposed
of by incineration. These incineration systems are constructed from
superalloys or ceramic materials which are designed for strength at
high temperatures and resistance to attack by oxygen. They may be
attacked much more rapidly in high temperature environments which also
contain chlorine. The mixed oxidation of several transition metals and
their alloys as well as several low-cost SiC based ceramics in high
temperature gas mixtures of argon, oxygen, and chlorine will be discussed.

1. INTRODUCTION

Many materials which are resistant to high temperature oxidation
because they form protective scales of oxide reaction products can be
attacked at much higher rates in high temperature environments where
more than one oxidizing species is present [1,2]. The mixed oxidation
of materials in environments containing oxygen and sulfur has received
considerable attention because sulfur is a common contaminant in fossil
fuels [3]. High temperature oxidation reactions can also be accelerated by the presence of chlorine contamination, and the economic
significance of corrosion by high temperature oxidation-chlorination is
increasing because of the presence of chlorine in many municipal and
industrial wastes which must be disposed of by incineration [4,5].
This paper addresses some of the fundamental problems which are encountered in these corrosion processes.

The structural materials which are used in high temperature systems are often metallic alloys based on iron, nickel, or cobalt which
contain aluminum and/or chromium for oxidation resistance, or non-oxide

495

H. Brodowsky and H.-J. Schaller (eds.), Thermochemistry of Alloys, 495–512.

ceramics such as silicon carbide or silicon nitride. All of these materials have resistance to high temperature oxidation because they contain components which form protective, adherent oxide layers when the materials are exposed to corrosive environments in service and these oxide layers separate the material from the environment as they grow [6]. In the presence of chlorine, all of these materials can also form chloride or oxychloride reaction products which are volatile at high temperatures [7]. The kinetics of corrosion of these materials in high temperature mixed O_2-Cl_2 environments are affected by the interactions between volatile and condensed reaction products during the corrosion process.

2. EXPERIMENTAL

The interpretation of thermogravimetric results in environments where both volatile and condensed species form is complicated because of the simultaneous progress of reactions which both increase and decrease the mass of the specimens. In this study, specimens have been subjected to thermogravimetric analysis in carefully controlled environments containing O_2 and Cl_2 at temperatures between 900 K and 1200 K. The condensed reaction products remaining on the specimens after corrosion have been examined by optical and scanning electron microscopy supplemented by structural and chemical analysis by X-ray diffraction and X-ray energy dispersive spectroscopy in order to facilitate the interpretation of the corrosion mechanisms.

The experimental procedure was essentially the same in all of the experiments described in this paper. During the high temperature exposures, the specimens were suspended from a Cahn microbalance and heated to the reaction temperature within a fused quartz reaction tube. Reagent grade gases consisting of Ar, O_2, and Cl_2 provided as premixed cylinders of Ar and Cl_2 were passed through columns to remove H_2O and CO_2, mixed in a packed column, and supplied to the reaction tube at controlled flow rates. Before each experiment, a specimen was cleaned and measured, and then was suspended within the reaction tube by a quartz fiber. The reaction tube was then purged with argon and the specimen was heated to the reaction temperature in this inert environment. When the temperature was stable, the experiment was initiated by replacing the argon by the corrosive gas mixture. The microbalance was used to monitor the mass of the specimen as a function of time during the exposure. After the experiment, the corrosive gases were replaced by argon, and the specimen was cooled and removed for microstructural examination.

3. RESULTS AND DISCUSSION

3.1 Corrosion of Oxides

All of the experiments in this study were performed under conditions where the oxide of the metal is the thermodynamically stable condensed phase. The vapor pressures of volatile chloride species can be substantial even under these conditions, and oxide specimens can be corroded in flowing gases containing chlorine by the formation of such chloride species. The kinetics of such corrosion processes were investigated in the Co-O-Cl system at 1000 K in gas mixtures containing Argon, oxygen, and 1% chlorine by volume using CoO specimens which had been produced by pre-oxidizing Co foil to completion in air at 1000°C [8]. The morphology of these specimens would be expected to be similar to that of oxide scales formed on metals in this temperature range.

Figure 1 shows the thermogravimetric results obtained in gas mixtures flowing at 1.5 cm s^{-1}. The mass of the specimens decrease with time indicating that volatile corrosion products are formed, and the rate of decrease increases as the partial pressure of oxygen in the environment decreases as would be expected if the reaction is between chlorine and an oxide species.

In many volatilization reactions, the rate of vaporization is controlled by diffusion in the gas phase with the partial pressure of the volatilizing species at the interface between the gas and the solid controlled by thermodynamic equilibrium considerations. Under these conditions, the rate of decrease in mass of the specimen is given by equation (1) [9]:

$$K_1 = h \ M/RT \ P_s \tag{1}$$

In equation (1), M is the molecular weight of the condensed species being consumed (g mole^{-1}), R is the gas constant (cm^3 mole^{-1} K^{-1}), T is the temperature in K, P_s is the partial pressure of the volatilizing species in equilibrium with the oxide and the corrosive environment (atm), and h is the mass transfer coefficient between the solid and the gas mixture (cm s^{-1}).

These experiments were performed under conditions of laminar flow and the specimens can be approximated as plates, so that the appropriate mass transfer coefficient is given by equation (2) [10]:

$$h - 0.664 \ D_{AB}^{2/3} \ v^{-1/6} \ (V_T/L)^{1/2} \tag{2}$$

In equation (2), D_{AB} is the diffusion coefficient of the volatilizing species in the gas mixture (cm^{2}s^{-1}), v is the kinematic viscosity of the gas mixture (cm^{2}s^{-1}), V_T is the superficial velocity of gas flow past the specimen (cm s^{-1}), and L is the length of the specimen (cm). Because the corrosive gas mixtures consisted largely of argon, v was taken as the kinematic viscosity of Ar and D_{AB} was approximated by the binary diffusion coefficient of the volatilizing species in Ar. D_{AB} can be calculated from the Chapman-Enskog equation using measured or estimated values of the Lennard-Jones parameters for the gas species. The other variables, V_T and L, are controlled experimentally.

The oxide phase in the Co-O system which is thermodynamically stable in environments containing 0.01 atm of O_2 or more at this temperature is Co_3O_4 [11]. Thus, the appropriate corrosion reaction would be expected to be reaction (3):

$$Co_3O_4(s) + 3Cl_2(g) = 3CoCl_2(g) + 2O_2(g) \tag{3}$$

The calculated rates of weight loss in gas mixtures containing 0.15 atm of O_2 were in good agreement with the experimental measurements over a range of gas velocities as shown in Figure 2, while calculations based on $CoO(s)$ or $CoCl_2(s)$ as the condensed phase seriously overestimated the rate of volatilization.

3.2 Simultaneous Oxidation-Chlorination of Metals

3.2.1 Paralinear oxidation/chlorination of cobalt.

The corrosion of a metal in a mixed O_2/Cl_2 environment is more complex than the corrosion of an oxide, because reactions with both oxidizing species are possible. The reaction with O_2 leads to the formation of condensed species and increases the mass of the specimen, while the reaction with Cl_2 leads to the formation of volatile species and decreases the mass of the specimen. The specific details of the kinetics and the resulting thermogravimetric plot depend on the order of the two reactions.

The simplest type of mixed oxidation in O_2/Cl_2 environments is that which obeys paralinear kinetics and occurs when the oxide corrosion products form before the chloride corrosion products. The equation describing this type of corrosion can be developed by considering the kinetics of the corrosion processes between the metal and the individual oxidizing species in uncontaminated environments [12]. If a metal forms a protective oxide scale when oxidizing in O_2, then the rate of growth of the oxide scale is controlled by transport through the oxide and is inversely proportional to the thickness of the scale. If this oxide scale is simultaneously attacked by chlorine, then the rate of consumption of the scale follows a linear rate equation. Summing these two processes leads to equation (4) for the thickness of the oxide scale:

$$dx/dt = k_p/x - k_1 \tag{4}$$

In equation (4) x is the thickness of the oxide scale, k_p is the parabolic rate constant for growth of the oxide ($cm^2 \ s^{-1}$), and k_1 is the linear rate constant for volatilization of the oxide ($cm \ s^{-1}$).

Equation (4) can be integrated with the initial condition that x=0 at t=0 to predict the variation of the thickness of the scale with time during mixed oxidation in O_2/Cl_2. It is apparent that the thickness of the scale increases with time and approaches a steady state value of k_p/k_1 when the rates of formation and consumption of the oxide are equal. When the integrated form of the equation is expressed using the rate constants written in the form of mass change per unit area as in equation (5), it can be used to predict the results of thermogravimetric experiments in O_2/Cl_2 environments.

$$t = \frac{K_p/2}{\delta^2 K_v^2} \left[\frac{-\delta K_v \frac{\Delta M}{A}}{K_p/2} - \ln\left(1 - \frac{\delta K_v \frac{\Delta M}{A}}{K_p/2}\right) \right] \tag{5}$$

In equation (5), K_p is the parabolic rate constant for growth of the oxide ($g^2 cm^{-4} s^{-1}$), δ is the weight fraction of oxygen in the metal oxide, K_v is the linear rate constant for volatilization of the specimen ($g\ cm^{-2} s^{-1}$), and M/A is the change in mass of the specimen per unit of exposed area ($g\ cm^{-2}$). When the specimens are exposed to corrosive environments under controlled conditions, equations (1) and (2) can be used to predict the value of K_v [13]. If the transport of oxygen through the oxide scale is not affected by the presence of Cl_2 in the environment, the value of K_p obtained in environments without chlorine contamination can also be used in equation (5) to describe the oxide growth in the mixed environment.

Equation (5) accurately describes the mixed oxidation of cobalt in gas mixtures containing 50% O_2 flowing at 1.5 cm s^{-1} at 1200 K as shown in Figure 3 [14]. When the specimens are oxidized in environments without Cl_2 contamination, they exhibit parabolic oxidation as expected. When chlorine contamination is added to the environment, less increase in mass is observed, and when the partial pressure of chlorine is sufficiently high, the mass of the specimen decreases with time after long enough periods of exposure. When the thickness of the oxide scale reaches its steady state value, the mass of the specimen decreases at a linear rate given by equation (1).

3.2.2 Ignition of nickel. Paralinear kinetics also describe the corrosion of nickel in gas mixtures containing 50% O_2 flowing at 1.5 cm s^{-1} at 1200 K as shown in Figure 4 [15]. Because NiO is closer to a stoichiometric oxide than CoO, the value of K_p for Ni is much smaller than that for Co so that the oxide scale on the specimen reaches its steady state value very quickly, and the thermogravimetric results can be approximated by a linear decrease in mass for each condition. The nickel oxide scale is corroded by reaction (6), and the rate of corrosion increases as the partial pressure of Cl_2 in the corrosive environment increases and as the partial pressure of O_2 in the corrosive environment decreases as is expected.

$$NiO(s) + Cl_2(g) = NiCl_2(g) + 1/2\ O_2(g) \qquad (6)$$

When nickel specimens are exposed to $Ar-O_2-Cl_2$ environments containing sufficiently high concentrations of chlorine at temperatures below 1200 K, the mechanism of corrosion can change drastically. This is illustrated in Figure 5 for gas mixtures containing 50% O_2 flowing at 1.5 cm s^{-1} at 1100 K. In gas mixtures containing 0.75% Cl_2 or less, the specimens exhibit the linear decrease in mass expected from equation (5), while in gas mixtures containing more than 1% Cl_2, the specimens exhibit a very rapid increase in mass. This rapid corrosion has been reported in earlier studies and is referred to as ignition [16]. The upper temperature limit for ignition was not recognized in earlier work because the specimens were heated in the corrosive environment and ignition occurred as the temperature of the specimen passed through the sensitive range. It is accompanied by an increase in the temperature of the specimen of up to 80°C and results in the complete consumption of the metal sample (typically, a disk 1 mm thick and 1.27 cm in diameter) in less than 20 minutes.

Ignition type corrosion is also observed in $Ar-O_2-Cl_2$ environments at 1000 K. The rate of the linear decrease in mass is too low to be accurately measured in these short term experiments at this temperature, but ignition can begin when the concentration of chlorine is as low at 0.25% as shown in Figure 6.

Figures 7 and 8 show the morphology of the corrosion products remaining on the specimens after linear volatilization corrosion and ignition corrosion, respectively. The non-ignited specimens consist of large grains of NiO which have been etched by Cl_2 The ignited specimens are covered by a layer of fine, approximately spherical grains of NiO which are much smaller than the grains on the non-ignited specimens. Energy dispersive X-ray analysis did not reveal the presence of chlorine on either of these specimens.

The transition from linear volatilization to ignition corrosion can be associated with the breakdown of the protective oxide scale on the specimen. The two corrosion mechanisms are shown schematically in Figure 9. When a protective oxide is formed on the specimen as in Figure 9-A, it quickly grows to its steady state thickness and the rate of overall corrosion is controlled by the rate of formation of $NiCl_2(g)$ from the NiO(s). This leads to linear volatilization type corrosion. If no protective oxide scale is formed, the chlorine in the environment can have direct access to the metallic Ni. The Cl_2 reacts with Ni to form $NiCl_2$ at a higher partial pressure than would be in equilibrium with NiO and the corrosive environment, and this $NiCl_2$ subsequently reacts with the O_2 in the environment to form the equilibrium product NiO(s) as shown in Figure 9B. Because the rate of the reaction by the mechanism shown in Figure 9B is not limited by a solid state diffusion step, it can proceed much more rapidly than the same reaction proceeding by the mechanism shown in Figure 9A. The rate of the reaction can be so high that the heat cannot be dissipated to the environment, so that the local increase in the temperature of the specimen accelerates the corrosion further, resulting in the rapid consumption on the specimen.

The transition from linear volatilization to ignition of nickel occurs when the oxide film on the metal surface is penetrated by chlorine. It is likely that a thinner oxide scale would be penetrated more easily than a thicker scale. The paralinear theory for mixed corrosion can be used to predict the steady state thickness of the protective oxide from K_p/K_v, which would be expected to be characteristic of the average thickness of the oxide scale on the nickel during volatilization type corrosion in any environment. K_p and K_v can be calculated by the methods described previously for each environment, and if penetration of the oxide scale is the necessary condition for ignition, then ignition should be expected when the value of K_p/K_v falls below a critical value. Table I shows the values of K_p/K_v tabulated for environments containing 20% O_2 and 50% O_2 at 1000 K, 1100 K, and 1200 K. Conditions which produced ignition of the specimens are indicated by asterisks in the table. At 1000 K and 1100 K, ignition is observed under conditions where K_p/K_v is less than 2 x 10^{-6} g cm^{-2}.

It is to be noted that this value corresponds to an average thickness of the NiO layer of only approximately 3.0 nm. This amounts to only about 5 unit cell dimensions for the NiO structure, and implies that a very small amount of the oxide is sufficient to passivate the surface. This dimension may correspond to the roughness of the interfaces on either side of the scale.

No ignition type corrosion was observed in specimens exposed at 900 K. The vapor pressure of $NiCl_2$ is low enough at 900 K that condensed $NiCl_2$ rather than volatile species would be expected to form at this temperature, so that the ignition mechanism would not be active. The absence of ignition type corrosion at 1200 K is partially due to the higher activation energy for K_p than for K_v, which leads to thicker oxides scales at higher temperatures in similar environments, and partially due to the shift in the equilibrium constant of reaction (6) with increasing temperature which leads to higher vapor pressures of $NiCl_2(g)$ in equilibrium with NiO at higher temperatures. Under these conditions, the $NiCl_2(g)$ formed at the metal surface in the absence of a protective NiO layer will not be converted to NiO so that volatilization type corrosion is favored.

3.2.3 <u>Active oxidation of Co</u>. Ignition type corrosion in O_2-Cl_2 gas mixtures in the temperature range of 1000 K to 1100 K appears to be unique to the corrosion of nickel, but cobalt also exhibits a form of accelerated corrosion in these environments as shown in Figure 10 for gas mixtures containing $0.2\%Cl_2$-$50\%O_2$-$49.8\%Ar$ [14]. Much more rapid corrosion is observed at 1000 K than at the higher temperatures 1100 K or 1200 K. Figure 11 shows the morphology of the oxide scale formed on Co after corrosion in Ar-O_2-Cl_2 gas mixtures at 1000 K [17]. This scale is composed of a porous layer of oxide particles similar to the scale formed on Ni during ignition with the exception of the large faceted grains at the gas-scale interface. These grains were identified to be Co_3O_4 by X-ray diffraction analysis. Similar faceted grains of Co_3O_4 were found on the walls of the reaction tube and on the weighing mechanism indicating that they were formed by a vaporization-condensation mechanism.

One fundamental difference between the corrosion of Ni and Co in gas mixtures containing 1% O_2 or more at 1000 K is the greater thermodynamic stablity of Co_3O_4 compared with Ni_3O_4. This leads to the formation of multilayered scales during the corrosion of cobalt by oxygen in this temperature range, and the transport through the scale is affected by this.

The volatilization and subsequent oxidation of chloride species may contribute to the accelerated oxidation of cobalt in a similar manner to its role in the ignition of nickel. The oxide scale formed on Co in these environments is porous and can be penetrated by $Cl_2(g)$. Chlorine may diffuse to the scale-metal interface and react with cobalt to form $CoCl_2(g)$ which then diffuses to the outside of the scale and reacts with oxygen to form Co_3O_4.

If the rate of growth of the scale is controlled by transport of Co through the porous scale in the form of $CoCl_2(g)$, and the partial pressures of $CoCl_2$ at the scale-metal interface and at the gas-scale interface are fixed by thermodynamic equilibria, then the kinetics of oxide formation would be expected to obey a parabolic rate law similar to that for normal oxidation. This corrosion process would differ from that in usual oxidation processes in that the transport process which controls the rate of growth of the scale would be vapor diffusion through pores in the oxide rather than solid state diffusion.

Under these conditions, the flux of Co atoms through the scale is related to the parabolic rate constant for oxidation and to the diffusion of $CoCl_2(g)$ through the scale by equation (7).

$$J_{Co} = C_{Co}k_p/\delta = D_{CoCl_2}/\delta * P_{CoCl_2}/RT * \varepsilon/\tau \tag{7}$$

In equation (7), C_{Co} is the molar density of Co in the CoO scale (mole cm^{-3}), k_p is the parabolic rate constant for growth of the scale (cm^2 s^{-1}), y is the thickness of the scale (cm), D_{CoCl_2} is the diffusion coefficient of $CoCl_2$ in the gas phase within the pores (cm^2 s^{-1}), P_{CoCl_2} is the pressure difference for $CoCl_2$ between the scale-metal and the gas-scale interfaces (atm). ε and τ are both dimensionless parameters which define the structure of the porous oxide scale. ε is the volume fraction of porosity in the scale and must have a value between 0 and 1. τ is a parameter called the tortuosity factor, which accounts for the additional length that the diffusion path must follow if the pores do not follow a straight path from the scale-metal interface to the gas-scale interface. τ must be greater than or equal to one and is usually less than 10. Equation (7) can be rearranged to express k_p in terms of the transport properties of the scale.

$$k_p = D_{CoCl_2}/C_{CoO} * \varepsilon/\tau * P_{CoCl_2}/RT \tag{8}$$

Figure 12 shows the thermogravimetric results obtained for Co at 1000 K in gas mixtures containing 50% O_2 and less than 0.2% Cl_2. In this range of concentrations of chlorine, the rate of increase in mass of the specimen increases as the partial pressure of chlorine increases. Apparent parabolic rate constants were extracted from the data in Figure 12 and are plotted in Figure 13 as a function of P_{Cl_2}. When converted to appropriate units, the data in Figure 13 can be expressed as a linear function of P_{Cl_2} given by equation (9).

$$k_p = 6.906 \times 10^{-6} P_{Cl_2} cm^2 s^{-1} \tag{9}$$

Because the partial pressure of oxygen at the scale-metal interface is very low, virtually all of the Cl_2 which diffuses to the scale-metal interface would be expected to react with Co and form $CoCl_2(g)$. Therefore, the vapor pressure of $CoCl_2$ at this interface would be expected to be approximately equal to the partial pressure of Cl_2 in the corrosive gas mixture. If the partial pressure of $CoCl_2$ at the gas-scale interface is controlled by the equilibrium constant of equation (3), then at this location, $P_{CoCl_2} = 0.0157 P_{Cl_2}$ and the difference in vapor pressure of $CoCl_2$ across the scale[2] is given by equation (10):

$$P_{CoCl_2} = (1-0.0157)P_{Cl_2} = 0.9843\ P_{Cl_2} \tag{10}$$

When equation (10) is used to define P_{CoCl_2}, equations (8) and (9) are equal, and can be rearranged to give ε/τ, the dimensionless parameter which characterizes the transport through the porous oxide scale.

$$\varepsilon/\tau = 6.906 \times 10^{-6}\ C_{CoO}/D_{CoCl_2}\ RT/P_{ClCl_2} \tag{11}$$

Equation (11) leads to a value of 0.071 for ε/τ. This value is consistent with the microstructure of the oxide scale shown in Figure 9. The porosity of the oxide scale is on the order of 0.1 to 0.2, and the tortuosity factor is greater than unity.

When the concentration of chlorine in the corrosive environment is greater than 0.2%, the thermogravimetric results for Co at 1000 K are not reproduceable, and generally show less weight increase than was observed in the 0.2% Cl_2-50% O_2 environment. Figure 14 shows the microstructure of the faceted grains of Co_3O_4 which form on the outside of the scale in environments containing 1% Cl_2 and 50% O_2 at 1000 K. The Co_3O_4 grains grow with a much more densely packed morphology in this environment than in the lower chlorine environments. Although the CoO layer below this faceted structure is porous in this environment, just as it is in the lower chlorine environments, the packing between the Co_3O_4 grains is sufficiently dense in this environment to interfere with the transport of vapor species through the scale, and as a result, the scale grows at a lower rate.

4. SUMMARY AND CONCLUSIONS

The presence of chlorine can accelerate the oxidation of metals by causing the formation of volatile corrosion products as well as condensed corrosion products. The effects of specific amounts of chlorine on the oxidation of particular metals in particular environments are determined by the morphology of the condensed oxide corrosion products and on the order in which the corrosion products are produced. When a protective oxide scale is formed on the metal, the formation of volatile chlorides leads to corrosion by paralinear kinetics described by the Tedmon equation. When the oxide scale is penetrated by chlorine, a porous oxide scale is produced and the rate of corrosion is generally accelerated. In the case of nickel, this leads to ignition of the sample. For cobalt, the kinetics are affected by the morphology of the oxide products formed by re-oxidation of the volatile chlorides.

504

5. ACKNOWLEDGEMENT

This research was sponsored by the Office of Basic Energy Sciences of the U.S. Department of Energy under grants Number 82 ER 12093 and 85 ER 45178.

6. REFERENCES

1. F. S. Pettit, J. A. Goebel, G. W. Goward, Corrosion Science, 9, pp. 903-913, 1969.
2. K. Holtke, P. Kofstad, Oxid. Metal., 14, pp. 919-936, 1980.
3. K. Natesan, S. J. Dapkunas, High Temperature Corrosion of Metals and Alloys, supplement to Trans. Jpn Inst of Metals, 24, pp. 411-418, 1983.
4. H. H. Krause, D. A. Vaughan, P. D. Miller, Journal of Engineering for Power, Trans., ASME,μ.bd 96,μpp. 216-222, 1974.
5. G. Marsh, P. Elliott, High Temperature Technology, 1, pp. 115-116, 1982.
6. N. Birks, G. H. Meier, Introduction to High Temperature Oxidation of Metals, Edward Arnold, Ltd., London, 1983.
7. P. L. Daniel, R. A. Rapp, Advances in Corrosion Science and Technology, 5, M. G. Fontana, R. W. Staehle, Eds., Plenum Press, NY, pp. 55-172, 1976.
8. M. J. McNallan, W. W. Liang, J. Amer. Ceram. Soc., 64, pp. 302-307, 1981.
9. G. H. Geiger, D. R. Poirier, Transport Phenomena in Metallurgy, Addison-Wesley, Reading, MA, 1973.
10. A. H. P. Skelland, Diffusional Mass Transfer, John Wiley and Sons, NY, 1974.
11. JANAF Thermochemical Tables, 1974 supplement, J. Phys. Chem. Ref. Data, 3, pp. 412-414, 1974, and 1975 Supplement, J. Phys. Chem. Ref. Data 4, pp. 69-71 and pp. 79-80, 1975.
12. C. S. Tedmon, J. Electrochem. Soc., 113, pp. 766-768, 1966.
13. N. S. Jacobson, M. J. McNallan, Y. Y. Lee, Metall. Trans., 17A, pp. 1223-1228, 1986.
14. M. J. Maloney, M. J. McNallan, Metall. Trans., 16B, pp. 751-761, 1985.
15. Y. Y. Lee, M. J. McNallan, Metall. Trans., 18A, pp. 1099-1107, 1987.
16. K. Hauffe, J. Hinrichs, Werkstoffe u. Korrosion, 21, pp. 954-965, 1970.
17. M. J. McNallan, W. W. Liang, J. M. Oh, C. T. Kang, Oxid. Metals, 17, pp. 371-389, 1982.

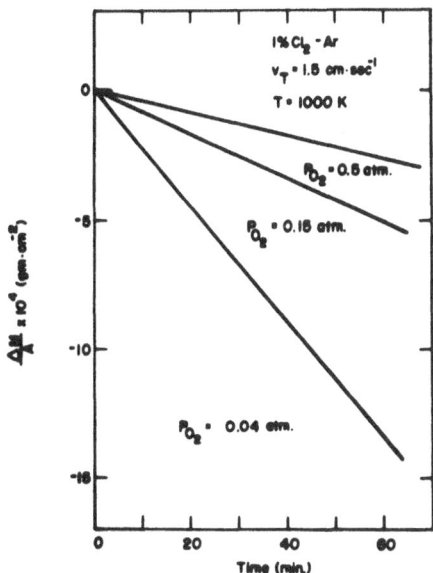

Figure 1. Thermogravimetric results for CoO in Ar-1%Cl$_2$-O$_2$ gas mixtures flowing at 1.5 cm s^{-1} at 1000 K.

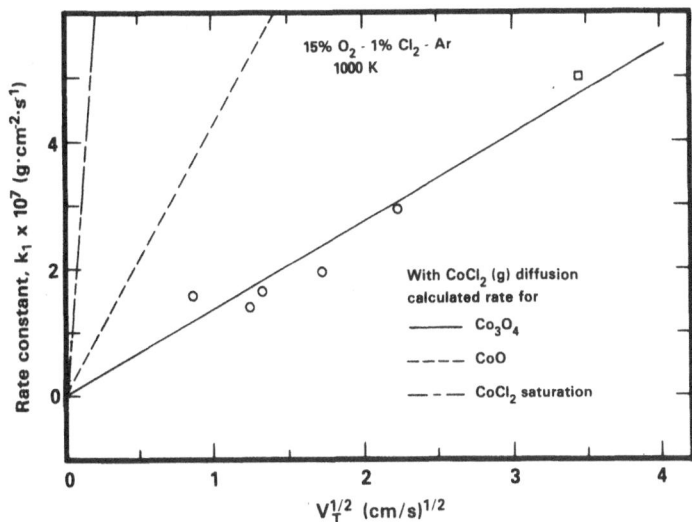

Figure 2. Experimental volatilization rate constants and theoretical volatilization rates for CoO corroding in Ar-1%Cl$_2$-15%O$_2$ showing agreement with calculations based on equilibrium with Co$_3$O$_4$.

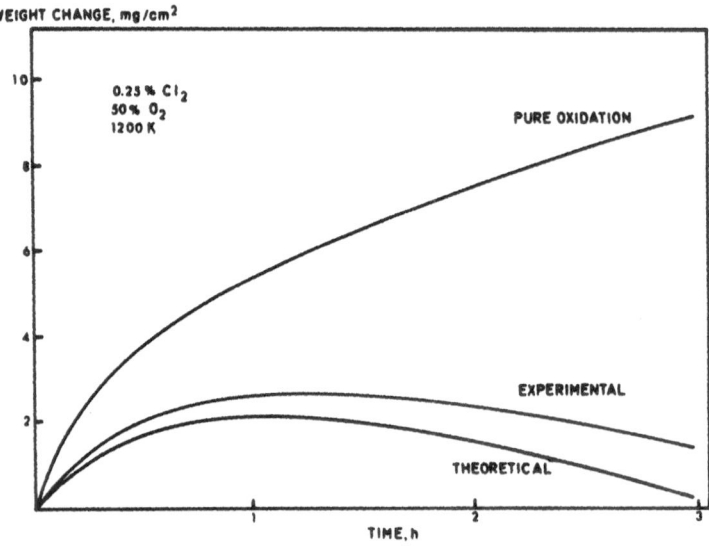

Figure 3. a) Thermogravimetric results for corrosion of Co in Ar-0.25%Cl$_2$-50%O$_2$ with results in uncontaminated O$_2$ and theoretical results from equation 5 shown for comparison.

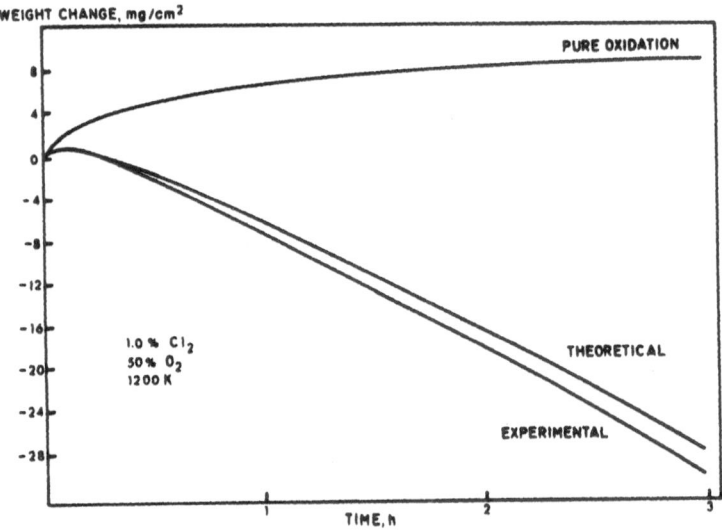

Figure 3. b) Thermogravimetric results for corrosion of Co in Ar-1%Cl$_2$-50%O$_2$ flowing at 1.5 cm s^{-1} at 1200 K with results in uncontaminated O$_2$ and theoretical results shown for comparison.

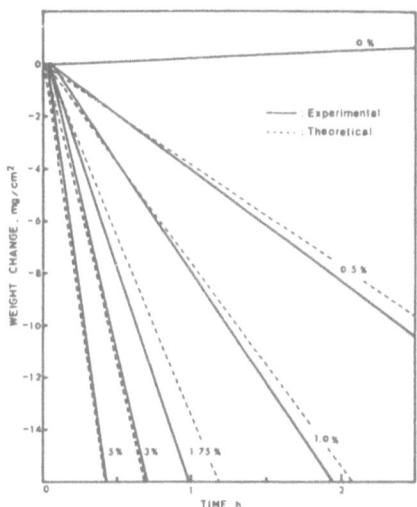

Figure 4. Thermogravimetric results for corrosion of Ni in Ar-Cl$_2$-50% O$_2$ gas mixtures flowing at 1.5 cm s^{-1} at 1200 K with theoretical results from equation 5 shown for comparison.

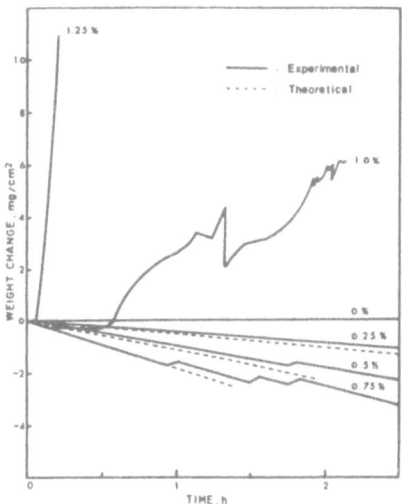

Figure 5. Thermogravimetric results for corrosion of Ni in Ar-Cl$_2$-50%O$_2$ gas mixtures flowing at 1.5 cm s^{-1} at 1100 K with theoretical results from equation 5 shown for comparison.

508

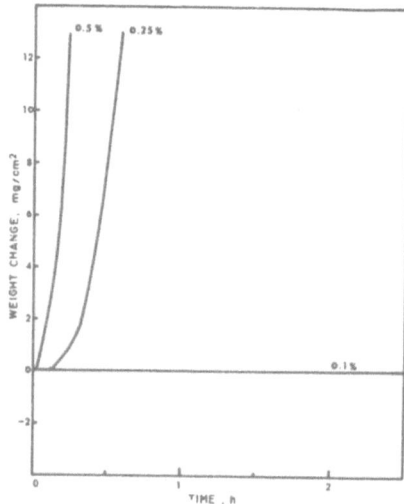

Figure 6. Thermogravimetric results for corrosion of Ni in Ar-Cl$_2$-20%O$_2$ flowing at 1.5 cm s^{-1} at 1000 K.

Figure 7. Microstructure of corrosion product scale formed on Ni after 150 minutes of corrosion in Ar-1.75%Cl$_2$-50%O$_2$ at 1200 K (no ignition) showing large, faceted grains of NiO.

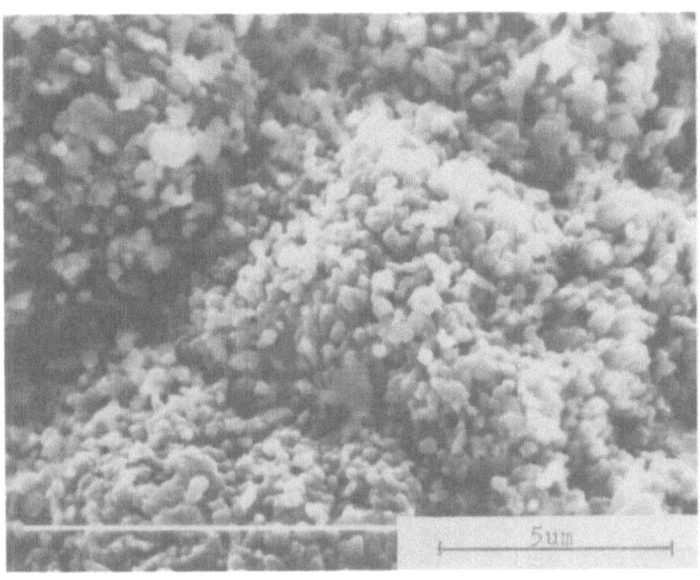

Figure 8. Microstructure of corrosion products formed on Ni after corrosion in Ar-0.75%Cl$_2$-20%O$_2$ at 1000 K (ignited specimen) showing fine spherical NiO particles.

Schematic Diagram of Mechanism

A. **Protective oxide scale corroded by Cl$_2$.**

B. **Oxide scale penetrated by Cl$_2$; Rapid oxide formation.**

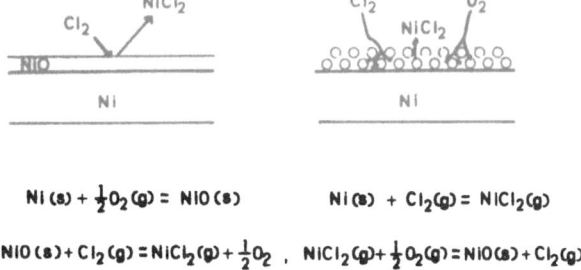

$$Ni(s) + \tfrac{1}{2}O_2(g) = NiO(s) \qquad Ni(s) + Cl_2(g) = NiCl_2(g)$$

$$NiO(s) + Cl_2(g) = NiCl_2(g) + \tfrac{1}{2}O_2 , \quad NiCl_2(g) + \tfrac{1}{2}O_2(g) = NiO(s) + Cl_2(g)$$

Figure 9. Reaction sequences for volatilization corrosion (A) and ignition (B) of nickel.

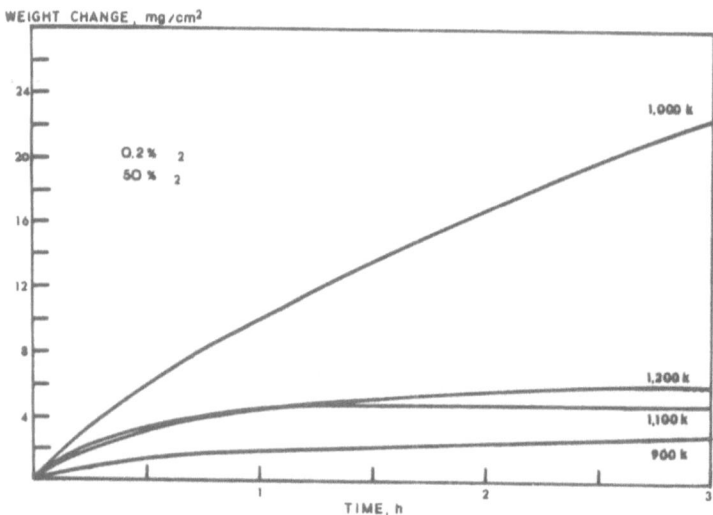

Figure 10. Effect of temperature on thermogravimetric behavior of cobalt specimens corroded in Ar-0.2%Cl$_2$-50%O$_2$ flowing at 1.5 cm s^{-1}.

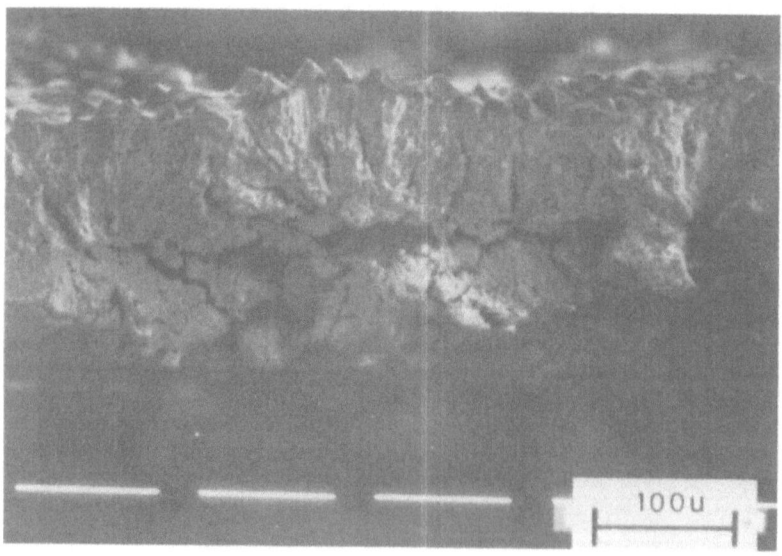

Figure 11. Morphology of oxide corrosion products formed on cobalt in Ar-1%Cl$_2$-50%O$_2$ at 1000 K.

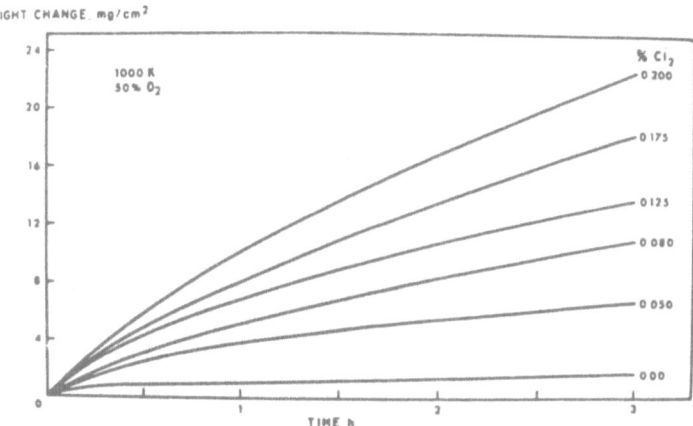

Figure 12. Thermogravimetric results for cobalt in environments containing 50% O_2 and less than 0.2% Cl_2 at 1000 K.

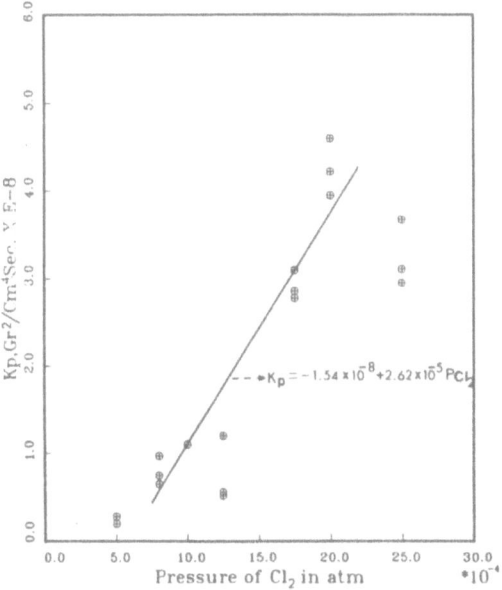

Figure 13. Effect of P_{Cl_2} on the parabolic rate constant for oxidation
of cobalt at 1000 K in environments containing 50% O_2 and less than
0.2 % Cl_2.

512

Figure 14. Morphology of outer surface of oxide scale formed on Co in Ar-1%Cl$_2$-50%O$_2$ at 1000 K.

THERMODYNAMICS OF CARBIDE-FLUORIDE SYSTEMS IN PACK CEMENTATION DEPOSITION

S. MAREELS, E. WETTINCK, A. P. VAN PETEGHEM
Laboratory for Non-Ferrous Metallurgy
Grote Steenweg Noord, 2
B-9710 Gent
Belgium

ABSTRACT.   Thermodynamical analysis becomes an important tool in
developping pack cementation processes, as will be demonstrated for the
case of carbide-fluoride systems. These systems are based on powder
mixtures containing at least one carbide and one fluosalt (e.g. SiC +
KBF_4 + ...). Other carbides and fluosalts can be added to the mixture,
as well as a filler material.
The thermodynamical analysis and background of these systems will be
discussed. The deposition reaction mechanisms will be presented, with
special emphasis on the role of oxygen in the deposition process. The
influence of impurities on the deposition process will be presented.
Finally, different coating types will be shown and the experimental
implementation of the coating process will be explained for each case.

1. INTRODUCTION.

Pack cementation is a deposition process whereby a substrate is coated
by heating it in a initially solid powder mixture. Thermally activated
reactions in the powder mixture will produce the desired deposition
atmosphere and deposition occurs mainly by gasphase reactions.
In conventional C.V.D. processes a reactant gas mixture (typically a
mixture of halide(s), hydrogen and a carrier gas) is passed through the
reaction chamber, where deposition occurs at the substrate by decomposition and/or reduction reactions.
C.V.D. systems are generally considered to be more flexible than pack
cementation processes, covering a much greater range of deposition
temperatures and pressures as well as a greater range of elements,
compounds and shapes that could be produced.
Carbide-fluoride systems are advanced pack cementation systems, with a
great potential of applications and possibilities. Their great flexi-
bility puts them in direct competition with modern C.V.D. processes.
Yet, until now, these systems were mostly unknown or greatly misunderstood. Lack of insight into the fundamentals and principles involved is

513

H. Brodowsky and H.-J. Schaller (eds.), Thermochemistry of Alloys, 513–526.
© *1989 by Kluwer Academic Publishers.*

the main reason for the underdevelopment of these systems. Indeed, their chemical complexity calls for very powerful thermodynamical analysis. The results of such an analysis are presented here, and will reveal why these systems work and how they work.

2. PACK CEMENTATION COMPARED WITH CONVENTIONAL C.V.D.

The basic deposition principles governing conventional C.V.D. and pack cementation are quite alike. In both systems deposition occurs by chemical interactions of various gasphase components with and at the substrate. Consequently, nucleation and growth, and so the deposition morphology will depend on the temperature and on the degree of super-saturation of the depositing species.

Unfortunately, the nature of these species is hardly known for most pack cementation systems, since only a few of the many - often uni-dentified - primary and secundary reaction products participate in the actual deposition steps of the process. Even the initial state and composition of the reactant system is all too often not fully known or misunderstood. The influence from the atmosphere, be it protective or not, and the influence of impurities are easily underestimated.

For C.V.D. however, the feed conditions are well known, and it is quite straightforward to determine the depositing species.

The major theoretical differences between C.V.D. and pack cementation are thus only related to the transport and production of the gases present in the deposition reactor. Practically spoken it is more a question of to know or not to know the gas phase composition. If one can determine the pack cementation atmosphere, one can influence it, optimize it and can calculate which type of coating will be formed, given certain initial conditions.

3. GENERAL DESCRIPTION OF THE CARBIDE-FLUORIDE SYSTEMS

Carbide-fluoride mixtures have been used for pack cementation boro-nizing [1]. However, we have shown already that these systems can be used for depositing boron, silicon [2,3], titanium [4,5], tungsten [5], chromium and berylium, for instance.

Generally speaking, the powder mixture contains a stable carbide (MC_x) of the depositing element (M), an unstable fluosalt of the same element $(MF_y)_m \cdot (XF)_n$ (with X = K, Na or NH_4) or eventually of another element (M'), a fillermaterial and atmospheric or other impurities.

The fluosalt should decompose at the deposition temperature, yielding the corresponding volatile fluoride, which will react with the carbide, thus creating subfluorides and oxyfluorides of M and or M'.

The filler materials are used to regulate the "depositioning power" [1] of the pack mixture and to avoid sintering of the powder. These filler materials - mostly oxides - can be partially or completely replaced by another carbide $M'C_{x'}$.

In this way, only one element can be deposited, or several elements can be deposited, either simultaneously or consecutively, depending on the interactions between the gasphase and the substrate.

One of the great advantages of these systems is their ability to deal with oxygen or water vapour. Consequently their is no need for any protective atmosphere. This makes the process more flexible and of course less expensive than other pack cementation or C.V.D. processes.

Direct hydrogen reduction of the fluorides is also circumvented since reduction takes place by disproportioning of the fluorides. This implies that there is no gas phase dilution by a carrier gas or a reductant, as in C.V.D. systems. It also means that the process can work, and preferentially is carried out, in open containers in direct contact with the outer (oxidizing) atmosphere.

Typical deposition temperatures are between 750 and 1300 K, depending on the decomposition temperature of the fluosalt and the reactivity of the carbide. Deposition times, normally 4 to 8 hours, eventually can range from 2 hours to more than 24 hours.

It is often possible to use the same class of powder mixtures in either fluidized bed or in salth bath processes, which will result in certain technological advantages. Unless otherwise specified, we will confine ourselves to the case of pure pack cementation processes.

4. CALCULATION OF THE EQUILIBRIUM STATE OF A REACTIVE SYSTEM.

The only situation that one can determine directly is the equilibrium state of the thermochemical system involved. Kinetic factors remain to be considered, since until today, no data are available for these complex systems. Yet, the equilibrium results for pack cementation can be trusted more than for comparable C.V.D. systems. Since there is no macroscopic gas flow, the dwelling time of the gaseous components in the reaction chamber is rather long (typically 500 to 1000 times longer than in corresponding C.V.D. reactors). Given the long dwelling times and the high temperatures, the asssumption of equilibrium generally introduces little or no error.

There are two different approaches for the calculation of the equilibrium state :

1. Minimization of the Gibbs free energy (or free enthalpy) of the system, with respect to the composition, considering the mass conservation requirements.

2. Solution of a system of non-linear equations, formed by the laws

$$\Delta G_r = \Delta G_r^0 + R\,T\,\ln K \qquad (1) \quad \text{for each reaction } r,$$

and the requirements for mass conservation.

The last approach is less straightforward, since one has to specify a whole set of r independent reactions, in which all expected components should appear at least once. Minimization of the free entalpy of the system does not require specifying any reactions, but only compounds, and this is the method we have choosen.

Due to the complexity of these pack cementation systems, the equilbrium has to been sought for numerically. We have written a computerprogram (EQUICALC) to evaluate very complex gas-solid equilibria at various temperatures and pressures. Until now, the activity of condensed phases (solid or liquid) is always assumed to be one, whereas for gaseous components the partial pressures are taken as activities.

The maximum number of compounds is principally only limited by the available free memory and practically by the consumed computertime. (For an average PC with MS-DOS[1] one can easily specify 30 to 70 components.) The data input consists of temperature and pressure, the required relative accuracy and/or the maximum number of iterations, and then for all compounds: the chemical formula, the initial molar quantity, the standard free enthalpy and the aggregation state (condensed or gaseous). The program then calculates the equilibrium quantities of all species involved. Care has been taken that one cannot make "false entries", that is to say that while specifying a (condensed) component in the input condition, that would not occur in the real equilibrium condition, the program will not erraneously produce this component, nor shift the calculated results in one way or another. This eliminates the necessity of recalculating a system over and over again, each time with altered data files.

Although the program performs very well, and the predicted results are generally in excellent agreement with the experiments, one should be aware of the following facts:

1. Due to the insensitivity of the free enthalpy function towards variations of the concentrations, one has to optimize this function to a normalized accuracy of at least 10^{-10}, eventually 10^{-14} or better, in order to realize the desired accuracy on the components.

2. The program cannot overcome inconsistencies of the input file (such as not considering H_2O, for the equilibrium calculations for the system C, CO, CO_2, O_2, H_2), yet it will warn the user that such results are absurd.

3. The calculations can never be more precise than the best accuracy that can be derived from the initial data; if the maximum error of these standard free enthalpy data can be estimated, one can perform a sensitivity analysis.

---

[1] MS-DOS is a deposited trademark of Microsoft Corporation.

4. Until now, only pure and stoechiometric phases and ideal gases are considered; this limitation can be overcome easily if sufficient data would be available to take deviations from the ideal behaviour into account.

Keeping these rules in mind, one can safely handle EQUICALC results. Furthermore, even besides from the experimental verification, there exists an easy control of the results, based on the phase rule. Indeed, the phase rule is derived, making exactly the same assumptions as are necessary for the proper application of this program. This implies that EQUICALC results always should coincide with the phase rule [6] predictions:

$$F = E - C + 1 \qquad\qquad\qquad (2)$$

with F: the number of degrees of freedom
C: the number of condensed phases
E: the number of elements in the system

The number of degrees of freedom gives the number of independent variables (temperature, pressure, element ratio 's,..) that can be choosen arbitrarily and which must be specified in order to fix the intensive state of the (reactive) system at equilibrium. If EQUICALC would falsly produce condensed phases, then F would become too small, eventually negative!

5. THE SiC - KBF$_4$ - O$_2$ SYSTEM.

We will now discuss the evaluation of a very (compositionally) simple pack cementation system, with only one carbide, SiC, and only one fluosalt, KBF$_4$, present. Since the actual process is carried out in direct contact with the atmosphere, we also take O$_2$ into account. For the top layers of the powder mixture, the initial amount of oxygen is greater than in the bulk of the mixture; moreover, the top layers receive a direct and/or diffusion supply of oxygen during the process. The diffusion flux towards the bulk is negligible as compared to the oxygen initially present in the intergranular space (typically there is 5 10^{-4} mole of oxygen per mole of potassium boron fluoride) of the mixture.
So we calculated the equilibrium composition of SiC-KBF$_4$ systems with initial quantities of O$_2$ ranging from 1 to 10^{-12} mole per mole of KBF$_4$, thus including the conditions from the top layers to the bulk as well as the situation that one would have when working under the utopial protective atmosphere.

We considered the following components:

A. compounds initially present: SiC, KBF_4, O_2

B. direct decomposition products: KF_1, BF_3

C. evaporation products: KF_g, KBF_{4g}

D. reactionproducts:

fluorides:	SiF_4, SiF_3, SiF_2, SiF
	BF_2, BF
	CF_4, CF_3, CF_2, CF
oxyfluorides:	$SiOF_2$, $SiOF$
	BOF_2, BOF
	COF_4, COF_2, COF
polymers:	$(KF)_2$, $(BF_2)_2$, $(CF_3)_2$,
	$(BOF)_3$
gaseous oxides:	CO, CO_2, SiO
oxides:	SiO_2, B_2O_3
slags:	K_2SiO_3, $K_2B_4O_7$, $K_2B_6O_{10}$
carbide:	B_4C
elements:	Si, B, C

Of course, not all these components will be present in the equilibrium mixture, nor could they. For instance, SiC can either be in equilibrium with Si or with C, not with both of them (see phase diagram). Further, since we have a six element system (B, C, F, K, O, Si), the maximum number of condensed phases C (2) will be 7 ($F=0$), and if we choose temperature and pressure ourselves ($F=2$) then $C_{max} = 5$. If C would become smaller than 5, then additional degrees of freedom will be found.

Table 1 gives the actual solid components (besides the solid components, KF_1 is the only liquid component, adding one to the total number of condensed compounds) as calculated for a starting mixture of 1 mole SiC^2, 1 mole KBF_4 and variable quantities of O_2. The components B, Si, B_2O_3 and $K_2B_6O_{10}$ will not be formed in these conditions. The component K_2SiO_3 is ony stable for a molar ratio (SiC, KBF_4, O_2) 1:1:1 and higher in oxygen, a ratio at which all the SiC is consumed.

From table 1 we can see that there are two degrees of freedom ($C=5$) for the highly oxygenated mixtures ($O_2 >= 0.005$ mole) and three degrees of freedom ($C=4$) for the other mixtures. We specified temperature and pressure, and so pinpointed two degrees of freedom. The third degree of freedom in the case of low oxygen mixtures does not allow us to specify yet another intensive variable; it is already fixed by the simple fact that the ratio $(K_g - F_g)/O_g$ is a constant, since condensation of K and F only occurs by formation of KF_1, and there is no condensation of O.

---

^{2}The actual molar quantities of SiC are between 7.5 and 60 per mole of KBF_4, but this becomes irrelevant for the calculations as soon as some SiC remains present in the equilibrium mixture.

Table 1. Molar quantities of solid components in the equilibrium state, for the system SiC (1 mole), KBF_4 (1 mole), O_2.

O_2 (mole)	$K_2B_4O_7$ (mole)	B_4C (mole)	C (mole)	SiC (mole)	SiO_2 (mole)
0.9	.1042	-	.9047	.0952	.5291
0.8	.1042	-	.8047	.1952	.4290
0.7	.1043	-	.7047	.2952	.3288
0.6	.1042	-	.6047	.3952	.2290
0.5	.1042	-	.5047	.4952	.1289
0.4	.1042	-	.4047	.5952	.0289
0.3	.0836	$1.6 \ 10^{-6}$	.3046	.6959	-
0.2	.0545	$1.5 \ 10^{-5}$	.2031	.7969	-
0.1	.0257	.0136	.1294	.8570	-
0.08	.0201	.0199	.1221	.8580	-
0.05	.0115	.0297	.1117	.8585	-
0.02	.0030	.0393	.1009	.8598	-
0.01	$1.2 \ 10^{-4}$	.0426	.0975	.8598	-
.008	$1.1 \ 10^{-5}$	.0434	.0970	.8596	-
.005	$8.9 \ 10^{-8}$	.0444	.0961	.8594	-
.002	-	.0455	.0954	.8591	-
.001	-	.0460	.0952	.8588	-
$8 \ 10^{-4}$	-	.0461	.0951	.8588	-
$5 \ 10^{-4}$	-	.0462	.0950	.8588	-
$2 \ 10^{-4}$	-	.0462	.0950	.8588	-
$1 \ 10^{-4}$	-	.0463	.0949	.8588	-
$8 \ 10^{-5}$	-	.0463	.0949	.8588	-
$5 \ 10^{-5}$	-	.0463	.0949	.8588	-
$2 \ 10^{-5}$	-	.0463	.0949	.8588	-
$1 \ 10^{-5}$	-	.0463	.0949	.8588	-
$8 \ 10^{-6}$	-	.0463	.0949	.8588	-
$5 \ 10^{-6}$	-	.0463	.0949	.8588	-
10^{-6}	-	.0463	.0949	.8588	-
10^{-7}	-	.0463	.0949	.8588	-
10^{-8}	-	.0463	.0949	.8588	-
10^{-9}	-	.0463	.0949	.8588	-
10^{-10}	-	.0463	.0949	.8588	-
10^{-11}	-	.0463	.0949	.8588	-
$10^{-1}2$	-	.0463	.0949	.8588	-

Table 2 gives the calculated equilibrium partial pressures of the boron and silicon fluorides. These fluorides are of the utmost importance tho the process, since they regulate the transport and deposition of boron and silicon, and are determinant for the equilibrium composition.

Table 2. Partial pressures of fluoride components in the equilibrium state for the system SiC (1 mole), KBF_4 (1 mole), O_2.

O_2 (mole)	BF_3 (mole)	BF_2 (10^{-6} mole)	BF (10^{-7} mole)	SiF_4 (mole)	SiF_3 (10^{-3} mole)	SiF_2 (10^{-5} mole)
.9	.59882	1.758	2.954	.37391	20.175	2.389
.8	.59869	1.755	3.540	.37404	20.174	2.388
.7	.59844	1.815	15.41	.37429	20.168	2.387
.6	.59867	1.756	2.972	.37407	20.170	2.389
.5	.59858	1.756	2.944	.37413	20.179	2.393
.4	.59859	1.787	4.371	.37410	20.191	2.396
.3	.67612	2.318	3.905	.29883	17.093	2.207
.2	.78257	3.340	5.833	.19547	12.737	1.952
.1	.84378	3.276	7.047	.13702	9.5713	2.073
.08	.84441	3.211	6.908	.13660	9.4895	1.855
.05	.84414	4.082	6.893	.13690	9.4928	1.544
.02	.84464	3.550	20.03	.13643	9.4677	1.479
.01	.84438	3.189	6.880	.13667	9.4814	1.443
.008	.84533	3.193	6.883	.13688	9.4854	1.939
.005	.84721	3.197	6.893	.13709	9.5071	3.275
.002	.84904	3.224	6.874	.13747	9.5170	1.656
.001	.84947	3.202	6.910	.13779	9.5366	2.517
.0008	.84959	3.202	6.873	.13783	9.5354	2.320
.0005	.84977	3.212	6.904	.13788	9.5442	3.405
.0002	.85015	3.200	6.878	.13779	9.5381	1.797
.0001	.85020	3.199	6.880	.13784	9.5410	2.119
.00008	.85021	3.199	6.878	.13785	9.5406	2.057
.00005	.85030	3.200	6.888	.13781	9.5373	1.700
.00002	.85049	3.205	6.877	.13767	9.5497	1.452
.00001	.85026	3.216	6.896	.13792	9.5449	1.450
$8\ 10^{-6}$	.85045	3.203	6.886	.13774	9.5369	1.450
$5\ 10^{-6}$	.85029	3.199	6.894	.13790	9.5448	1.449
$1\ 10^{-6}$	.85038	3.201	6.884	.13782	9.5400	1.450
10^{-7}	.85041	3.200	6.879	.13780	9.5392	1.450
10^{-8}	.85051	3.210	6.909	.13769	9.5499	1.450
10^{-9}	.85039	3.200	6.880	.13782	9.5397	1.450
10^{-10}	.85043	3.202	6.878	.13778	9.5373	1.450
10^{-11}	.85042	3.200	6.879	.13779	9.5387	1.450
10^{-12}	.85042	3.200	6.879	.13779	9.5387	1.450

Table 3 gives the calculated equilibrium partial pressures of the boron and silicon oxyfluorides and of CO.
Besides the components given in the tables, KBF_{4g} ($\pm 5\ 10^{-8}$ mole), KF_1 ($\pm .99717$ mole), KF_g ($\pm 1.5\ 10^{-3}$ mole), $(KF)_2$ ($\pm 7.2\ 10^{-4}$ mole), $(BF_2)_2$ ($\pm 3.3\ 10^{-6}$ mole) and CO_2 ($< 10^{-7}$ mole) are also formed

Table 3. Partial pressures of oxygenated components in the equilibrium state for the system SiC (1 mole), KBF_4 (1 mole), O_2.

O_2 (mole)	CO (mole)	BOF (mole)	BOF_2 (mole)	$(BOF)_3$ (mole)	SiOF (mole)	$SiOF_2$ (mole)
.9	$5.62E^{-5}$	$6.63E^{-4}$	$4.08E^{-9}$	$4.11E^{-3}$	$5.61E^{-8}$	$2.58E^{-6}$
.8	$5.66E^{-5}$	$6.63E^{-4}$	$3.67E^{-9}$	$4.12E^{-3}$	$6.58E^{-8}$	$2.12E^{-8}$
.7	$5.60E^{-5}$	$6.65E^{-4}$	$3.49E^{-9}$	$4.12E^{-3}$	$1.51E^{-8}$	$2.10E^{-8}$
.6	$5.61E^{-5}$	$6.63E^{-4}$	$3.49E^{-9}$	$4.12E^{-3}$	$1.53E^{-8}$	$2.82E^{-8}$
.5	$5.62E^{-5}$	$6.64E^{-4}$	$3.95E^{-9}$	$4.13E^{-3}$	$1.75E^{-8}$	$2.07E^{-8}$
.4	$5.63E^{-5}$	$6.66E^{-4}$	$3.51E^{-9}$	$4.14E^{-3}$	$1.51E^{-8}$	$2.09E^{-8}$
.3	$4.84E^{-5}$	$7.04E^{-4}$	$3.48E^{-9}$	$4.91E^{-3}$	$1.29E^{-8}$	$1.64E^{-8}$
.2	$4.35E^{-5}$	$7.92E^{-4}$	$4.13E^{-9}$	$6.09E^{-3}$	$1.48E^{-8}$	$9.43E^{-9}$
.1	$3.31E^{-5}$	$7.91E^{-4}$	$3.15E^{-9}$	$6.48E^{-3}$	$1.25E^{-8}$	$1.37E^{-8}$
.08	$2.85E^{-5}$	$7.69E^{-4}$	$4.73E^{-9}$	$6.41E^{-3}$	$1.81E^{-8}$	$1.49E^{-8}$
.05	$2.88E^{-5}$	$7.68E^{-4}$	$6.10E^{-9}$	$6.41E^{-3}$	$7.59E^{-9}$	$2.00E^{-8}$
.02	$2.88E^{-5}$	$7.78E^{-4}$	$4.95E^{-9}$	$6.40E^{-3}$	$1.63E^{-6}$	$1.87E^{-8}$
.01	$4.04E^{-5}$	$7.72E^{-4}$	$3.86E^{-9}$	$6.40E^{-3}$	$2.10E^{-7}$	$5.52E^{-7}$
.008	$3.36E^{-5}$	$7.25E^{-4}$	$2.98E^{-9}$	$5.28E^{-3}$	$6.09E^{-9}$	$1.32E^{-7}$
.005	$2.82E^{-5}$	$6.32E^{-4}$	$3.83E^{-9}$	$3.26E^{-3}$	$9.19E^{-9}$	$4.97E^{-9}$
.002	$2.01E^{-5}$	$4.43E^{-4}$	$4.99E^{-7}$	$1.23E^{-3}$	$4.88E^{-6}$	$4.30E^{-9}$
.001	$1.27E^{-5}$	$3.45E^{-4}$	$1.27E^{-8}$	$5.76E^{-4}$	$2.28E^{-6}$	$1.09E^{-8}$
.0008	$1.53E^{-5}$	$3.17E^{-4}$	$3.17E^{-9}$	$4.46E^{-4}$	$3.79E^{-6}$	$2.24E^{-6}$
.0005	$1.10E^{-5}$	$2.63E^{-4}$	$1.05E^{-7}$	$2.56E^{-4}$	$7.91E^{-9}$	$1.69E^{-7}$
.0002	$8.73E^{-6}$	$1.76E^{-4}$	$7.20E^{-8}$	$7.73E^{-5}$	$1.68E^{-8}$	$1.87E^{-7}$
.0001	$4.50E^{-6}$	$1.24E^{-4}$	$7.81E^{-8}$	$2.67E^{-5}$	$4.32E^{-8}$	$2.63E^{-8}$
.00008	$4.30E^{-6}$	$1.08E^{-4}$	$1.04E^{-8}$	$1.80E^{-5}$	$1.37E^{-8}$	$1.12E^{-8}$
.00005	$2.94E^{-6}$	$7.98E^{-5}$	$4.37E^{-9}$	$7.16E^{-6}$	$1.35E^{-8}$	$1.51E^{-7}$
.00002	$1.76E^{-6}$	$3.76E^{-5}$	$1.18E^{-8}$	$7.83E^{-7}$	$6.3E^{-10}$	$5.27E^{-9}$
.00001	$7.17E^{-7}$	$1.96E^{-5}$	$5.78E^{-9}$	$1.66E^{-7}$	$2.51E^{-9}$	$1.28E^{-9}$
$8\ 10^{-6}$	$6.01E^{-7}$	$1.56E^{-5}$	$9.0E^{-10}$	$1.06E^{-7}$	$9.0E^{-10}$	$2.13E^{-9}$
$5\ 10^{-6}$	$4.82E^{-7}$	$8.97E^{-6}$	$8.92E^{-7}$	$2.89E^{-8}$	$6.3E^{-10}$	$6.4E^{-10}$
$1\ 10^{-6}$	$9.42E^{-8}$	$1.99E^{-6}$	$9.9E^{-12}$	$2.5E^{-10}$	$1.1E^{-10}$	$6.5E^{-10}$
10^{-7}	$1.06E^{-8}$	$1.81E^{-7}$	$1.75E^{-8}$	$3.9E^{-12}$	$1.6E^{-11}$	$7.7E^{-12}$
10^{-8}	$1.07E^{-9}$	$1.98E^{-8}$	$9.0E^{-12}$	$3.9E^{-13}$	$6.2E^{-13}$	$4.9E^{-12}$
10^{-9}	$7.6E^{-10}$	$2.01E^{-9}$	-	-	$1.7E^{-13}$	$1.9E^{-12}$
10^{-10}	$1.0E^{-11}$	$2.0E^{-10}$	-	-	-	-
10^{-11}	$8.0E^{-13}$	$2.2E^{-11}$	-	-	-	-
10^{-12}	$2.0E^{-13}$	$3.6E^{-12}$	-	-	-	-

The results of the optimization procedures can be used in two ways:

1. Determination of the reaction mechanism.
2. Determination of the feasibility of deposition, by "contacting" the calculated gas phase with a substrate and calculating this new equilibrium.

6. THE SiC - KBF$_4$ - O$_2$ REACTION MECHANISM.

The most important gas-solid reactions are those between SiC and BF$_3$, or SiC, BF$_3$ and O$_2$. The complete set of reactions is as follows:

(1) KBF_4 <=> KF_1 + BF_3

(2) KF_1 <=> KF_g (*)

(3)   2 KF_g <=> $(KF)_2$ (*)

(4) KF_g + BF_3 <=> KBF_4 (*)

(5)   2 BF_3 +   SiC   <=> 2   BF   + SiF_4 +   C   (*)

(6)   2 BF_3 +   SiC   <=> 2 BF_2 + SiF_2 +   C

(7)   2 BF_3 +   SiC   <=> BF_2 + SiF_4 +   C   +   B

(8)   ½ O$_2$ + BF_3 +   SiC   <=>   BOF   + SiF_2 +   C

(9)   O$_2$ +   SiC   <=> SiO_2 +   C

(10)   ½ O$_2$ +   C   <=>   CO

(11)   4 B   +   C   <=> B_4C

(12)   1½ O$_2$ +   2   B   <=> B_2O_3

(13)   3 B_2O_3 +   2 KF_1 <=> 2   BOF   + $K_2B_4O_7$

(14)   2 SiO_2 +   2 KF_1 <=> $SiOF_2$ + K_2SiO_3

(15)   2 BF_2 <=> BF_3 +   BF   (*)

(16)   2 SiF_3 <=> SiF_4 + SiF_2 (*)

(17)   2 BF_2 <=> $(BF_2)_2$ (*)

(18)   3 BOF   <=> $(BOF)_3$ (*)

(19)   BOF   + SiF_3 <=>   SiOF   + BF_3 (*)

(20)   BOF   + SiF_2 <=> $SiOF_2$ + BF   (*)

(21)   BOF   + SiF_4 <=> SiF_3 + BOF_2 (*)

(22)   2 CO   <=> CO_2 +   C   (*)

Reactions marked with (*) deal with subsystems that are at equilibrium on their own. These subsystems are really driving the global system to its equilibrium state and determine its concentrations. It happens that all interfluoride systems are at local equilibrium, which results in a certain "buffer" effect, keeping the partial pressures of these components at a constant value over a wide range of oxygen input levels.

The reactions with oxygen, or other oxygenated compounds, only proceed in a measure more or less correlated with the initial amount of oxygen. The only gaseous components, directly linked with the O_2 concentration, are BOF and CO; other oxygenated gases are formed by further reaction of these compounds. The slag formation reactions (13),(14), and also reaction (9) occur only at very high initial oxygen levels, well above the levels that usually prevail in pack cementation systems.

Direct verification of such a complex reaction mechanism is very difficult. Yet, some gases, such as CO and $(BOF)_3$, were directly detected. Other components, such as boron and silicon fluorides, were detected as a group. The actual deposition process of course, is also a verification.

7. DEPOSITION REACTIONS.

The reactions between the gasphase and the powder mixture are only of secondary importance as compared to the interactions with the substrate. Three kind of interactions are to be considered:

1. Oxidation of the substrate or of the deposited species.

2. Fluoride induced corrosion of the substrate.

3. Deposition of B or Si, or both.

So, taking the calculated gas atmosphere of the carbide-fluoride equilibrium in contact with the substrate (Fe or Ni), we consider the oxide, fluoride, boride and silicide compounds that could be formed with the substrate element. Special attention is paid to the iron-boron system; conventional deposition processes (C.V.D., P.V.D. and pack cementation alike) produce twophasic coatings (FeB and Fe_2B) - mostly microcracked as a result of too high stresses at the interface - , whereas monophasic Fe_2B are technologically preferable.

With high initial oxygen levels, oxidation of the substrate is of course possible; also there are then considerable losses of boron and silicon as oxygenated compounds, both condensed and gaseous. However, this situation is only possible in the very top layer of the powder mixture, where there is free contact with the surrounding air.

In the bulk of the mixture, no oxidation nor fluoride attack of Fe or Ni takes place. Even if slightly oxidized specimen were used, or if the samples would oxidize during the initial stages of the process (when the protective atmosphere only starts developping), the oxide would be reduced in the deposition stage of the process:

$$(23) \quad Fe_2O_3 + 3\ CO \qquad \Longleftrightarrow 2\ Fe + 3\ CO_2$$

$$(24) \quad Fe_2O_3 + 3\ BF \qquad \Longleftrightarrow 2\ Fe + 3\ BOF$$

Deposition of boron occurs by:

$$(25) \quad 2\ BF \qquad \Longleftrightarrow \quad B + BF_2$$

$$(26) \quad 3\ BF_2 \qquad \Longleftrightarrow \quad B + 2\ BF_3$$

Formation of iron boride can proceed by direct reaction (27),(28) with the fluorides as long as free iron is present at the surface; in a later stage (29) takes place at the growing interface:

$$(27) \quad 2\ Fe + 2\ BF \qquad \Longleftrightarrow \quad Fe_2B + BF_2 \qquad\qquad (*)$$

$$(28) \quad 2\ Fe + 3\ BF_2 \qquad \Longleftrightarrow \quad Fe_2B + 2\ BF_3 \qquad\qquad (*)$$

$$(29) \quad 2\ Fe + B \qquad \Longleftrightarrow \quad Fe_2B$$

Again, the fluoride subsystems are at local equilibrium! From this, it is clear that the driving force for deposition is the difference in activities of two fluoride subsystems, one in equilibrium with the carbide environment, the other in equilibrium with the substrate.
Since the oxygen influence is very effectively buffered over a wide range of initial concentrations (from about 0.1 mole to $1\ 10^{-12}$ mole), it follows that deposition is feasible in this whole range. This also implies that there is no necessity for any protective atmosphere.
The EQUICALC results also show that the only stable compound is Fe_2B; so no FeB or iron silicides are formed.
If nickel is choosen as a substrate, the same can be said about oxygen or fluoride attack. However, this time silicon will be deposited, Ni_2Si being now the stable phase:

$$(30) \quad 2\ SiF_2 \qquad \Longleftrightarrow \quad Si + SiF_4$$

$$(31) \quad 4\ SiF_3 \qquad \Longleftrightarrow \quad Si + 3\ SiF_4$$

With the silicide formation reactions:

$$(32) \quad 2\ Ni + 2\ SiF_2 \qquad \Longleftrightarrow \quad Ni_2Si + SiF_4 \qquad\qquad (*)$$

$$(33) \quad 2\ Ni + 4\ SiF_3 \qquad \Longleftrightarrow \quad Ni_2Si + 2\ SiF_4 \qquad\qquad (*)$$

$$(34) \quad 2\ Ni + Si \qquad \Longleftrightarrow \quad Ni_2Si$$

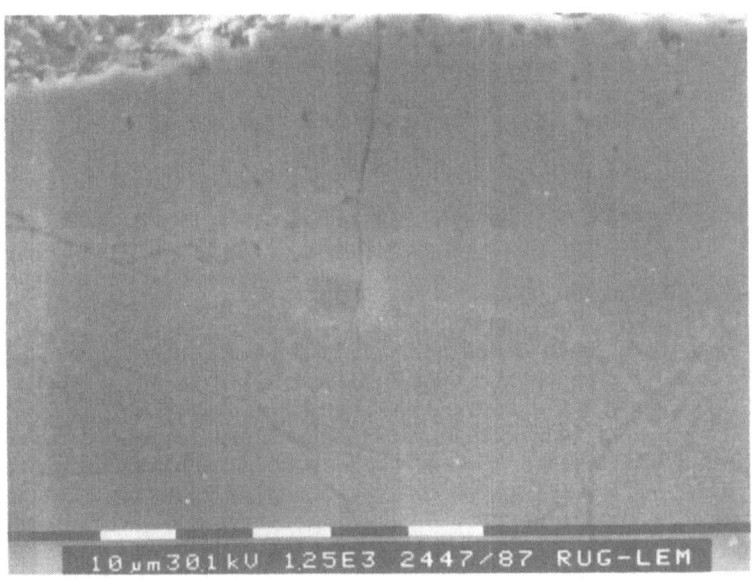

Figure 1. Conventionally produced twophasic boride coating.

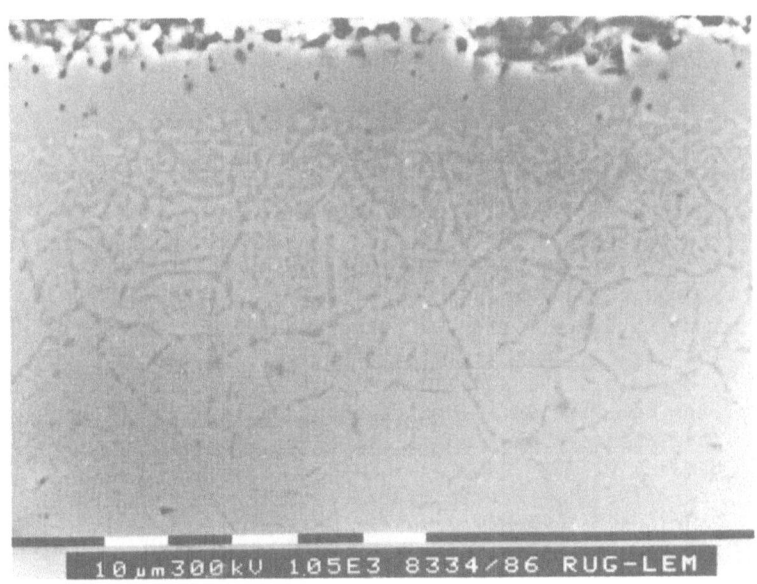

Figure 2. Carbide-fluoride produced monophasic Fe_2B coating.

The above reactions, as predicted through EQUICALC analysis, are fully consistent with the deposition experiments. On nickel, indeed mono-phasic Ni_2Si coatings are formed, while on ferrous substrates, mono-phasic Fe_2B coatings are deposited. This is a major breakthrough, since it was for the first time that monophasic boride coatings were formed on alloyed steels, with any process! Figure 1 shows a conventional-microcracked - twophasic coating on 304 stainless steel and figure 2 shows the new - crackfree - monophasic coating, also on 304 steel.

8. DISCUSSION AND CONCLUSIONS.

We have shown that carbide-fluoride systems are capable of depositing elements - even those with a great affinity for oxygen - without the use of a protective atmosphere. These systems are able to buffer the influence of oxygen over a very wide concentration range. Even slightly oxidized samples can be treated with good results, since the atmosphere in the pack mixture becomes strongly reducing (this is in sorts comparable to the operation of a blast furnace). Although these systems are chemically very complex, EQUICALC permits to analyse, with high accuracy, the entire process. Consequently it becomes possible to optimize these processes, and to develop new processes, with these simulations.
In this way, we designed new processes to deposit chromium, titanium [4],[5] and tungsten [5].

9. REFERENCES.

[1] Kunst, H.; Schaaber, O. : 'Beobachtungen beim Oberflächenborieren von stahl, III. Borierverfahren.' Härterei-Tech. Mitt. 22 (1967) 4, p. 275-292.
[2] Mareels, S.; Wettinck, E. : 'Borieren und Silicieren mit aktivier-ten Carbiden, I. Diffusionsverfahren und Reaktionsmechanismus.' Härterei-Tech. Mitt. 40 (1985) 2, p. 73-76.
[3] Mareels, S.; Wettinck, E. : 'Borieren und Silicieren mit aktivier-ten Carbiden, II. Diffusionsschichten - Aufbau und Zusammen-setzung.' Härterei-Tech. Mitt. 40 (1985) 4, p. 168-177.
[4] Mareels, S.; Wettinck, E. : 'Boronizing: C.V.D. compared with Pack Cementation.' Oberflächentechnik : Proceedings of the 3rd International Congress for Surface Technology. (1985), VDE-Verlag, Berlin, p. 199-206.
[5] Mareels, S.; Wettinck, E. : 'Pack Cementation of TiN and TiB_2 Coatings.' Internal Report. Presented at the 1986 Fall Meeting of the Electrochemical Society. To be published.
[6] Van Zeggeren, F., Storey, S. : The computation of Chemical Equilibria. Cambridge University Press, (1970).

CORROSION AND HYDROGEN PERMEATION STUDIES IN METAL ALLOY FILMS

D. A. Stevenson
Department of Materials Science and Engineering
Stanford University
Stanford, California 94305
USA

ABSTRACT. The deposition of metal alloy films usually leads to non-equilibrium structures that have significantly different properties from the bulk equilibrium phases of the same composition. We describe studies relating to the preparation, structure and some chemical properties of selected binary transition metal alloy films. The principal method of preparation is by planar magnetron sputter deposition. A number of experimental methods are used to characterize the composition and structure of the films, with particular focus on distinguishing amorphous from crystalline structures. The passive properties are established with electrochemical methods and the hydrogen behavior is established with an electrochemical double cell method. Significant differences between the amorphous and the crystallized structure are observed in the passive behavior, the hydrogen solubility, and the hydrogen diffusivity. These observations are correlated with the structure and microstructure of the alloy.

1. INTRODUCTION

There are substantial variations in the properties of thin metal alloy films, when compared with the corresponding bulk alloys. Furthermore, there can be significant differences between alloy films of the same composition prepared by different methods. There is a great deal of technological interest in these films for such applications as magnetic media and magnetic heads for memory storage, electrical interconnects for integrated circuits, and for modifying corrosion, friction and wear properties. A variety of methods are used to prepare thin alloy films: vapor quench methods (VQ) such as thermal evaporation and a variety of sputter and chemical vapor deposition methods; liquid quench methods (LQ) such as piston and anvil and splatt rolling; rapid thermal cycling using laser and arc annealing methods; and electrodeposition. Ion implanation also develops similar changes in the surface properties in some systems.

The influence of the preparation method and the alloy composition on the properties is well documented, but a detailed correlation with the structure of the films is usually not made. This discussion specifically addresses the relationship between the composition and structure of metal alloy films and some chemical properties of the films: the electrochemical passive properties; and the hydrogen permeation. Emphasis is placed on documenting

H. Brodowsky and H.-J. Schaller (eds.), Thermochemistry of Alloys, 527–546.

differences between amorphous and crystallized structures.

The systems selected for study are binary systems with components from the early transition-late transition metals. These systems often produce amorphous structures over a range of composition when prepared by rapid quench methods and they usually consist of two elements with significant differences in passive behavior. As a consequence, they provide ideal systems for studying the influence of composition and structure on electrochemical passivity. In addition, we have explored the hydrogen permeation in selected systems to provide insight into the amorphous structure.

2. CORROSION PROPERTIES

The corrosion properties of amorphous alloys have elicited considerable interest due to early reports of impressive corrosion resistance of metal-metalloid alloys, particularly the Fe-Ni-P-C alloys [1-4]. As a consequence, the amorphous state has gained a reputation for superior corrosion resistance under all circumstances. The main features of the amorphous state that may influence the corrosion behavior are: the absence of physical imperfections, such as grain boundaries, stacking faults, and dislocations; and homogeneous chemical composition, not restricted by equilibrium solubility limits. The latter feature is particularly significant, since rapidly quenched films allow homogeneous compositions to be studied that are not otherwise accessible. The metal-metalloid systems have been most extensively studied; however, the metal-metal class are inherently less complex and provide a simpler basis for comparing the amorphous and crystalline state. Binary pairs were selected that, in most cases, form amorphous structures over a range in composition and have substantially different passive properties for the two components. Studies were made in the following binary systems: Cu-Zr; Cu-Ti; Ni-Ti; Co-Zr; and Fe-Ni. Samples were prepared by planar magnetron sputter deposition (PMS) and analyzed by potentiodynamic anodic polarization (PAP) in various electrolytes.

2.1. EXPERIMENTAL APPROACH

Films of the alloys in question were prepared by planar magnetron sputter deposition (PMS) using two geometrical configurations [5]: a "composition spread" mode, consisting of two overlapping coplanar sources with a static substrate located parallel to the source plane at various source-substrate distances; and a "rotating substrate" mode, with substrates mounted on a turntable rotated at 800 rpm. The "composition spread" samples had a continuously varying composition and were used to survey the influence of composition upon various properties, whereas the samples from the "rotating substrate" mode were used for more detailed studies of specific compositions and to contrast the behavior with the "composition spread" samples. As-deposited and crystallized films (24 hours at 500°C in vacuum ampoules) were characterized by several techniques. The film compositions were determined with electron probe microanalysis and further characterization was performed using differential scanning calorimetry (DSC), X-ray diffraction (XRD), and transmission electron microscopy (TEM). The corrosion behavior was evaluated by potentiodynamic anodic polarization (PAP) using a specially designed small area anode (~4 mm^2) corrosion cell. The electrolytes were mainly 1 to 0.1 N H_2SO_4, HNO_3 and HCl. The potentials and currents were controlled and monitored with a Princeton Applied Research Model 173 Potentiostat and Model 376 log current convertor. Scan rates were either 0.2 or 2 mv/sec. Special care was taken in the surface pretreatment (surface polishing and cleaning) to establish consistent surfaces at the initiation of the potentiodynamic scan. The PMS preparation method was found to produce amorphous structures over a somewhat broader composition range than for the LQ technique.

2.2 . RESULTS AND DISCUSSION

When a metal electrode is polarized to potentials more positive than the corrosion potential, there are four major possibilities: the alloy may anodically etch at increasing rates, in accord with an activation or concentration polarization law; it may form a macroscopic oxide, such as in the anodization of Al or Ta; oxygen may evolve from the metal; or, the metal may form a thin passive film, with a dramatic reduction in the anodic current density. The latter phenomenon is of interest in the present study. Figure 1 displays an idealized response of a passive metal to PAP. The significant feature is the large decrease in current density (i) at the primary passive potential (E_{pp}) as the potential is increased into the passive region, with the passive current density designated as i_{pass}. When the corrosion potential of a freely corroding metal is in the passive region, then the corrosion current density will equal i_{pass}. In most cases, the decrease in current density for the active-passive transition is several orders of magnitude and the corrosion rate in the passive region is very low. When a passive metal is alloyed with a less passive or active metal, there is usually a monotonic decrease in i_{pass} with increasing composition of the passive metal, with the major decrease occuring over a relatively small composition range (10-20%).

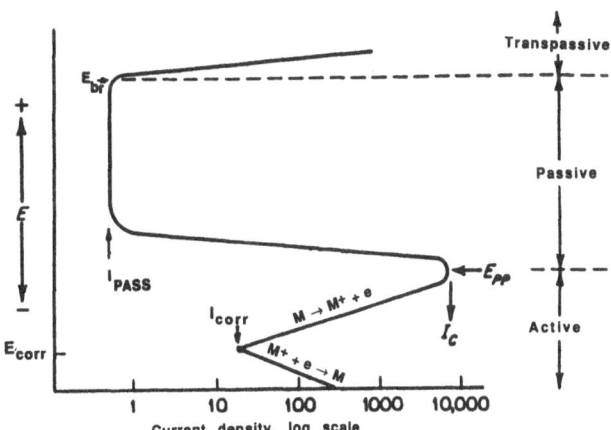

FIG.1 Idealized anodic polarization behavior of an active-passive metal

These behaviors and trends are clearly established from a family of polarization curves for PMS Co-Zr alloys (Fig. 2). Pure cobalt is active, showing logarithmic (Tafel Law polarization) and saturation (concentration polarization) regimes. The 12 and 15 a/o Zr alloys show active-passive transitions, with i_{pass} decreasing with increasing a/o Zr. Similar trends are seen in the Cu-Zr system (Fig. 3) and in the Ni-Ti and Fe-Ni systems. To establish trends with composition, it is appropriate to compare the most significant parameter, i_{pass}, for the different alloys. Since i_{pass} usually varies somewhat with potential in the passive region, it is necessary to select a specific potential that is most meaningful for comparing the respective alloys. Such a plot is given in Fig. 4 for the PAP behavior for Cu-Ti alloys in 0.1 M H_2SO_4. The expected trends are observed: there is a monotonic decrease in i_{pass} with a/o Ti, with a rapid decrease up to ~30 a/o Ti and then a more gradual decrease. This trend is typical for active-passive alloy systems, for

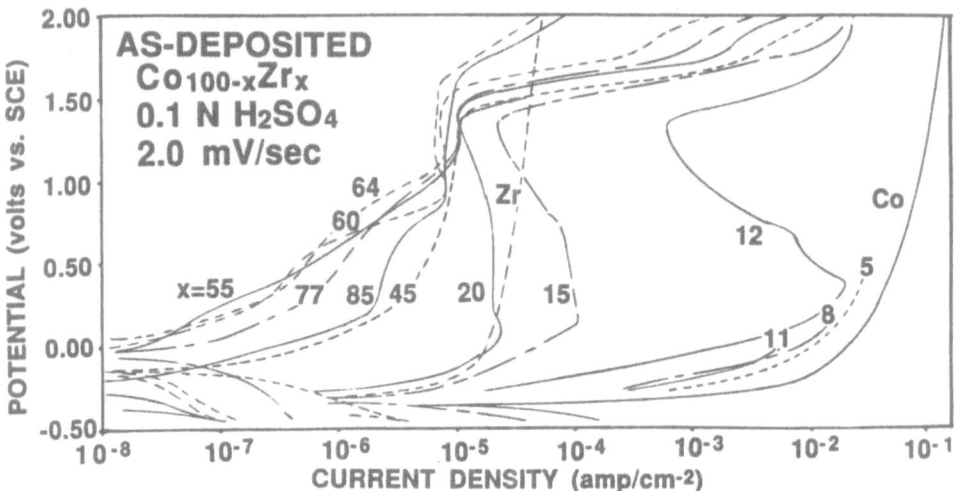

FIG.2 Anodic polarization scans for a series of Co-Zr amorphous alloy sputtered films. The X value is the a/o Zr in Co.

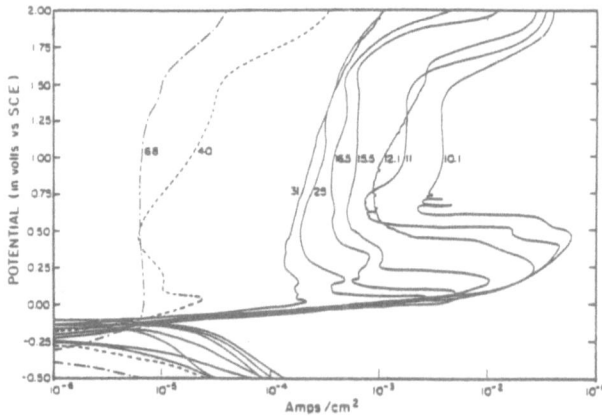

FIG. 3 The Potentiodynamic Anodic Polarization behavior of amorphous Cu-Zr "composition spread" alloys. The numbers indicate a/o Zr. The electrolyte was 0.1N H_2SO_4 and the scan rate was 2 mV/sec.

example, Fe-Cr [6,7] and Cu-Ni [8] alloys. Below 20 a/o Ti, the i_{pass} values are significantly lower for the amorphous alloys than for the crystallized alloys, whereas above 20 a/o, there is no significant difference. According to the equilibrium phase diagram,

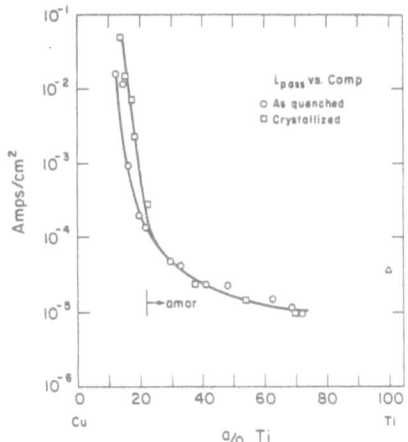

FIG. 4 Comparison of passive current density (@1.0 V, SCE) in 0.1 N H_2SO_4 between as-deposited and crystallized Cu-Ti alloys .

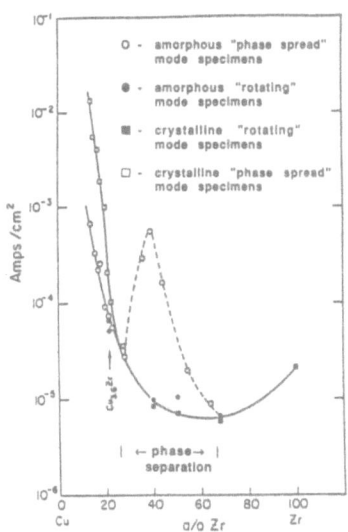

FIG. 5 Current density in the passive region at the arbitrary potential, 0.75V, is plotted versus composition for the following Cu-Zr alloys in 0.1N H_2SO_4.

alpha-Cu and Cu_4Ti are the equilibrium phases at the lower compositions. Thus, an active alpha-Cu is expected to coexist with a more passive Cu_4Ti phase and develop local active-passive cells with a resulting high current density. The presence of these phases was confirmed by TEM [10]. A similar plot is given for the Cu-Zr system in Fig. 5 which shows similar trends for samples prepared by the "rotating substrate" mode of deposition and for crystallized alloys, but there is an anomalous behavior for films prepared by the "composition spread" mode; the value of i_{pass} first decreases with increasing Zr content to ~25 a/o Zr, and then increases, with a maximum value at ~40 a/o Zr. The values of i_{pass} for these alloys are actually higher than for either the "rotating substrate" samples or the crystallized samples. This unusual behavior reflects a significant difference in structure for films prepared by the composition spread mode. Fig. 6 shows the first diffraction maximum for a "composition spread" $Cu_{60}Zr_{40}$ alloy deposited at different source-substrate distances. The development of composition fluctuations during synthesis is evident from the increasing extent of peak separation as the source-substrate distance is decreased. The peak width at half-height of alloys deposited at 6 1/2 inches is quite broad and indicates composition inhomogeneity for this sample also. The diffraction results for films deposited on rotating substrates are qualitatively similar to those obtained for LQ alloys of similar composition.

Further evidence for amorphous phase separation in the "phase spread" samples was provided by TEM and DSC. TEM observations on thin sections revealed a microstructure that was mottled on a ~150 Å scale [11] and electron diffraction and dark field microscopy established that the structure was noncrystalline. Energy dispersive analysis confirmed the compositional inhomogeneity of this structure. In addition, dramatic differences in the DSC behavior between "composition spread" and "rotating substrate" specimens of the same composition provided further evidence of the differences in the as-deposited structures; the composition spread samples gave two DSC peaks, one representative of a lower Zr content and the other representative of a higher Zr content.

The Co-Zr system behaves similarly to the Cu-Ti system, but with a few anomalies that suggest possible phase separation, but verification has not yet been established with structural studies. By contrast, there is only a modest change of i_{pass} for the Ni-Ti system as Ti (the more passive component) is added to Ni (Fig. 7). In contrast to the Cu and Co containing systems in which Cu or Co are always active, both Ni and Ti are passive in non-activating acids. For the Ni-Ti system, there is little difference in the as-deposited (amorphous) and the crystallized state, with Fig. 8 showing typical trends for a 44 a/o Ti in Ni. Films of Fe-Ni alloys prepared by PMS "composition spread" were always crystalline and showed an interesting trend in i_{pass} versus composition; at ~70 a/o Fe in Ni, there is an abrupt increase in i_{pass} with increasing Fe content in dilute acid electrolytes. This is related to the appearance of a BCC second phase in the films for ~> 70 a/o Fe, in contrast to a pure FCC structure at lower Fe content.

Halogen ions are particularly deleterious to the integrity of passive films. The potential range for passivity and the passive current density may be significantly influenced by halide ions. In particular, the transition to a transpassive region by film breakdown (pitting) usually tends to lower potentials with increasing halide ion content. An example of the effectiveness in degrading passive behavior is illustrated by comparing the behavior of pure Ni in HNO_3 and in HCl (Fig. 9). In the former acid, classic active-passive behavior is shown, whereas in HCl, only active behavior is exhibited.

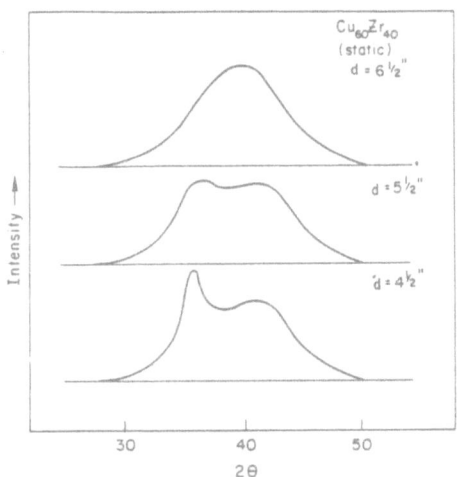

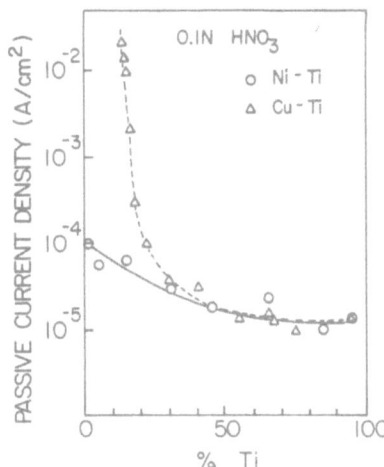

FIG.6 X-Ray diffraction patterns (Cu K alpha radiation) of Cu-Zr specimens with nominally identical compositions deposited by phase-spread mode sputter deposition at 3 different source substrate separations, d.

FIG. 7 Passive current densities (@ 1 V, SCE) versus Ti concentration of Ni-Ti and Cu-Ti alloys deposited by the "concentration mode" in 0.1 N HNO_3.

To evaluate the influence of halide ions on the passive properties and to establish any differences between the amorphous and crystallized state, a number of PAP scans were performed with chloride additions to the electrolyte. For example, the anodic polarization behavior of amorphous and crystallized $Cu_{73}Ti_{27}$ in sulfuric acid-chloride solution (0.1 M H_2SO_4 + 0.03 M NaCl) are shown in Fig. 10. The amorphous $Cu_{73}Ti_{27}$ is spontaneously passivated while the crystallized alloy shows a multiple-step active-passive process. The addition of small amounts of chloride shifts the corrosion potentials of both amorphous and crystalline alloys to more active potentials and increases i_{pass}, but i_{pass} is about an order of magnitude lower for the amorphous state, a much greater difference than when the chloride ion is absent. The improvement in the passive properties for the amorphous structure versus the crystallized structure is most dramatic for the Ni-Ti system in HCl. Figure 11 shows the behavior of a $Ni_{56}Ti_{44}$ alloy in 1 N HCl; only active behavior is observed for the crystallized alloy; for a homogeneous microcrystalline alloy there is an active-passive transition, but with a small passive region; and for the amorphous alloy, there is self-passivation, with i_{pass} several orders of magnitude lower than i_{active}. This behavior may be contrasted with the behavior in HNO_3 and H_2SO_4 (Fig. 8).

534

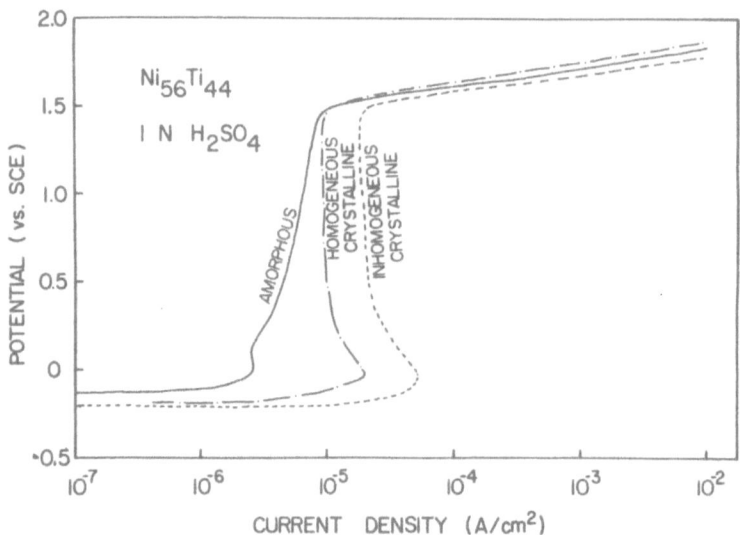

FIG. 8 Comparison of potentiodynamic anodic polarization curves of amorphous, crystallized single phase and two phase $Ni_{56}Ti_{44}$ alloys in 1 N H_2SO_4.

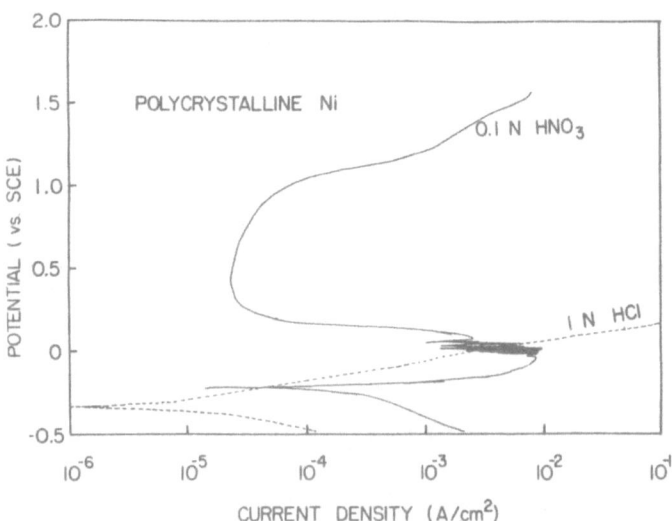

FIG. 9 Potentiodynamic anodic polarization curve of polycrystalline Ni in 0.1 N HNO_3 and 1 N HCl.

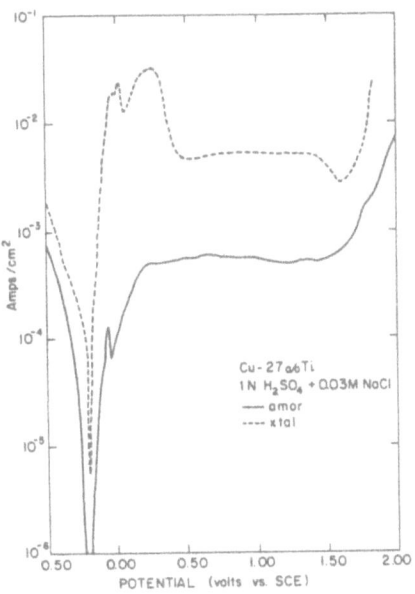

FIG. 10 Comparison of the polarization behavior of amorphous and crystallized Cu$_{73}$Ti$_{27}$ alloys in a chloride-containing electrolyte (1 N H$_2$SO$_4$ + 0.03 M NaCl).

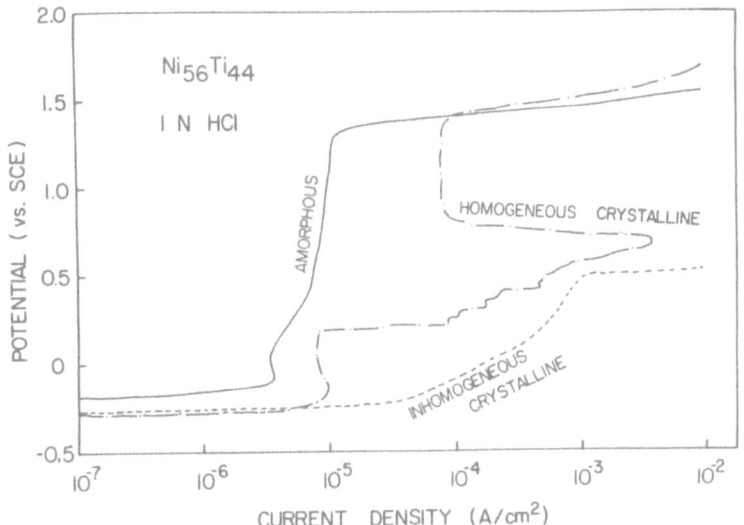

FIG. 11 Potentiodynamic anodic polarization curves of amorphous, and crystallized single phase and two phase Ni$_{56}$Ti$_{44}$ alloys in 1 N HCl.

2.3. SUMMARY AND CONCLUSIONS

There are significant trends in the passive properties of amorphous alloy films, as compared to the corresponding crystallized state. One observes the expected trend of decreasing i_{pass} with increasing composition of the more passive component. For non-activating acid electrolytes, there is relatively little difference in the two states for the higher concentration of the more passive elements (a/o Zr or Ti ~> 20), but self-passivation occurs at a somewhat lower concentration. For the lower concentration in Cu alloys (a/o Zr or Ti ~<20), there is a substantially higher i_{pass} for the crystallized alloys which is associated with the appearance of an active Cu-rich phase in the microstructure. For Cu-Zr alloys prepared by the composition spread mode of deposition, there is an anomalously large i_{pass} between 25 and 70 a/o Zr in Cu. This behavior is attributed to an amorphous-amorphous phase separation, which is confirmed by evidence from XRD, TEM and DSC. In this structure, the inhomogeneous amorphous state has a substantially inferior resistance to corrosion than either the crystallized state or the homogeneous amorphous state. From a practical standpoint, the most impressive improvement in the passive behavior of the amorphous state concerns the behavior in activating acid environments; there are significant improvements in the tendency toward self-passivation and low i_{pass} for the amorphous alloys. The differences in the passive behavior is ascribed to the chemical and physical homogeneity of the amorphous state.

3. HYDROGEN PERMEATION IN ALLOY FILMS

The behavior of hydrogen in amorphous alloys is of interest for fundamental and practical reasons. The substantial differences in the diffusivity and solubility of hydrogen between the amorphous and the crystallized state indicate that hydrogen behavior is extremely sensitive to these structural differences and may be a valuable reflection of the structure [12,13]. Hydrogen incorporation also influences the mechanical properties [14], magnetic properties [15-18], and electrical properties [19] of amorphous alloys. Recently, attention has centered on the fundamental question of the local structure of hydrogen atoms in amorphous alloys. This has been studied by X-ray and neutron scattering [20,21,26], nuclear magnetic resonance (NMR) [22,23,38], small angle scattering [24], Mossbauer spectroscopy [25], internal friction [27-29], the Gorsky effect [30], and photoelectron spectroscopy [25] and various combinations of these techniques.

We describe electrochemical studies of the solubility and diffusivity of hydrogen in amorphous Cu-Ti, Pd-Si and Ni-Ti alloys prepared by planar magnetron sputtering. The Ti containing systems are of interest because of the high hydrogen storage capacity of Ti [31,32], and Pd based systems are classic ones for the study of hydrogen behavior because of the high solubility and diffusivity of hydrogen.

3.1. EXPERMENTAL APPROACH

Hydrogen diffusion and solubility in metals is often studied by gas-phase permeation methods in which the time dependence of the permeation rate of hydrogen through a metal membrane is measured. The electrochemical double-cell methods are attractive because of their versatility and sensitivity. However, there are some important restrictions in applying these techniques: surface reactions should be fast compared with the rate of bulk diffusion; and the metal surface should be free from an impedance layer, such as an oxide, which prevents the penetration of hydrogen. This causes difficulties in applying electrochemical methods to systems other than noble metal alloys. In the present work, the solubility and diffusivity of hydrogen in Cu-Ti, Ni-Ti and Pd-Si amorphous alloys are studied using a double cell electrochemical method.

The following cathode reaction may proceed with appropriate electrolytes and applied potentials:

$$H^+ + e^- = H \qquad (1)$$

where $\underline{H}$ is atomic hydrogen dissolved in a metal cathode. Under carefully controlled conditions, the Nernst equation may be used to relate the emf (E) of the electrode interface (with respect to a suitable reference electrode) to the dissolved hydrogen activity and the pH, and Faraday's Law may be used to relate the amount of hydrogen introduced to the integrated cell current. The reverse reaction (anodic) may proceed by applying an appropriate potential. Thus, with simple measurements, one can control and monitor the amount of hydrogen introduced at a specific electrode interface and the corresponding hydrogen activity at that interface. A thin metal membrane is mounted between two identical cells, with each cell containing a reference electrode (SCE; standard calomel electrode) and a Pt counter electrode. The electrolyte is a 1 to 2 solution of phosphoric acid to glycerin by volume [33,34]. At one side of the membrane (the hydrogen input side), the hydrogen concentration is increased electrochemically causing the hydrogen to diffuse through the foil and to be detected at the other side (the detection side). Measurements may be performed at different temperatures by thermostating the cell.

A variety of boundary conditions may be imposed at both the input and the detection sides of the sample by appropriate instrumentation [35]. The input side, for example, can be polarized cathodically by a constant current or by a fixed potential and the total current can be integrated by a coulometer. At the detection side, the potential of the membrane can be maintained at specific values by means of a potentiostat and the anodic current simultaneously recorded, or, the potential can be measured with respect to a reference electrode/buffer amplifier combination with no anode current, so that the variation of hydrogen activity may be monitored as a function of time. Several methods have been used for diffusion measurements and are well described in the literature [35]. Two methods were used in the present study:,the galvanostatic method and the pulse method.

In the galvanostatic method, a constant hydrogen flux is imposed on the input side by a constant cathodic current, i_o, while the hydrogen concentration is maintained at zero at the detection side by applying a sufficiently positive potential (500 mV vs. SCE) and the resulting anodic current is recorded versus time. The chemical diffusivity (D) of hydrogen can be calculated either by the break-through time, t_b, from the relation [35]

$$t_b = 0.76 L^2 / \pi^2 D \qquad (2)$$

or by the time lag t_l,

$$t_l = L^2 / 2D \qquad (3)$$

where L is the film thickness and the break-through time, t_b, is defined as the intersection time of the tangent at the inflection point of the i versus time curve at the detection side with the initial level, and the time lag, t_l, is the intercept on the time axis of the extrapolation of the straight part of the total quantity of hydrogen emerging from the detection side and this is calculated to be the time to reach $0.63 i_o$ at the detection side.

In the pulse technique, a short cathodic current pulse is applied to the input side of the membrane which initially has a uniform hydrogen concentration. The potential change is measured at the detection side, which arises from the diffusion of hydrogen from the input side. The break-through time, t_b, is defined in a similar way as in the galvanostatic method and is related to the diffusion coefficient as follows:

$$t_b = 0.05\, L^2\, /D \qquad (4)$$

The galvanostatic method provides information on the permeability through the membrane and, by comparing the cathodic and steady state anodic current, can be used to check if the applied cathodic current exactly corresponds to the hydrogen absorbed into the membrane. However, this method gives diffusivity values for the average hydrogen concentration in the membrane and does not provide information on the dependence of the hydrogen diffusivity and activity on the hydrogen concentration. The pulse method can provide this information by applying successively small current pulses.

The pressure-composition-temperature (P-C-T) isotherms in this work are determined by the cathodic introduction of hydrogen at the input side of the specimen, and the equilibrium hydrogen pressure determined by measuring the equilibrium potential at the detection side. The hydrogen concentration is defined as the ratio of the number of hydrogen to metal atoms and is determined by Faraday's law:

$$C = \frac{\int I_c(t)\ dt\ /F}{AL\, /V_m} \qquad (5)$$

where $I_c(t)$ is the cathodic charging current (amps), F is the Faraday's constant, A is the sample area, L is the sample thickness and V_m is the molar volume of the sample.

Hydrogen pressure is related to the electrochemical potential ($\emptyset$) and to the chemical potential (μ) by the Nernst equation (with hydrogen gas at one atmosphere taken as a reference state):

$$\mu = (\emptyset - \emptyset^o) = \frac{RT}{2F}\ \ln P_{H2} \qquad (6)$$

At each constant concentration level, the pulse method may be used to determine the hydrogen diffusivity by applying a short current pulse and monitoring the time response. The P-C-T isotherms were also determined by immersing the sample into the electrolyte and measuring the equilibrium potentials for different concentrations and the results are consistent with those of the bielectrode method.

Amorphous $Ni_{68}Ti_{32}$, $Ni_{56}Ti_{44}$, $Ni_{28}Ti_{72}$, $Cu_{70}Ti_{30}$, and $Cu_{48}Ti_{52}$ alloys were prepared by the rotating substrate mode of magnetron sputter deposition and amorphous $Pd_{83}Si_{17}$ and $Pd_{81}Si_{19}$ alloys were prepared by the piston-anvil LQ method. For the Ni-Ti and Cu-Ti alloys, both sides of the films were coated with a thin layer of Pd (50-100 nm thick) *in situ*, to eliminate the surface barrier (oxide) for hydrogen entry and exit. Specially prepared substrates (electronic grade glass slides from the Hoya Company) were used in order to avoid pin holes in the films arising from unintentional incorporation of extraneous particles during the deposition. Thicknesses of the films were about 5-8 µm. The structure of as-deposited films was examined by X-ray diffractometry (XRD) giving broad amorphous peaks superposed on the Pd peaks. The only crystalline peaks observed were from Pd.

The $Ni_{68}Ti_{32}$ and the Pd-Si alloys were crystallized by annealing the as-deposited amorphous alloys in evacuated and sealed quartz tubes at 600°C for 5 hrs and at 550°C for 2 hrs, respectively. When $Ni_{56}Ti_{44}$ and $Ni_{28}Ti_{72}$ alloys were annealed by this method, the surfaces were oxidized, presumably because interdiffusion of the alloy with the Pd surface layer developed a high Ti surface composition. By contrast, the Ti content at the surface of $Ni_{68}Ti_{32}$ alloys may be sufficiently low after crystallization so that a barrier oxide layer does not form. When the Cu-Ti alloys were crystallized, they became too brittle to handle.

3. 2. RESULTS

Hydrogen permeation through Pd-coated amorphous Ni-Ti and Cu-Ti alloys was initially studied by the galvanostatic technique to evaluate the effectiveness of Pd coatings for eliminating surface barriers for hydrogen incorporation. This was achieved by comparing the steady state cathodic and anodic current; the discrepancy between the cathodic and the observed steady-state anodic currents (current loss) for different charging current densities was usually well below 5%. Assuming that the discharge of atomic hydrogen is the only anodic reaction, the steady state values of the anodic and cathodic currents will be identical if all of the charged hydrogen diffuses through the specimen and is oxidized at the anodic surface. The small differences observed in the two currents indicate that these processes are dominant. The small hydrogen loss may arise from diffusion in a radial direction in the specimen, oxygen reaction, or hydrogen recombination reaction on the cathodic surface, non-steady state conditions, or some combination of these effects. The discrepancy in the anodic and cathodic current decreases as the charging current density (and thereby the diffusivity) increases. The transient curves shift toward shorter time as the applied cathodic current increases, indicating that hydrogen diffusivity increases as the applied current (and the hydrogen concentration) increases. Quantitative data on diffusivity and solubility are obtained by the pulse technique as described below.

Solubility and diffusivity of Pd-Si, Cu-Ti, and Ni-Ti alloys were measured by the pulse technique. Solubility at very low hydrogen concentrations was also measured by immersing the whole sample into the electrolyte, as described earlier. The solubility isotherms (hydrogen concentration versus hydrogen chemical potential using Eqs. 5 and 6) for selected amorphous alloys are shown in Figs. 12-14. No plateau pressure was observed for any of the amorphous alloys studied and Sievert's law was not obeyed. For the Ti containing alloys, the hydrogen solubility at a given hydrogen pressure increases with the Ti concentration. This is consistent with the stability of TiH_2, which forms at low hydrogen pressures at 25°C, in contrast to the hydrides of Cu and Ni which are relatively unstable (NiH requires 6 Kbar hydrogen pressure at 25°C [12]).

The change in the hydrogen solubility for the amorphous and the crystallized alloys at room temperature is shown in Figs. 12 and 15. The hydrogen solubility in amorphous Pd-Si alloys is substantially higher than in the crystallized alloy, as seen in Fig. 12, but for the Ni-Ti alloys, the amorphous form has a higher solubility at low hydrogen pressure but a lower solubility at higher hydrogen presssures (Fig. 15). The crystalline alloys obey Henrian behavior (Sievert's law) at low concentration but deviate at higher concentrations, whereas the amorphous alloys do not obey Sievert's Law at

540

any concentration studied. Amorphous alloys show hysteresis for the charging-discharging cycle for hydrogen over a broad range of pressure, rather than the narrow range near plateau pressures for crystalline alloys.

The diffusion coefficient of hydrogen in the amorphous alloys show a significant increase with increasing hydrogen concentration, as illustrated in Figs. 16-18. This is in contrast to the relative independence of the hydrogen diffusivity values on the hydrogen concentration for the crystallized alloys, which are shown for comparison in Figs. 16 and 18. Diffusivity measurements were performed for different temperatures and analyzed with Arrhenius plots to obtain activation energies, as illustrated in Fig. 19 for a Ni-Ti alloy. At all temperatures, the diffusivity increases with increasing hydrogen concentration while the activation energy decreases.

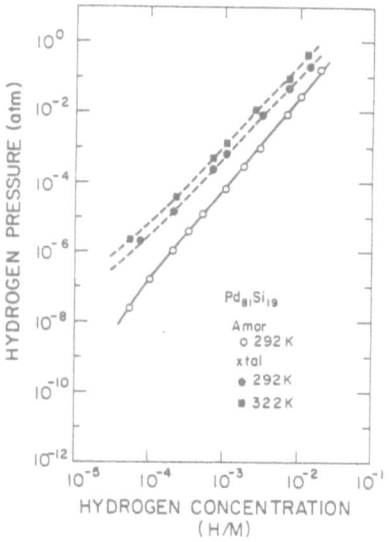
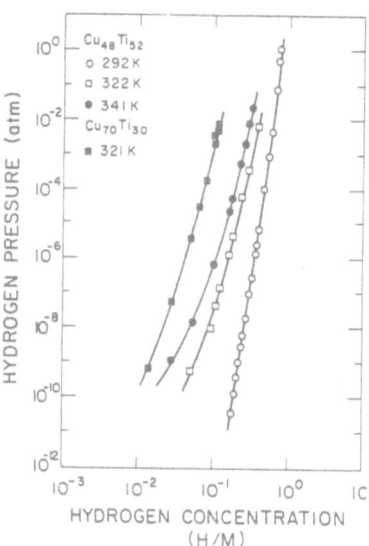

FIG. 12 Pressure-composition-temperature (P-C-T) diagram for amorphous and crystallized Pd$_{81}$Si$_{19}$ alloys.

FIG. 13 Pressure-composition-temperature (P-C-T) diagram for amorphous Cu$_{70}$Ti$_{30}$ alloys.

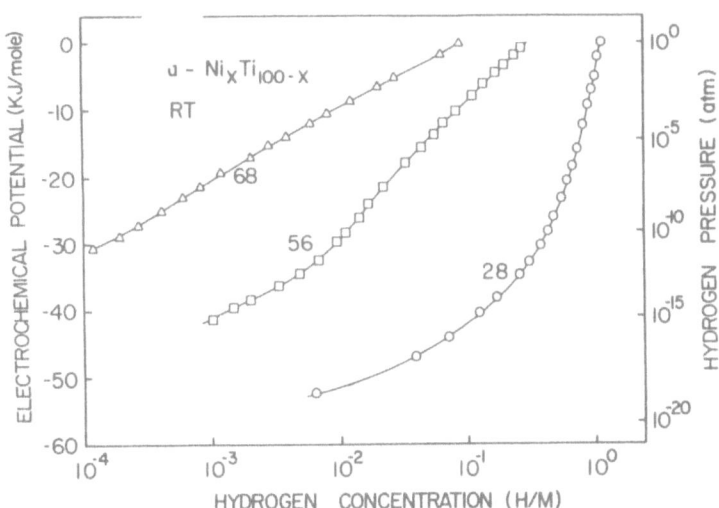

FIG.14 Pressure-Composition-Temperature (P-C-T) diagram for amorphous $Ni_{28}Ti_{72}$, $Ni_{56}Ti_{44}$ and $Ni_{68}Ti_{32}$ alloys.

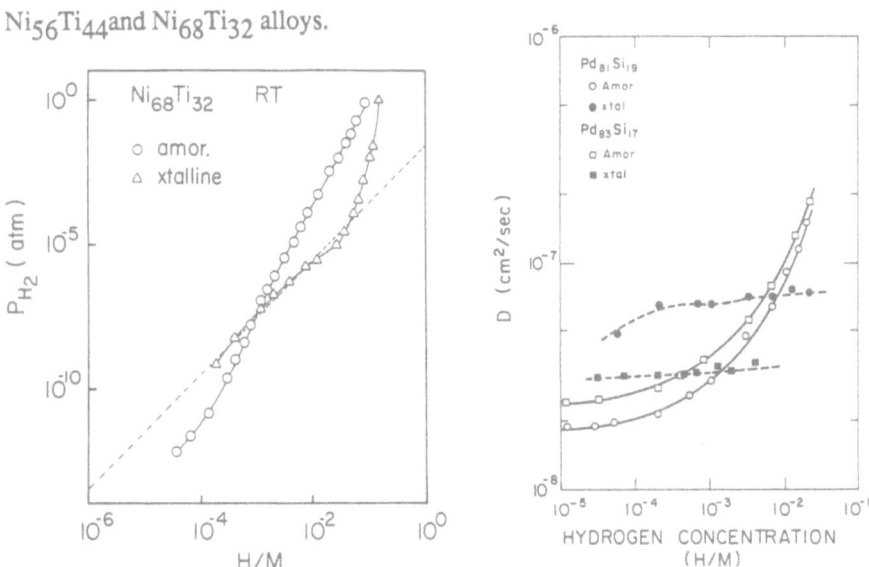

FIG.15 Comparison of Pressure-Composition-Temperature (P-C-T) diagram between amorphous and crystallized $Ni_{68}Ti_{32}$ alloys at room temperature.

FIG.16 Comparison of the hydrogen diffusivity at 292 K between amorphous and crystalline Pd-Si alloys.

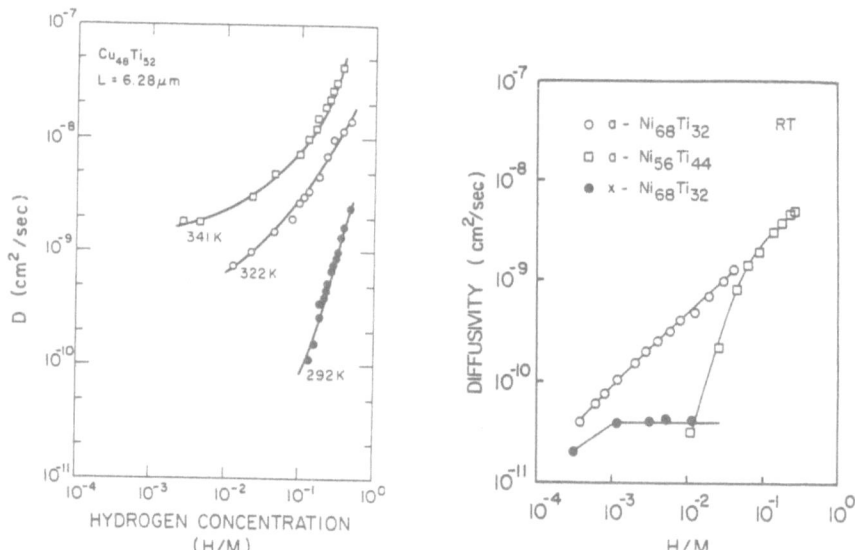

FIG.17 Hydrogen diffusivity, measured by the pulse method, as a function of hydrogen concentration for an amorphous $Cu_{48}Ti_{52}$ alloy.

FIG.18 Hydrogen diffusivity, measured by the pulse technique, as a function of hydrogen concentration at room temperature for amorphous $Ni_{56}Ti_{44}$ and $Ni_{68}Ti_{32}$, and crystalline $Ni_{68}Ti_{32}$ alloys.

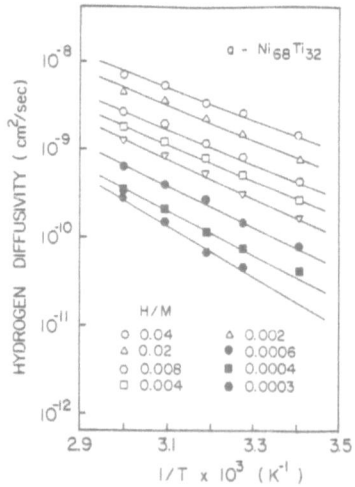

FIG. 19 Arrhenius plot of the diffusivity of hydrogen at different hydrogen concentrations in an amorphous $Ni_{68}Ti_{32}$ alloy

3.3. DISCUSSION

The solubility of hydrogen in the amorphous alloys increases monotonically with increasing hydrogen pressure without showing a plateau pressure; none of the P-C-T isotherms of the amorphous alloys showed a pressure plateau. XRD measurements after hydrogenation up to 1 atm at room temperature indicated that these alloys remain amorphous. All these facts indicate that the amorphous alloys are stable and do not form a hydride upon hydrogenation, at least up to 1 atm of hydrogen pressure at room temperature. A plateau region in the P-C-T isotherm for an amorphous metallic alloy has not yet been reported. The diffusivity of hydrogen increases significantly as the hydrogen concentration increases, typically an increase of more than a two orders of magnitude for the range of hydrogen activities studied. Similar trends are reported for amorphous Pd-Si [34] and Fe-Ni-B [36] alloys. This behavior implies that there is a broad density of (energy) states for hydrogen in the amorphous state; as the hydrogen concentration increases, progressively higher energy sites are occupied with a correspondingly higher diffusivity and lower activation energy. By contrast, interstitial sites in crystalline alloys have comparable energies with relatively little dependence of the hydrogen diffusivity on the hydrogen concentration at lower concentrations.

Insight into the distribution of sites available for occupation by the hydrogen in the amorphous alloys is provided by analyzing the solubility isotherms with lattice statistical models. Assuming a Fermi-Dirac distribution function for the hydrogen occupation of available sites (the occupation number is 0 or 1) and assuming the distribution function can be approximated with a step function equal to 1 or 0 for the site energy less than or greater than the hydrogen chemical potential, respectively, it is shown that the density of sites for hydrogen atoms at a given chemical potential of hydrogen, $g(\mu)$, can be determined from the following relation [37]:

$$g(\mu) = C \frac{d \ln C}{d\mu} \qquad (8)$$

The assumption of a step function for the distribution law is reasonable if the spacing between successive energy states is small compared to RT. The density function, $g(\mu)$, obtained using the data for the Ni-Ti system in Fig. 14 with Eq. 8, is shown in Fig. 20. The hydrogen density of states for amorphous $Ni_{28}Ti_{72}$ alloy shows a bimodal distribution with peak positions at -37 and -18 kJ/mole, respectively. The amorphous $Ni_{56}Ti_{44}$ alloy shows a low intensity broad peak around -30 kJ/mole and an exponential increase above -25 kJ/mol up to 0 kJ/mole and amorphous $Ni_{68}Ti_{32}$ alloys show an exponential increase above -20 kJ/mole with no peaks below that energy. All the calculated distributions were broad, with the energy spacing between successive energy states being large compared to RT. Thus, the assumptions made in deriving Eq. 8 are justified.

The bimodal distributions for amorphous $Ni_{28}Ti_{72}$ and $Ni_{56}Ti_{44}$ alloys within the 0-1 atm range of hydrogen pressure implies that there are two different kinds of interstitial sites in both alloys, each of them with its own energy distribution. It is not possible, however, to identify the site from solubility measurements only. The mean energy of the each peak and the crystallization path [38] can provide information on the characteristics of density of states.

544

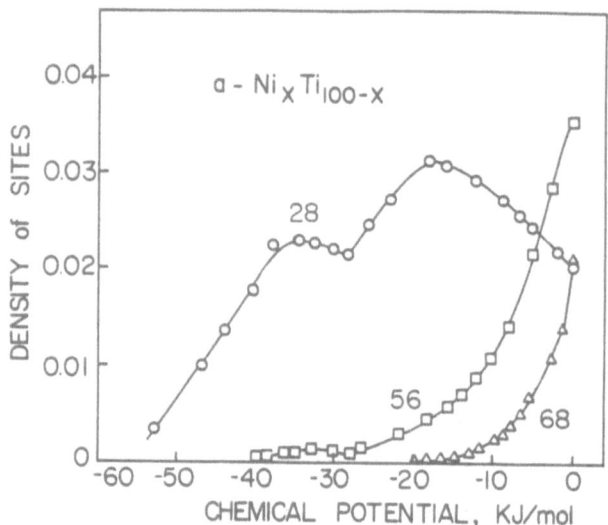

FIG.20 Density of sites for hydrogen, g(μ), as a function of the hydrogen chemical potential, μ, for amorphous $Ni_{28}Ti_{72}$, $Ni_{56}Ti_{44}$ and $Ni_{68}Ti_{32}$alloys calculated from the P-C-T diagram in Fig. 13.

The peak positions in the density of states of the lower energy peaks are -35 and -30 kJ/mole for amorphous $Ni_{28}Ti_{72}$ and $Ni_{56}Ti_{44}$ alloys, respectively, which are close to the formation energy of TiH_2 of -36.6 kJ/mole [39]. This implies that the lower energy peak is related to the dissolution of hydrogen in a local structure similar to the interstitial sites in crystalline TiH_2, where hydrogen occupies the tetrahedral sites of the FCC structure. The lower intensity for $Ni_{56}Ti_{44}$ arises from the lower Ti concentration and the shift in the peak to successively higher energies arises from the interaction of hydrogen with nickel atoms as the hydrogen concentration increases.

Crystallization studies on the hydrogenated amorphous $Ni_{28}Ti_{72}$ alloys [39] indicate that the hydrogen occupies tetrahedral sites surrounded by four Ti atoms at low hydrogen concentration (H/M < 0.33), i.e., at low energy states, since the crystallization products consist mainly of TiH_2 and $NiTi_2$. At higher concentrations (H/M>0.5) amorphous alloys crystallize to form TiH_2+NiTi. The intermetallic compound $NiTi_2$ decomposes into TiH_2 + $NiTiH_x$ upon hydrogenation [40] which implies that hydrogen atoms occupy sites similar to the interstitial sites of crystalline $NiTi_2$ along with the sites surrounded by four Ti atoms. It is noteworthy that the transition from one distribution to another in amorphous $Ni_{28}Ti_{72}$ alloy takes place at around H/M=0.38.

These results are in general agreement with recent X-ray [25], neutron diffraction [20,26,41], Mössbauer and photoelectron spectroscopy [21] and small angle scattering and neutron diffraction studies [24] on the hydrogen sites in other amorphous intertransition metal alloys. These studies provide evidence that hydrogen occupies mainly the tetrahedral sites surrounded by four hydride forming atoms, such as Zr or Ti. NMR studies indicate that the occupation of tetrahedral sites by hydrogen in amorphous $Cu_{50}Ti_{50}$ are similar to that in crystalline CuTi phase from [23]. Mössbauer and photoelectron spectroscopy studies on the hydrogen in amorphous $Zr_{76}Fe_{24}$ [21] lead to similar conclusions (hydrogen atoms preferentially occupy sites which are very similar to those in crystalline ZrH_2) and also indicate that hydrogen begins to occupy less favorable sites (eg., with one iron nearest neighbor) after the sites with only Zr nearest neighbors are filled. Anomalous small angle scattering and neutron diffraction [24] of hydrogenated $Cu_{67}Ti_{33}$ amorphous alloy indicate the presence of small clusters of Ti hydride having the TiH_2 composition. The Young's modulus of hydrogenated Pd-Si amorphous alloys [42] with the hydrogen below a critical composition have a constant Young's modulus, E, whereas above this critical value, E decreases linearly with increasing hydrogen concentration. This behavior is similar to the change in thermal stability with hydrogen concentration observed in the present study.

3.4. SUMMARY AND CONCLUSIONS

The hydrogen solubility and diffusivity in several Ni-Ti, Cu-Ti and Pd-Si amorphous alloys were investigated and the behavior compared to that in the crystallized alloys. The hydrogen activity - hydrogen solubility isotherms for the amorphous alloys do not show a Henry's Law region or plateau regions, in contrast to the behavior of the crystallized alloys. For the Pd-Si alloy, there is a higher solubility in the amorphous state than in the crystallized state at a given alloy composition, but for Ni-Ti alloys there is the opposite behavior. The solubility of the Ti containing alloys increases with the Ti content and substantial hydrogen solubility (up to H/M ~ 1.0) is observed for some of the alloys. For all of the amorphous alloys, there is a large increase in the diffusion of hydrogen and a decrease in the diffusion activation energy with increasing hydrogen content whereas in the crystallized alloys, the diffusivities and activation energies are relatively insensitive to the hydrogen concentration. This implies that there is a broad distribution of sites for hydrogen occupation in the amorphous alloys. The solubility isotherms are analyzed with a simple statistical model to yield a density of states for hydrogen in the amorphous alloys. The density of states is interpreted in terms of the amorphous structure and the interaction between the incorporated hydrogen and the component atoms in the alloy.

REFERENCES

1. M. Naka, K. Hashimoto and T. Masumoto, J. Japan Inst. Metals, 38 (1974) 835.
2. M. Naka, K. Hashimoto and T. Masumoto, Corrosion, 32 (1976) 146.
3. T.M. Devine, J. Electrochem. Soc., 124 (1977) 38.
4. R.B. Diegle and J.E. Slater, Corrosion 32 (1976) 155.
5. T.W. Barbee, Jr. and D.L. Keith, in Synthesis and Properties of Metastable Phases p. 93, eds. E.S. Machlin and T.J. Rowland Conf. Proc. AIME (1980).
6. H. Uhlig, Corrosion and Corrosion Control, J. Wiley and Sons, New York (1971)
7. P. King and H. Uhlig, J. Phys. Chem. 63 (1959) 2026.
8. H. Uhlig, Z. Electrochem 62 (1958) 700.
9. R.G. Walmsley, Ph.D. Thesis, Stanford University, Stanford, CA (1983).
10. Y.S. Lee, Ph.D. Thesis, Stanford University, Stanford, CA (1981).
11. A.F. Marshall and R.G. Walmsley, 39th Ann. Proc. EMSA, Atlanta GA (1981)

546

274.

12. A.J. Maeland, L. E.Tanner and G.G. Libowitz, J. Less- Common Met. 74 (1980) 279

13. Y. S. Lee and D. A. Stevenson, J. Non-Crystalline solids 72 (1985) 249

14. B. S. Berry and W. C. Pritchet, Proc. 3rd Intl. Conf. rapidly quenched Metals 2 (1978) 21

15. J. M. D. Coey, D. H. Gignoux, A. Lienard and J. P. Rebouillat, J. Appl. Phys. 53 (1982) 7804

16. Y. Boliang, D. H. Ryan, J. M. D. Coey, J. O. Ström-Olsen and F. Razavi, J. Phys. F: Met. Phys. 13 (1983) L217

17. D. W. Forester, P. Lubitz, J. H. Schelleng and C. Vittoria, J. Non-Crystalline Solids 61&62 (1984) 6857.

18. D. J. Sellmyer, C. G. Robbins and M. J. O'Shea, ibid, pg. 655

19. B. Chelluri and R. Kirchheim, J. Non-Crystalline Solids 54 (1983) 107

20. K. Suzuki, N. Hayashi, Y. Tomizuka, T. Fukunaka, K. Kai and N. Watanabe, ref 6, p.637

21. B. Grzeta, K. Dini, N. Cowlam and H. A. Davies, J. Phys. F : Met. Phys. 15 (1985)2069

22. R. C. Bowman Jr., W. L. Johnson, A. J. Maeland and W.-K. Rhim, Phys. Letters 94A (1983) 181

23. R.C. Bowman Jr., J. A. Maeland and W.-K. Rhim, Phys. Rev. 26 (1982) 63622

24. P. Goudeau, A. Naudon, B. Rodmacq, P. Mangin and A. Chamberod, Proc. Intl. Conf. , Grenoble, July 8 - 12 (1985)

25. S. M. Fries, H-G. Wagner, S. J. Campbell, U. Gonser, N. Blaes and P. Steiner, J. Phys. F: Met. Phys. 15 (1985) 1179

26. K. Kai, T. Fukunaka, T. Nomoto, N. Watanake and K. Suzuki, Proc. 4th Intl. Conf. Rapidly Quenched Metals V2, Sendai, Japan Inst. Met. (1982) 1609

27. B. S. Berry and W. C. Prichet, Scripta Met. 15 (1981) 637

28. K. Agyman, E. Armbruster, H. U. Künzi, A. Das Gupta and H.-J. Güntherodt, J. de Physique 42 (1981) C5-535

29. L. E. Hazelton and W. L. Johnson, J. Non-Crystalline Solids 61&62 (1984) 667

30. B. S. Berry and W. C. Prichet, Phys. Rev. B 24 (1981) 2299

31. M. A. Gutjahr, H. Buchner, K. D. Beccu and H. Saeuffer, Power Sources 4 (1973) ed. D. H. Collins, Oriel Press, Newcastle, pg. 79

32. V. H. Buchner, M. A. Gutjahr and K. D. Buccu, Z. Metallkde 63 (1972) 497

33. R. Kirchheim, F. Sommer and G. Schluckebier, Acta Met. 30 (1982) 1059

34. R. Kirchheim, Acta Met. 30 (1982) 1069

35. N. Boes and H. Züchner, J. Less- Common Metals, 49 (1976) 223

36.Y. Sakamoto, K. Baba, W. Kurahashi, K. Takao and S. Takayama, J. Non-Crystalline Solids 61&62 (1984) 691

37. J. J. Kim and D. A. Stevenson , in preparation

38. J. J. Kim and D. A. Stevenson, submitted for publication

39. J. Barin, O. Knacke and O. Kubaschewski, Thermodynamical Properties of Inorganic Substances, Vol. 1, P750, Springer-Berlag, Berlin (1977)

40. K. Yamanaka, H. Saito and M. Someno, J. Japanese Chem. Soc. 8 (1975) 1267

41. J. J. Rush, J. M. Rowe and A. J. Maeland, J. Phys. F : Met. Phys. 10 (1980) 1283

42. R. S. Finocchiaro, C. L. Tsai and B. C. Giessen, J. Non-Crystalline Solids 61&62 (1984) 661

THE SIMULTANEOUS CHROMIZING-ALUMINIZING COATING
OF AUSTENITIC STAINLESS STEELS

D.M. Miller, S.C. Kung, S.D. Scarberry, R.A. Rapp
Department of Metallurgical Engineering
The Ohio State University
116 West 19th Avenue
Columbus, Ohio   43210

ABSTRACT.   Chromium and aluminum were simultaneously codeposited by diffusion into austenitic stainless steel substrates, by a single step pack cementation process.   The mechanism for the formation of diffusion-coated products on   304 and 316 stainless steels and on Incoloy 800 are discussed.   The morphologies of the phases formed at the surface, namely, and external beta layer and an underlying multiphase interdiffusion zone, are presented.   The formation of the brittle beta outer layer was minimized by variations in the pack composition and activator.   The coated 304 and 316 steels exhibited excellent scaling resistance upon oxidation in air at 1000 C.

INTRODUCTION.   An increase of the metal temperature in electric utility boilers from about 550 C to 650 C would greatly enhance energy conversion efficiency.   Operating at higher temperatures requires that materials have the necessary mechanical strength, as well as high temperature corrosion resistance.   The mechanical strength of austenitic stainless steels have been found suitable for these applications and one can consider enhancing their corrosion resistance by enriching the surface with chromium and aluminum via a single step pack cementation process.

Surface coatings can be applied using chemical vapor deposition, plasma spraying, and electron beam evaporation techniques etc.   Pack cementation offers the advantage of coating complex shapes economically.   A possible application of pack cementation would be the coating of both the inner and outer walls of heat exchanger tubes.   Commercially, the deposition of only aluminum is the most widely used pack cementation technique.   However, if one were to only aluminize austenitic stainless steels, a resulting depletion of the chromium content at the surface would occur, and this may decrease the corrosion resistance of the alloy.   Chromium depletion could be quite problematic in environments supporting hot corrosion.   Likewise, aluminization results in the formation of a brittle external beta NiAl layer which spalls readily.   Therefore, it is desirable to

547

H. Brodowsky and H.-J. Schaller (eds.), Thermochemistry of Alloys, 547–558.

simultaneously deposit both chromium and aluminum into austenitic alloys and to avoid the formation of the external NiAl layer. This not only avoids chromium depletion, but as will be shown later, the addition of the ferrite stabilizing element, chromium, also increases the kinetics of the coating process.

Bangaru and Krutenat (1,2) recently characterized and evaluated aluminum diffusion coatings on 316, 310, and Incoloy 800H stainless steel substrates. Each coating was observed to consist of two distinct layers, an outer layer comprised of a continuous beta NiAl intermetallic phase, and an underlying layer termed the interdiffusion zone which contained a chromium-rich alpha ferrite matrix with both NiAl and Ni_3Al precipitates distributed throughout. While both diffusion layers had better corrosion resistance than the austenitic substrate alloy, the outer beta layer tended to be brittle and cracked and spalled upon thermal shock. The underlying multiphase interdiffusion zone, however, remained essentially intact and oxidation resistant after prolonged exposure to a wide range of industrial environments. (1,2)

Bangaru and Krutenat propose that the austenite stability of the substrate alloy controls both the thickness and phase distribution of the diffusion coating. The multiphase interdiffusion zone forms by aluminum combining with the austenite-stabilizing element nickel, locally depleting the alloy of nickel, and in turn, causing a local phase transformation from fcc austenite to bcc alpha ferrite. The bcc ferrite then acts as a "short circuit" path for more rapid aluminum diffusion into the alloy.(1,2) Bangaru and Krutenat suggested that the aluminum activity could be adjusted in order to produce diffusion coatings that consisted of only the ductile and corrosion resistant interdiffusion zone without the brittle outer beta NiAl intermetallic layer. Also, it was recommended that the ferrite be replenished with Cr in order to further enhance the corrosion resistance.

Based on the aforementioned studies and halide activator studies of Gupta and Seigle (3), this investigation was undertaken to produce pack cementation diffusion coatings composed of only the interdiffusion zone, while simultaneously replenishing the chromium content of 304, 316, and Incoloy 800 stainless steels. This was achieved by varying both the Cr/Al masteralloy composition and the halide activator until a suitable aluminum activity was attained in the pack. The diffusion temperatures and pack compositions were not specified in any way in the previous work (1,2), despite an impressive characterization of the coated morphologies. Therefore, this study also provides the experimental details necessary to achieve the simultaneous pack cementation deposition of chromium and aluminum into austenitic alloys.

EXPERIMENTAL PROCEDURE. The austenitic alloys used in this investigation were received in sheet form and substrate samples were cut from the sheet to approximate dimensions of 1cm x 1cm x 2mm. The nominal composition of each alloy and the relative austenite stability index used by Bangaru and Krutenat (1,2) are given in Table 1. Prior to the cementation process, each sample was polished to 600 grit,

ultrasonically cleaned, and the initial weight and thickness were
measured. The pack consisted of the following :
   (a) 75 wt.% of alpha alumina powder, 80-200 mesh.
   (b) 23 wt.% of Cr-Al masteralloy powder, 100 mesh.
   (c) 2 wt.% of halide activator, reagent grade.

Table 1: Nominal compositions of alloys and their relative austenite
stabilities.

Substrate	Cr	Ni	Mn	Si	C	Fe	Other
304SS	19	10	1.5	.5	.05	bal	- - - - -
316SS	17	12	1.5	.5	.05	bal	2.5 Mo
Incoloy 800	21	33	.75	.008	.05	bal	.38 Cu
							.38 Al
							.38 Ti

Substrate	Cr* Equivalent	Ni Equivalent	Stability(+) Index
304SS	19.75	12.25	.98
316SS	20.25	14.25	.90
Incoloy 800	21.77	35.26	.44

* See for example reference 5.
(+) Cr Eq. - 4.99/Ni Eq. +2.77
     (lower values indicate a higher austenite stability)

TAble 2: Masteralloy/activator combinations used for each substrate at
1000C for 27 hours.

Substrate	Masteralloy	Activator
304SS	80Cr-20Al (wt%)	NaCl
304SS	90Cr-10Al	NaCl
304SS	95Cr- 5Al	NaCl
304SS	95Cr- 5Al	NaF
304SS	95Cr- 5Al	AlF$_3$
316SS	90Cr-10Al	NaCl
316SS	90Cr-10Al	NaF
316SS	95Cr- 5Al	NaF
316SS	95Cr- 5Al	AlF$_3$
Incoloy 800	70Cr-30Al	NaF
Incoloy 800	90Cr-10Al	NaCl
Incoloy 800	95Cr- 5Al	NaF

In order to produce the desired interdiffusion zone with chromium enrichment without the outer beta aluminide layer, several masteralloy/activator combinations were tested. Table 2 lists the masteralloy/activator combinations used for each alloy.

The substrate samples and pack materials were placed in alumina crucibles (3.8 cm. dia. x 6.3 cm.) which were sealed with an alumina lid using an alumina-base cement. The dried pack was placed into an electrical tube furnace and pure argon was introduced into the furnace at an approximate flow rate of 100 ml./min.. Once the inert atmosphere had been established, the packs were heated at 1000 C ($\pm$ 2 deg. C) for 27 hours.

After coating, the packs were furnace cooled in the same atmosphere. The samples were removed from the pack and ultrasonically cleaned to remove any loosely adherent pack material; thickness and weight measurements were again made. Each sample was cut in cross-section with a low speed diamond saw, mounted in epoxy, and polished for metallographic examination. The microstructural features of the coatings were revealed by etching in a solution containing 3 parts hydrochloric acid, 2 parts nitric acid, 2 parts acetic acid, and 1 part glycerol. The samples were then examined using the optical microscope and SEM. Both x-ray diffraction and EDS analysis were used to characterize the resulting diffusion coatings.

The oxidation resistance of the desired coatings was later tested by TGA measurements in air at 1000 C for 30 hours. Weight changes were recorded using a Cahn precision microbalance.

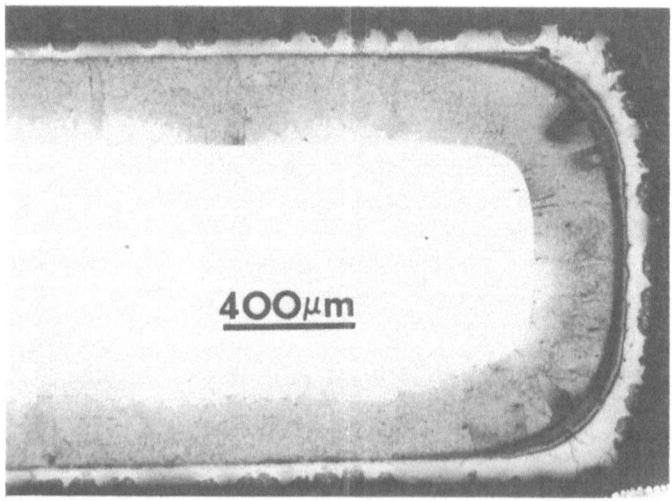

Fig. 1. Optical micrograph of a 304 stainless steel coated with 80Cr/20Al masteralloy and NaCl activator at 1000 C for 27 hours, showing multiphase interdiffusion zone and outer beta aluminide layer.

CHARACTERIZATION OF COATINGS. Figure 1 shows the morphology of the diffusion coating obtained on a 304 stainless steel substrate resulting from a pack containing a 80Cr/20Al masteralloy and a NaCl activator. The pack was heated 27 hours at 1000 C in forming gas. The diffusion coating consists of an outer outward grown layer and an underlying multiphase layer. The dark regions in the outer layer shows this layer has incorporated some of the pack powders. X-ray diffraction has verified earlier characterizations (1,2) and confirmed that the outer diffusion layer consists of the brittle beta NiAl intermettalic compound, while the matrix of the underlying polycrystalline interdiffusion zone was found to be bcc alpha ferrite.

At the present time, a definitive characterization of the precipitates found in the interdiffusion zone has not been completed. As expected, rather crude EDS analysis has shown these precipitates to be rich in both Ni and Al. The previous characterization performed by Bangaru and Krutenat (1,2) using TEM and microdiffraction techniques identified three types of precipitates in the interdiffusion zone. The largest of these precipitates was determined to be Ni-rich (NiAl) aluminides, while the smaller secondary precipitates were also tentatively identified as the NiAl phase. A third precipitate identified as the Ni_3Al gamma prime phase was observed to form inside the larger NiAl precipitates.

The results of the present investigation show no evidence of the formation of the gamma prime phase within the larger precipitates. However, as shown in Figure 2, barely resolvable secondary precipitates are dispersed throughout the interdiffusion zone with the larger precipitates. A more detailed description of the diffusion coatings obtained on each substrate is presented in the following.

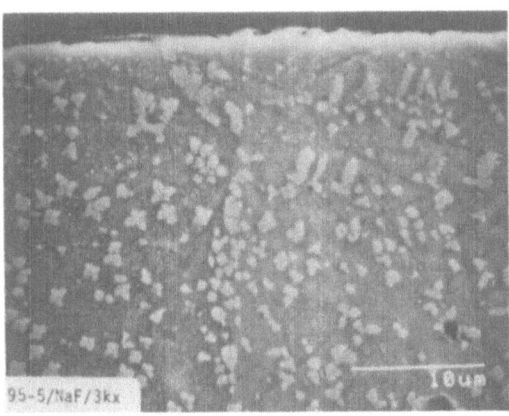

Fig. 2. SEM micrograph showing the two types of precipitates (NiAl and perhaps Ni_3Al)present in the inter-diffusion zone on aluminized 304 SS.

304 STAINLESS STEEL.   Initially, a pack consisting of a 80Cr/20Al masteralloy and a NaCl activator in Ar was attempted.   As shown in Figure 3, a relatively thick (approximately 50 micron) outer aluminide layer which formed by outward growth incorporated some of the pack powders.   Underneath the outer aluminide layer a thick (approximately 250 micron) inward grown interdiffusion zone was observed. Obviously, the aluminum activity of this pack was too high, as evidenced by the formation of the outer aluminide layer.   A pack with a lower aluminum activity which consisted of a 90Cr/10Al masteralloy and a NaCl activator was then tested.   This pack yielded a morphology quite similar to Figure 3 except that the interdiffusion zone was slightly thinner.

The diffusion coating resulting from a pack containing a 95Cr/5Al masteralloy and NaCl in Ar shown in Figure 4 shows that the aluminum activity was reduced too much and the resulting coating was very thin (0-100 microns) and irregular over the entire sample.   In this case, the aluminum activity was reduced to such a low level that the austenite to ferrite phase transformation could not occur at a sufficiently rapid rate.   The thermodynamic activities of aluminum in chromium-aluminum alloys depend sharply on the aluminum content in chromium-rich alloys.(4)   Therefore, efforts were concentrated on using a pack containing a 95Cr/5Al masteralloy while varying the halide activator.

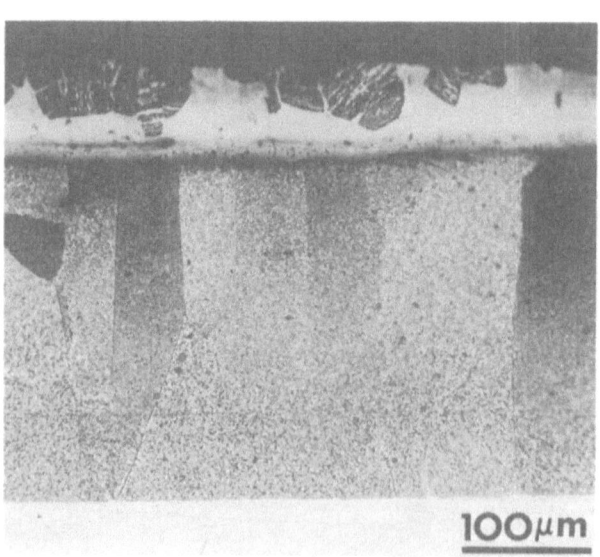

Fig. 3.   Optical micrograph of 304 SS coated in a 80Cr/20Al + NaCl pack at 1000 C for 27 hours.

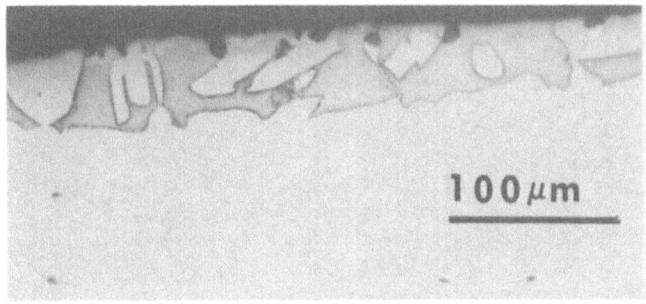

Fig. 4. Optical micrograph of 304 SS coated in a 95Cr/5Al + NaCl
pack at 1000 C for 27 hours.

To increase the aluminum activity in 95Cr/5Al packs, the NaCl
activator was replaced with either AlF$_3$ or NaF. Figure 5a shows a
95Cr/5Al pack with an AlF$_3$ activator in Ar. As can be seen, the
replacement of NaCl with AlF$_3$ increased the aluminum activity too
much, as seen by the presence of the outer aluminide layer. However,
a 95Cr/5Al + NaF pack in Ar yielded a uniform interdiffusion zone
with the nearly complete elimination of the outer aluminide layer as
illustrated in Figure 5b. Therefore, for the given temperature and
diffusion time, this pack composition seems to be optimum for coating
Type 304 stainless steel.

316 STAINLESS STEEL. Since 316 SS has a slightly higher austenite
stability than 304 SS, as shown in Table 1, the initial pack
compositions were chosen to have slightly higher aluminum activities
than the optimum pack for 304 SS. Therefore, packs consisting of a
90Cr/10Al masteralloy and either NaCl or NaF as activators were tested
in Ar. Both packs produced outer aluminide layers and hence the
aluminum activities were determined to be too high. Figure 6a shows
the diffusion coating obtained from a 90Cr/10Al + NaCl pack. The NaF
pack did, however, yield a slightly thicker interdiffusion zone than
did NaCl, consistent with the activator studies of Gupta and
Siegle.(3)
The aluminum activity of the pack was then reduced by using a
95Cr/5Al masteralloy with AlF$_3$ or NaF used as activators. The AlF$_3$
pack still produced the outer aluminide layer, while the NaF pack in
Ar nearly eliminated the outer layer as shown in Figure 6b. However,
the interdiffusion zone obtained on the 316 SS using a 95Cr/5Al + NaF
pack in Ar showed a marked reduction in thickness compared to 304 SS
coated in the same pack. This is attributed to the higher austenite
stability of the 316 SS.

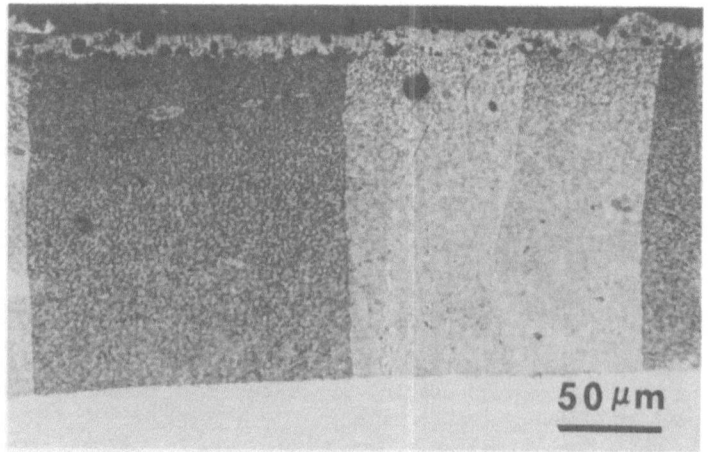

(a)

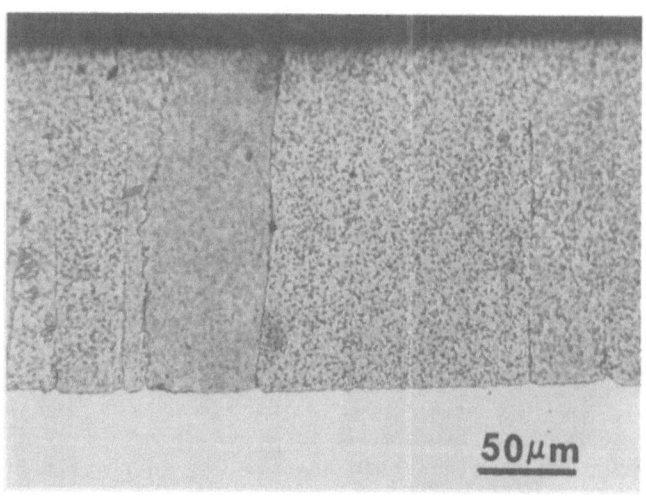

(b)

Fig. 5. Optical micrographs of (a) 304 SS coated in a 95Cr/5Al + AlF$_3$ pack, (b) 304 SS coated in a 95Cr/5Al + NaF pack, at 1000 C for 27 hours.

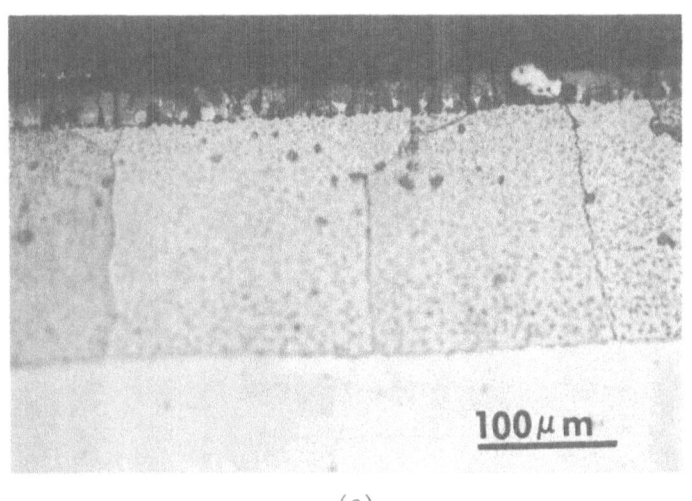

(a)

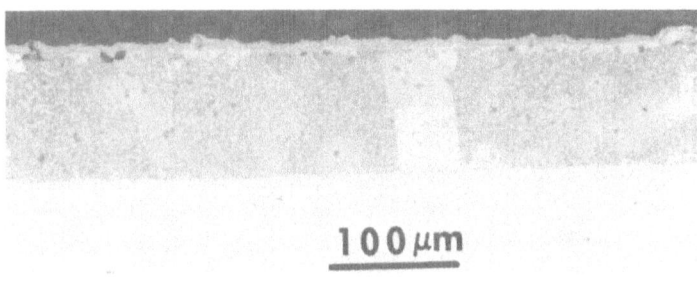

(b)

Fig. 6.   Optical micrographs of (a) 316 SS coated in a 90Cr/10Al
          + NaCl pack, (b) 316 SS coated in a 95Cr/5Al + NaF pack, at
          1000 C for 27 hours.

INCOLOY 800.   After only a few trial packs, an optimum pack
composition for the elimination of the outer aluminide layer has not
been achieved.   However, several interesting aspects in the coating
morphology were observed.   Figure 7 shows the diffusion coating
produced from a pack containing a 90Cr/10 Al masteralloy and a NaCl
activator.   This pack composition produced a relatively thin
interdiffusion zone without the elimination of the outer aluminide
layer.   As seen in Figure 7, the interdiffusion zone is comprised of
elongated aluminide particles in an alpha ferrite matrix.   This type
of cellular growth morphology was observed earlier by Bangaru and

Krutenat(1,2) on Incoloy 800H and was attributed to a match of the aluminide growth rate to the migration velocity of the ferrite/austenite transformation front. Further reduction of the aluminum activity by using a 95Cr/5Al + NaF pack in Ar resulted in little aluminum diffusion or transformation in the substrate matrix. However, such difficulty with the more stable austenitic alloys should not be discouraging because the potential applications offering the greatest improvement in corrosion resistance would involve effective coatings on the less expensive 304 and 316 alloys.

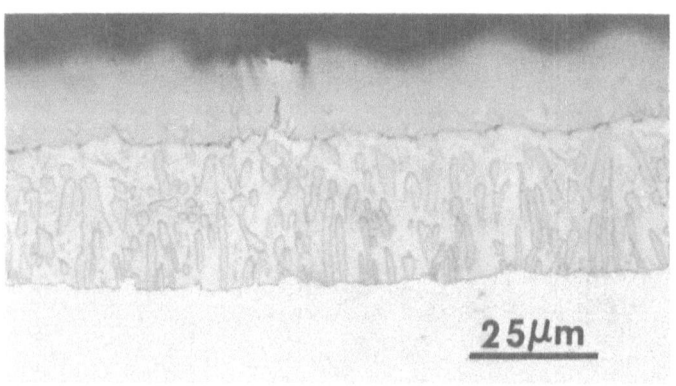

Fig. 7.   Optical micrograph of Incoloy 800 coated in a 90Cr/10Al + NaCl pack, at 1000 C for 27 hours.

TGA RESULTS.   Figure 8 is a TGA plot of the oxidation kinetics for the coated 304 and 316 stainless steels.   The weight-gain per unit area increased rapidly during the initial stages of oxidation but reached a steady-state value of approximately 1 mg/cm^2 for 304 SS and about 2 mg/cm^2 for 316 SS.   In an uncoated condition, the kinetics of oxidation at 1000 C for these stainless steels are extremely rapid. Clearly, a protective alpha-Al$_2$O$_3$ scale was formed.   After exposure to air at 1000 C, the coupons were air quenched to room temperature, whereby spalling of the protective oxide scale occurred.   Indeed, the service application of the coated stainless steels at temperatures as high as 1000 C would not be expected, since the desirable interdiffusion zone would be degraded rather rapidly by diffusive loss to the interior.

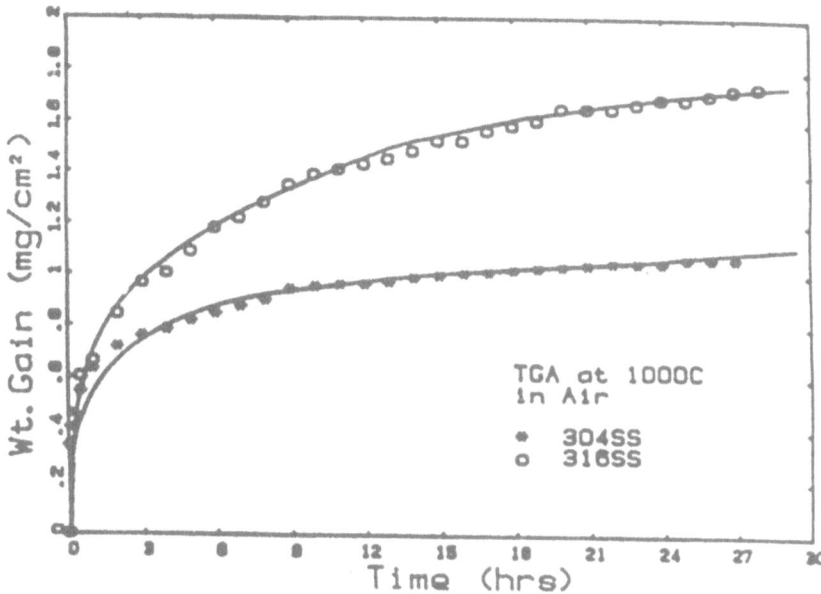

Fig. 8. TGA results at 1000 C for 304 SS and 316 SS coated in 95Cr/5Al + NaF packs.

CONCLUSIONS. In this study, chromium-aluminum masteralloys and various alkali halide activators were used to produce an interdiffusion zone enriched in chromium and aluminum for Type 304, 316, and Incoloy 800 stainless steels. Pack compositions that minimize the formation of the brittle outer aluminide layer during a single step pack cementation process have been obtained for Types 304 and 316 stainless steels. A mixture of a 95Cr/5Al masteralloy powder and NaF as the activator was shown to be the optimum pack composition for both steels. But the difference in austenite stability for the two steels caused the interdiffusion zone for Type 304 SS to be substantially thicker than that for 316 SS.

A suitable pack composition for the elimination of the outer beta aluminide layer on Incoloy 800 was not found. Previous results, however, suggest that a 90Cr/10Al masteralloy and an activator which results in a lower aluminization rate than NaCl, e.g. NaI or NH_4Cl, may produce the desired result.

Furthermore, in this study, the aluminum activity was decreased by the use of a chromium-aluminum masteralloy known to codeposit chromium. The codeposition of chromium in this way not only assists the corrosion resistance of the coating but also supports the austenite to ferrite transformation and the restriction of the growth of the outer beta phase. However, if the aluminum activity is reduced too much, the aluminum halide vapors are inadequate to support sufficient reaction of aluminum and nickel. Hence the austenite to ferrite phase transformation would not occur, or else the advance of the austenite/ferrite interface could be too slow. Obviously, the optimum coating pack compositions for the elimination of the outer aluminide layer would differ for various austenitic steels, depending on the austenite stability of the substrate.

TGA oxidation kinetics measurements performed on both 304 and 316 stainless steels coated in 95Cr/5Al + NaF packs produced slow scaling kinetics, but scale spalling occurred during cooling. The spalling problem may be reduced by introducing a small amount of yttria (Y_2O_3) into the Al_2O_3 matrix powder(6). The serious diffusive degradation of the interdiffusion zone observed at 1000 C would be greatly reduced for service applications at significantly lower temperatures.

ACKNOWLEDGEMENTS. The financial support of EPRI on project RP 2278-1 (J. Stringer, project monitor) is appreciated. The stainless steel coupons were provided by R. John of Shell Development Corporation.

REFERENCES.

1. N.V. Bangaru and R.C. Krutenat, J. Vac. Sci. Technol., B2, (1984) 806-815.

2. N.V. Bangaru and R.C. Krutenat, NATO ASI Series No. 85, (1984) 427-451.

3. B.K. Gupta and L.L. Seigle, Thin Solid Films, 73, (1980) 365-371.

4. W. Johnson, K. Komarek and E. Miller, Trans. Met. Soc. AIME, 242, (1968) 1685-1688.

5. A.L. Shaeffler, Iron Age, 162, (1948) 72.

6. R.A. Rapp, D. Wang, and T. Weisert, "Simultaneous Chromizing-Aluminizing of Iron and Iron-Base Alloys by Pack Cementation" accepted for publication in Metallurgical Coatings, Eds. R. Krutenat and M. Khobaib, TMS-AIME, Warrendale, PA, (1986).

LIST OF CONTRIBUTORS

Anik, S.
Institut für Metallurgie
Allgemeine Metallurgie
Technische Universität Berlin
Joachimstaler Str. 31/32
1000 Berlin FR Germany

Bennett. L.H.
National Bureau of Standards
Gaithersburg, MD 20899 USA

Bessoud, A.
Laboratoire de Thermodynamique
et Physico-Chimie Métallurgiques
ENSEEG, BP 75
38402 Saint Martin d'Hères
Cédex France

Bros, J.P.
Thermodynamique des Systémes
Métalliques
Université de Provence
3, Place Victor-Hugo
13331 Marseille Cedex 3 France

Castanet, R.
Centre de Thermodynamique et
Microcalorimétrie du CNRS
26, rue du 141e R.I.A.
13003 Marseille France

Chang, Y.A.
Department of Metallurgical
Engineering
University of Wisconsin-Madison
1509 University Avenue
Madison, WI 53706 USA

Colinet, C.
Laboratoire de Thermody-
namique et
Physico-Chimie Métallurgiques
ENSEEG, BP 75
38402 Saint Martin d'Hères
Cédex France

Davenport, J.W.
Dept. of Physics, Building 510A
Brookhaven National Laboratory
Upton, NY 11973 USA

Egan, J.J.
Brookhaven National Laboratory
Department of Applied Science
Upton, NY 11973 USA

Fickel, G.
Institut für Physikalische Chemie
Universität Kiel
Olshausenstr. 40
2300 Kiel FR Germany

Frohberg, G.
Institut für Metallurgie
Allgemeine Metallurgie
Technische Universität Berlin
Joachimstaler Str. 31/32
1000 Berlin 15 FR Germany

Gambino, M.
Thermodynamique des Systèmes
Métalliques
Université de Provence
3, Place V. Hugo
13331 Marseille Cedex 3 France

Gronvold, F.
Department of Chemistry
University of Oslo
Blindern, 0315 Oslo 3 Norway

Hayes, F.H.
Metallurgy and Materials
Science Dept.
University of Manchester/UMIST
Grosvenor Street
Manchester M1 7HS UK

Hehenkamp, T.
Institut für Metallphysik
und SFB 126
Universität Göttingen
Hospitalstr. 3/5
3400 Göttingen FR Germany

560

Hertz, J.
Laboratory of Metallurgical
Themodynamic
University of Nancy I
U.A. C.N.R.S. 1108
B.P. 239
54506 Vandoeuvre-lès-Nancy
Cédex France

Hillert, M.
Div. of Physical Metallurgy
Royal Institute of Technology
100 44 Stockholm Sweden

Ipser, H.
Institute of Inorganic
Chemistry
University of Vienna
Währinger Str. 42
1090 Wien Austria

Jaggy, F.
Max-Planck-Institut für
Metallforschung
Institut für Werkstoff-
wissenschaften
Seestr. 92
7000 Stuttgart 1 FR Germany

Kieninger, W.
Max-Planck-Institut für
Metallforschung
Institut für Werkstoff-
wissenschaften
Seestr. 92
7000 Stuttgart 1 FR Germany

Kirchheim, R.
Max-Planck-Institut für
Metallforschung
Institut für Werkstoff-
wissenschaften
Seestr. 92
7000 Stuttgart 1 FR Germany

Kofstad, P.
Department of Chemistry
University of Oslo
P.B.1033
0315 Oslo 3 Norway

Komarek, K.L.
Institute of Inorganic
Chemistry
University of Vienna
Währinger Str. 42
1090 Wien Austria

Krachler, R.
Institute of Inorganic
Chemistry
University of Vienna
Währinger Str. 42
1090 Wien Austria

Kung, S.C.
Department of Metallurgical
Engineering
The Ohio State University
116 West 19th Avenue
Columbus, Ohio 43210 USA

Lin, R.
Institut für Metallurgie
Allgmeine Metallurgie
Technische Universität Berlin
Joachimstaler Str. 31/32
1000 Berlin 15 FR Germany

Maaz, A.
Institut für Physikalische Chemie
Universität Kiel
Olshausenstr. 40
2300 Kiel FR Germany

Mareels, S.
Laboratory for Non-Ferrous
Metallurgy
Grote Steenweg Noord 2
9710 Gent Belgium

McHugh, G.
Metallurgy and Materials
Science Dept.
University of Manchester/UMIST
Grosvenor Street
Manchester, M1 7HS UK

McNallan, M.
GEMM Department, m/c 246
University of Illinois at Chicago
Department of Civil Engineering
P.O. 43 48
Chicago, IL 60680 USA

Miedema, A.R.
Philips Research Laboratories
5600 JA Eindhoven Netherlands

Mikula, A.
Institut für Anorganische Chemie
Universität Wien
Währinger Str. 42
1090 Vienna Austria

Miller, D.
Department of Metallurgical
Engineering
The Ohio State University
116 West 19th Avenue
Columbus, Ohio 43210 USA

Mütschele, T.
Max-Planck-Institut für
Metallforschung
Institut für Werkstoff-
wissenschaften
Seestr. 92
7000 Stuttgart 1 FR Germany

Neckel, A.
University of Vienna
Institute for Physical Chemistry
Währinger Str. 42
1090 Vienna Austria

Niessen, A.K.
Philips Research Laboratories
5600 JA Eindhoven Netherlands

Pasturel, A.
Laboratoire de Thermodynamique
et Physico-Chimie Métallurgiques
ENSEEG, BP 75
38402 Saint Martin d'Hères
Cédex France

Peteghem van, A.P.
Laboratory for Non-Ferrous
Metallurgy
Grote Steenweg Noord 2
9710 Gent Belgium

Predel, B.
MPI f. Metallforschung,
Inst. f. Werkstoffwissenschaften
und Institut für Metallkunde
der Universität Stuttgart
Seestr. 75
7000 Stuttgart 1 FR Germany

Rapp, R.A.
Department of Metallurgical
Engineering
The Ohio State University
116 West 19th Avenue
Columbus, Ohio 43210 USA

Rebouillon, P.
Thermodynamique des Systèmes
Metalliquès
Université de Provence
3, Place V. Hugo
13331 Marseille Cedex 3 France

Scarberry, S.D.
Department of Metallurgical
Engineering
The Ohio State University
116 West 19th Avenue
Columbus, Ohio 43210 USA

Schaller, H.-J.
Institut für Physikalische Chemie
Universität Kiel
Olshausenstr. 40
2300 Kiel FR Germany

Schmid, D.
Institut für Metallphysik
und SFB 126
Universität Göttingen
Hospitalstr. 3/5
3400 Göttingen FR Germany

Schmid-Fetzer, R.
Technical University
Clausthal
Metallurgical Center
Robert-Koch-Str. 42
3392 Clausthal-Zellerfeld
FR Germany

Stevenson, D.A.
Department of Materials Science
and Engineering
Stanford University
Stanford, California 94305 USA

Stølen, S.
Department of Chemistry
University of Oslo
Blindern
0315 Oslo 3 Norway

Ter Minassian, L.
Laboratoire de Chimie Physique
Université Pierre et Marie Curie
11, rue P. et M. Curie
75231 Paris Cedex 05 France

Tomiska, J.
Institute of Physical Chemistry
University of Vienna
Währinger Str. 42
1090 Vienna Austria

Watson, R.E.
Department of Physics
Brookhaven National Laboratory
Upton, NY 11973 USA

Weinert, M.
Department of Physics
Brookhaven National Laboratory
Upton, NY 11973 USA

Wettinck, E.
Laboratory for Non-Ferrous
Metallurgy
Grote Steenweg Noord 2
9710 Gent Belgium

Zahra, A.-M.
Centre de Thermodynamique et
Microcalorimétrie du CNRS
26, rue du 141e R.I.A.
13003 Marseille France

LIST OF PARTICIPANTS

Albertsen, K.
Institut für Physikalische Chemie
Universität Kiel
Olshausenstr. 40
2300 Kiel FR Germany

Alqasmi, R.
Brookhaven National Laboratory
Department of Applied Science
Upton, NY 11973 USA

Anik, S.
Institut für Metallurgie
Allgemeine Metallurgie
Technische Universität Berlin
Joachimstaler Str. 31/32
1000 Berlin FR Germany

Becker, D.
Institut für Experimentalphysik
Universität Kiel
Leibnitzstr. 19
2300 Kiel FR Germany

Bessoud, A.
Laboratoire de Thermoeynamique
et Physico-Chimie Metallurgiques
ENSEEG, BP 75 38402 Saint
Martin d'Hères Cédex France

Borzone, G.
Universita di Genova
Instituto di Chemica Generale
Viala Benedetto XV C.A.P.
16132 Genova Italy

Bredesen, R.
Department of Chemistry
University of Oslo
P.O. Box 1033 Blindern
0315 Oslo 3 Norway

Bretschneider, Th.
Institut für Physikalische Chemie
Universität Kiel
Olhausenstr. 40
2300 Kiel FR Germany

Brodowsky, H.
Institut für Physikalische Chemie
Universität Kiel
Olshausenstr. 40
2300 Kiel FR Germany

Bros, J.P.
Thermodynamique des Systémes
Métalliques
Université de Provence
3. Place Victor-Hugo
13331 Marseille Cedex 3 France

Cacciamani, G.
Universita die Genova
Instituto di Chemica Generale
Viala Benedetto X C.A.P.
16132 Genova Italy

Cakir, O.
Ondokuz Mayis University
Faculty of Art and Sciences
Dept. of Chemistry
Kurupelit-Samsun Turkey

Castanet, R.
Centre de Thermodynamique et
Microcalorimétrie du CNRS
26, rue du 141 R.I.A.
13003 Marseille France

Chang, Y.A.
Department of Metallurgical
Engineering
University of Wisconsin-Madison
1509 University Avenue
Madison, WI 53706 USA

Colinet, C.
Laboratoire de Thermody-
namique et
Physico-Chimie Métallurgiques
ENSEEG, BP 75
38402 Saint Martin d'Hères
Cédex France

564

Egan, J.J.
Brookhaven National Laboratory
Department of Applied Science
Upton, NY 11973 USA

Fickel, G.
Institut für Physikalische Chemie
Universität Kiel
Olshausenstr. 40
2300 Kiel FR Germany

Ghohs, G.
Dept. of Metallurgy and
Mincral Engineering
de Croylaanz
3030 Heverlee Belgium

Hämäläinen, M.
Helsinki University of Technology
Vuorimientie 2K
02150 Espoo Finnland

Hayes, F.H.
Metallurgy and Materials
Science Dept.
University of Manchester/UMIST
Grosvenor Street
Manchester, M1 7HS UK

Hehenkamp, T.
Institut für Metallphysik
und SFB 126
Universität Göttingen
Hospitalstr. 3/5
3400 Göttingen FR Germany

Hertz, J.
Laboratory of Metallurgical
Thermodynamic
University of Nancy I
U.A. C.N.R.S. 1108
B.P. 239
54506 Vandoeuvre-lès-Nancy
Cédex France

Heyer, H.
Philipps Universität Marburg
Institut für Physikalische Chemie
Hans Meerweinstr.
3550 Marburg FR Germany

Hillert, M.
Div. of Physical Metallurgy
Royal Institute of Technology
100 44 Stockholm Sweden

Humbeeck van, J.
Katholieke Universiteit Leuven
Department Metaalkunde en
Toegepaste
Materialkunde
de Croylaan 2
3030 Leuven (Heverlee) Belgium

Ipser, H.
Institute of Inorganic
Chemistry
University of Vienna
Währinger Str. 42
1090 Wien Austria

Jorgensen, P.W.
Department of Metallurgy
The Technical University of
Denmark
Building 204
100 Limdtoftevey
2800 Lyngby Denmark

Karlsson, A.
Kemisk Laboratorium IV
Fysik Kemi
H.C. Orsted Institutet
Universitetsparken 5
Kobenhaven Denmark

Kofstad, P.
Department of Chemistry
University of Oslo
P.B. 1033
0315 Oslo 3 Norway

Konetzki, R.A.
Metall. and Mineral Engineering
University of Wisconsin-Madison
1509 University Avenue
Madison, WI 53706 USA

Küppers, H.
Mineralogisch-Petrographisches
Institut
Universität Kiel
Olshausenstr. 40
2300 Kiel FR Germany

Lin, R.
Insitut für Metallurgie
Allgemeine Metallurgie
Technische Universität Berlin
Joachimstaler Str. 31/32
1000 Berlin 15 FR Germany

Magill, M.
Institut für Physikalische Chemie
Universität Kiel
Olshausenstr. 40
2300 Kiel FR Germany

Mareels, S.
Laboratory for Non-Ferrous
Metallurgy, Rijksuniversiteit Gent
Grote Steenweg Noord 2
9710 Gent Belgium

McNallan, M.
GEMM Department, m/c 246
University of Illinois at Chicago
Department of Civil Engineering
P.O.43 48
Chicago, IL 60680 USA

Mens van, R.
Philips Research Laboratories
P.O. Box 80.000
5600 JA Eindhoven Netherlands

Miedema, A.R.
Philips Research Laboratories
P.O. Box 80.000
5600 JA Eindhoven Netherlands

Mikula, A.
Institut für Anorganische Chemie
Universität Wien
Währinger Str. 42
1090 Wien Austria

Miller, D.
Department of Metallurgical
Engineering
The Ohio State University
116 West 19th Avenue
Columbus, Ohio 43210 USA

Mütschele, T.
Max-Planck-Institut für
Metallforschung
Institut für Werkstoff-
wissenschaften
Seestr. 92
7000 Stuttgart 1 FR Germany

Nagarajan, K.
Max-Planck-Institut für
Metallforschung
Institut für Werkstoff-
wissenschaften
Seestr. 92
7000 Stuttgart 1 FR Germany

Neckel, A.
University of Vienna
Institute for Physical Chemistry
Währinger Str. 42
1090 Vienna Austria

Peteghem van, A.P.
Laboratory for Non-Ferrous
Metallurgy
Grote Steenweg Noord 2
9710 Gent Belgium

Pfalzgraf, B.
Institut für Experimentalphysik
Universität Kiel
Leibnitzstr. 19
2300 Kiel FR Germany

Predel, B.
MPI f. Metallforschung,
Inst. f. Werkstoffwissenschaften
und Institut für Metallkunde
der Universität Stuttgart
Seestr. 75
7000 Stuttgart 1 FR Germany

Prepeneit, J.
Institut für Physikalische Chemie
Universität Kiel
Olshausenstr. 40
2300 Kiel FR Gemany

Press, W.
Institut für Experimentalphysik
Universität Kiel
Leibnitzstr. 19
2300 Kiel FR Germany

Rapp, R.A.
Department of Metallurgical
Engineering
The Ohio State University
116 West 19th Avenue
Columbus, Ohio 43210 USA

Rasmussen, S.
Department of Metallurgy
Technical University of Denmark
Building 204
100 Limdtoftevey
2800 Lyngby Denmark

Rebouillon, P.
Thermodynamique des Systèmes
Metalliquès
Université de Provence
3. Place V. Hugo
13331 Marseille Cedex 3 France

Schaller, H.-J.
Institut für Physikalische Chemie
Universität Kiel
Olshausenstr. 40
2300 Kiel FR Germany

Schindler, R.N.
Institut für Physikalische Chemie
Universität Kiel
Olshausenstr. 40
2300 Kiel FR Germany

Schmid, D.
Institut für Metallphysik
und SFB 126
Universität Göttingen
Hospitalstr. 3/5
3400 Göttingen FR Germany

Schmidt-Fetzer, R.
Metallurgical Center
Robert-Koch-Str. 42
3392 Clausthal-Zellerfeld
FR Germany

Sohége, J.
Drägerwerk AG
Grundlagenentwicklung
Postfach 1339
2400 Lübeck FR Germany

Skibowski, M.
Institut für Experimentalphysik
Universität Kiel
Leibnitzstr. 19
2300 Kiel FR Germany

Stevenson, D.A.
Department of Materials Science
and Engineering
Stanford University
Stanford, California 94305 USA

Stolen, S.
Department of Chemistry
University of Oslo
Blindern
0315 Oslo 3 Norway

St.Pierre, G.
Department of Metallurgical
Engineering
141 Fontana laboratories
116 West 19th Avenue
Columbus, Ohio 43210-1179 USA

Theesen, F.
Institut für Physikalische Chemie
Universität Kiel
Olshausenstr. 40
2300 Kiel FR Germany

Tomiska, J.
Institute of Physical Chemistry
University of Vienna
Währinger Str. 42
1090 Vienna Austria

Watson, R.E.
Department of Physics
Brookhaven National Laboratory
Upton, NY 11973 USA

Yörük, S.
Firat Universitesi
Mühendislik Fakültesi
Kimya Mühendisligi Bölümü
Elazig Turkey

Zahra, A.-M.
Centre Thermodynamique et
Microcalorimétrie du CNRS
26, rue du 141e R.I.A.
13003 Marseille France

Index